Patterns of Vertebrate Biology

E. W. Jameson, Jr.

Patterns of Vertebrate Biology

With 164 Figures

Springer-Verlag
New York Heidelberg Berlin

E. W. Jameson, Jr.
Department of Zoology
University of California
Davis, California 95616
U.S.A.

Sponsoring Editor: Philip Manor
Production: Kate Ormston

Library of Congress Cataloging in Publication Data

Jameson, E. W., Jr. (Everett Williams), 1921–
Patterns of vertebrate biology.

Bibliography: p.
Includes index.
1. Vertebrates. 2. Biology. I. Title.
QL605.J33   596   81-696
AACR2

ISBN 0-387-90520-0   Springer-Verlag New York Heidelberg Berlin
ISBN 3-540-90520-0   Springer-Verlag Berlin Heidelberg New York

To my wife, Sumiko

Preface

This book grew from a series of lectures on vertebrate natural history. The topics have been developed over a period of nearly 30 years, and today scarcely resemble the original subject matter. The progress is primarily technical. Some concepts provide a synthetic framework for viewing much modern research, but many of these concepts either date from Darwin or have developed from observations of later students.

Animal science courses follow a sequential pattern in which there are three discrete levels of undergraduate instruction. Initially, students study subject matter contained in such courses as biology and general zoology. These courses introduce students to animal phylogeny, basic plans of morphology and certain physiological aspects; incidental to these subjects the student acquires a broad zoological vocabulary.

At the other end of the academic spectrum are courses that emphasize synthesis and theory: evolution, zoogeography, behavior and ecology are important courses whose role is to explore the relationships of various aspects of the physical and biological world. In these courses theory and analysis prevail. They are not, however, essentially "subject matter" courses with distinct bodies of knowledge. For example, a student can relate information from many fields to concepts in evolution or ecology, but one does not in fact "study evolution" or "study ecology."

Between these two types of course are several discrete fields, each with its own subject matter which is generally distinct from that of another field. A student emphasizing biological sciences, such as zoology, botany or bacteriology, will pursue knowledge in such fields as genetics, morphology, physiology, cell biology and cytology, as well as taxon-oriented fields, such as entomology or mammalogy.

Although natural history has long been a popular subject, the limits between natural history on the one hand and ecology on the other are nebulous. Ecology is, or originally was, an outgrowth of natural history. It has since become an activity that emphasizes analysis and theory and bases its concepts on information from many separate fields. Natural history remains a study of natural events in the life cycles of wild animals. In writing this textbook on natural history, I have tried to preserve the concept of three levels of courses: (1) introductory, (2) advanced subject matter and (3) analysis and theory.

I have assumed the reader to have a background normally derived from one or two introductory courses, and I have proceeded to more detailed and specific discussions of vertebrate biology. This has demanded occasional references to anatomy, physiology, geology and other disciplines, but the student need not have had advanced work in such fields to follow the discussions in this book. Rather, a background in natural history will assist him or her in an appreciation of other aspects of biology. Because this textbook is primarily descriptive, comprising a collection of discussions of field biology of vertebrates, I have tried to avoid the modern theories that have developed within ecology, evolution and behavior. Nevertheless these fields derive much of their foundation from natural history, and hence overlap is unavoidable.

The subject matter is subdivided so as to assist the reader in finding specific topics. I have attempted to organize examples from as many parts of the earth as possible so that the reader can derive some feeling for the similarities and differences that are founded on contrasting environments as well as on diverse taxa.

Stated most simply, my aim is to pursue the study of vertebrate life cycles, to explore their patterns and variations in terms of the influences of (1) their biotic associates and (2) the abiotic (physical) influences. My ultimate goal is the pleasure of satisfying this aim. The assembling of one's own data and elucidation of specific patterns of behavior and physiology provide the sole base from which general concepts can be generated. To study population cycles of voles, one should study voles, or, if one wishes to explore adaptations of an amphibian, one needs to consult the amphibian. As Louis Agassiz expressed it several generations ago: "Study nature, not books."

In addition to establishing the originality and authenticity of one's own data, one must analyze the information so as to establish or reject all possible relationships. Special responses to the activities of other species, abiotic factors or endogenous patterns ultimately rest on the analysis of original data. The separation of different factors and their aggregate effects on biological patterns constitutes a major challenge to the field biologist. Although simultaneously exerted forces can be clearly isolated by meticulous statistical procedures, cautiously designed analyses succeed only with unimpeachable data.

The most realistic as well as the most difficult approach is the analysis of an activity as a part of the whole, and not in isolation. This *Gestalt* approach, however painstaking and slow, eventually yields patterns that are unequivocally valid. Biological events are rarely exactly the same in every place and in every year; eventually concepts are modified and qualified to accommodate the continual variations in space and time.

Please do not interpret my enthusiasm for the *Gestalt* as a criticism of laboratory studies. Many phenomena are virtually impossible to monitor in the field, and controlled observations of isolated factors greatly aid in their identification and description. But the laboratory, however essential and productive, is a departure from the animal's real world. In this context one should heed Professor Einstein's gentle admonition: "Everything should be made as simple as possible, but not simpler," or the more acrimonious comment of H. L. Mencken, "For every problem there is a simple solution, which is usually wrong."

One example will illustrate the pitfall of drawing hasty conclusions from isolated studies. In laboratory populations of ground squirrels *(Citellus lateralis)*, females enter hibernation before males. In the field, however, adult females are active long after the males have become dormant. The reason for this difference may reflect the failure of captive females to breed; wild females spend some weeks nursing young, while males are already accumulating fat. It is not known for certain that lactation accounts for the delayed dormancy of female squirrels in nature, but the patterns of captive and wild squirrels are nevertheless different.

In this volume are included many investigations by students who have found lifetimes of satisfaction in pursuing segments of the natural history of vertebrates. These examples are of many sorts, but they all have the underlying genesis of providing joy to the investigator. I hope that their effects will both stimulate you to develop your own distinctive program of study and also contribute to a more profound understanding of vertebrate patterns.

Davis, California   E. W. Jameson, Jr.

Acknowledgments

In the early development of this volume I was assisted by many students who took great care in reviewing several sections. I am indebted to them for reflecting the viewpoint of a student reader. Prominent among these students are Hannah Carey, Christine Carter, Janet Dole, Rebecca Gilliland, Susan Hoffman and Kelly Stadille.

Numerous colleagues and friends have generously provided their time and special competence toward the improvement of the following discussions. While I accept responsibility for any errors and ambiguities that may still survive, the value contained in the specialized topics results from the effort provided by these many friends. It is a pleasure to acknowledge their help. They include Allen Allison, James W. Atz, Daniel I. Axelrod, George A. Bartholomew, Ernest C. Bay, Jacques Berger, Duncan M. Cameron, Jr., Robert L. Carroll, James F. Case, Simon Conway-Morris, Richard Cowan, Joel Cracraft, Alan L. Dyce, Perry W. Gilbert, Kenneth Gobalet, Richard C. Goris, Richard Haas, Jack P. Hailman, Richard M. Hansen, Walter Heiligenberg, Alfred A. Heusner, Milton Hildebrand, William H. Hildemann, D. L. Ingram, John S. Lake, L. C. Llewellyn, William M. Longhurst, Charles P. Lyman, R. M. McDowall, Rodney A. Mead, L. E. Mount, Peter Moyle, William T. O'Day, Oliver P. Pearson, Hans J. Peeters, John A. Phillips, J. Robb, Ranil Senanayake, Arthur M. Shapiro, Arnold Sillman, Martin P. Simon, Judy Stamps, P. Quentin Tomich, Teru Aki Uchida, Miklos D. F. Udvardy, David B. Wake, Marvalee H. Wake, Kenneth E. F. Watt and A. H. Whitaker.

For meticulous care in typing the several versions of this manuscript I wish to thank June Elliott, Susan Kortick, Theda Strack and Ellen Tani.

The original illustrations in this book are the work of Suzanne Black, and it is a pleasure to thank her. I am deeply grateful, also, to Philip C. Manor for editorial guidance during the final stages of the preparation of the manuscript.

Contents

Part I

Phylogenetic Development

Fundamental to exploring the biology of various species, it is essential to examine their genesis. Although phylogeny and evolution are not the message of this book, the biology of living entities reflects their ancestry, their phylogenetic history and their past geographic movements, just as past events may aid in our comprehension of the biology of modern forms.

A phylogeny is a postulated relationship of past forms to each other and to modern taxa. The mechanics of evolution, based largely on experimental genetics and population biology, attempt to explain speciation and its role in the diversity of modern faunas. The evolution of vertebrates from the earliest fish-like forms to present aquatic and terrestrial species occurred against a background of seafloor spreading and movement of continental plates, faunal isolation and climatic changes. It is difficult to imagine organic evolution in the absence of these biogeographic events.

This section finally includes the observable and documented movements of vertebrates today. Some movements are local and microgeographic and may be repeated daily; they lead to the concept of home range. More extensive movements involve greater distances with an eventual return to the place of origin; they are migrations. Although most migrations are annual, some are more frequent, even daily, whereas others consume five to eight years and occur only once. Very likely these observable movements, over geologic time, contribute to the faunal movements that form the basis of zoogeography.

1. An Overview of Vertebrate Phylogeny

The first question in vertebrate biology is the basic one: who are the vertebrates and where did they come from. So we begin this book with a survey of vertebrate phylogeny, a complex history constructed by generations of naturalists, systematic biologists and paleontologists on the basis of the fauna surrounding us and the fossil record (Table 1.1).

Vertebrate Progenitors

There is no fossil evidence that unambiguously indicates the origin of the first vertebrates, but possible ancestors occur in several groups of marine animals, including Cephalochordata and Urochordata (Fig. 1.1).

The most likely vertebrate progenitors seem to be lancelets (subphylum Cephalochordata) or creatures close to the lancelets. Not only do they closely resemble vertebrates in form, but also the muscular segmentation (myomeres) is very much like that of modern fishes. The presence of an apparent cephalochordate resembling *Amphioxus* from Middle Cambrian strata supports their candidacy as vertebrate ancestors.[181]

Tunicates (subphylum Urochordata) are chordates by virtue of their notochord, dorsal hollow nerve cord and perforated pharynx. The three groups of tunicates suggest a means by which fish-like chordates may have evolved. The sea squirts (Ascidiacea) have tadpole-like larvae, but as adults they are sessile. They are most common along rocky coasts near the low tide line, and they extract plankton by filtering water through gill slits: microscopic food, trapped by mucus, is diverted to the gut. The notochord is lost in the adult sea squirt. The chain-tunicates (Thaliacea), like the sea squirts, begin life as planktonic larvae and later form pelagic, drifting, colonies. The thaliaceans, like the ascidians, resemble other chordates only in their free-living larval stage. In contrast, members of the third

Table 1.1. Correlation of major geologic events and evolutionary development of vertebrates from early Paleozoic.

YBP[a]	Epoch	Geologic Events	Faunal Changes
10,000 to present	Holocene	Melting of glacial ice	Faunal movements and isolation with climatic changes
3 mil to 10,000	Pleistocene	Ice ages; several Bering connections; land bridges to continental islands; tropical aridity in glacial periods	Many faunal shifts in elevation and latitude; extinctions of many large mammals; movement of human populations into New World
	Pliocene	Closure of Panamanian Portal; accelerated mountain building	Exchange of NA and SA faunas
5 mil	Miocene		Gradual development of grassland mammalian faunas
25 mil	Oligocene	Increased elevation and aridity of land masses	Radiation of birds and placental mammals; spread of tropical forests
35 mil			
55 mil	Eocene	Final separation of Europe and North America, and Antarctic from South America and Australia	Appearance of placental mammals; radiation of marsupials
	Palaeocene		

Q^b

C

YBP[a]	Era[b]	Period	Geological events	Biological events
63 mil				
	M	Cretaceous	Separation of North America from South America and Africa; Separation of Africa from Antarctica	Pinnacle of large reptile development; Spread of angiosperms; first snakes
136–140 mil	M	Jurassic		Early mammals and early birds; Radiation of dinosaurs
220–225 mil	M	Triassic		Appearance of dinosaurs; Expansion of gymnosperms
240 mil	P	Permian		Expansion of reptiles
	P	Carboniferous		Seed ferns, horsetails and early conifers; first reptiles
290 mil	P	Devonian		First amphibians
330 mil	P	Silurian		Early jawed fishes
	P	Ordovician		Ostracoderms
360 mil	P			
420 mil	P	Cambrian		First cephalochordates

[a] YBP, Years before present.
[b] Q, Quaternary; C, Cenozoic; M, Mesozoic; P, Paleozoic.

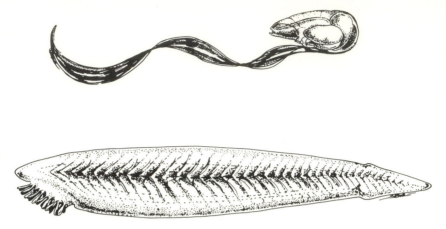

Fig. 1.1, a–b. Two marine chordates that are possible ancestors of vertebrates. Upper: larvacean urochordate; Lower: cephalochordate (Suzanne Black).

group of urochordates, the Larvacea, not only retain tails and notochords but also remain free living as adults. Indeed the larvaceans (or appendicularians) begin to look like vertebrates, for they are *neotenic,* or sexually mature, in their chordate-resembling larval forms.

Both larvaceans and cephalochordates are filter-feeders, extracting food and oxygen as water passes over the perforated pharynx. The persistent notochord provides longitudinal firmness against which lateral muscles can bend, effecting directional movement. Among living chordates, they most closely resemble a hypothetical vertebrate ancestor.

Jawless Fish

The earliest clearly vertebrate species are represented by four types of jawless fish called Cephalaspidomorphi and Ostracodermi (literally "shell-skin," in reference to their armored skin) (Fig. 1.2). These jawless fishes were not sequential in appearance but existed more or less simultaneously: the Osteostraci and Anaspida are found in fresh-water or brackish-water strata from Silurian to Devonian in age; the Heterostraci were mostly marine and lived from the Late Cambrian to the Devonian age, whereas the least known, Coelolepida (which may even have been immature heterostracans), are known from brackish-water strata from Silurian to Devonian in age.

Although the first vertebrates were very probably marine creatures living in warm, shallow seas of the Cambrian and Lower Ordovician 450 to 500 million years ago, adaptation to fresh water occurred early in their history (Table 1.1). Possibly a rich food supply in the form of many aquatic arthropods drew early jawless fishes up-river. No doubt the invasion of fresh-water habitats was assisted

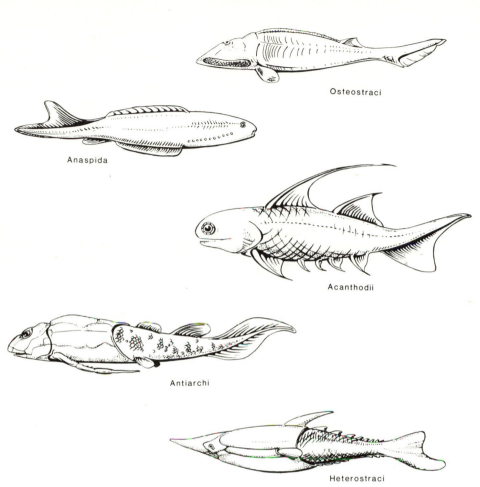

Fig. 1.2. Representatives of some early fishes. The jawless Osteostraci and Anaspida occurred in fresh or brackish waters from the Silurian to Devonian. The Heterostraci, also jawless, are known mostly from marine strata from Late Cambrian to Devonian. Acanthodians, or "spiny sharks," were active predators from the Silurian to the Permian, and the antiarchs (a type of placoderm) are known from the Devonian. Thus these groups overlapped broadly through geologic time (Suzanne Black).

by the development of the vertebrate kidney. The flow of fresh water into a fish requires a kidney that can discharge large amounts of water.

The earliest jawless vertebrate had a terminal mouth and a concentration of sensory tissue above the mouth, the beginning of cephalization. The breathing apparatus consisted of a series of gill slits separated by stiffened gill arches, which in turn were provided with striated muscles. They had a nasal passage, undoubtedly an olfactory sense, and both a median pineal body and paired eyes.

The Osteostraci, or cephalaspids, were dorsoventrally flattened with a ventral mouth and must have been bottom-dwelling detritus feeders. Although they had no paired appendages, most forms had posteriorly diverging lateral projections from the armored head, and these may have served as balancers. As in the present-day lampreys, osteostracans had a single median nostril, two (paired) semicircular canals, a pair of dorsal eyes, a pineal opening and a series of gill openings. They have been proposed as ancestors of the lampreys.

The Anaspida, unlike the flattened osteostracans, were terete but resembled the Osteostraci in many other features. They had a single nostril and a pair of pectoral appendages or spikes, and their mouth was terminal. Their tail was hypocercal (with supporting tissue in the ventral lobe), perhaps providing an upward thrust to the head in swimming.

The Heterostraci (or pteraspids) date from the Devonian period, unless Ordovician fragments from Central North America and Russia represent heterostracans. The eyes were lateral (in contrast to the dorsal eyes of the osteostracans). Like the hagfish, they were marine (or mostly so), and had a single pair of gill openings.

The Coelolepida (not illustrated) resembled heterostracans in most features but were unique in their lack of heavy armor and in the abundance of small scales.

Living Jawless Fish

The 30 or more species of lamprey (Petromyzontidae) and the 15 species of hagfish (Myxinidae) are the only living jawless vertebrates. Although some systematists believe these two groups to be closely related, their biology tends to refute such a concept. Lampreys are anadromous (spawning in fresh water and migrating to the sea), with a buried, filter-feeding larval stage, whereas hagfish are totally marine and show direct development. Hagfish have a terminal nostril through which water passes, via the pharynx, through a series of openings to the gill chamber and the gill opening(s). In the unattached, nonfeeding lamprey, water passes from the mouth, via a single orifice, to the branchial chamber and out through a series of gill openings.

Jawed Fish

Jawed fishes appeared while the early jawless forms flourished, but there is no clearly apparent line of descent from any jawless group, and no transition type has been found. Indeed differing gill formation has been advanced as an argument against such descent.[723] Although many types of jawed fishes are known from Devonian fossil beds, they fall into two major groups: Acanthodians and Placoderms. Although there is no obvious ancestor for these groups, one can suggest lines of descent from them.

Acanthodians, or "spiny sharks," were predatory fish with lateral fins and large lateral eyes (Fig. 1.2). They were contemporary with the heterostracans in the Silurian and survived into the Permian. Distinctive not only in the possession of terminal jaws, they also had a series of paired lateral fins supported by stout spines; the pectoral and pelvic fins were enlarged and developed skeletal support, and sometimes smaller spines occurred between them. There were both marine and fresh-water forms. Most students believe they evolved into bony fishes.

Placoderms (Placodermi), the second major group of jawed fishes, include arthrodires, antiarchs (Fig. 1.2) and some minor groups. They are known from fresh-water Devonian strata, where most of them were bottom-dwelling predators. The skull was articulated by a distinctive ball-and-socket neck joint with the trunk shield, and some groups had dermal armor. One line of arthrodires bears a strong resemblance to a living group of cartilaginous fishes, the chimeras.

Living Jawed Fish

Living jawed fish include several major groups that are old and fairly distinct from each other. The elasmobranchs (cartilaginous fishes) consist of sharks (Selachii) and rays (Batoidei), almost all marine predators. A third group of cartilaginous fishes, the chimeras (Holocephali), are also marine and nocturnal, but are not well known.

Elasmobranchs, the sharks and rays (Chondrichthyes), appeared in the Devonian period after the first bony fishes. They have been among the most conspicuous marine fishes ever since, and today about 575 species of these predatory fishes exist. The heterocercal tail and broad pectoral fins support these fish swimming in the buoyancy provided by salt water, but such an arrangement does not promote efficient maneuverability; the heterocercal tail provides, among other movements, an upward thrust to the tail. Their large size also reduces their ability to make sudden turns. That their origin is unknown is partly due to their cartilaginous skeleton; only their hard teeth provide well-preserved fossils. There was a broad variety of types after the Devonian, including a few fresh-water species, but the elasmobranchs today are few in contrast to the abundance of bony fish species.

Bony Fish

The bony fishes (Osteichthyes)—some of which (the chondrosteans) have cartilaginous axial skeletons, first appearing in the Devonian period—fall into two major groups: the lobe-finned and lung fishes (collectively referred to as the Sarcopterygii) and the ray-finned fishes (the Actinopterygii). The further primary subdivisions and relationships of bony fishes are not uniformly agreed on by systematists. Differences exist in interpretation of fossilized structures, and actual affinity is not always clearly separable from convergence.

Among *living* bony fishes today there are three groups of ray-finned fishes (Fig. 1.3). The major group is the Teleostei, comprising about 25,000 species in both marine and fresh water. Two other groups are remnants of earlier lines, the Chondrostei and Holostei fishes. There is no discrete break between the primitive chondrosteans (sometimes called palaeoniscoid fishes) and the relatively modern chondrostean fishes, but it is nevertheless useful to make a distinction.

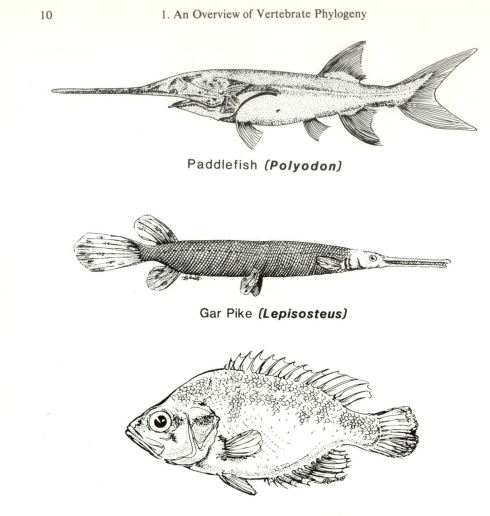

Paddlefish *(Polyodon)*

Gar Pike *(Lepisosteus)*

Sunfish *(Archoplites)*

Fig. 1.3. The three living groups of ray-finned fishes (Actinopterygii). Most living fishes are teleosts. Gars are survivors of the holostean fishes (Suzanne Black).

Among the chondrosteans in Africa are the bichirs and the rope-fish, which live in slow-moving rivers. They have a well-developed bilobed air bladder and can breathe air when their water becomes oxygen-poor. In the Northern Hemisphere the sturgeons (Acipenseridae) and paddlefish (Polyodontidae) are relics of the chondrosteans that flourished from the late Mesozoic.

The holostean fishes, dating from the late Paleozoic, exist today as gars (Lepisosteidae) and the sole-surviving bowfin (Amiidae). They were widespread in the Jurassic and now live only in eastern North America, testifying to the geologic and climatic stability of that region. Fossil holosteans possessed ganoid scales (which created a heavy inflexible armor). In more advanced and modern holostean fishes the skeletally heterocercal tail is functionally homocercal, and one, the bowfin (*Amia*), has cycloid scales. The loss of lift or upward thrust that would have been provided by a heterocercal tail is compensated for by the well-developed air bladder, which is both a hydrostatic and a respiratory organ in this fish.

The vast majority of modern fishes are teleosts, a group that includes many orders and families. Teleost fishes developed in fresh waters of the Triassic and existed together with holosteans and palaeoniscoids, both of which declined as the teleosts diversified and expanded. Earlier forms present most of the features, but teleost species are distinguished more by a combination of characters than by new structures.

Teleost morphology facilitated their invasion of fresh water. Apatite (a strong, light mineral of calcium, phosphorus and carbon) is the main constituent of bone in teleosts, and creates a rigid, light and flexible framework for muscle. Fresh water being lighter than sea water, early invaders from the ocean needed some compensation for the density of their tissues and bone. The air bladder also developed in the early fresh-water jawed vertebrates as a hydrostatic organ. The scales of teleost fishes vary in shape and size, but are usually smaller and thinner than those of holostean fishes and provide teleosts with a more flexible covering. In addition, the anterior tooth-bearing bones of the teleost jaw are somewhat movable, and this feature has supplied many teleosts with partly protrusible mouthparts, which increases their versatility in feeding. The efficient kidney, which has led some physiologists to suggest that fishes evolved in fresh water,[947] can be considered to have made it possible for marine forms to enter fresh water at an early era (Silurian).

These characteristics define an extremely diverse and abundant group of fresh-water and marine fishes, both large and very small. They have an extensive repertoire of reproductive, feeding, breathing, migratory and other major features of the life history. The teleosts are by far the largest, most varied and least appreciated group of vertebrates.

Amphibians

From one of these early jawed bony fishes amphibians developed, and the earliest unequivocal amphibians are tailed four-legged forms known from fresh-water fossil strata of the Upper Devonian age. In these same strata occur rhipidistian crosspterygian fishes, air-breathing lobe-finned fishes of fresh waters. The two groups lived at the same time and in the same place, and their physical similarities suggest that their relationship is phylogenetic as well as ecologic. One of the more interesting aspects of amphibian origins is the proposal that the class Amphibia is diphyletic,[498] the anurans (frogs) and Caudata (salamanders) having evolved separately from two different stocks of rhipidistian fishes. The argument for the diphyletic stem of the Amphibia lies in the similarity of certain cranial features of anurans and Caudata to those of two groups of rhipidistian fishes. The rejection of the proposal for separate origins for frogs and salamanders is based not on the evidence, but rather on its interpretation.

Earliest Amphibians: *Ichthyostega*

The earliest amphibian, *Ichthyostega* (of the order Stegocephalia) (Fig. 1.4), shared many characters with some Devonian rhipidistians: they both had a single occipital condyle, a single gill cleft on each side, some small scales on the skin, a caudal and dorsal fin supported by rays, internal nares, a lateral line system, teeth with folds of enamel and the separation of the skull by a transverse suture. This feature of the teeth provided these early amphibians with the name Labyrinthodontia. *Ichthyostega* was fish-like, scaled and probably mostly aquatic. The stegocephalians were nevertheless bona fide amphibians: their neck vertebrae had high neural spines, indicating muscular support for the head, and the limbs were pentadactyl (with a stout bone in the upper limb and a pair of bones in the lower), suggesting that they could bear the weight of the animal on land. Because of the many morphologic features shared by the rhipidistian lobe-finned fishes and the labyrinthodont amphibians, one is led to the conclusion that a common ancestor occupied their environment in a pre-Devonian time.

Environment of Early Amphibians

One is justified in making several assumptions about the origin of terrestrial vertebrates. The environment must have been both warm, or mild, and humid. An amphibian could not readily leave water and move about on land if the atmosphere were dry and desiccation a serious threat. The plant associates (ferns, lycopods, psilophytes and horsetails) of *Ichthyostega,* moreover, indicate a climate that was neither cold nor arid. The locality of *Ichthyostega* is Greenland, which in the Devonian period was in subtropical or tropical latitudes. If this Devonian swampland resembled humid tropics or subtropics of today, one may also assume that the waters underwent periods during which dissolved oxygen was scarce. Oxygen reduction could have been seasonal or daily or both, so that occupants of such waters would be forced to develop means of breathing air. Finally, if breathing of Devonian fish and amphibians resembled that of present forms, they

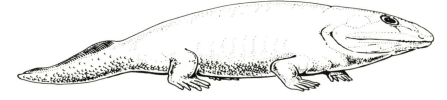

Fig. 1.4. *Ichthyostega,* the earliest amphibian, was intermediate between rhipidistian fishes and modern amphibians. Although it has fine scales, an opercular-like gill cover and fin rays, it also had high neural spines on the neck vertebrae and pentadactyl limbs with heavy bony elements (Suzanne Black).

extracted oxygen not only by lungs (or an air bladder) and gills but also through the skin. When leaving water and entering air, *Ichthyostega* would be deprived of breathing through the gills but would still be able to breathe through lungs and skin. Skin breathing is dependent upon a moist epidermis, which, in turn, requires a humid atmosphere. Small scales in early amphibians might have retarded, but would not have prevented, dermal breathing.

If this description is accurate, the movement from water to land may have been caused by a deprivation of dissolved oxygen in water. On land, oxygen is ample, and breathing would depend on adequate atmospheric humidity. Some students have proposed that increased terrestriality in early vertebrates was a result of seasonal aridity and drying of ponds. Seasonal disappearance of ponds, however, is always accompanied by atmospheric aridity, which would certainly cause the death of an amphibian or early reptile not resistant to desiccation. It seems far more plausible that the transition from water to land occurred at times of reduced oxygen in ponds and high atmospheric humidity. The abundance of completely terrestrial amphibians in humid tropics today supports this latter suggestion.

Terrestrial Egg-Laying

The move from water to land was a major ecologic response affecting evolution of terrestrial vertebrates. The transition to terrestrial life was a multifaceted process; deposition of eggs on land was critical. Terrestrial breeding has developed many times in different families of amphibians, and numerous frogs and salamanders are totally terrestrial.

Although the majority of midlatitude amphibians lay their eggs in water, terrestrial breeding is very common in the humid tropics, and there are even terrestrial tadpoles in some species. Among Neotropical leptodactylid frogs there are many that have reached various degrees of independence from ponds or streams. Many leptodactylid frogs build nests of foam or froth in the water (Fig. 1.5). This

Fig. 1.5. A pair of leptodactylid frogs in the process of creating a foam nest. Foam nests are created both in water and on land. Terrestrial foam nests tend to retard desiccation (Suzanne Black).

is apparently an adaptation to xeric habitats, and seems to facilitate the transition to terrestrial egg-laying; some terrestrial breeders build froth nests on land.[431] *Leptodactylus albilabris, L. marmoratus* and *L. pentadactylus* lay eggs in a nest of froth placed in a subspherical cavity in the soft earth. As the larvae begin to swim, the frothy covering is destroyed and metamorphosis takes place in nearby standing water. In still less aquatic species (e.g., *L. labialis*) the male digs a subterranean burrow and nest chamber from which he calls. In some Neotropical leptodactylids, terrestriality is also seen in adaptations of the larvae: tadpoles of *L. pentadactylus* and *L. poecilochilus* survive drying in holes and crevices and under vegetation, and experimentally they survived desiccation in petri dishes for 100 hours after tadpoles of species from permanent ponds had died.[1041] The Leptodactylidae in the Australian Region show a similar series of steps from aquatic to terrestrial egg-laying.[647]

There are terrestrial tadpoles also in the Neotropical Hylidae.[622] Two species of *Thoropa* breed on moist cliffs where the larvae remain, never entering water. Similarly, species of *Oocormus* breed in terrestrial streamside habitats, and the larvae are quite terrestrial.

Within the Ranidae, land breeders occur most prominently in South Africa.[1053] The ranid *Arthroleptis wagneri* deposits its eggs under damp leaves, and *A. crusculum* lives on grassy montane environments where there is no water for oviposition. The tadpole stage is passed within the egg and small frogs emerge. The hogsback frog *(Anhydrophryne rattrayi)*, another South African ranid, lays a small clutch of eggs in depressions in the soil; tadpoles transform within the fluid-filled nest about four weeks after hatching. Another African ranid *(Arthroleptella lightfooti)* deposits its eggs in damp moss, and the partly metamorphosed tadpoles (with hind legs) remain on land. Similarly, *Nannophrys ceylonenses,* a ranid frog of Ceylon, lays eggs on land and the tadpole is entirely terrestrial, living in damp moss, much like the Neotropical hylid, *Oocormus.*

Terrestrial egg-laying is common in the large and diverse family Microhylidae,[694] in the Old World tree frogs (Rhacophoridae), the New Zealand Leiopelmidae, the Middle American Centrolenidae and others. Many species of tropical and subtropical frogs lay their eggs in trees: some attach their eggs to leaves that overhang water into which the tadpoles drop, whereas many Neotropical frogs deposit eggs in water that collects in the axils of bromeliads, epiphytic plants of humid tropical forests.

Among the species of Caudata, most are aquatic or at least are so during the breeding season.[92] Members of the Hynobiidae, Cryptobranchidae, Salamandridae (except the live-bearing *Salamandra*), Amphiumidae and Proteidae all lay eggs in ponds, lakes or streams. In the Ambystomidae, *Ambystoma cingulatum* and *A. opacum* lay eggs on land in autumn in sites that are annually inundated by winter precipitation; the gilled larvae are washed into the pond. Other ambystomid salamanders lay eggs in water.

In the Plethodontidae there is a series of genera that exhibits a transition from aquatic to terrestrial and even arboreal.[266] Typical are the species of totally aquatic *Gyrinophilus* and *Pseudotriton,* which lay eggs in water. There is some divergence in egg-laying among species of *Desmognathus: D. phoca* and *D. quadramaculatus* lay eggs attached to the undersides of rocks and other structures in streams, whereas *D. fuscus* and *D. ochrophaeus* lay eggs in litter on damp hillsides, and the larvae enter the stream below. The four-toed salamander, *Hemidactylium scutatum,* deposits its eggs in clumps of moss overhanging ponds, and the larvae drop to the water below. The species of the genera *Plethodon, Aneides, Hydromantes* and *Ensatina* all lay eggs on land, as do those species of plethodontid salamanders that have invaded the tropics.

Terrestrial egg-laying may well have been as frequent in Paleozoic amphibians as it is in most modern, tropical forms. Breeding on land has occurred independently, so many times that there seem to be no great adaptations needed. Most probably, development of early reptilian features occurred among wholly terrestrial amphibians: that is to say, reptilian evolution followed terrestriality in the Amphibia.

The earliest known amphibians were much larger than most modern forms, they were generously covered by scales, and many species were terrestrial. Most of them disappeared with the development of reptiles at the close of the Paleozoic era. Apparently the demise of these larger land-dwelling amphibians was related to reptilian expansion, for the more aquatic amphibians persisted, and some very small terrestrial amphibians seemed unaffected by early reptilian competitors or predators.

These terrestrial amphibians resembled modern salamanders in some features: their small size favored integumentary breathing and exploitation of a burrowing or semifossorial way of life. They probably occupied a forest-floor habitat where the humidity was high and where the fauna of small arthropods provided an abundant source of food. The earliest modern amphibian (Lissamphibia) is *Doloserpeton annectens,* of the Lower Permian of Oklahoma.[107] This species was small, probably completely terrestrial, and presumably fed on soil arthropods.

Modern Amphibians

The three orders of modern amphibians [Anura, Caudata and Apoda (caecilians)] are distinct from their earliest fossil records: anurans (frogs) and caudates (salamanders) date from the Jurassic, and caecilians are known from the Paleocene. Although *Doloserpeton* is clearly on the lineage of modern amphibians, it is not closely allied to any one order. In addition, another Permian amphibian, *Goniorhynchus,* resembles the Apoda in several aspects of skeletal anatomy.

The Anura and Caudata have long had separate histories and remain today fairly distinct from one another. The remaining order, the Apoda (caecilians), has unique features and is not close to either frogs or salamanders. The relatively heavy cranial structure in caecilians may be an adaptation to a fossorial life (many frogs have massive bones in the roof of the skull); however, some students have suggested that they represent a lineage long distinct, and are descended from microsaurs, small amphibians from the Permian and Carboniferous.[156]

Reptiles

Although the climate of the late Paleozoic included widespread seasonal aridity and desert or semidesert conditions in some regions, the birth of the Reptilia occurred in such permanently swampy regions as produced coalbeds, regions of ferns and horsetails and other moisture-loving plants. It is very likely that, with the development of a genuinely terrestrial amniotic egg as exists today, reptiles moved into areas of seasonal dryness, but it is very unlikely that they originated there.

Early Reptiles

Early reptiles diverged from their amphibian ancestors in the form and ossification of the vertebra, the position and form of the limbs and the form of the skull. The ancestral amphibians were already terrestrial. Body support and movement on land are easier in small than in large animals, and the most primitive reptiles were small, with small heads. Fossil reptiles can be distinguished from most fossil amphibians by the single occipital condyle (double in all but primitive amphibians) and the sacral attachment, which articulates with at least two vertebrae (instead of one, as in amphibians).

The dentition of reptiles varies but is usually intermediate to that of amphibians and mammals. Tooth generation is polyphyodont (with continuous replacement) as in amphibians and usually homodont (without extensive differentiation in tooth shape and size). Turtles lack teeth. In Paleozoic amphibians some large teeth were on the roof of the mouth (on the pterygoid, palatine and vomer). In tetrapod reptiles the roof of the mouth may have small, or no, teeth, but snakes have well-developed pterygoid and palatine teeth. In some snakes the maxillary bones bear anterior fangs carrying a toxin from labial glands, and in some other venomous species, called rear-fanged snakes, a pair of enlarged teeth arise at the rear of the maxillary bones.

The integument of modern reptiles presents many features that separate them from modern amphibians. Reptiles have scales that are epidermal in origin, and which are usually shed periodically; as they grow, they are replaced. Thus they gradually increase in size but not in number. Reptilian skin has few glands and is dry and rather resistant to desiccation. Nevertheless reptilian skin is not impervious to gases, and in some turtles, snakes and lizards the skin allows some passage of oxygen and carbon dioxide. An important feature of the dermal layer of some lizards, and possibly some other reptiles, is the occurrence of fluid melanin, providing the basis for color change.

Reptile Eggs

The amniotic egg is a consistent character separating modern oviparous reptiles from living amphibians. Even viviparous reptiles, which do not lay down a shell over the egg, still retain the same fetal membranes found in eggs of oviparous reptiles. The amniotic egg retards the passage of water and is thus resistant to desiccation, especially when covered with a calcareous or coriaceous shell (the cleidoic egg, from "cleis" = key). This enables reptiles to breed on dry land.

A terrestrial egg, surrounded by air, has a much greater oxygen supply than has an aquatic one, permitting an increase in size with consequent decrease in surface-to-mass ratio. The large amniotic egg contains much more yolk than do most amphibian eggs, enabling the growing embryo to develop beyond the larval (or gill cleft) stage before hatching.

Egg Fertilization

Because their amniotic egg has a coriaceous shell, fertilization in reptiles is necessarily internal. The means for achieving internal fertilization varies among the four living orders. There is no intromittent organ in the tuatara, *Sphenodon* (Rhynchocephalia). Turtles (Chelonia) and crocodilians (Crocodilia) have a median penis (of different tissue), and in lizards and snakes (Squamata) a paired structure, the hemipenes, provides for transmission of sperm, but only one-half of the hemipenes of the squamates is used at any one time. None of these structures appears to be homologous with any other, and such penile diversity suggests that reptilian intromittent organs evolved independently at least three times. Internal fertilization must have preceded the development of an eggshell, and there is no reason to assume that internal fertilization developed only once in reptiles.

Possibly the different primitive reptilian features developed separately, and not necessarily in the same sequence, in different amphibian (and transition) lineages. Indeed there is no evidence even that the *first* amniotic egg was laid by a reptile, or that the amniotic egg developed only once. Although the origin and final clas-sification of reptiles may not be agreed on for many years, one should realize that there are arguments for and against a polyphyletic origin of reptiles, and that the final decision will depend on the interpretation of the evidence as much as on the evidence itself.

Birds

It is ironic that some of the central unifying features of terrestrial vertebrates are soft structures that usually disappear during fossilization. Although dinosaur eggs have been preserved, their amniotic structure is not apparent in their fossil condition. The features with the skeletal imprints of *Archaeopteryx* are unique; probably other avian fossils without preserved feathers are indistinguishable from reptilian fossils. Thus, because these important diagnostic features are virtually lacking in the fossil record, it is very difficult to recognize the first bona fide birds.

Birds have been called "glorious reptiles" and "hot reptiles," and the similar-ities between the two taxa outnumber the differences. Birds are much more like reptiles than are mammals, and possibly some very early avian fossils are erro-neously assigned to the Reptilia. For example, both birds and reptiles have a single occipital condyle—a lower jaw of several bones attached to a movable quadrate bone. Most birds have the ribs reinforced by an uncinate process, and the same structure is found in the tuatara. Then, birds have a system of air sacs, which also happens to be present in chameleons; their function in these lizards is unknown but may be thermoregulatory.

Most characteristic aspects of birds relate to either flight or endothermy, two distinctive avian features. Both features relate to the fact that all birds are more or less completely covered with feathers. The body and legs are clothed with both an undercoating of fluffy down and a firmer overlying layer of contour feathers. Although these contour feathers add virtually nothing to bird flight, they do retard heat loss, and also give the bird much of its characteristic color. The wing feathers (remiges) and tail feathers (rectrices) also contribute to coloration, but more importantly they provide the air-foils for flight and aerial maneuver.

Bird skeletons exhibit extreme modifications for flight and dynamic balance. The bones of the "hands" and feet are reduced in both size and number so that as the wings move in flight there is minimal shifting of the center of gravity. The long bones are hollow and connect to the air sacs and lungs; this results in a stronger, lighter skeleton and extensive connection between air sacs, lungs and the outside air. Similarly the loss of teeth reduces the weight of the head, and the extreme abbreviation of the tail (and its replacement by the retrices) also saves weight and concentrates bone closer to the center of gravity.

Corresponding to the displacement of bone toward the interior of the body is the central placement of muscles. The two large pairs of pectoral muscles attach to the keeled sternum and extend onto the wings as reduced strands and tendons, another concentration of weight. The keel on the sternum develops in response to enlargement of the pectoral muscles: absent in nonflying birds (ratites), it appears in bats and moles.

Early Birds

The earliest known remains that are definitely avian are imprinted forms in a slate-like granite, dating from the Jurassic. It is not known if this bird, *Archaeopteryx lithographica,* was ancestral to all living birds, or if modern birds are from two or more lines. The possible origin of feathers from reptilian scales, having occurred once, could have happened more than once; there are also some fundamental differences between some main stocks of birds. The earliest birds probably descended from small, fast-running coelurosaur dinosaurs.

Few birds are known from Cretaceous formations. *Hesperornis* was a large, loon-like bird, with legs placed far in the rear. It possessed teeth, but lacked a keel on the sternum and was certainly flightless. *Ichthyornis* was gull-like and, with a keeled sternum and hollow bones, may have been one of the earliest flying birds. There was probably a substantial divergence by the close of the Cretaceous, for early Cenozoic fossils were of many types. Many of the well-differentiated orders of today were presumably established in the Paleocene and Eocene, but small and delicate arboreal birds would seldom have remained intact for preservation as fossils. The known early Cenozoic birds were thus large forms.

From Eocene strata of North America and Europe came the seven-foot ratite named *Diatryma*. At this time Europe bridged the North Atlantic and was joined to eastern North America. Other ratites occurred in Africa and Asia, and the initial stock must have existed in the Cretaceous, for ratites spread to Australia and New Zealand. Some of the large ratites existed until the Pleistocene: *Dinornis* of New Zealand and *Aepyornis* of Madagascar were both more than three meters tall.

Modern Birds

Today there are 20 or more living orders of birds, as diverse as loons, hummingbirds, chickens and owls; yet all are birds, with many features in common, and all, even today, are rather reptilian.

The distinctive avian features of endothermy and flight have behavioral as well as structural implications. Migration to and survival of birds in polar regions cannot be equaled by reptiles. Much avian courtship display involves color and conformation of feathers, and color is a sexual trait in many groups of birds. A mass movement into trees throughout the Cenozoic radiation of birds presented an opportunity for a dietary radiation unparalleled in reptiles. Perhaps the greatest divergence is the sequence of nest-building and parental care, which provides the basis for the education of offspring.

Mammals

Mammary tissue, the central unifying element among mammals, does not occur in the fossil record. Early mammals are recognized by skeletal changes that are presumed to have occurred with the development of mammae. Some other mammalian characteristics occur in certain reptiles: ichthyosaurs, for example, were viviparous, and some modern squamates have a chorioallantoic placenta. Also, although endothermy is difficult to establish for any Mesozoic reptiles, it may have developed in early paramammals (mammal-like reptiles), and probably existed prior to mammary tissue.

The presence of sweat and sebaceous glands in the skin of mammals contributes to their distinctive biology, and such glands are not known in any living reptile. The occurrence of skin glands in mammals is intimately associated with thermoregulation, behavior, milk production and hair; probably skin glands and pelage in paramammals constituted one of the early steps from which endothermy and lactation naturally followed. The simplest glands, *sweat glands,* are tubular ducts that discharge water, salts and urea, which are collectively known as sweat. They are important in thermoregulation through evaporative cooling of the skin onto which the water is discharged. *Mammary tissue* consists of concentrations of highly specialized glands similar to sweat glands. Complex, branched glands, called *sebaceous glands,* occur not only with hair follicles but also in several other locations, such as over the breasts. Sebaceous secretions are oily, and apparently prevent excessive drying of the structures or surfaces over which they flow. All these glands and their associated tissue and structures must have been present in the earliest mammals, and their traces may never be found in the fossil record.

Other mammalian features are also not apparent in the fossil record. Highly distinctive mammalian characteristics, such as metabolic rate, four-chambered heart, non-nucleated erythrocytes and the muscular diaphragm, are not preserved. The identification of the first mammal may therefore always remain equivocal.

In tracing mammalian origins, certain skeletal characteristics suggest some of the above changes. Teeth became increasingly pointed and differentiated (heterodont) for special functions, and their origin became restricted to the premaxilla, maxilla and dentary. The dentary became the sole bone of the lower jaw. Dental replacement became limited to two generations of teeth (diphyodont); because teeth became permanent, deeper roots and alveoli suggest the diphyodont condition. Certain foramina in the skull, allowing for passage of nerves and blood vessels, suggest movable lips, a feature necessary for sucking milk from nipples.

Some noncranial skeletal features changed as some Triassic reptiles became increasingly mammalian. The appendicular skeleton is more vertical than lateral, so that not only is the body held higher but the limbs move in a plane parallel to the axis of the body. The movement of the hind limbs is reflected in the increasingly lateral position of the head of the femur as the latter assumes a vertical position. As a result of these changes, running speed is increased and (with the elevation of the head) vision improved. These changes frequently occurred with elongation of the cervical vertebrae; although almost all mammals have seven cervical vertebrae, they are elongated in most mammals with long legs. To provide attachment for the muscles that hold the head, the cervical or thoracic vertebrae have high neural spines.

Early Mammals

Possibly the first mammals were nocturnal endotherms. They are known to have been small, with well-developed olfactory and auditory systems, suggesting nocturnality.[199] Almost all small mammals today are nocturnal, possibly for thermoregulatory reasons. With incipient endothermy, nocturnal mammals can leave their burrows for short periods and return to their tunnels for a stable ambient temperature. Thus they avoid both hypo- and hyperthermia, even if they are somewhat thermolabile. Incipient endothermy, however, probably enabled early small mammals (Fig. 1.6) to become nocturnal insectivores. Nocturnality would also enable them to avoid most diurnal predators, and they could forage for insects as easily at night as in the daytime.

Viviparity, together with lactation, produced an important behavioral change, which probably accounted for the great success of mammals in the Cenozoic. With lactation, there was a period of the newborn young's depending on the mother, with a decrease in early mortality (compared to reptiles). Parental care, only in its incipiency in a few reptiles, became elaborate and prolonged in some mammals, presenting an opportunity for parental influence on their young.

Fig. 1.6. Reconstruction of *Megazostrodon rudnerae,* one of the earliest known mammals (Crompton, Taylor and Jagger, 1978, courtesy of Nature).

This brief résumé of vertebrate phylogeny points out some of the features that have differentiated the major groups, and suggests the backdrop against which these changes may have occurred. Skeletal features, apparent in the fossil record, indicate some aspects of the biology of ancient forms, but, for the most part, lives of prehistoric forms remain obscure. Their diversity was extensive, and their biology may in many cases be inferred from the biology of living vertebrates. The living groups are the subject of the balance of this book.

Suggested Readings

Andrews SM, Miles RS, Wallar AD (eds) (1977) Problems in vertebrate evolution. Linnean Society Symposium Series No. 4

Carroll RL (1969) Problems of the origin of reptiles. Bio Rev 44:393–432

Carroll RL (1970) The ancestry of reptiles. Philosophical transactions of the Royal Society of London, ser. B. 257:267–308

Halstead LB (1968) The pattern of vertebrate evolution. Freeman, San Francisco

Hildebrand M (1974) Analysis of vertebrate structure. Wiley, New York

McFarland WN, Pough FH, Cade TJ et al (1979) Vertebrate life. MacMillan, New York

Schmalhausen JL (1968) The origin of terrestrial vertebrates. Academic, New York

Stahl BJ (1974) Vertebrate history: problems in evolution. McGraw-Hill, New York

2. Mechanics of Evolution

What forces account for the tremendous diversity and profound biologic adaptations seen in modern vertebrates? Although certain major climatic changes are known to have altered past environments, the evolutionary changes of ancient groups must be inferred largely from genetic changes in living forms. Much of the experimental investigation of evolutionary trends involves insects, especially fruit flies *(Drosophila),* but generally evolutionary patterns are assumed to apply to all animals.

Requirements for Selection

Evolution through natural selection depends on several conditions. There must be individual and random genetic variation, including mutations, many of which are recessive and therefore hidden. These alleles must differ in survival value, so that the more desirable traits tend to increase at the expense of those with a lesser survival value. In species maintaining their numbers after reduction by prereproductive mortality, there is a continuous loss of individuals of all ages. In most species, mortality is greatest before sexual maturity. Selection can affect the frequency of alleles through nonrandom mortality of the prereproductive individuals.

Over geologic time, environmental change alters selective forces. Changes in climate or land form can cause movements of animals from one region to another—with exposure to new species with which to compete—and also present new physical factors to which the fauna must adapt. The total effect of these processes alters the genetic composition of species of animals. These changes are apparent in the vertebrate fossil record, and a brief review of evolutionary mechanisms will suggest how genetic composition is altered by selection in nature.

Rates of Evolution

Within a given phylogenetic line of fossil material, there are sometimes rather sudden changes of morphologic traits, and if the fossil record is rich enough, each taxon may be represented by few or many specimens. The dating of such fossils is often not sufficiently precise to determine whether rate of change per unit time is constant or whether change proceeds in brief spurts. Differences in rates can be seen, however, by comparing evolutionary change of two or more taxa through the same geologic strata: that taxon showing the greater difference is the one having the more rapid rate of evolution. For example, during late glacial (Würm or Wisconsin) and postglacial times, some European mammals have become markedly smaller, and a few have become larger (Fig. 2.1).

Actual dating of changes may be inferred from timing of other events. For example, terrestrial animals isolated on continental islands since the last glaciation may be compared morphologically and the differences taken as their changes over 10,000 years. This evaluation assumes that they were similar when the islands were connected to the parent land mass. Many other kinds of faunal movements occurred in the Pleistocene, and probably accelerated rates of morphologic changes. In the forested regions of tropical South America and Africa today are numerous vertebrates that were isolated during glacial arid periods in the Pleis-

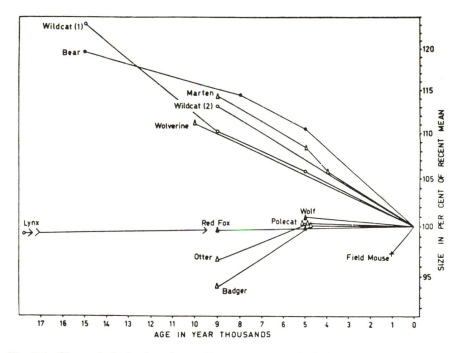

Fig. 2.1. Change in body size of some European mammals during late glacial and postglacial times (Kurtén, 1960, courtesy of Cold Spring Harbor Laboratory).

tocene. Today they occur as subspecies with allopatric but contiguous geographic ranges; the morphologic differences that distinguish these subspecies are presumed to have developed during the last glacial epoch, or between 30,000 and 10,000 ybp (years before present).

Morphologic changes over a longer interval can be seen in some South American mice (Cricetidae) that entered from North America following the closure of the Panama Portal at the end of the Pliocene, about 3,000,000 ybp. Of 39 genera of Neotropical cricetid mice, seven occur north to Central America and only three extend to North America; this suggests that over 3 million years changes had reached the generic, but not yet the familial level. Unfortunately it is not known how many Nearctic genera gave rise to these 39 Neotropical genera. Over the same period there seems to have been less differentiation (i.e., a slower rate of evolution) among the cricetid mice in North America. Invasion into a new region with new faunal associates probably accelerated the evolutionary rate of the mice invading South America.

Selection and Adaptation

As selection tends to eliminate some individuals sooner than other (better adapted) individuals, selection slowly alters the overall genetic composition of the population. This process requires that such elimination occur before the end of the individual's reproductive life, and elimination will have a greater evolutionary effect if it occurs early in the reproductive life or before sexual maturity. The genetic composition of the population is a composite of the genotypes of its individuals, and these genotypes are usually close to an optimum for the conditions at any given time. As conditions change, however, selection operates and the genotype tends to shift toward a new optimum. A change of conditions is commonly a climatic change or a faunistic change. Such changes occur constantly. Some of them are slight and temporary, such as a 2–3-year decline in annual rainfall. Other major changes, such as glacial epochs or climatic changes, are usually accompanied by faunal movements.

An adaptation that benefits the individual is ipso facto beneficial to the species. In addition, for many years there has existed the concept of altruism. It involves the postulated occurrence of a trait deleterious to the individual but, at the same time, advantageous to the species. The existence of hereditary altruism is difficult to prove or disprove, but has provided the basis for much recent discussion.[224,664]

Preadaptation

Preadaptation is a concept that, like "altruism," means different things to different people; it may indeed be a concept that occupies a greater role in literature than in nature. Typically, preadaptation refers to a trait that is nonadaptive in one generation but which may become adaptive in a later generation that occupies a different environment. For example, lungs in some air-breathing fishes are sometimes said to have been a preadaptation to terrestrial life. This statement overlooks the fact that lungs are essential to some fishes and that some definitely terrestrial amphibians (some Plethodontidae) lack lungs. The loss of flight in some insects on oceanic islands is occasionally cited as an example of preadaptation because winds tend to blow insects off islands. Most insects clearly were not flightless before reaching the island, and the adaptation *followed* their establishment on the island. In one sense, virtually all adaptations are preadaptations, for, if they are lacking when the need arises, the individual faces a high risk of early demise. "Preadaptation" is part of the verbal armament of many students of evolution.

Extinction and Survival

The prevalence of extinction overshadows that of long-term survival. The vast majority of known taxa has disappeared and left no descendents. Certainly extinctions occur for a great variety of reasons. First one may assume that an environmental change is either so sudden or of such magnitude that there is no time for adaptation or accommodation. A fairly obvious example is the extinction of marsupial carnivores in South America following the establishment of the Isthmus of Panama, which allowed placental carnivores to enter South America. The disappearance of many terrestrial vertebrates during the Pleistocene could have resulted from sudden changes of climate, shifts in fauna or predation by aborigines ("Pleistocene overkill").[650,651] Whatever the cause, for many species the rate of adaptation was too slow.

At any given time, two conditions characterize an environment: (1) the climate and other abiotic factors and (2) faunal and floral elements. Scarce species have relatively small gene pools and are poorly equipped for rapid genetic change. An abundant species has both greater real and potential variability, and it can suffer great mortality and still survive as a newly adapted remnant. Generally, in a given area, a large species has fewer individuals than has a small species, and there are usually fewer individuals of a predatory than of a prey species. Thus population size (the density of numbers and the geographic range) in itself is an important factor in survival. Variability (and mutation rate) can differ within populations of the same size, and greater variability is tantamount to greater adaptability.

Extinction or survival are determined by factors other than the genetic aspects of a population. Some environments are inherently stable, whereas others are notoriously variable; the latter impose severe demands on its inhabitants. The two most important environmental factors are temperature and precipitation; in some regions they are extremely variable, both over short and long periods, and in other regions relatively stable. In eastern North America and eastern Asia stability has preserved many evolutionary lines of both animals and plants, some of which have occupied these areas since or before the Mesozoic era. These ancient lineages include the paddlefish *(Polyodon),* alligator *(Alligator),* giant salamanders (Cryptobranchidae) and pit vipers *(Agkistrodon),* all groups that no longer occur in intervening and less stable regions.

By way of contrast, western North America has seen a constant reduction of humidity and retraction of forested areas throughout the Cenozoic, as well as extensive mountain-building at the close of the Cenozoic and during the Pleistocene.[34] The greatest stability exists on ancient islands, such as New Zealand and Madagascar. They are the homes of Mesozoic faunas that have disappeared on continents. Even the small and isolated Seychelles Islands in the Indian Ocean have surviving caecilians and a relict family of frogs (Sooglossidae). Thus population size and variability are not the only factors affecting extinction. Relatively small populations can survive for extremely long periods in a favorable environment.

Geography and Subspecies

Much of the phenotypic variation within a species is correlated with its geographic distribution. The differences (as in bill-shape of birds, color, size and scale counts) may be minor, but they may also be sufficiently consistent within a segment of the overall geographic range that such a variant may be designated a subspecies or geographic race. These differences are usually genetic and presumably adaptive; they seem to occur in all but the most geographically restricted species. Populations of fishes, distinguished by subtle meristic differences, are clearly distinct, morphologically and geographically, but are called *races* and are not usually given scientific names. Among most vertebrates, however, a race refers to a subspecies that *does* bear a scientific name.

The study of geographic variation from earliest times has suggested, in one way or another, the role of geographic isolation in speciation. In most cases closely allied taxa, especially at the subspecific or specific level, occupy adjacent geographic regions, with or without contiguous boundaries: less closely related, but perhaps congeneric, species may occur in the same region but in different habitats. This concept is an old one and is sometimes called Jordan's law. The concept that closely allied but sympatric species occur in different habitats and thus avoid direct competition is commonly called competitive exclusion. This is a logical corollary of Jordan's law. It follows that closely related species are morphologically and physiologically similar; when sympatric, competition is minimized if species are biologically dissimilar.

The unit of the taxonomist is the species, a group of interbreeding individuals with similar morphologic and physiologic features. Most vertebrate species are bisexual, and fertility is assumed to exist among mature members of the opposite sex. A species consists of one or more breeding assemblages called *populations* or *demes;* any individual of a deme is, presumably, spatially close enough to every other (of the opposite sex) for reproduction. Such a population or deme might range from hundreds confined to an island to thousands on a continent. In some species, such as ducks that form pair bonds on their wintering grounds, a deme may be very large. A species may consist of one or many demes.

Considerable conspicuous genetic variation may be seen within a species: an individual variant, or morph (such as a color variant), may have a survival value close enough to the prevalent form so that both of them occur commonly. Some vertebrates are dimorphic or even polymorphic, and the most frequent morph varies from one region to another. In southeastern Canada the redbacked vole *(Clethrionomys gapperi)* is sometimes a chocolate color, and this morph was once thought to be a different species, *Clethrionomys fuscodorsalis.* Such color phases are not uncommon in many animals. Regional variation is extremely common, especially in widely distributed species, and such variants are frequently named as geographic forms or subspecies.

Ecogeographic Rules

Together with the recognition of speciation, subspecies and races came the awareness of frequent patterns of variation associated with one or more environmental features. Common types of geographic variation have been formalized in several ecogeographic rules. Among teleosts, for example, populations reared in cold waters tend to have more vertebrae than those from warmer waters. This phenomenon has been observed in nature and reproduced experimentally, and has been referred to as Jordan's rule (not to be confused with Jordan's law, referring to geographic speciation).

Among many birds and mammals, those from high latitudes are frequently larger than their temperate or tropical counterparts. This apparently reflects the decrease of surface-to-mass ratio with an increase in mass, reducing the surface from which heat is lost. This increase in body mass with a decrease in environmental temperature is Bergmann's rule; it is seen in many species of endotherms, but logically not in fossorial or hibernating mammals. Not only does body size increase, but appendages decrease in length, with a decrease in temperature. The significance lies in the appendages, such as ears in desert hares, being important for the dissipation of body heat and, whereas this geographic variation of appendage size is sometimes called Allen's rule, it is the same phenomenon as that subsumed by Bergmann's rule. In birds, heat loss occurs through the bill, and birds in warm regions frequently have longer bills than their boreal fellows.

These generalizations have exceptions, but Bergmann's rule is widely applicable, and the exceptions usually have some obvious environmental basis. Among reptiles, smaller forms sometimes occur in colder regions, but there are exceptions. A small heliotherm acquires heat more rapidly than a large one, but other factors are involved. Ectotherms have indeterminate growth, and a small population of predators may result in greater mean longevity and greater mean size. Ecogeographic rules have stimulated a large amount of research, frequently with conflicting results. One should bear in mind that the environmentally correlated features are not necessarily genetic.

Sexual Selection

Sexual selection can be intra- or intersexual. In some mammals aggression between males is more important than mate selection by females (females seemingly being nonselective), and in such species large size contributes to the dominance of the male. In sea lions (Otariidae) there seems to be a greater attachment of a female to a particular rookery and a tendency to return annually to the same site for parturition. Mating promptly follows parturition, and the females may stray at this period. In many groups of terrestrial mammals, including some artiodactyles in which males exceed females in size, females are active in mate selection.

Sexual dimorphism in size follows patterns only within certain groups of mammals, and these patterns are difficult to apply to mammals as a class. Although males of polygynous species tend to be large, as do many pinnipeds, there is also marked dimorphism in many weasels (Mustelidae) and in whales which, while they may or may not be polygynous, never develop harems as do sea lions (Otariidae). There is marked sexual dimorphism in small mammals as well as in large species, and there is certainly no single underlying determinant of sexual dimorphism in mammals.

In birds there is sometimes a conspicuous sexual difference in form and color of plumage, and in some species one sex is larger than the other. In birds in which the male mates with several or many females (polygyny), he is usually larger than his mates. Among raptorial birds females may be much larger, a phenomenon for which there are many possible explanations, with no one being generally accepted.

Although sexual colorations among many lizards and fishes vary, among most amphibians and many reptiles males and females closely resemble each other; in these groups sexual recognition (and presumably selection) is more commonly based on odor and behavior.

Whatever the means of sexual recognition, there seems to be frequent strife among males. This struggle, whether mild or violent, is for territory and mates, and the distinction is not always clear. The most vigorous males presumably claim the best territories and, whether the female selects a territory or a mate, the result is the same. Females are not always passive in mate selection; nor does fidelity always prevail.

Frequently specific features apply to only one sex, and the greater differentiation is usually seen in the male. Such secondary sexual characteristics can be in form, colors, size, voice and sometimes odor. Sexual characteristics are selected by other males or by females of the same species, and probably some characters are selected by both sexes.

A secondary sexual trait may be a voice or song to which both sexes respond. Males of many species of birds initiate the reproductive season by establishing territories, at which time females are frequently absent. Display and song are for other males, but the same display and song serve for sexual recognition when females arrive. Sexual selection is mostly a product of responses of the same sex in some species and also of the opposite sex in others. There is no reason why both sexes cannot assist in building and reinforcing sexual selection. In reality it is difficult to determine if a sexual feature serves a purpose beyond sexual and specific recognition. Putting the question differently, does a female entering or passing through a series of established territories select a territory or a mate?

Genetic Stability

Wild populations are generally polymorphic at a large number of loci. In large populations (free from the effects of genetic drift) in which random mating occurs, the frequency of alleles remains constant from one generation to another. This concept, known as the Hardy-Weinberg law, assumes no selection of any of these alleles and no introduction of additional alleles, conditions that probably never exist in nature. The significance of the Hardy-Weinberg law in natural populations lies in the tendency for the perpetuation of an extremely polymorphic gene pool so that, when environmental conditions change, there is an ability, previously latent, for the population to give rise to a different allele frequency, one better suited to the changed conditions.

The conditions required by the Hardy-Weinberg law effectively preserve the genetic status quo, but several natural events alter gene frequency. (1) Movement of individuals into and from a population is a common event, especially pronounced in the prereproductive period of animals, and this movement produces *gene flow.* Emigration (i.e., dispersal) occurs in virtually all populations; dispersing individuals change the allele frequency of the species in the area they leave as well as in the area they enter. (2) Mutations appear continuously, but the rate varies; unless the appearance of an allele (by mutation) balances its disappearance (by dispersal or selection or both), mutation alters allele frequency. (3) In small populations *genetic drift,* the random fixation of some alleles (and the elimination of others), produces a marked and sometimes rapid change in allele frequency. (4) Selection, by a nonrandom elimination of individuals before (or during) reproduction, produces a different allele frequency in the subsequent generation.

Expression of alleles is always partly an effect of the environment, and some characters are extremely flexible. Although, for example, many circadian and circannian rhythms are genetically fixed, their appearance is scheduled by environmental cues. For example, in some ground squirrels (*Citellus* spp.) that hibernate, spring activity is determined partly by soil temperatures and varies from year to year. If the timing of vernal arousal were rigidly fixed, the squirrels would be forced to forage in deep snow and cold temperatures in years of late winters; this would cause a drastic depletion on lowered stores of body fat. Both brood and clutch size are also genetically fixed and environmentally responsive to food supplies.

The goshawk *(Accipiter gentilis)* commonly rears from two to four young annually, but, in years when their prey species are depressed, goshawks may not breed. Some voles (*Microtus* spp.) are also responsive to diet: substances in their food can increase litter size, presumably by affecting ovulation. Neoteny, in many salamanders, is a highly variable response to environmental conditions, although the ability to become neotenic is most certainly genetic.

This flexibility in expression of alleles is, on the one hand, an important buffer between the environment and the genotype: it provides for survival of a genotype under unfavorable conditions, while permitting full expression under the most favorable conditions. On the other hand, this phenotypic flexibility reduces the selective effect of the environment. Probably the expression of many alleles varies with external factors, and in this sense these loci are partly immune to selection by these factors. This elasticity is obviously genetic, but the specific phenotypic expression (brood or litter size, timing of hibernation, and so on) is both adaptive and nongenetic. These alleles include not only various aspects of reproduction but also growth and development.

Genetic Drift

Changes may occur more quickly in a small population or in a series of isolated populations, and nonadaptive characters tend either to be fixed or lost. This is genetic drift, or the Sewall Wright effect, and is characteristic of small isolated populations, or demes. Presumably much Pleistocene evolution, which has resulted in subspecies, reflects differences that may have started as genetic drift.

Genetic drift can arise from several different situations characterized by a small population at one point. A small population possesses an effective gene pool of its breeding adults; the next generation, the product of the gametes of this gene pool, is even more limited in variability. Because the gametes are but a small sample of an initially small gene pool, some genes are of necessity never passed on but instead disappear with the parent generation. The effect is to concentrate genes in a random (nonselective) manner.

Genetic drift can result from a population's being continuously held to a low level. About the periphery of its range in northeastern United States, the red-backed vole, for example, occurs in small isolated pockets of swamps of spruces and other conifers. These populations are restricted and local and have probably been so for several thousand years. Some of these populations have been named as subspecies; many other unnamed populations occur, and a scarcity of named taxa does not indicate a rarity of genetic drift.

The establishment of a species to an oceanic island may be associated with an extremely small gene pool, perhaps a single gravid female. This particular example of genetic drift is called the *founder effect*.

Genetic drift may follow fluctuations in population as well as deme isolation. When a population is low, subsequent generations will necessarily result from a restricted gene pool, a small part of which existed prior to the population's decline.

Much of our knowledge of natural selection comes from carefully monitored breeding experiments, and by inference this information is applied to observable natural situations. One can, for example, produce genetic drift in the laboratory, but one can only *assume* that certain isolated and distinctive wild populations constitute natural genetic drift. One can observe the result of 10,000 years of isolation of a mouse of a coastal island, but one cannot observe the process by which this population became distinctive. Ironically, much of the testing of genetic theory is conducted in the laboratory, whereas the significance of such work lies in nature. Although there may be few people of sufficient intellectual breadth to be genuinely competent in both population genetics and natural history, it is incumbent on each investigator to become familiar with the activities of the other.

Speciation versus Phyletic Evolution

Evolutionary change involving geographic isolation is called speciation, in distinction to phyletic evolution, or the gradual modification of a line over time but within a given locality or area. With isolation and stability, a continental area or a very large island is somewhat buffered from the forces that stimulate genetic change; over geologic time such an area will have a fauna of (1) endemic species and sometimes genera and (2) a rather low ratio of species to genus, such a fauna having been derived mostly from phyletic evolution.

Among large islands, New Zealand and Madagascar are reasonably well isolated with a high degree of endemicity. Their isolation dates from the Cretaceous, and their early history saw them much closer to continental faunas than they are today. The southern beech *(Nothofagus)* forests of Patagonia have an uncommon fauna, with many endemic species and some endemic genera of vertebrates. These *Nothofagus* forests have been isolated (by arid barriers) from the rest of South America since perhaps as early as the Miocene age. The trend in evolution in these isolated regions is phyletic and occurs with time, and this does not produce an increase of species at any given period.[1061]

Suggested Readings

Cracraft J, and Eldredge N (1979) Phylogenetic analysis and paleontology. Columbia University Press, New York

Grant V (1977) Organismic evolution. Freeman, San Francisco

Johnson C (1976) Introduction to natural selection. University Park Press, Baltimore

Mayr E (1963) Animal species and evolution. Harvard University Press, Cambridge, Massachusetts

Rensch B (1959) Evolution above the species level. Methuen, London

Simpson GG (1953) The major features of evolution. Columbia University Press, New York

3. Zoogeography

As we have seen, the genetic composition of a population responds to both physical and biotic shifts in the environment. Ultimately these genetic changes bring about more profound phylogenetic changes, such as those briefly described in the first chapter. Zoogeography and evolution are opposite sides of the same coin, and the mechanics of natural selection explain their relationship.

Zoogeography deals with the ancient and modern geographic distribution of faunas. It involves two major factors: dispersal and vicariance. Dispersal is the movement or extension of a species into a new region. Prior to the mid-1960s it was thought to account for most present distributions. However, evidence firmly establishing the concept of plate tectonics has resulted in a major revision of the role assigned to vicariance, the movement of an entire biota with the drift of the continental plate that it occupies.

We therefore begin with the geologic history of the continents in the framework of the zoologic consequences. Dispersal and the role it plays will then lead us to describe today's faunal distribution.

Disjunctions and Endemism

An important measure of the distinctiveness of a fauna is the frequency of unique or *endemic* taxa, those that are native and confined to the area in question. They can be either young groups that are expanding or old taxa that persist in one place and are extinct in former parts of their range. Today, numerous species of kangaroo rats and pocket mice (Heteromyidae) occur in North America; they are evolutionarily vigorous and endemic to that continent. The lemurs (Lemuridae) are endemic to both Madagascar and the nearby Comoro Islands and are an ancient stock. Some endemics consist of single species, once widely distributed, that exist today only in restricted areas. The tuatara *(Sphenodon punctatus)* remains now only on several small islands off the coast of New Zealand but formerly occurred on continental areas in the Mesozoic era. This sort of endemic, the ancient survivor, is a *relict*. It is no accident that relicts occur in *refugia,* areas that have been relatively stable for long periods of geologic time. New Zealand is a refugium for many taxa that probably occurred in other regions in the Mesozoic. The three species of kiwis *(Apteryx),* now only in New Zealand, must have occurred on continental areas before the separation of New Zealand from Antarctica.

Many modern plants and animals occur in areas separated by unoccupied habitat. Such *disjunct* distributions are often at the family or generic level, and less commonly one species lives in two distant areas. Disjunctions are often inexplicable, considering the distances that sometimes lie between these populations, but others clearly reflect the climate or the geologic history of the regions, or both.

Many intercontinental disjunctions involve a single species, and these are almost always birds or mammals. In the far north, North America and Eurasia are close together, and arctic taxa are usually Holarctic; technically they occur on different continents and by that token are disjunct (Fig. 3.1). The breeding range of the snow bunting *(Plectrophenax nivalis)* is genuinely Holarctic but actually occupies a rather small part of the earth. This species certainly had much wider distributions during the Pleistocene and probably occurred over the Bering Land Bridge. Farther south are some birds that occupy the great boreal coniferous forests; the pine grosbeak *(Pinicola enucleator)* is a breeding bird in most of the spruce-fir forests of the Northern Hemisphere. Still farther to the south, the wren *(Troglodytes troglodytes)* is a dweller of mixed coniferous and broad-leaved forests.

Some animals today exist in widely separated areas and presumably were once continuously distributed. Frogs of the family Ascaphidae (= Leiopelmidae) today occur in New Zealand (where there are three species of *Leiopelma*) and in extreme western North America (where there is one species, the bell toad, *Ascaphus truei*). Because one cannot imagine that a small frog, especially one that inhabits cool damp forests, has *dispersed* overseas and crossed the equator to span the distance between its two areas, one must assume that: (1) its evolutionary history dates from at least the Cretaceous, when New Zealand was rafted from Antarctica and (2) it once had a continuous terrestrial distribution. For these two conditions to be met, probably its present homes have not been greatly modified since the Mesozoic, and the areas in which ascaphid frogs now survive are refugia. These frogs are relicts, survivors of an ancient fauna.

Other disjunctions are the various *pantropical distributions,* in which there are more examples among the older classes of vertebrates. There are pantropical fishes, such as lungfish, which occur in Africa *(Protopterus),* South America *(Lepidosiren)* and Australia *(Neoceratodus).* The caecilians (Apoda) are legless amphibians, which survive in a humid equatorial belt, including islands such as the Seychelles and the Philippines. An ancient group of fossorial reptiles sometimes called worm lizards or graveyard snakes (which are neither lizards nor snakes), the amphisbaenians, are found today on not only all tropical continents

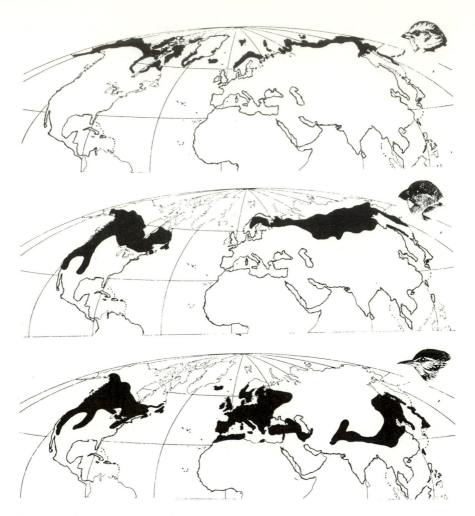

Fig. 3.1. Three Holarctic birds. Top: the snow bunting *(Plectrophenax nivalis)* nests in the far north, in an almost continuous distribution in the arctic tundra. Middle: farther to the south the pine grosbeak *(Pinicola enucleator)* occupies forests of spruce and fir. Bottom: the wren *(Troglodytes troglodytes)* is more widely disjunct in mixed forests of pine and broad-leaved trees of temperate latitudes (Suzanne Black).

but some islands, such as the Greater Antilles (in the West Indies) and Socotra (in the Indian Ocean). There are fewer pantropical birds, but parrots, trogons and the ratite birds are perhaps the best examples of nonmigratory birds inhabiting all the tropics.

There is an apparent correlation between modern pantropical disjunctions in different classes of vertebrates and the position of the continental blocks during various stages of vertebrate evolution. Continental continuity was extensive during development of families of fishes, amphibians and reptiles, but most avian and mammalian evolution occurred after the modern continents had parted.

Background to Plate Tectonics

Since the days of early marine exploration and the first charts of coastal outlines, geographers and geologists have noted the congruent aspects of the Atlantic shores of Africa and South America. Similarity of the coastal outlines of these two land masses suggested that in early geologic time these two continents were one and that their separation created the Atlantic Ocean. Subsequently maps of the seafloor showed that the two submerged continental shelves matched even more closely than do the coastal outlines. Early paleontologists produced fossil evidence that Europe and North America were once connected.

Geologic evidence for continental drift came also from the Southern Hemisphere. Eduard Suess, an Austrian geologist, related outlines and some geologic features of certain southern land masses, and suggested the name Gondwanaland for southern land masses and India. The name Gondwana came from an area in India, for the Indian subcontinent was presumed to have moved northward from the southern region. In various parts of Gondwanaland is found tillite, a glacial rubble left by extensive glaciation in the Permian and Carboniferous. With tillite occur numerous fossils of the distinctive "seed-fern," *Glossopteris.*

Species of *Glossopteris* are important indicators of land-mass relationships because, unlike true ferns, they had heavy seeds that were not windborne. Thus their fossil occurrence on lands widely separated today strongly suggests that an ancient land connection existed in the Permian and Carboniferous. Tillite deposits have been found in India, southern Africa, Madagascar, South America and Antarctica. These isolated lands all have members of the *Glossopteris* flora, with the same species often being represented in two or more areas that are now widely separated.

In the Northern Hemisphere a second ancient supercontinent, Laurasia, was proposed for the origin of North America, Eurasia and Greenland. In the first half of this century a great deal of evidence was prepared by Alfred L. Wegener of Germany and Alex L. du Toit of South Africa. From time to time, biologists, mostly entomologists, argued for continental drift, but for decades they were in the minority. Through the first half of the 20th century the discussion went both ways: one group considered relative movement of land masses obvious, whereas their opponents regarded the subject as absurd. Vertebrate zoologists were conspicuous in the latter group.

The mid-1960s produced new and convincing evidence for continental movement: paleomagnetism, benthic sedimentation, global coordination of seismic studies, bathypelagic temperatures, detailed maps of submarine topography, magnetic anomalies on the ocean floor, and the increasing age of islands away from the midocean ridges. Simultaneously there were new and very convincing kinds of fossil evidence. Today, few serious scholars question continental drift, but its significance to the distribution of recent vertebrates is still subject to varied interpretation.

Mechanics of Plate Tectonics

The concept of plate tectonics provides for continental land masses floating on a molten magma, each major land mass constituting a plate. Within the magma are slow convection currents rising along discrete lines in the ocean basins, these lines being marked by submarine ridges, such as the Mid-Atlantic Ridge. As the magma moves upward, the seafloor moves away from the ridge. The seafloor may raft the continental plates away from the ridge, or the floor may buckle under the plate in what is called a subduction zone. In the latter case the downward movement of the floor may elevate the plate, causing mountain-building or volcanic activity, or both.

It is appropriate to review evidence supporting this concept. In the western hemisphere, the Mid-Atlantic Ridge fits very nicely with the two opposing continental outlines. This evidence is by itself circumstantial, but related studies explain the situation. First, the Mid-Atlantic Ridge not only parallels the opposite continental shelves but lies halfway between them. Samples of sedimentation of the ocean floor are shallowest near the Ridge and increasingly deeper away from the Ridge and toward the continents, suggesting greater age of the deposits nearer the continents and youthfulness of the sedimentary layers nearer the Ridge. Close to the Ridge the strata are about 280 m thick and an estimated 18 million years old, but 800 km from the Ridge the strata are more than 472 m thick and 85 million years old. Second, temperatures of the ocean itself are cooler away from the Ridge than they are close to it, suggesting warmer crustal temperatures below the Ridge. These two features suggest that the Mid-Atlantic Ridge overlies an upward movement of magma and that the new ocean floor moves away from the Ridge. Third and most important is the pattern of paleomagnetism on the seafloor. Periodically the earth's polarity has changed; this has occurred an estimated 170 times in the past 75 million years. As molten rock solidifies, it assumes the magnetism obtaining in the area at that time and, when the direction of magnetism changes, subsequent molten rock, as it cools, adopts the new magnetism. Careful examination of paleomagnetism on either side of the Mid-Atlantic Ridge shows that bands of magnetized seafloor are parallel to, and at equal distances from, the Ridge are equal in direction and width, suggesting that the seafloor at equal distances from the Ridge is of equal antiquity.

The concept of continental drift, then, can be explained by a lateral displacement of the ocean floor from central ridges toward continents, this movement tending to raft continents away from the ridges. Thus the vague and inaccurate expression "continental drift" is replaced by the more precise and descriptive "plate tectonics and seafloor spreading."

Establishment of seafloor spreading constitutes proof of separation of the continents. Both du Toit and Wegener emphasized the similarity of the rocks and geologic formations on the opposing coasts of west Africa and Brazil: matching rock of 600 million and 2000 million years of age on both continents line up precisely on each side of the ocean. Dates and details were established with much greater precision in the 1960s. By evaluating various sorts of evidence, geologists have calculated times and rates of continental separations. The zoogeographic significance of plate tectonics rests primarily on the stage of evolution of the various taxa of terrestrial, fresh-water and littoral animals at the times of continental separations.

The ocean floor, according to the concept of seafloor spreading, is oldest nearest the continental shelves, and the oldest marine deposits in any given basin probably date the beginning of the separation of the adjacent continents. In the South Atlantic Ocean, the oldest marine deposits suggest that South America and Africa began to separate at the southern region in the late Jurassic or early Cretaceous of the Mesozoic, and that the two continents since then have been moving apart at an annual mean rate of two centimeters. The separation was not complete (and a barrier for terrestrial animals) between Brazil and West Africa until the middle Cretaceous.

Separation of Continental Plates

The concept of the existence of the ancient supercontinent, Pangaea, involving the major land masses of the earth, is now accepted by most students. Pangaea persisted intact to at least the end of the Paleozoic. Fragmentation of Pangaea began in a number of areas around the close of the Triassic. The sequence of separations of the major land areas and the consequent zoogeographic implications have been outlined, discussed and reviewed by many scholars.[32,36,132,192,989,990]

The initial fracture of Pangaea resulted in a northern Laurasia and a southern Gondwana, but numerous rifts appear to have occurred at different sites at about the same time. Somewhat different patterns are outlined by different workers, but there is general agreement about the early Mesozoic relationships of the continental plates, the order in which they split apart and the direction in which they moved. Although the breakup started around the mid-Mesozoic, some connections (e.g., North America to Europe, South America to Antarctica and Australia to Antarctica) remained until the Eocene, by which time many families of terrestrial vertebrates were well-differentiated. This situation confirmed the relationship of "continental drift" to modern vertebrate distribution.

North America was connected both to northwest Africa and to Europe, and the initial separation was from Africa. Rifting of the two plates started in the Triassic, about 180–200 mybp (millions of years before present) when Africa and South America were still connected. To the north, North America began to part from Europe along two rifts on either side of Greenland in the early Mesozoic. A connection persisted between northern North America and Europe into the Eocene. North America on the west shares its plate with northeastern Asia, and in the Bering region the two areas were connected intermittently in the Cenozoic.

The initial opening between Africa and South America began in the south, in the Lower Cretaceous, 130–140 mybp, and advanced northward to effect a complete parting of Africa and South America in the Upper Cretaceous, about 100 mybp. North and South America parted when the former separated from Africa.

Central America was an archipelago from the Cretaceous and did not form a continuous connection to North and South America again until late Pliocene. As North and South America were rafted westward, the land segment that was Mesozoic Central America remained, broke up and formed the Greater Antilles.

Antarctica occupied a central position from which other land masses broke off and moved away, during which time it moved southward from the Mid-Antarctic–Indian Rise toward its present position. The original position of India is equivocal; it parted from Antarctica (or possibly Africa) in the Cretaceous, and moved toward Asia from the Paleocene to early Eocene. The former attachment of Madagascar is also uncertain, but it probably separated from Africa in the late Cretaceous. Some students maintain that there was an early connection between India and Australia. The Seychelles and Comoro Islands presumably parted from India as the latter moved north. India joined Eurasia in the region of the future Himalayan Range sometime in the Eocene, and, as India underrode the plate of Asia, the Himalayas began to rise (in the Miocene) (Fig. 3.2).

When Africa and South America began to part in the south, the latter was still connected to Antarctica. The Scotia Arc joined these two plates until approximately the Eocene. Australia and Antarctica also remained connected until the Eocene (50 mybp). At the time these plates were contiguous, Antarctica was roughly 20° north of its present position. New Zealand moved away from Antarctica in the late Cretaceous (80 mybp).

For the biogeographer several important features of this sequence are apparent. (1) As late as the Eocene there were connections between South America and Antarctica, Australia and Antarctica, and between North America and Europe. (2) Northwestern North America and northeastern Asia were sporadically connected by dry land in the Cenozoic. (3) There was no continuous land connection between South and North America from some time in the Cretaceous until the late Pliocene.

Paleontologic Confirmation of Plate Tectonics

Important paleontologic discoveries, especially in the Southern Hemisphere, followed closely the major advances relating to seafloor spreading in the 1950s and 1960s. The ultraconservative critics of continental drift received their severest setback with the discovery of early Triassic amphibians and reptiles in the Coalsack Bluff deposits in Antarctica. These fossils include labyrinthodonts, salamander-like amphibians three or four feet long. The presence of such large amphibians in Australia, Antarctica and the Northern Hemisphere points to a connection of Antarctica to other land masses up to about the early Triassic, and coincides with the calculated rate of the Gondwanaland breakup. The presence in lower Triassic Antarctic strata of the reptilian genus *Lystrosaurus,* also known from India, Sinkiang (in China) and South Africa, validates the relationship of the drift concept to vertebrate distribution.[176] Other lower Triassic reptiles are common to South Africa, Antarctica and Australia, and some of these characterize the known lower Triassic fauna of India.

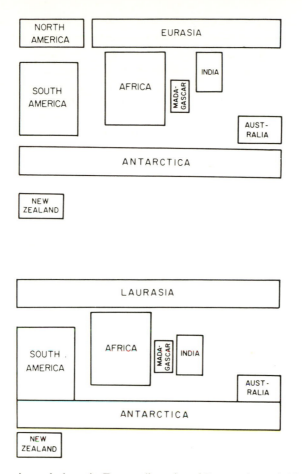

Fig. 3.2. The continental plates in Eocene (lower) and Present (upper) (Cracraft, 1974; reproduced, with permission, from the Annual Review of Ecology and Systematics).

Modern Vertebrate Distributions and Plate Tectonics

Insofar as the continental drift concept has applications to modern zoogeography, distributions of recent vertebrates with long histories must reflect some of the geologic events of the Mesozoic. Had there been no evolution, extinctions or overseas dispersal since the Eocene, we would find orderly Laurasian and Gondwanan patterns of distribution. After many millions of years, however, modern fauna include many taxa that do not reflect ancient continental relationships. Only some elements of fauna distribution relate to plate tectonics. With these limitations in mind, one can find many distributional patterns that are best explained in terms of Laurasian or Gondwanan history.

Although austral disjunctions had been noted for many years, most zoogeographers believed that the taxa concerned had spread southward from their original distributions to their present positions on the southern continents. Most investigators agreed that movements to the southern continents occurred in the distant past, variously pegged "some time in the Mesozoic" or "early in the Cenozoic." The separation of the southern land masses occurred when a number of living families of fishes, amphibians and reptiles had already evolved. In other words, some of the taxa (families and even genera) that are today disjunct existed prior to the rifting of the major southern land masses. These disjunctions can be explained as having resulted from the movement of land masses, rather than from a dispersal of animals.

The Southern Hemisphere has some conspicuous examples of disjunctions involving forms now living in Africa, South America, Australia (including New Guinea and Tasmania) and New Zealand, or any two of these areas. Some interesting examples are found among insects. The primitive hemipterans of the family Peloridiidae have been discussed by several workers: they are minute, flightless fragile creatures living in and feeding on moss in forests of *Nothofagus;* today they occur in South America, New Zealand, Tasmania, southern Australia and Lord Howe Island.[214] The exclusively austral and more-or-less flightless carabid beetles of the tribe Migipondini have a distribution very much like that of the peloridiids; they also have no close relatives in the Northern Hemisphere. There are also conspicuous affinities of some South American aquatic midges to those in Australia and New Zealand, whereas the taxa in the latter two areas are not especially close.[130,131]

Among terrestrial vertebrates are included all primary fresh-water fishes ("terrestrial" in this context), amphibians, reptiles exclusive of marine turtles and snakes, birds exclusive of pelagic families and mammals except cetaceans and pinnipeds.

Fish

Fresh-water fish are restricted to fresh water, although a few can move through brackish or marine water. The latter are presumably derived from marine forms and are called secondary fresh-water fish. This is not to say that the latter do not disperse as do primary fresh-water fish, but simply that there is a greater possibility for movement through marine waters.

Gondwanan disjunctions include two groups of primary fresh-water fish found today in Africa, South America and Australia. The lungfish (Dipnoi) are confined to Africa, South America and Australia. The Osteoglossidae also occur in fresh water on these three continents and also in southeast Asia. There are fossils (Eocene) from fresh-water deposits in North America and Asia, but the major part of the modern distribution of the family is on Gondwanaland masses. Lungfish and osteoglossids are apparently the only primary fresh-water fish in Australia. The remaining Gondwanan disjunctions (Characidae, Cichlidae and Nandidae) are found in South America and Africa. The Characidae include the various species of tetras in tropical fish culture, and the Cichlidae also have many members commonly kept in aquaria. The Dipnoi, osteoglossids, characids and cichlids must all have had a broad Gondwanan distribution in the late Cretaceous.

Laurasian disjunctions, at the family level, are more numerous than are Gondwanan disjunctions, probably reflecting the more recent continuity in the north. Sturgeons (Acipenseridae) occur in many major rivers of the Northern Hemisphere, and the family has had a wide distribution throughout North America and Eurasia since the Mesozoic. Similarly gars (Lepisosteidae) and bowfins (Amiidae) have had fossil records in Laurasia since the Cretaceous, although both groups are confined to eastern North America today. The paddlefish (Polyodontidae), another ancient group, are today represented by two species, one in the Mississippi drainage of North America and the other in the Yangtze River in China; it is clearly Laurasian. Suckers (Catostomidae) and minnows (Cyprinidae) probably had a Laurasian origin, although the minnows occur in northern and central Africa today. The mud minnows (Umbridae) today occur in the Danube and Dniester drainages of Europe and in eastern North America, with a broadly disjunct genus *(Novumbra)* in northwestern United States (Washington). Additional Laurasian fresh-water fish include the pike (Esocidae), perches (Percidae) and blackfish (Dalliidae).

Amphibians

Amphibians spread over land masses much more rapidly than do fishes, and some amphibians seem to have been rafted over salt water. Nevertheless certain amphibian groups illustrate ancient distributional patterns. The caecilians (Apoda) comprise two pantropical families. They are aquatic or fossorial and today occur in Africa, South America, India and Sri Lanka (Ceylon), the tropical Orient and the Seychelles. They were clearly widely distributed in Pangean times.

In contrast to the Pangean distribution of the caecilians, salamanders (Caudata) are a Laurasian group and have probably dispersed through northern land connections prior to the final fracture in the Eocene. Four families (Plethodontidae, Cryptobranchidae, Salamandridae and Proteidae) are all in Eurasia and North America. The Plethodontidae has extended its range into northern South America. Of the remaining families of salamanders, the Ambystomidae occurs widely throughout North America, and the sirens (Sirenidae) and the "Congo eels" (Amphiumidae) are confined to the southeastern Unites States and the lower Mississippi drainage; the latter two are distinct and undoubtedly relicts.

Frogs (Anura), on the other hand, however confusing current taxonomic arrangements, had an extensive early history in the southern continents. The Pipidae are aquatic anurans of South America and Africa; it is extremely difficult to imagine overseas dispersal for this group. Tree frogs (Hylidae) occur in the New World (including the West Indies), Australia, New Guinea and some nearby islands. There is one (or several) species of *Hyla* in Eurasia, almost certainly a rather recent immigrant from the New World across a Bering connection. The Leptodactylidae are similarly distributed in the Australian Region and South America, and barely extend north to the United States; one genus occurs in South Africa, and an Eocene fossil is known from India. The Microhylidae is a very large and widely distributed pantropical family that must have been established before the breakup of Pangaea. Additional anuran families endemic to South America (and some including the Greater Antilles and Central America) clearly show that frogs have had a very long history in the New World tropics: neotropical endemics include the Pseudidae, Dendrobatidae, Centrolenidae, Atelopodidae, and the strange frog-eating frogs, Ceratophryidae.

The paucity of endemic anuran families in the Northern Hemisphere might suggest separate geographic origins for the two groups. Most frogs in the Northern Hemisphere are old groups that are widespread in the world. The Bufonidae is nearly cosmopolitan, being absent in Madagascar and Australia; only species of *Bufo* are Laurasian, and the movement into North America quite probably was through a Cenozoic Bering connection. The Ranidae is virtually cosmopolitan, but only the genus *Rana* is in Eurasia and North America; South American species may represent a post-Pliocene movement. The genera *Sooglossus* and *Nesomantis* (Sooglossidae is sometimes placed in the Ranidae) are endemic to the Seychelles. The movement of the Ranidae into the New World was very probably through a Cenozoic Bering connection. The Pelobatidae (including the spadefoot toads) are clearly Laurasian. The Discoglossidae is distributed throughout Eurasia, and occurs also in the Philippines; there is a North American Cretaceous fossil assigned to this family. The bell toad *(Ascaphus truei)* of western North America and the New Zealand frogs (*Leiopelma* spp.) are the only surviving members of the family Leiopelmidae. Both are very ancient and clearly relicts of a once widespread line.

In summary, one can see that amphibians reflect continuity of continental plates in the Mesozoic. Also (1) caecilians were, and are today, pantropical; (2) most frog families seem to have developed in the Southern Hemisphere where they are still most diverse; and (3) salamanders are, and perhaps always have been, essentially boreal in distribution.

Reptiles

Reptiles were the major element of Mesozoic tetrapod faunas. In view of their antiquity, it seems surprising that there are not more pantropical families than there now are, but some modern families are very clearly associated with Gondwanaland masses.

Many families of reptiles are *incompletely* pantropical: i.e., they are absent from one low-latitude area or another but nevertheless have a modern distribution pattern which suggests that they had wide tropical occurrence in the Mesozoic. Such incompletely pantropical groups are often absent from Africa, where some probably once occurred, and also from Australia, which they may never have reached.

The side-necked turtles, Chelyidae and Pelomedusidae, have probably attained their modern ranges through Gondwanaland connections. The chelyid side-necks today occur in South America, Australia and New Guinea, and their fossil history is confined to the Southern Hemisphere. The pelomedusids were widely distributed in the Cretaceous, and can be regarded rather as austral relicts than bona fide Gondwanan elements. Today pelomedusid turtles are found in South America, Africa, Madagascar, the Cape Verde Islands (in the Atlantic) and Mauritius and the Seychelles (in the Indian Ocean).

Several families of turtles seem to have originated and spread through North America and Eurasia before the Eocene. The soft-shelled turtles (Trionychidae) may have dispersed through northern routes but today occur in Eurasia, North America, Africa and through the Greater Sundas to New Guinea. The family is absent in Australia and South America. Other boreal turtles are either restricted, as are the snapping turtles (Chelydridae), or widespread (Testudinidae). Possibly the latter dispersed from the north into Africa, for they are absent in Australia. Testudinidae is an old family and occurs on the Seychelles, Madagascar and the Galápagos.

The tuatara *(Sphenodon punctatus)* of New Zealand is the last of a great variety of reptiles of the order Rhynchocephalia that was widely distributed in the Mesozoic. It is a Mesozoic relict from Pangaea.

There remain several reptilian families that are ancient and (in most cases) were apparently widespread before the initial breakup of Pangaea. This situation would provide for both a Gondwanan and a Laurasian dispersal.

Amphisbaenians (suborder Amphisbaenia) are burrowing reptiles that superficially resemble snakes. Today they are pantropical (and are also on the Greater Antilles), but in the Cenozoic they occurred across Laurasia. One genus *(Trogonophis)* is on Socotra Island (off the coast of Arabia), but the family is absent from both Madagascar and the Australian Region. Their fossorial habit suggests that their occurrence in the Greater Antilles and on Socotra reflects vicariance and not dispersal.

The lizard family Iguanidae presents one of the most interesting Gondwanan distributions because of insular forms in the Pacific, its presence on Madagascar, and the apparently subsequently evolved family Agamidae. Their occurrence is complementary: the agamids occupy temperate and tropical Eurasia, including the East Indies, parts of the Australian Region and Africa, except Madagascar. The most plausible explanation for the agamid–iguanid distribution is that iguanids were ancestral in Africa, and, after the separation of South America and Madagascar from Africa, the African stock evolved into or was replaced by agamids. Southeast Asia is rich in agamid diversity, and the family may have originated there as well. Iguanids were most certainly rafted to the Galápagos, and perhaps also to the Fiji and Tonga Islands, continental fragments that today have endemic iguanid lizards.

Chameleons (Chamaeleontidae) are a typical Old World Gondwanan family and today occur in Africa, Madagascar, India, Ceylon and the Seychelles. Geckos (Gekkonidae) are pantropical, and their arboreal habit has enabled them to disperse to many oceanic islands. New Zealand geckos are related to those in Australia and undoubtedly were rafted to New Zealand long after its departure from Antarctica.

Snakes constitute a fossorial offshoot from reptiles, and apparently arose during the lower Cretaceous. Nevertheless they frequently reflect plate tectonics in their modern distributions.

Two families of small fossorial snakes are called "worm snakes": The Typhlopidae is probably the more ancient and occurs throughout the warm regions of the world, including Madagascar and the Australian Region. The presence of typhlopids in the West Indies could be attributed, partly at least, to these small snakes' hiding under loose bark or logs; they could passively disperse by rafting. Some kinds are very small and have inadvertently been carried in the soil of potted plants. The Leptotyphlopidae is superficially similar to the Typhlophidae and occurs in parts of Eurasia, Africa and the tropical New World, even reaching the southcentral United States, but is absent from the Australian Region and Madagascar.

Some other reptiles have disjunct tropical distributions today. The primitive snake family Aniliidae occurs today in the tropical orient and northern South America. Another primitive group of colubrid snakes, the snail-eating Pareinae (including Dipsadinae), is represented by species in two genera in southeast Asia and in three genera in the neotropics. Boid snakes are pantropical and ancient. As are pelomedusid turtles and iguanid lizards, the boas of Madagascar are most closely related to those in South America. The lizard family Xenosauridae contains two species: one in southern China, the other in southern Mexico. There is no fossil record for this family. Crocodilians date from the Cretaceous and are pantropical today.[926]

Birds

Birds and mammals reflect Mesozoic geography much less clearly than do fishes, amphibians and reptiles. Although the two classes of endothermic vertebrates were well-established by the late Cretaceous, most modern families evolved during the Cenozoic.

Examples from birds are limited to several that illustrate Mesozoic or early Cenozoic dispersal. Even though birds are capable of long overseas movements, and some of them fly thousands of miles annually, they are generally restricted to a geographic range that is smaller than the apparently suitable or potentially available range.

Some students of phylogeny and geographic distribution of birds remain adamant that plate tectonics has had little or no effect on the modern distributional patterns of birds. The occasional tenacity of outmoded concepts testifies that science is not always free from subjective evaluations.

At least 20 families of birds existed by the close of the Mesozoic, and the actual diversity was perhaps much greater; early Cenozoic birds were of even more types. Eocene faunas included ratite birds, hawks, owls, parrots, trogons and others, at a time when Eurasia was still connected to North America (through Greenland), and South America and Australia were still part of a temperate Antarctica. The Bering connection apparently persisted up to the middle of the Cenozoic (Miocene), at which time plant fossils indicate a warm humid climate for northeast Asia and northwest North America.[36]

Several groups of birds have distributions today that must reflect late Mesozoic or Eocene connections in the Southern Hemisphere. Two of these, penguins and ratite birds, are flightless. Ratite birds are large, and there is no real evidence that they are not a monophyletic group. One allied family, the tinamou, is capable of flight. Ratites include the Rheidae of South America, Apterygidae and Dinornithidae of New Zealand, Casuariidae and Dromiceidae of the Australian Region and Struthionidae of Africa. Fossil ratites include elephant birds (Aepyornithidae) from Madagascar and the eastern Canary Islands as well as from Africa, and some fossil ostriches from Eurasia.[192]

Parrots (Psittacidae) constitute another major and primarily austral group. They are pantropical today but occurred in northern continents in the Cenozoic. Avian antiquity in Australia, New Zealand and South America is suggested by the rich and diverse parrot fauna in these areas, and a Gondwanan origin probably accounts for much of the Southern Hemispheric distribution of parrots today. Other essentially Gondwanan elements include galliform birds, pigeons, trogons and cuckoos.

Mammals

Mammals in the Southern Hemisphere are often endemic at the family level (without transoceanic disjunctions), and intercontinental relationships are rather difficult to establish. Most endemic African and South American mammals are fairly distinct from those of northern regions today, and they are distinct from each other as well. The fossil record and apparent relationships do not indicate a general exchange of mammalian stocks between these two land masses.[934] South America differed also in having a diverse marsupial fauna, which is not known to have occurred either in Asia or Africa; fossil marsupials are known from the Eocene of Europe. The South American fauna is further complicated by the sudden arrival of platyrrhine monkeys and caviomorph (= hystricomorph) rodents in the Oligocene. Although on geographic grounds a North American origin for these two groups seems plausible, North American fossils do not include probable ancestors for these mammals. On phylogenetic grounds, the platyrrhines and caviomorphs must have come from Africa, necessarily over a marine barrier, when Africa and South America were much closer than they are today.[446] Marsupials offer the one unequivocal example of mammalian distribution throughout Gondwanaland. Their fossil record is old (Cretaceous) in North and South America, and their lineage is a long one in Australia. Thus, although the origin and direction of dispersal of marsupials are still undetermined, it is clear that Antarctica was involved.[533]

This brief review is sufficient to illustrate how certain Mesozoic and Cenozoic geographic changes effected some patterns of modern vertebrate distribution. In the Southern Hemisphere disjunctions are more impressive because of the broad oceans that separate related taxa.

Dispersal

Evolution, speciation, extinction and especially *dispersal* have modulated the zoogeographic distribution due to vicariance, often to a large extent. Modern biogeographic patterns, as we shall see, are consequences not only of the slow workings of tectonic forces but also results of the dispersal of species.

The movement of an individual away from its birthplace, never to return, is called *dispersal*. At the periphery of the species' geographic range dispersing animals may actually extend the range, but usually such range extensions are abortive. Beyond the geographic boundary of the range, environmental conditions are presumably unsuitable for the species' survival. At the edge of the range some sort of barrier may prevent the permanent expansion of given species; the barrier may be a river, an ocean, a mountain range or an ecologic requirement.

Geographic ranges tend to be dynamic, and two events can allow the permanent expansion of the range into new areas. First, environmental changes may create conditions favorable to the establishment of the dispersing animals, and environmental changes are more the rule than the exception. Second, genetic changes from one generation to another may endow a dispersing individual with attributes (preadaptations) needed for survival in the new area.

Corridors, filters and other dispersal routes resemble one another, differing only in degree. For example, during the early Cenozoic, and until some time in the Eocene, a broad connection between northeastern North America and northwestern Europe allowed the exchange of animals. These two regions were farther south than they are today, and this connection enjoyed a mild climate inhabited by a large and diverse fauna. Such a connection is a *corridor*. The Pleistocene Bering connection, in contrast, was a cool grassland, inhabited by a reduced steppe fauna in the late Cenozoic; because the faunal movements across the more recent Bering land bridges were environmentally restricted, it was a *filter*.

Dispersal to islands is unlikely for most vertebrates except birds and bats. Rafting, or waif dispersal, has probably allowed some kinds of rodents, lizards and frogs to be passively dispersed over a marine barrier. Dispersal over ocean water is either by rafting, as on floating vegetation, or windborne. Occasionally strong storms carry birds or bats out to sea, quite apart from those migratory species that regularly visit oceanic islands. Some tropical regions are subjected to torrential rains and erosion of riverbanks. During extremely heavy rainstorms, trees—including their roots with some soil and the arboreal inhabitants—are transported to sea. If the ground has been water soaked, many ground dwellers may have taken refuge in such a tree before its journey to the sea. In other words, waif dispersal has a real basis and can account for much of the flora and fauna on oceanic islands.

Oversea dispersal is affected by two factors. First, the ability to travel over a marine barrier is virtually absent in many amphibians, fossorial animals such as moles and caecilians, many large mammals and flightless birds. Other forms, such as a few snakes and lizards and some rodents, are able to survive a journey by rafting. Oversea dispersal is impossible for some kinds of vertebrates, and possible but difficult for others. Second, for those groups that can somehow or other occasionally reach oceanic islands, success is largely a matter of chance, such chance being affected by the size of the island and its distance from the mainland. That is, a small island will have fewer introductions than a large one, and proximity to the mainland will increase the likelihood of introduction. In an archipelago, chance alone will favor more species on those islands closer to the continent, accounting for the phenomenon known as the *filter effect,* or the decrease in faunal diversity with increase in distance from the mainland. The filter effect is seen in the faunas of many islands in the Pacific, and has been most carefully documented for insects.

Faunal Distributions

We now turn to an examination of the combined effects of vicariance, dispersal and evolution as reflected by modern observations. Zoogeographic data dealing with the question of which species inhabit(ed) what areas may be dealt with either taxonomically or regionally. With a taxonomic approach one selects a species and determines regions in which it is found. With a regional approach one first selects an area and then characterizes the fauna. Although the methods are equivalent ("Bats live in Madagascar" vs. "Madagascar is a habitat for bats"), each method has its advantages. The regional approach is more suitable for practical biology and fieldwork; the taxonomic is more useful in illustrating long-range vicariance and dispersal patterns.

Zoogeographic data of course have dimensions not only of place and species but also of time. In the final section of this chapter we shall examine the important changes that occurred during the ice ages.

Regional Zoogeography

Early systems designed to distinguish different geographic ranges resulted in the description of zoogeographic regions such as those characterized by Alfred R. Wallace and Philip L. Sclater. Regions are characterized by endemic groups, usually vertebrates, and by the absence of taxa (usually families) in neighboring regions (Fig. 3.3). Sclater's regions, modified by subsequent authors, reflect not only geography, evolution and history but also current ecologic conditions.

Classic Zoogeographic Regions

In considering the classic vertebrate zoogeographic regions, the *Palaearctic* and *Nearctic* regions are sometimes combined into a broad *Holarctic* region. They occupy most of the land in the Northern Hemisphere. Eastern North America was continuous with northwestern Europe until the Eocene, and northeastern Asia and northwestern North America were joined periodically from the Cenozoic to the Pleistocene by lowered sea levels. Consequently there are numerous birds and mammals common to both the Nearctic and Palaearctic regions, and these similarities are more conspicuous at high than at low latitudes.

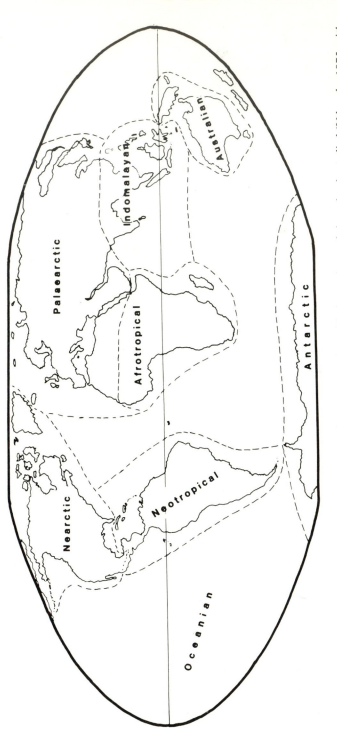

Fig. 3.3. Biogeographic regions of the world. The boundaries vary somewhat with the worker and the animals studied (Udvardy, 1975; with permission of the International Union for Conservation of Nature and Natural Resources).

Southeastern Asia has an increasingly tropical aspect with decreasing elevation and latitude. The *Oriental* or *Indomalayan* region is the home of distinctive tropical groups, many of which replace more boreal taxa of the Palaearctic region. The Oriental region extends to the Greater Sundas (Java, Sumatra and Borneo), which lie on the continental shelf of Asia.

The Palaearctic region of the southwest occupies much of northern Africa; the Palearctic aspects of the fauna of northern Africa are attributed to land connections around the Mediterranean Sea. Africa south of the Sahara becomes the *Ethiopian* or *Afrotropical* region.

The *Neotropical* region includes South America north to at least the area of the Isthmus of Panama. Although there has been an intercontinental faunal exchange between North and South America since the closure of the Panama Portal (in the Pliocene), the Neotropical region possesses a distinctive vertebrate fauna. The affinities of the Neotropical fauna with elements of both the Ethiopian and Australian regions have produced diverse theories of land bridges and parallel evolution.

The *Australian* region is distinguished not only for the radiation of such groups as marsupials but also by the absence of many others. Included in the Australian region are Tasmania to the south and New Guinea to the north.

The biogeographic realms (Table 3.1) include the vast area of Oceania, of which only 1/500 consists of land areas, and the Antarctic realm, which has a meager but distinctive fauna. In both Oceania and Antarctica most of the important faunal elements are invertebrates, and most of the vertebrates are birds.

The classification of the faunistic regions or biogeographic realms neglects the hundreds of large or small areas, mostly islands, that do not fit logically or clearly into the major categories. Some intermediate regions, such as Wallacea (between the Australian and Oriental regions) and Mesoamerica, are areas of extremely complex mixtures. Oceania includes a great variety of islands that differ in age, origin, topography, climate and distance from the mainland, and their inclusion under one heading is simply for convenience.

Table 3.1. Faunistic regions, floristic kingdoms and biogeographic realms.[a]

	Faunistic Regions	Floristic Kingdoms	Biogeographic Realms
Holarctic	Palaearctic Nearctic	Boreal	Palaearctic Nearctic
	Ethiopian Oriental	Paleotropical	Afrotropical Indomalayan
	Australian	Australian	Australian
	Neotropical	Neotropical	Neotropical
	--	--	Oceanian
	--	--	Antarctic

[a] The classical faunistic regions are those of Sclater (1858), the floristic kingdoms were described by Engler (1879) and the biogeographic realms are those of Udvardy (1975).
--, Not relevant.

Areas of Integration

Although faunistic regions or biogeographic realms have clearly defined boundaries at most junctures, some areas are intermediate or otherwise ambiguous. Two of these will illustrate the difficulty of drawing faunal boundaries in some instances: Mesoamerica and the East Indies.

Mesoamerica. Numerous scholars have discussed faunal exchanges between North and South America.[428,793,847,880,881,931,933,1079] The Panama Portal, a seaway from the Cretaceous, was closed in the late Pliocene, allowing passage of terrestrial animals in both directions. Prior to the late Pliocene a few terrestrial vertebrates were apparently rafted across the Portal: these included iguanid and teiid lizards, anurans of the genera *Bufo* and *Hyla,* and one procyonid mammal. To appreciate the passage of vertebrates through the Isthmus, it is necessary to consider the apparent geologic events in Mesoamerica during the Cenozoic, the region between southern Mexico and Colombia. As the plates supporting North and South America moved westward, from the Cretaceous to the present, Mesozoic Mesoamerica "trailed behind," and eventually became the Greater Antilles (Cuba, Jamaica, Hispaniola and Puerto Rico) and the island arc of the Lesser Antilles (Fig. 3.4). This separation occurred during the incipiency of true mammals, when there were already marsupials and insectivores in North and South America.

Numerous stocks of amphibians and reptiles were well-established. Some living vertebrates of the Greater Antilles may be survivors of an original Cretaceous fauna: amphisbaenians, some anurans and insectivores (*Solenodon* and *Nesophontes*) had an ancient North American origin, and may have existed on the Greater Antilles when they became insular.[861] Such invertebrates as onycophorans and terrestrial annelid worms were probably also part of the original Antillean faunas. The elements in this archipelago that remained and moved westward with North and South America gradually joined each other (as a result of volcanic activity) during the Cenozoic, and became contiguous by the late Pliocene (Fig. 3.5).

Apart from the creation of the West Indies, the important aspect of pre-Pliocene history of Mesoamerica was its insular nature. This area is poor in vertebrate fossils and its Cenozoic fauna is unknown. There was probably a broad diversity of small creatures (especially amphibians and reptiles) with few sizable animals, such as large herbivores and carnivores.

Most modern zoogeographers, considering past and present ranges of mammals in North and South America, designate Mesoamerica as intermediate between the Nearctic and Neotropical Regions. The Mesoamerican avifauna is

Fig. 3.4. The North and South American plates migration. The westward drift in the late Cretaceous left fragments that were to become the Greater Antilles and also created a marine barrier between North and South America (Malfait and Dinkelman, 1972).

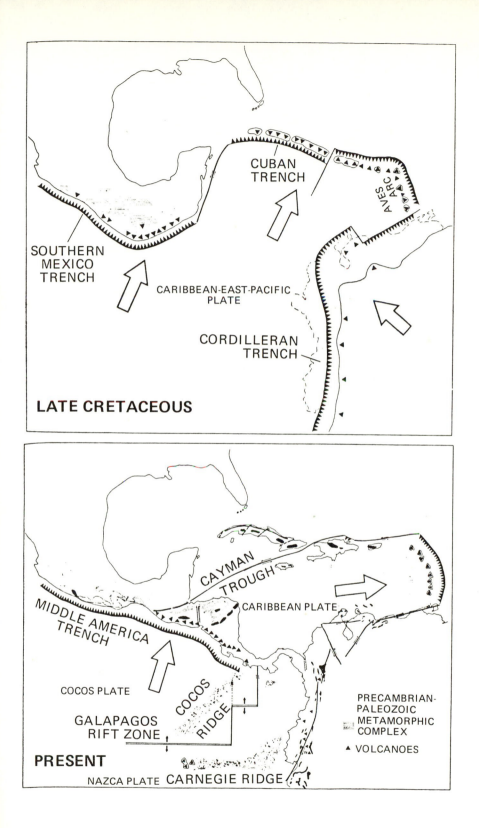

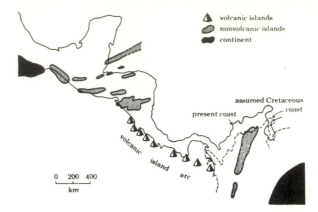

Fig. 3.5. Reconstruction of Mesoamerica, primarily by volcanic action, during the Cenozoic. From the late Cretaceous until the late Pliocene Mesoamerica was an archipelago (Raven and Axelrod, 1975, courtesy American Scientist, journal of Sigma Xi, The Scientific Research Society; after Dengo, 1973).

not distinctive, and the area is poor in fresh-water fishes. Thus modern Mesoamerica has little that is unique in its fresh-water fishes, birds or mammals.

The amphibians and reptiles of Mesoamerica, on the other hand, are in a large measure unlike the Nearctic and Neotropical herpetofaunas and constitute a distinct (not intermediate) fauna.[881] The Mesoamerican species of these two classes are placed in 159 genera, of which 44 percent are centered in Mesoamerica: species in 29 genera are endemic to this area, and another 18 genera are essentially restricted (almost endemic) to the region. In contrast, only 23 genera (or 14 percent) of the Neotropical genera extend north of Costa Rica. Thus the herpetofauna of Mesoamerica is not a transition of the fauna to the north and south but rather a separate and distinct fauna with many species of endemic Mesoamerican genera.

The mammals, which have been emphasized (or overemphasized), do indeed indicate Mesoamerica to be a region of mixture of Nearctic and Neotropical elements. At the close of the Pliocene, several northern groups of mammals invaded South America: These included not only placental carnivores (dogs, cats, bears and weasels), but deer, camels, horses, peccaries, tapirs and cricetid rodents. This invasion seems to have occurred when characteristic Neotropical taxa, such as marsupial carnivores, condylarths and ground sloths, became extinct; the invaders from the north may have hastened the disappearance of the southern forms. From South America some armadillos, generalized marsupials and hystricomorph rodents moved north. The opossum, *Didelphis,* appeared in North America in late Pliocene, punctuating the return of marsupials to that continent, where they had been extinct since the Miocene.

Why do Mesoamerican amphibians and reptiles present such a different picture from that of the mammals? Is the picture affected by the land area needed to sustain a population of larger mammals, on the one hand, and "herptiles" (amphibians and reptiles), on the other? The smaller size of ectothermic amphibians and reptiles permits them to exist in greater numbers than would be possible for the endothermic (and energy-demanding) mammals. It does seem thoroughly plausible that the Cenozoic Mesoamerican archipelago may have been the site of radiation of the many distinctive amphibians and reptiles that give this area its characteristic herpetofauna today.

East Indies or Wallacea. Another area of complex faunal mixture lies between Java and Borneo (the southeastern limit of the Oriental Region) and western New Guinea and northern Australia (the northwestern limit of the Australian Region). Within this area lie 13,000 islands known as the East Indies. They have probably had very different geologic histories, for, although they lie in a highly volcanic belt, vulcanism does not preclude the origin of some by fragmentation or seafloor elevation. This area was named Wallacea by Elmer D. Merrill in 1926. Wallacea consists of a few large islands, the best known being the Celebes, Halmahera, Timor and Ceram, but the smaller islands are important, and have probably contributed to the faunal complexity of Wallacea (Fig. 3.6).

In the mid-19th century, Alfred R. Wallace noted relatively sudden and conspicuous faunal changes when traveling from one island to another, east of the Oriental Region. These marked differences in faunal composition were most readily observed in birds and most conspicuous at the eastern end of Borneo, Java and Bali. The distinction is greatest between the faunas of Bali and the next smallest island to the east, Lombok, because the distance is so short, but the faunal break is just as real between Borneo and Celebes, across the Makassar Strait.

Clearly Bali, Java, Borneo and Palawan were connected to each other and to the Malay Peninsula during the Pleistocene; logically they have modern, continental faunas appropriate to their size and topography. To the east of these islands lie the many islands of Wallacea, mostly surrounded by deep water and not connected to the continental shelf of southeast Asia. The faunal boundary between the Oriental Region and Wallacea is Wallace's Line. It runs between Bali and Lombok, north through the Makassar Strait and south of Mindanao. (As originally proposed by Huxley, Wallace's Line went north to include the Philippines.)

The eastern boundary of Wallacea is not nearly so clear. Although New Guinea is unequivocally in the Australian Region (with respect to the distribution of vertebrates), there is no clear-cut point at which one travels west and suddenly leaves the Australian Region and enters Wallacea. The sharp distinction does not exist. An eastern boundary is required, however, if Wallacea is to be acknowledged as a reality, which indeed it is.

To understand this dilemma, one must consider the events that led to the relative clarity of Wallace's Line. The Pleistocene continuity of the Asiatic continental shelf permitted free movement of vertebrates to Bali and Borneo but not beyond. Although not all vertebrate families of the Malay Peninsula occur on the Greater Sundas, all of those on the Greater Sundas also live on the mainland. Although the extension of the continental shelf did not join Borneo to Celebes, the distance in the Pleistocene was reduced to roughly 45 km, increasing chances

Fig. 3.6. Wallacea and surrounding lands. Shaded marine areas indicate continental shelves, which were probably exposed in glacial periods of the Pleistocene (Mayr, 1944).

for waif dispersal from the mainland. To be sure, some vertebrates dispersed beyond Bali and Borneo, but the frequency of Oriental taxa beyond Wallace's Line decreases eastward, displaying a typical filter effect. Probably this was the path used by murid rodents, some squirrels, skinks and agamid lizards; microhylid frogs probably entered the Australian Region through Wallacea. Several mammalian orders do not cross Wallace's Line at all. Of those that do, some were almost certainly carried by aboriginal man: these include cattle, pigs, some viverrid carnivores and possibly deer. Birds, although capable of flying the short distances involved, sometimes do not move freely across Wallace's Line; those that do, show the filter effect, with fewer Oriental taxa to the east.

Some taxa are notoriously poor at oversea travel, and cross Wallace's Line with difficulty, if at all. Fresh-water fish are abundant on the Greater Sundas and very scarce on Celebes and throughout Wallacea. Those that do occur in Wallacea are species that can breathe air, and some of these have been transported, covered by damp cloth, by aborigines for food. They survive for weeks in a basket, and are carried from one island to another. Some mammals enter Wallacea: on Celebes are shrews, mice and squirrels and two species of monkeys. Among amphibians, the ranid and rhacophorid frogs, caecilians and salamanders do not enter Wallacea.

The boundary between Wallacea and the Australian Region depends on the taxon being considered.[665] The eastern limit is described by Weber's Line, defined as the point at which the group in question has one-half of its representatives from the Oriental Region and one-half from the Australian Region. Because Weber's Line lies at somewhat different positions for different taxa, the eastern limit to Wallacea is a broad area. The position of Weber's Line, rather close to the continental shelf of the Australian plate, can be explained by a much greater faunal movement eastward than westward, and this in turn seems predicated on the much greater faunal diversity in the Oriental Region.

These two intermediate areas, Mesoamerica and Wallacea, are not exactly comparable. Mesoamerica represents an area of mixture for birds and mammals but is a distinct faunal entity for amphibians and reptiles. Wallacea, on the other hand, is populated by groups from both neighboring regions and lacks a distinctive character of its own.

Islands

Island biotas are isolated from many forces that continuously mold the size and composition of continental life. The vast and diverse environments of continents experience more rapid and drastic climatic changes than do islands. A classification of island types is necessary to an understanding of the development of insular faunas (Table 3.2). Traditionally islands have been described as either *oceanic* or *continental*. Oceanic islands result from either volcanic action, or some other geologic event, that raises the seafloor above sea level. Continental islands lie on a continental shelf and are separated from the mainland by rather shallow

Table 3.2. Island types and characteristic features.

Type	Geologic Origin	Biotic Characteristics	Examples
Continental	On continental shelf	Harmonic	British Isles, Sri Lanka, Taiwan
Continental fragment	Rafted continental plate, isolated by deep water	Harmonic but ancient	Madagascar, New Zealand, Seychelles
Oceanic	Isolated vulcanism from ocean floor	Disharmonic	Hawaiian Islands, Galápagos

water (about 150 m or less); they were connected to the mainland during the Pleistocene. A third, important type of island is the *continental fragment;* this is a continental crust that has been rafted away from the mainland and is surrounded by relatively deep water.

Oceanic Islands

Many oceanic islands succumb to erosion and sink shortly after birth. But larger and more substantial islands persist for tens of thousands of years, and sometimes much longer, and develop typical oceanic faunas. Volcanic islands may also experience cycles of growth and subsidence. Although this process may be prolonged, as in the example of the Hawaiian Islands, it may also be rapid. Falcon Island, in the Tonga group, has been elevated to heights of 50 to 200 m and alternately submerged three times in less than 100 years, and this example may not be exceptional.[368]

The fauna of an oceanic island is distinctive, not only for the types that occur but also for those that do not. Most terrestrial vertebrates are birds, and the most numerous mammals are bats. Reptiles and amphibians occur according to their dispersal powers, but amphibians and soil-dwelling forms seldom travel far over salt water. Such faunas then are not representative of the adjacent mainland and are *disharmonic* or *unbalanced*. The disharmony is both taxonomic and ecologic, i.e., some taxa can disperse over marine barriers and others cannot, and this feature results in some habitats being left unexploited and some taxonomic groups being left absent.

Oceanic islands' faunas reflect the island's age, size, topography and distance from the mainland, and these factors are paramount in the development of oceanic biota. Also, if forms are introduced by rafting, the direction of an ocean current is critical to the development of the island fauna. The direction of prevailing winds, especially storm winds, can deflect the normal paths of bats and birds. This complex of factors is difficult to simplify and analyze, for each pulsates independently of the others, and attempts to unravel their roles tend to be artificial and contrived.

Foremost among the well-studied oceanic islands are the Hawaiian Islands. They lie 3200 km from Alaska, 5400 km from Japan and 3200 km from the continental United States. Closer are many small islands to the south and west, also of oceanic origin; the abundance of guyots—flat-topped seamounts, the summits of which have been eroded to sea level—attests to the earlier presence of many other mid-Pacific islands in the past. These islands may well have served as "stepping-stones" in the development of the Hawaiian biota. Most of the plants and invertebrates have their affinities to the south and west and are apparently oriental in origin. Of the two native land mammals, the Polynesian rat *(Rattus exulans)* is from the southwest; it is widespread among other Pacific islands, and was presumably transported only by aborigines. The only other land mammal is a bat *(Lasiurus cinereus semotus)* of New World origin; species of *Lasiurus* are strong fliers and highly migratory.[1016] Reptiles and amphibians of the Hawaiian chain were brought there by man. The native avian fauna, on the other hand, is composed mostly of species whose relatives occur in North America or which are Holarctic. One endemic family, the honeycreepers (Drepanididae), surpasses all other vertebrate families in the degree and rapidity of its adaptive radiation.[10]

The Galápagos, an archipelago of oceanic islands 880 km from the coast of South America, are also volcanic and have had no connection to the mainland. They are products of a series of late Cenozoic flows of basalt from the Galápagos Platform. The Cocos Ridge and Carnegie Ridge have never joined these islands to the mainland (Fig. 3.4), but they may have been elevated in the Miocene and perhaps deflected currents toward these islands. The oldest exposed rocks on the Galápagos are Pliocene, but lava may cover older strata. The 16 islands total about 4500 km². They lie on the equator but are cooled by the Humboldt (or Peru) Current and the Southeast Trades. The land is generally semiarid, with heavy rainfall only at higher elevations (1500 m). The soil is thin and porous and there is very little standing water. Both the nearness of the mainland and the Humboldt Current probably account for the presence of Neotropical rats, iguanid lizards, land turtles and penguins on these islands.

The fauna of these islands has been extensively studied. "Darwin's Finches," a group of four genera of Neotropical finches, is one of the best known small group of wild birds. They illustrate adaptive radiation and are especially distinctive in divergence of diet, bill shape and structure of digestive tract.[110] This radiation apparently resulted from a single introduction. The other land birds are all from Neotropical stocks and include a cuckoo, a warbler (Parulidae), martin, pigeon and mockingbird. The reptiles include two large iguanid lizards, geckos and colubrid snakes. The giant land tortoises are the best known large vertebrates on the islands, and gave them their name. The endemic bat *(Lasiurus brachyotis)* is allied to the bat on the Hawaiian Islands, and *Lasiurus c. cinereus* (from the mainland) visits the Galápagos on migration. There are neither fresh-water fish nor amphibians. The vertebrate invasion of the Galápagos must have been by flight or rafting; the currents from the mainland move about 120 km daily, and rafting would take about one week, not a difficult journey for a small reptile or a plant-eating rodent.

Other oceanic islands are Iceland in the Atlantic and numerous small volcanic islands in the Pacific. In each case they lack ancient connections to continental lands, and their biotas developed by oversea dispersal. Most vertebrates are birds, bats and geckos, but scincid lizards are widely distributed on oceanic islands. Both geckos and skinks are prone to hide under loose bark of fallen trees; here, rafting is a logical means of dispersal.

Continental Islands

From a faunistic point of view, it is important to distinguish oceanic from continental islands, for the latter have usually had repeated connections with the mainland, and have faunas very similar to those on the mainland. Because the intervening water is shallow, most continental islands were joined to the adjacent mainland one or more times in the Pleistocene, when there were worldwide lowerings of sea levels. Continental shelves were exposed and many continental islands became coastal mountains; opportunities for movements of animals in such regions were ample. Melting of Pleistocene ice raised sea levels, inundating coastal lowlands; coastal mountains once again became isolated and today are recognized as continental islands.

Because of their recent separation from the continents, faunas of continental islands are harmonic in composition; they include a near-complete representation, both taxonomically and ecologically, of the fauna on the nearby continent. Continental islands are usually the homes of fresh-water fishes, amphibians and fossorial vertebrates, all creatures that travel seldom, if at all, over salt water. After a continental island is separated from the mainland, the fauna can lose taxa by extinction and gain taxa by oversea dispersal. In time, then, a continental island can develop some aspects of an oceanic island but not to the point of losing the harmony typical of continental islands.

The largest continental island is Greenland; it is still in a Pleistocene-like condition and has a meager fauna of arctic birds and mammals. The British Isles, continental islands of Europe, have continental vertebrate elements (fresh-water fish, amphibians and moles) that contribute to the harmonic aspect of their fauna. The well-known absence of snakes in Ireland is attributed to its post-Pleistocene separation, which must have occurred before the climate became warm enough for snakes; England, Scotland and Wales remained connected to the mainland until the climate was warm enough to support snakes.

The four main islands of Japan seem to have become isolated from each other at different times: the faunas of the three "older" islands (Honshu, Shikoku and Kyushu) resemble one another closely and are somewhat different from those on both Hokkaido and the adjacent mainland (Korea). Hokkaido, on the other hand, has a fauna similar to that on the mainland today, and was apparently joined to the mainland, together with Sakhalin (or Karafuto), until late in the Pleistocene. Honshu, Shikoku and Kyushu have many species of mammals (e.g., a dormouse and a water shrew) that do not now occur on Hokkaido or in Korea, but whose relatives live in Southern China.

The offshore island of Taiwan is clearly continental, as are the Greater Sundas (Borneo, Sumatra and Java) and Sri Lanka. The depth of the water separating them from the nearby mainlands is 100 m or less, and these islands were all joined to continental Asia during parts of the Pleistocene. New Guinea is another major continental island; during the Pleistocene a large part of the Arafura Sea was dry, and New Guinea was then joined to Australia. Today the New Guinea terrestrial vertebrate fauna is rather like that of northern Australia.

Continental Fragments

Continental fragments include not only such large islands as New Zealand and Madagascar but also some small ones, such as the Seychelles and Socotra. They often have many faunal elements that one would expect on continental but not on oceanic islands: they may have amphibians and fresh-water fishes, fossorial animals (such as caecilians and annelid worms), large mammals and a balanced fauna. For example, on the Seychelles there is an endemic family of frogs (Sooglossidae) as well as caecilians. Socotra has an amphisbaenian, and on Madagascar there is a diverse mammalian fauna, including many forms that do not readily disperse over marine waters. New Zealand, although lacking native land mammals (except bats), has frogs (of an ancient lineage), the archaic rhynchocephalian *(Sphenodon)* and ratite birds.

Many continental fragments are quite small but may be important zoogeographically. Continental fragments in the Indian Ocean include not only Madagascar and the Seychelles but also the Crozet Plateau, with Prince Edward Island and Crozet Island, the Kerguelen–Heard Plateau, with Kerguelen Island and Heard Island, as well as the isolated islands of Amsterdam, Reunion, Comoro, Aldabra and Socotra, and, in the Atlantic, the eastern Canary Islands. Continental fragments in the Pacific include Auckland Island and Campbell Island south of New Zealand, Norfolk Island, New Caledonia and Fiji to the north, and New Zealand. The Greater Antilles are also continental fragments. Many continental fragments have faunas that existed on continents in the Cretaceous or early Cenozoic.[331]

Certain invertebrates seldom or never disperse successfully over salt water and are accurate indicators of continental soils. Earthworms occur in most continental soils not glaciated in the Pleistocene, and their occurrence on certain continental fragments supports geologic evidence of their origin. Endemic genera of annelid worms link Australia with New Zealand and New Caledonia. There is an endemic genus on the Seychelles, and a number of endemic genera on such islands as the Auckland and Campbell Islands and Puerto Rico. There are comparable distributions of free-living soil nematodes.[291]

The Theory of Island Biogeography by MacArthur and Wilson[631] stimulated interest in the development and dynamics of insular faunas. These authors codified the generally accepted features affecting natural dispersal to oceanic islands, including the cumulative effects of island size and topography, distance from the mainland, island age and other aspects familiar to students of historical biogeography. A major part of their thesis is that stability, or the constancy in number of species, represents a balance or equilibrium between rates of introduction and rates of extinction. Their "theory" has been both praised and attacked. A major problem is that it is extremely simplistic, and fails to consider the profound differences in the origin and composition of the faunas of oceanic islands, continental islands and continental fragments.

A conspicuous aspect of the many thousands of oceanic islands is their disharmonic faunas. An introduction to a disharmonic fauna may represent a previously unfilled niche and need not result in a compensatory extinction. A continental island, on the other hand, has a fauna fairly representative of the adjacent mainland; virtually all continental islands date from the Pleistocene and have modern faunas. An introduction into such a fauna would be expected to disturb the balance just as it would on a piece of contiguous mainland. A continental fragment presumably had a harmonic fauna at its inception, which may have been in the Mesozoic or early Cenozoic. Introductions to continental fragments may have effects not predicated on a simple one-to-one relationship.

MacArthur and Wilson proposed that the avifauna of the Hawaiian Islands illustrates their concept of equilibrium between introductions and extinctions. The introduced birds, which are supposed to have accounted for the retreat and decline of the native species, remain mostly in urban areas, where they are associated with the introduced flora. The Japanese white-eye *(Zosterops japonica)* is one of the few alien birds that has invaded the native forests. In reality, however, the native birds have declined with the large-scale removal of native vegetation: they do not accept vast areas of sugarcane and pineapple as substitutes for native cover. Also the introduction of alien mosquitoes brought with it avian malaria, to which the native birds are not resistant. Hence, MacArthur and Wilson's discussion of the Hawaiian avifauna is unfortunately laced with speculation of what they presume to have occurred over the past geologic time in these islands.

The concepts of the MacArthur and Wilson equilibrium "theory" are more ecologic than biogeographic in substance. There is a detailed discussion of this topic in Pielou (cited at the end of this chapter).

Deserts

Deserts are important not only for the adaptations shown by vertebrates that live there but also because they present interesting biogeographic assemblages that some workers have compared to island faunas. Deserts differ from islands, however, in two important aspects: (1) desert boundaries are not as discrete as the separation of an island by sea and (2) deserts reflect a climatic condition (aridity) that can change in severity over geologic time.

At least some deserts are relatively recent in area and geographic position. Using the North American Sonoran Desert as an example, relatively small and local arid localities were the sites of some early (Paleocene–Eocene) pockets of dry-adapted vegetation in Mexico; they did not, however, prevail in parts of southern California, where the northern end of the Sonoran Desert lies today. Instead, at least along creek beds, there was a flora that included *Araucaria, Taxodium* and *Magnolia,* none of which is native near California today, suggesting substantial rainfall evenly distributed throughout the year.[35] The area of southeastern California, northwestern Arizona and adjacent Mexico now covered by the Sonoran Desert developed its xeric features gradually, from the Eocene through a succession of increasing arid floras, to the desert seen today (Fig. 3.7). The final and major drop in rainfall probably occurred as a result of mountain-building from the late Pliocene, punctuated only by temporary increases of rainfall during the interglacials. Other desert regions today do not seem to be ancient.[35]

Desert vertebrates seem to be heterogenous in their ecologic origins and include neither isolated ancient vertebrates nor major faunal groups. In the Sonoran Desert there are no genera of mammals or birds confined to deserts; most desert vertebrates are species of genera that are widespread in western North America. Thus, except for a small number of reptiles, desert vertebrates are recently adapted from less xeric types, and deserts—despite their climatic conditions—cannot be regarded as "ecologic islands" comparable to oceanic islands and continental fragments.

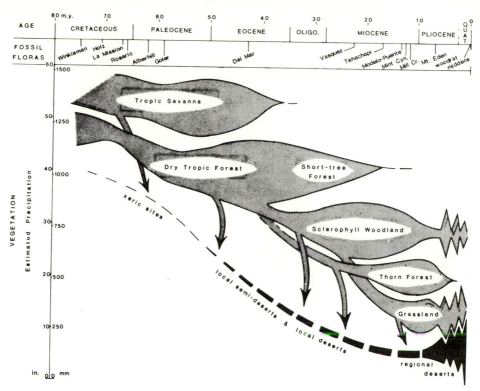

Fig. 3.7. The gradual development of the Sonoran Desert from the Cretaceous to the present. As estimated annual precipitation decreased, the desert biota developed (Axelrod, 1979).

The Pleistocene

Temperate latitudes in the latter part of the Cenozoic became cooler and drier. Gradually the coniferous forests of the Pliocene were shifted to lower latitudes and elevations as the cooler climate of the Pleistocene appeared. Most scholars agree that there were four major periods of glaciation, both in Eurasia and North America, and between each major glaciation was a mild or warm interglacial period. The end of the Pliocene, around 3 million years ago, was the start of the first glacial period. This is known as the Blancan (in North America) or Villafranchian (in Europe); the last glaciation is called the Wisconsin or Würm, in North America and Europe, respectively. The exact timing and number of glacial and interglacial periods are not firmly established; superimposed on the four major epochs may be as many as 17 glacial and interglacial periods over the past 1.7 million years. This fine division of the Pleistocene is based on ^{18}O dating of oceanic sediments and does not invalidate the concept of four *major* glacial periods.

Today, about 10 percent of the earth's land surface is covered by glacial ice, but in the early Pleistocene more than 30 percent was glaciated. At present more than 11 million km² of ice cover Antarctica, an area more than one-and-a-half times that of the United States and seven times that of Greenland.[249]

The ice covered much of Canada, Greenland, Alaska and the northern United States, with isolated montane glaciers in parts of the Rocky Mountains, Cascade Range and the Sierra Nevada. In Europe the ice blanketed Scandinavia and the northern countries, and there were montane glaciers in the Alps, Pyrenees and Caucasus. In eastern Eurasia glaciers existed in both the Himalayas and the high ranges of Japan and Korea. Glacial ice also occurred in Tasmania, New Guinea, southern New Zealand and Antarctica. There was some glacial ice on the Hawaiian peak of Mauna Kea and an isolated glacier on Mt. Itaziaia in Brazil.

Large parts of eastern Asia escaped glaciation, apparently as a result of insufficient precipitation. Because of lowered sea levels during the glacial ages, the Bering land bridge stopped the cold waters of the Arctic Ocean from entering the Bering Sea; this in turn allowed the warm Kurshio current to reach the land bridge and ameliorate the climate from Kamchatka to Alaska. The absence of glacial ice on the Bering land bridge itself permitted movement of cold-adapted animals between eastern Asia and Alaska.

Water locked in glacial ice lowered sea levels, exposing many continental shelves and connecting many islands to their adjacent mainlands. In North America many coastal islands along the west coast of Canada were contiguous with the mainland, and in the Gulf of St. Lawrence Anticosti Island was part of the mainland. In Europe the British Isles were united to the continent, Spain was joined to northwest Africa, and Sardinia, Corsica and Sicily were continuous with the European continent. To the east, Ceylon was joined to India, the Greater Sundas were joined to Malaysia, and New Guinea and Tasmania were connected to Australia. Some of the modern islands of the Galápagos and Hawaiian Islands were joined to each other as were the main islands of New Zealand. The four main islands of Japan were joined to Sakhalin in the north and Korea in the south, and Taiwan was also part of the mainland.

As a consequence of the climatic and geographic changes during the Pleistocene, parts of the biota (1) were destroyed beneath the ice, (2) slowly retreated before the advancing ice or cold climate or both and became isolated in favorable areas in lower latitudes and lower elevations or (3) moved onto temporary peninsulas, which later became post-Pleistocene islands. Some aquatic forms later became isolated in remnants of once-extensive pluvial or meltwater lakes. Such arctic species as the muskox and the pika (*Ochotona* sp.) ranged south to at least Pennsylvania, and the muskox moved far south in Eurasia in the Pleistocene.

As entire faunas retreated from the glaciers in the north, many species moved into separate isolated habitats to the south, and some such populations gradually differentiated. When this isolation persisted for extended periods, speciation developed to the point that the isolated populations became genetically incompatible. When their ranges once again became geographically contiguous, interbreeding did not occur. With the melting of ice sheets, some animals moved not only back in a northerly direction but also to higher elevations in mountains, whereas still others became stranded in scattered lakes or meadows, and rising sea levels left others on continental islands. In this manner many relict populations of vertebrates now persist on islands, mountain ridges, sphagnum bogs or remnants of Pleistocene pluvial lakes.[467]

Bering Land Bridge

Although the Bering Land Bridge is very important for understanding modern vertebrate distributions, one should remember that there were pre-Pleistocene faunal exchanges between North America and Eurasia. Evidence indicates several (perhaps six or seven) times in the Cenozoic when vertebrates moved between eastern Asia and Alaska.[930] Periods of faunal passage are suggested by Holarctic distributions of faunas today. In Miocene and Oligocene flora were forests of hardwoods not greatly unlike the modern broad-leaved forests in eastern Asia and eastern North America. They included species of *Carya, Fagus, Quercus, Corylus* and *Juglans,* trees that must have supplied mast for mid-Cenozoic mice and squirrels; *Nyssa* and *Liquidamber* now occur only in lowland areas in eastern United States and eastern Asia.[590]

The similarity of floral elements persisted on both continents through the Pliocene, but conifers gradually dominated broad-leaved hardwoods. These floral shifts probably indicate late-Cenozoic drops in mean temperature, resulting in a drift of the broad-leaved forests to lower latitudes. Fossils of both plants and animals point to milder climates in the earlier Cenozoic, times when the Bering connection was forested with broad-leaved hardwoods. Presumably the animals that moved through this region during mild periods now occur in temperate regions far to the south. For example, a primitive or ancestral procyonid moved into Asia to give rise to the modern pandas, and an early or mid-Cenozoic tapir probably crossed the Bering Land Bridge into the New World. One should remember, however, that some early students of animal distribution were unaware that eastern North America and western Europe remained joined by Greenland until the Eocene. Thus Holarctic occurrences of early Cenozoic vertebrates, such as the early horses, do not necessarily reflect movement through the Bering region.

The unglaciated region of eastern Siberia and western Alaska, together with the nearby continental islands and continental shelf, is called Beringia. It was a flat treeless plain in the Pleistocene, cool-temperate but not as frigid as it is today. As the climatic pulsations raised and lowered sea levels, faunas moved up and down mountains and across and between continents, and Beringia was the region in which critical faunal exchanges occurred. Bering connections occurred several times in the Pleistocene, and there were also several marine transgressions during which the sea moved far inland. Just as a warm Bering Sea and intercontinental dry land connections characterized glacial times, waters of cooler temperatures and a Bering Strait were features of the Pleistocene interglacial periods.

Evidence of these temperature changes are seen in deposits of marine fossils at the bases of wave-eroded terraces that are today far inland from the present coast and 30–70 m above the modern sea level. Such evidence is known from the Chukot (or Chukotski) Peninsula in Asia and the Seward Peninsula of Alaska, and may indicate not only changes in sea levels but also tectonic uplift. During the interglacials, when there was a Bering Strait (as it exists today), marine animals could pass between the Atlantic and the Pacific Oceans. The puffin *(Fratercula arctica)* and seals (*Pusa* spp.) have closely allied but now disjunct populations in the North Atlantic and North Pacific (Fig. 3.8), testifying to a prior continuous range.[1035]

The passage of Pleistocene mammals between Siberia and Alaska is documented by numerous Pleistocene fossils, many of which have been dated by radiocarbon. Birds can move easily over short marine barriers, but mammals were the conspicuous vertebrates of the far north, and many of them were large. Ectotherms must have been a trivial element in Beringia then as now.

An inventory of arctic mammals would show a general similarity between the faunas of the Old and New Worlds, and also that the preponderance of those in North America is of Old World origin. The more boreal species in each continent are closely allied to each other, frequently conspecific, whereas those in more temperate lower latitudes, although clearly similar, are more often of different species or even genera. Holarctic species include carnivores such as the brown, or grizzly,

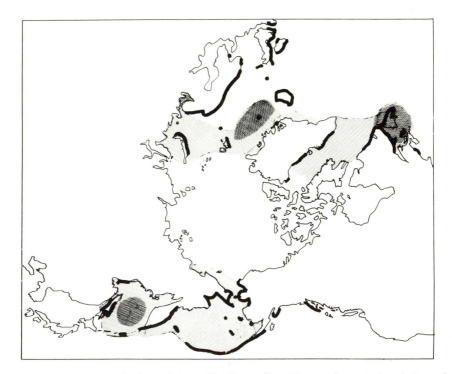

Fig. 3.8. Present distributions of the puffin (heavy line; *Fratercula arctica*) and the seals (shaded areas; *Pusa groenlandica* and *P. fasciata*). Presumably the Atlantic and Pacific populations intermingled during Pleistocene interglacials (Udvardy, 1963).

bear *(Ursus arctos)*, arctic fox *(Alopex lagopus)*, ermine *(Mustela erminea)* and lynx *(Felis lynx)*, arctic ground squirrel *(Citellus undulatus)* and voles (*Microtus oeconomus* and *Clethrionomys rutilus*).

Farther to the south, in more temperate New World latitudes, dwell more distant relatives of Asiatic mammals, species that probably moved across an older Bering connection when the climate was warmer. Some typical examples of Nearctic mammals representing earlier movements are the black bear *(Ursus americanus)*, long-tailed weasel *(Mustela frenata)*, bobcat *(Felis rufus)*, red-backed vole *(Clethrionomys gapperi)* and the ground squirrel *(Citellus columbianus)*. There was also some movement from Alaska into the Old World. Early in the Pleistocene horses (*Equus* sp.) moved into Asia and later the muskox *(Ovibus moschatus)* returned to the Old World (from whence it had come at an earlier date). There were also some mammals that entered North America across the Panamanian isthmus in the late Pliocene and moved as far north as Alaska without crossing into Siberia. The ground sloth *(Megalonyx)* reached Alaska, and the porcupine *(Erethizon dorsatum)* today occurs in the northern coniferous forests, but neither is known to have crossed into Asia.[469]

Latitudinal Shifts

Dramatic biotic shifts in distributions resulted from changes between the pluvial glacial periods and the warmer, drier interglacial times. Tropical forms such as the jaguar *(Panthera onca)*, peccary *(Tayassu)* and tapir *(Tapirus* sp.) ranged far to the north of where they live today. In Europe, interglacials saw northward movements of such species as rhinoceros, hippopotamus and the large cats.

Post-Pleistocene Refugia

Amphibians and reptiles were also affected by the climatic changes of the Pleistocene, and many species probably extended their ranges far to the north and south of their current distributions. The panhandle of northern Idaho is a refugium for some amphibians that occur elsewhere only along the Pacific coast. The giant salamander *(Dicamptodon ensatus)*, a newt *(Taricha granulosa)* and the tailed frog *(Ascaphus truei)* dwell in cool mountain brooks, and the lack of differentiation suggests that the disjunction is probably relatively recent. The spotted frog *(Rana pretiosa)* occurs in northern Idaho and western Wyoming, and a plethodontid salamander *(Plethodon vandykei)* is represented along the Pacific coast and in northern Idaho by distinct subspecies.

Northern Idaho is also a refugium for a number of trees characteristic of Pacific coastal forests: notably there is hemlock *(Tsuga heterophylla)*, fir *(Abies grandis)* and cedar *(Thuja plicata)* as well as shrubs, such as Oregon grape *(Mahonia nervosa)*. Today the intervening area is dry: a high steppe of sagebrush, with an annual precipitation of 15 cm. During the Pliocene or Pleistocene these five amphibians and associated plants must have been continuously distributed between northern Idaho and the coast.

Some of these disjunctions predate the Pleistocene. The western United States in the Cenozoic had coniferous forests of fir *(Abies)*, larch *(Larix)*, spruce *(Picea)*, cedar *(Chamaecyparis),* redwood *(Sequoia),* bigtree *(Sequoiadendron)* and Douglas fir *(Pseudotsuga).* In the Oligocene and Miocene these forests extended to Utah and, in some cases, east to Colorado and south to Arizona. Heavy summer rains at this time are indicated by the presence of such broad-leaved trees as chestnut *(Castanea),* madrone *(Arbutus),* persimmon *(Diospyros),* sweetgum *(Liquidambar),* sassafras *(Sassafras)* and elm *(Ulmus)* in the northern Great Basin in the Miocene.[34] Decreasing precipitation (together with increasing extremes in temperature) in the later Cenozoic probably accounted for the elimination of many of these trees, except for montane relict stands or those that survive today along the coastal belt. Some of these broad-leaved trees occur today only in the eastern United States and, in some cases, eastern Asia.

Many vertebrates have similar histories and are now absent from the central United States. The turtle genus *Clemmys,* for example, is restricted to the extreme western United States (one species) and the eastern states (three species), with a fossil record from the Paleocene to the Pliocene in the Great Basin (Fig. 3.9).[140] Many birds have similar east–west disjunctions in North America: these include conspecific or closely allied taxa of orioles *(Icterus),* flickers *(Colaptes),* juncos *(Junco)* and others. Although these modern disjunctions may have begun before the Pleistocene, post-Pleistocene drying of central North America probably accentuated their patterns.

Southern relicts exemplify the southerly movement of biotas during glacial periods. Today the wood frog *(Rana sylvatica)* and the Canadian toad *(Bufo hemiophrys)* occur in small pockets far south of the general distribution of these spe-

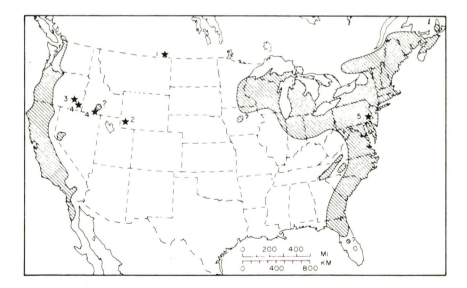

Fig. 3.9. Present disjunct distribution (shaded) of the nearctic turtle genus *Clemmys;* Cenozoic fossils are indicated by stars (Bury and Ernest, 1977).

cies. One of the more interesting boreal islands is in the Turtle Mountains of North Dakota and Manitoba. This is an area of less than 800 m in elevation and, although the flora lacks the dominant coniferous trees of the north, the relict mammalian fauna is characteristic of the northern coniferous forests and includes caribou *(Rangifer tarandus)*, moose *(Alces alces)*, arctic shrew *(Sorex arcticus)*, snowshoe hare *(Lepus americanus)*, lynx *(Felis lynx)*, wolverine *(Gulo luscus)* and others.

The Great Plains of the west-central United States and Canada is semiarid today, with woody vegetation clustered along water courses; however, at the close of the Wisconsin glaciation, about 13,000–9000 ybp, heavier rainfall permitted the existence of coniferous and, later, broad-leaved forests and the associated fauna. Today isolated populations of such mammals as the lemming-mouse *(Synaptomys cooperi)* remain in small favorable pockets. Similarly the big-eared bat *(Plecotus townsendii)*, apparently widespread at the close of the Wisconsin, is now fragmented in several areas (Fig. 3.10).

After the glacial retreat in Europe, some of the fauna moved to the north and some elements moved into higher elevations in mountains. Postglacial movements to the British Isles were partly limited by the gradual refilling of the English Channel seven to six thousand years B.C. Among the family of newts (Salamandridae), three species (*Triturus vulgaris, T. helveticus* and *T. cristatus*) extended

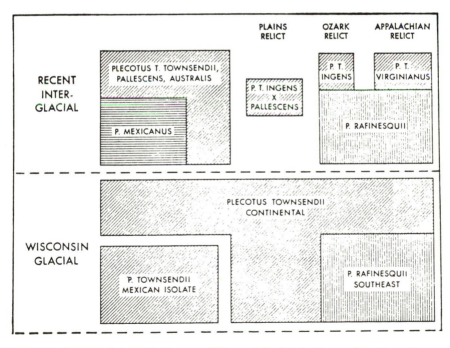

Fig. 3.10. Suggested late Pleistocene (Wisconsin) distribution and modern (Recent Interglacial) of the big-eared bats (*Plecotus* spp.) (Humphrey and Kunz, 1976).

their ranges to England and Scotland before the reestablishment of the marine barrier. Presumably *T. vulgaris* arrived first, for it alone reached Ireland before the country was separated from Scotland and England. Other newts remain in refugia in mountains in Europe: *Salamandra atra* is confined to the Alps, *Triturus montandoni* occurs only in the Carpathians and Tatras and *Euproctus asper* is relict in the Pyrenees. There are also glacial salamandrid relics in eastern Asia, but in this region the refugia are the coastal islands of Hong Kong, Taiwan and Japan.

Unglaciated Pluvial Regions

The cool rainy periods had an estimated annual precipitation of 90 cm in the Great Basin of North America, and comparably heavy rains fell in other temperate regions. Thus it was that in unglaciated areas south of the cross-continental ice sheets there was a great accumulation of ground water, together with the creation of some very large Pleistocene lakes. Much of the flatland of the Great Basin was flooded by lakes 300 m or more deep, and the level of Pleistocene Lake Chad of North Africa was 340 m above modern Lake Chad. In central Eurasia there was a large West Siberian lake dammed at its northern border by glacial ice. Glacial ice also closed the Baltic Sea twice in the last 10,000 years, creating a fresh-water lake.

Many of our modern deserts are post-Pleistocene, and there is much faunal evidence to show that these past pluvial lakes furnished broad corridors for the passage of fresh-water fishes as well as for some of the more aquatic amphibians. Scattered in Death Valley of North America are small creeks, lakes and pools, refugia for a number of species of small cyprinodont fishes called pupfish (*Cyprinodon,* spp). They are the remnants of what was undoubtedly a much more extensive distribution when that region was well-watered.[701] Isolated fishes occur today in the area of the Pleistocene lakes Lahonton and Bonneville. In the Sahara Desert is evidence of pre-Pleistocene forests. Fishes and crocodiles occur in isolated lakes of the Sahara, as well as fossil remains of large mammals that now dwell only on the plains to the south, implying a moister, more genial climate in the Pleistocene.[376]

The Tropics in the Pleistocene

In lower latitudes, in both Africa and South America, mean temperatures dropped $2°-3°C$ during glacial periods. Although vegetational zones descended to lower elevations in the tropics during times of glacial maxima, tropical lowlands simultaneously experienced a *decrease* in rainfall. In South America there was an expansion of nonforested habitat during the Pleistocene arid phase and a concurrent expansion of the fauna of steppe and savannah (Fig. 3.11). In the last glacial period (25,000–12,000 ybp) extensive regions of tropical Africa were more arid than they are today. There was a contraction of tropical rain forests and an expansion of the Sahara southward, with desert replacing grassland and savannah. Tropical rivers became intermittent and shallow desert lakes shrank. As in higher

Fig. 3.11. Movements of Neotropical biota during the glacial arid phase of the Pleistocene (Müller, 1973).

latitudes, montane forests descended by as much as 1000 m below their interglacial elevations. As a consequence, in low latitudes forest biota tended to become
disjunct and replaced by grassland biota.[390,714] This was the opposite of the peri-
glacial changes at midlatitudes.

In some tropical rain forests today the topography is hilly and rolling, and the
soil sandy. Such sand dunes are preserved by the roots of woody vegetation that
developed after the last glacial epoch; the presence today of "fossil dunes" sub-
stantiates the other evidence for former arid conditions in regions of modern tropical rain forests.[878]

Warm ocean water tends to increase wind flow and convection currents, augmenting the moisture in the overlying air and increasing precipitation over the adjacent land. When surface waters cool, convection is reduced and trade winds carry less moisture.[938] This relationship of cool surface waters to adjacent terrestrial aridity is seen today in regions where upwelling brings cool waters to the surface. Along the Guinea coast of west Africa, for example, a strong upwelling lies off Accra (at 5° N), a region normally of scant rainfall; in years when currents bring the surface sea temperature above the norm, however, rainfall increases. On either side of Accra, where there are no upwellings, there are tropical rain forests. Similarly heavy rainfall in the Cera area of northeastern Brazil is correlated with, and apparently results from, an increase in offshore sea–surface temperature, and droughts occur when the temperature drops.[640] Apparently this phenomenon accounts for tropical aridity during times of glacial ice formation at high latitudes. Precise dating of pollen cores by ^{14}C confirms that xeric plant communities in the tropics spread during 15,000–12,000 ybp, during the last (Wisconsin or Würm) glacial epoch.[390]

Pleistocene Extinctions

During the latter part of the Wisconsin glaciation, or in the 5000–8000 years following this period, many of the large mammals of the Holarctic region disappeared. In North America these extinctions included not only the original horses, camels and elephants, but also the giant bison, some antelopes, and the relatively recent arrivals from South America—ground sloths and glyptodonts. It is remarkable that these extinctions occurred so close together and from no clearly established causes. One school of thought contends that predation by man hastened the Pleistocene departure of these large mammals.[651] This belief is predicated on the somewhat earlier arrival of man in the New World but does not account for the parallel extinctions in Eurasia and Australia, where man has had a much older history. These extinctions may not have been nearly so sudden as they appear to have been. The Pliocene to late Pleistocene fossil record consists of a series of strata neither evenly spaced nor evenly represented, so that extinction rates may be distorted. Thus the apparently sudden (post-Wisconsin) extinctions may be partly an artifact of the record.

Concurrent with these Pleistocene extinctions of many of the larger species of mammals was the development of stocks of many smaller species. This has been explained by selective hunting of larger taxa by aboriginal man for spearheads and other artifacts sometimes occur with Pleistocene fossils of large game. On the other hand, some Pleistocene mammals were smaller than their modern counterparts.[467,567] The early Pleistocene (Villafranchian) glutton *(Gulo schlosseri)* was much smaller than the living species. Several Pleistocene bears (*Ursus etruscus* and *U. minimus*) were small; *U. etruscus* may have given rise to the larger *U. thibetanus*. Some insular Pleistocene elephants were very small; on the island of Malta the minute *Palaeoloxodon falconeri* stood about 1 m high, and dwarf proboscidians occurred also in the Greater Sundas and on Santa Cruz Island, California. The mid-Pleistocene *Dicerorhinus etruscus* was about the size of the smallest modern rhinoceros. Thus, throughout the Pleistocene some mammals increased in size, whereas others decreased, and many forms, both large and small, have not survived to the present.

Summary

The overall effects of two to three million years of glaciation have been numerous and diverse, a genuine kaleidoscopic event. Water levels changed, connecting and separating islands from continents, creating large inland lakes and resulting in the movement of floras and faunas. The influence of glacial ice was greatest at high latitudes and high elevations, but heavy precipitation altered the environments of temperate regions. The Pleistocene changes in tropical regions were drastic and profound and caused major shifts in deserts and forests.

Subsequent to the last glacial advance, prevailing temperatures exceeded those of today, giving credence to the suggestion that we are still in the Pleistocene. The prospect for future climatic and biotic changes is one of tantalizing uncertainty: the prognosis is for change but the direction is unpredictable. Within historic times, Viking agriculture has been frozen out of the coast of Greenland, and aboriginal settlements in some areas of the southwestern United States have been dried out. Our modern climate tends more toward change than stability.

The geographic distribution of animals, whether by dispersal or by vicariance, exposes populations to almost constantly varying climatic conditions. Also, as animals move, each deme encounters other species to which adjustments must be made. Zoogeography and evolution interact to maintain the pace of natural selection. Thus abiotic and biotic factors, acting together, mold the biology of each species.

Suggested Readings

Banarescu P (1975) Principles and problems of zoogeography. (Principii si Probleme de Zoogeografie). Translated by the NOLIT Publishing House, Terazije 27/II, Belgrade, Yugoslavia

Croizat L (1962) Space, time, form: the biological synthesis. Privately published. Caracas, Venezuela

Darlington PJ Jr (1957) Zoogeography: the geographical distribution of animals. Wiley, New York

Engler A (1879–1882) Der such einer entwicklungsgeschichte der pflanzenwelt. Engelmann, Leipzig

Good R (1964) The geography of flowering plants. Longmans, London

Hopkins DM (1967) The Bering land bridge. Stanford University Press, Stanford, California

Mayr E (1944) Wallace's line in the light of recent zoological studies. Q Rev Biol 19:1–14

Pielou EC (1979) Biogeography. Wiley, New York

Sclater, PL (1858) On the general geographical distribution of the memebers of the class Aves. J. Linn. Soc. (Zool.) 2: 130–145

Tarling DH, Tarling M (1971) Continental drift. Doubleday, Garden City, New York

Udvardy MDF (1969) Dynamic zoogeography. Van Nostrand, New York

Udvardy MDF (1975) A classification of the biogeographical provinces of the world. Occasional Paper No. 18. International Union for Conservation of Nature and Natural Resources. Morges, Switzerland

Zeuner FE (1959) The pleistocene period. Hutchinson, London

4. Movements and Migration

As has been indicated, dispersal is a permanent exodus from place of birth and may lead to an expansion of a species geographic range. Most movements, however, involve a return to the place of origin and may be confined to a small area. The concepts of territory, home range, movements within a home range, dispersal and migratory movements are generally considered to be distinct. In nature the distinctions are sometimes temporary and difficult to define. The overlap of these concepts in theory and their actual overlap in nature require that they be discussed together.

Movements and migrations within regular, defined areas are typical vertebrate activities. They occur for reasons related to age, sexual activity, season, population density and abundance of food. Home range, territory, movements and migration reflect temporary, secular or seasonal responses to the environment and constitute areas of major interest to the field biologist. They are the topic of this chapter.

Territory and Home Range

In understanding the relationship of an individual to habitat, territory and home range are fundamental concepts. An area that an individual defends is usually considered a *territory*, and an area normally occupied by an individual—and frequently not defended—a *home range*. Territories are usually mutually exclusive, but home ranges may overlap. With birds, the territory is most clearly apparent during the nesting season, for then males are often conspicuous in singing or chasing other males. Territoriality seems to weaken markedly after young leave their parents; the adults then forage over a home range that may not be defended. Vertebrates defend territories using various means and signals. Mammals may employ either visual or olfactory signs, while birds communicate mostly visually. In any group, combat or mock combat may occur between defending individuals.

Territoriality, because it is directed primarily at other individuals of the same species, affects *intraspecific* relationships, and probably does not play a major role in interspecific competition. Territorial aggression is regarded as being directed against the same sex and species, but this conception is predicated on sex recognition. In species in which the sexes are similar, males are aggressive toward all other individuals until sex is established by typical behavioral responses. In many species of birds males and females are similarly colored, and males attempt to exclude all intruding individuals. Sexual recognition is based on a fight-or-flight response: namely, males fight or leave, whereas females indulge in elusive flight but neither fight nor leave.

The needs for a home range remain more or less constant, although the *quality* of the home range (i.e., its ability to produce food) may fluctuate seasonally; therefore, the home range of an individual may fluctuate from month to month. The defended territory, on the other hand, depends mostly on the hormonal state of the individual, and therefore usually becomes markedly altered between the breeding and nonbreeding seasons.

Males of the middle-American frog, *Dendrobates granuliferous,* call from an elevated site, on a tree, and engage in combat with invading males.[201] In some other species of *Dendrobates,* however, females defend their territory.[201] Male frogs of the Nearctic *Centrolenidae* call from vegetation overhanging water. The female attaches her eggs to some plant over water and the male remains nearby.[261] The male remains at the site, sometimes mating a second time with another female and defending his territory from intruding males (Fig. 4.1). In the bullfrog *(Rana catesbeiana),* vigorous males compete for the most attractive territories (Fig. 4.2), and may mate with several females in one season.[454]

Territorial defense in birds is usually the responsibility of both sexes when males and females both care for the young; in most raptors both sexes defend the territory. In polygynous species the male defends a courting area but may be unaware of where the female has nested. In some monogamous birds, especially in species with conspicuously colored males, the male is more vigorous in territorial defense. Some nonmigratory birds have winter territories, but they may not be identical to those that are defended during the nesting period. Winter home ranges are commonly larger than the breeding territories, and some birds, such as some chickadees, tits (*Parus* spp.) and woodpeckers (Picidae), forage in groups over the winter, and the group may share the feeding area.

As with avian species, some species of shrews (*Sorex* spp.) tend to have territories with adjacent but nonoverlapping borders.[411] Female shrews seem not to tolerate other females, although males move more freely between territories of females. On the other hand, female sea lions and females of some other polygynous pinnipeds wander from one territory to another, and may mate with several males in a single season, although the males defend the territory. Beavers *(Castor canadensis)* indicate territory by mud mounds on which they deposit castoreum or castor, a pungent brownish secretion from paired inguinal sacs. All beavers in a colony contribute to these scent mounds, and castor from an alien beaver is recognized and strongly resented.[4] The buck pronghorn *(Antilocapra americanus)*

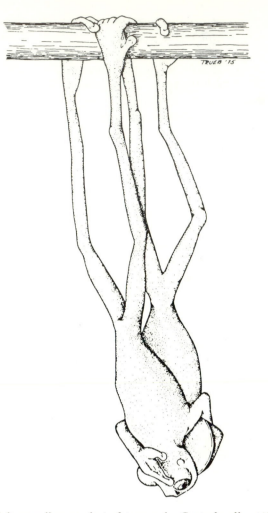

Fig. 4.1. Territorial grappling combat of two male *Centrolenella griffithsi* (Duellman and Savitsky, 1976).

defends his territory only if an intruding buck displays his side to the resident buck. An intruding buck is not allowed to display to females, indicating that the display serves for courtship as well as for territorial defense.[124]

Territorial defense among lizards is more conspicuous among insectivorous species. Aggressive behavior toward invaders is prompt in most insectivorous iguanid and agamid lizards but, in some plant-feeding species, group feeding is frequent. The New Zealand gecko, *Hoplodactylus pacificus,* feeds in groups on nectar.[1088] In most plant feeders food is unlimited, and territoriality is most probably a competition for mates.

Fig. 4.2. Territorial combat of male bullfrogs *(Rana catesbeiana)* (Suzanne Black, from a photograph in Howard, 1978).

Territory size must represent a profitable balance between the cost of defense and the rewards of excluding potential competitors. When more time is spent in defense, less time is available for foraging. Thus a large territory may be a luxury from which the resident obtains little profit at great cost. Territory size appears to represent interactions of such factors as needs of the animal and abundance and concentration of resources in the territory, as well as such variables as density of competitors and the degree to which both sexes participate in territorial defense.

Changes in Home Range

Size of a home range may change with (1) the productivity of the area utilized or (2) the number of individuals in the area. The quality of a home range changes with seasons and with the pressures placed on it; lack of rain or drastic lowering of temperatures render the environment temporarily unsuitable for many vertebrates. Although a home range may be fairly constant over a season or for the lifetime of an individual, small variations in source of food and vegetative features are reflected in temporary deviations from usual home range. In a temperate region in late winter, food is frequently at an annual low level of abundance; this scarcity is one factor (among others) for vigorous territorial defense in the spring. In late summer, on the other hand, food may be both relatively abundant and irregularly concentrated in certain areas. This may account for concentrations of individuals, greater local wandering and reduced territorial defense in late summer.

Selective use of an entire home range varies with the sudden appearance of food items. The chipmunk *(Tamias striatus)* feeds heavily on forbs in the spring and can satisfy its nutritional needs within a small area, but in autumn it forages more widely in search of mast, some of which it stores for later use.[1115] In Montana, chipmunks *(Eutamias amoenus* and *E. minimus)* temporarily invaded an area of oat grass *(Danthonia unispikata)* when the seeds matured in midsummer. At such times the animals had two areas of activity, connected by a path that was virtually linear and of negligible area, and the total of these two areas amounted to the individual's home range. Conversely, as an animal moved permanently from one den site to another and took up a new area (or center) of activity, the old and new areas would not be used simultaneously but would be considered separately in calculating home ranges.[653]

Determination of Home Ranges in the Field

Home ranges of mammals have been studied intensively since the introduction of mass-produced live traps. Students have calculated home ranges for many kinds of small mammals, and have proposed numerous procedures for summarizing field data. Translation of trapping records to a definition of a home range requires some assumptions about the individual's movements. Some workers set a grid of evenly spaced traps; others place them at sites most likely to be visited by small mammals. When trapping covers a brief period in the life of an animal, the worker assumes that the capture sites fall within its normal area of movement, namely, its home range. Long-term studies show that many small mice and shrews sometimes leave one home to assume residence a few hundred yards or more distant. By plotting capture sites on a chart and connecting adjacent outermost points, one can derive the minimal area covered by an individual. With only a small amount of data the worker cannot usually detect irregularities in the home range, but many capture sites may reveal details in its outline. Some students calculate home ranges from as few as two capture sites. From a polygon determined in this manner, one can calculate the home range. Within two possible perimeters of roughly the same size and shape, a large number of records reveals whether the individual moves at random within its home range or concentrates its activity in certain areas (Fig. 4.3). Some animals, such as pocket gophers, move within a sharply delineated area (their tunnels), and require special treatment (Fig. 4.4).

By observing movement of one or both of a pair of nesting birds and plotting the observed perches on a map, one can develop an outline of the area they use. By connecting only several outer points at an arbitrarily determined maximal dis-

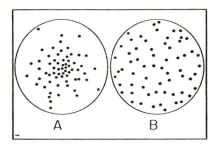

Fig. 4.3. Hypothetical home ranges of two different individuals based on the same sample size (number of captures) and area covered (Fitch, 1958).

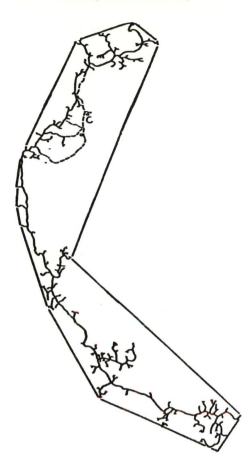

Fig. 4.4. The rather lineal home range of a pocket gopher, as illustrated by its burrow system (Mohr, 1947).

tance from the center, a maximum territory is outlined, and this may include areas not visited by the pair. A more accurate indication of *utilized territory* can be found by connecting all peripheral points; such an outline tends to exclude areas not visited.

Another measure of home ranges is the "probability density function," which is determined by a series of recapture radii from the geometric center to every recapture point. Recapture points normally include sites well within the "minimum polygon," and, when plotting data from animals captured in live traps in a regular grid, one can assume that each animal could move halfway between its capture point and the next outermost trap. A mean recapture plus two standard deviations will determine a theoretical circle which includes 95 percent or more of the animal's movement.[238]

Size of Home Range

The size of a home range is proportional to the resources it provides: a few square meters may be adequate for a mouse, whereas a shrew of equal size may regularly forage over an area several times as large. The home range, then, is predicted on the availability of food, its nutritional value and the energy needs of the animal. A shrew, for example, has a greater energy expenditure than has a lizard of the same weight, for the latter obtains much solar energy in the daytime and experiences a great reduction in body temperature and energy demands at night. Thus a lizard has a relatively small home range compared with that of a shrew. Also, all other aspects being equal, larger animals have larger home ranges. Within a major taxon, such as a family, size of home range is predicated mostly on abundance of food, suitability of cover and distribution of home or nest sites. There is a general relationship between body weight and territory size in birds, and predators occupy greater areas than do herbivores (Fig. 4.5).

The size of either a home range or a defended territory reflects an adjustment between the needs of the animal and the abundance of these resources in the hab-

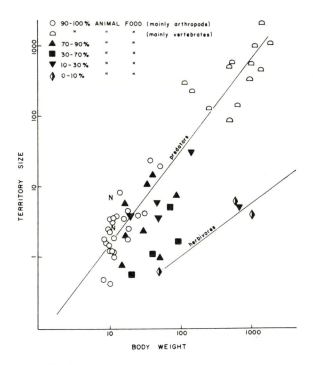

Fig. 4.5. The relationship between individual body mass and territory size in birds. Open symbols are predators; solid symbols are omnivores (eating 10–90% animal food); half-filled symbols are herbivores (eating less than 10% animal food) (Schoener, 1968, copyright 1968 by the Ecological Society of America).

itat. If the home range is not defended (and it usually is not), it may not be density-dependent. Many herbivores, such as deer, move no more than necessary during a day; when forage quality declines, however, the individual may expand its home range.[525] Expansion of a home range increases the cost of foraging, and if a herbivore must cover a larger area in periods of forage deterioration, the increase in food thus obtained may not justify the expense of obtaining it. In a given habitat, a vole *(Microtus)* may be almost totally herbivorous and satisfy its needs close to its nest, whereas another, such as a deer mouse *(Peromyscus)* or wood mouse *(Apodemus),* may feed on small seeds and nuts as well as on small arthropods, and may need to forage a greater distance from its nest. The prairie vole *(Microtus ochrogaster),* on the Natural History Reservation of the University of Kansas, is a typical herbivore, and has a mean home range of 0.22 hectares. In contrast, the white-footed mouse *(Peromyscus maniculatus),* a smaller species that feeds extensively on seeds and insects, has a mean home range of 1 hectare and the short-tailed shrew *(Blarina brevicauda),* almost entirely insectivorous, has a mean home range more than twice the area of the white-footed mouse.[297] The pika *(Ochotona princeps)* defends an area about the pile of hay it has gathered in autumn; each individual, regardless of age, gathers its own hay.[529]

The dusky salamander *(Desmognathus fuscus)* has a riparian home range of about three lineal meters,[66] and the milk snake *(Lampropeltis triangulum)* moves over an area of about 20 hectares.[297] Thus the needs of one species for food, nest sites, cover for young and other environmental features may differ from the environmental requirements of another species, and the totality of these needs for different species is found in areas of different size.

Home ranges of fish and amphibians in streams assume a special dimension, for food is carried by the flow of water, and therefore foraging need not consume as much time as it does in quiet waters or on land. With an increase of food, emigration may be reduced, and there may be a slight increase in density and biomass resulting from growth and reproduction. This does not affect the increase in territoriality of substrate spawners during their reproductive seasons. When young salmon leave their nest, or redd, at age 16 weeks, they defend feeding territories and wait for the moving water to bring food to them.[162]

Certain bathypelagic fishes make extensive vertical movements daily. Some of these, such as lanternfish (Myctophidae), rise and fall 900 m or more twice a day.[252] Because this is a daily movement, one might define their "home range" as the volume of ocean included within this 900 m column of water. On the other hand, this is a regular movement with a return to the place of origin, and thus fits the definition of a migration as well as falling within the concepts of a home range.

For many vertebrates the home range seems to consist of a general area, not necessarily bounded by landmarks. Very commonly males occupy a somewhat larger home range than do females. They are prone to make occasional shifts in their home ranges, whereas females tend to remain in specific areas for extended periods. Young animals commonly disperse from their natal home before reaching sexual maturity.

Migration

When an animal moves away from its home range to reside in a new area (or wander over a broad region) and later returns to its place of origin, this movement is called *migration.*

Migration requires that the returning individual locate its area of origin. The distance covered in movement to the winter home is usually far beyond the animal's summer home range, so that the ability to return, called *homing,* requires mechanisms of *orientation* and of *navigation.* In this context, homing is the drive to return to the general area of the birthplace, orientation is the ability to determine one's position along the way, and navigation is the capacity to recognize direction—to plan the route from one place along the way to the destination. Homing, orientation and navigation are prerequisites of migration.

Migrations are so adjusted that the animal is in the proper environment at the right season; sometimes one speaks of a spawning or breeding migration when the destination is the breeding area, and there are specific segments of a migratory path that are primarily for feeding and overwintering. Extensive migrations are characteristic of some fishes, a few amphibians, reptiles, many birds and a few kinds of mammals. Migrations are regular annual events in the lives of many species that inhabit regions where there are marked seasonal variations in temperature and rainfall, and usually in food as well. Some migrations take more than one year but nevertheless eventually end in a return to the place of origin for breeding. Pacific salmon (*Oncorhynchus* spp.) move to the sea in the first year of their lives and do not return for several years; they breed once and die. Most species, however, move annually, and any given individual may make two or more migrations in its lifetime. A few vertebrates move in synchrony with tides, and these short trips are technically migrations.

A theoretically important advantage of homing is that young are born in the specific locality to which they are adapted, by virtue of their parents having been born and reared there. The successful transplanting of many migratory animals, such as salmonid fishes, suggests that their environmental demands are flexible and/or duplicated in many areas.

Environmental Influences on Migration Routes

Marine and aerial migrators may employ water or air currents in movement; this not only relieves the animal of some work but affects the speed and direction of movement. Larval fish, such as larval herring or leptocephali larvae of eels, may drift passively with a current, but cetaceans may seek out currents that assist them, or birds may modify their route and benefit from winds. Marine currents are slow and fish carried in them have no reference to stationary objects, so that travel is by passive drift; some downstream migrants, however, swim actively with the currents. Fish moving either actively or passively with a current are said to be *denatant,* whereas those moving against the current are *contranatant;* the direction of both types is dependent upon that of the current. Because water is a heavy medium through which to move, denatant passage saves energy for growth and reproduction. If a gyral (large eddy in the ocean) is of the proper size and location, a species could complete its entire migration by denatant movement within the gyral; the problems of homing, orientation and navigation then would not exist. Perhaps some fish swim actively with the current of a gyral, but such details are not as well known for fish as for migratory birds. Some stocks of Atlantic herring and Pacific salmon may migrate in this manner; currents provide a directional clue as well as transport.

Oceanic migrations may be closely associated with ocean currents, both horizontal and upwellings. Young cod in the western Atlantic arctic seem to follow complex currents between Greenland and Iceland. A northern offshoot of the Gulf Stream, the Irminger Current, carries warm water to Iceland, where it moves both west to Greenland and east to northern Iceland (Fig. 4.6). Larvae from spawning on the south of Iceland make a denatant in a clockwise path about the north and others are carried passively to the east coast of Greenland; the movement of mature cod is in the opposite direction, a contranatant journey against the currents that effected the passive drift of the larvae (Fig. 4.7).

Migrations of Fish. Fish are *diadromous* when they migrate between salt and fresh water; when they move from the sea to rivers and spawn in fresh water the species is *anadromous,* and the reverse movement with spawning in the ocean is *catadromous* migration. Other words have been created to describe migrations

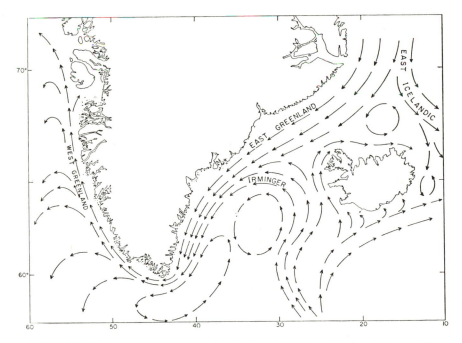

Fig. 4.6. Surface currents in the Iceland–Greenland area (Harden Jones, 1968).

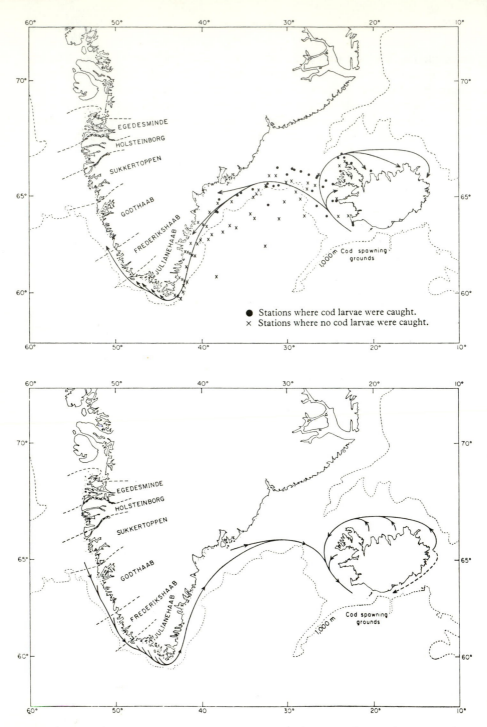

Fig. 4.7. Upper: the passive drift of eggs and larvae of cod *(Gadus morhua)* from the spawning ground south of Iceland (Harden Jones, 1968). Lower: the active movement (against the current) of cod to the spawning ground (Harden Jones, 1968).

within an ocean, a lake or a river. So many fishes are anadromous that spawning in fresh water must constitute a great advantage. Perhaps abundance of insect food for the fry, high oxygen content of the water or a relative scarcity of predators justifies the expenditure involved in swimming two or three thousand miles. On the other hand, one can imagine that an anadromous fish is simply returning to the ancestral, fresh-water home of the species, and that the major benefit in an anadromous migration is the abundance of food in the ocean.

One should assume, however, that there is a biologic advantage to the individual in both movements to the sea and the return to fresh waters. The anadromous pattern persists in species that become landlocked: the Atlantic smelt *(Osmerus mordax)*, an anadromous salmonoid fish introduced into the Finger Lakes of New York, ascends small creeks to spawn and returns to the deep lakes to feed; the salmon (e.g., *Oncorhynchus nerka*) spawns in streams or along the lakeshore when landlocked. Less clear-cut but generally anadromous migrations are made by the striped bass *(Morone saxatilis):* the fry move downstream and may remain near the mouth of the river of their birth, or move a short distance into the sea, only to return upsteam when sexually mature.

The eel (*Anguilla* spp.) is the best known catadromous fish. In the North Atlantic Ocean eels breed near the Sargasso Sea, and the larvae, drifting clockwise with the Gulf Stream, enter rivers in eastern North America and in western Europe, swim to the eastern Mediterranean Sea where they enter the Nile and penetrate to the Azores. Each group is not known to be genetically the same. Several years later, eels return to the sea, spawn and die. The original pattern was established by the painstaking effort of Johannes Schmidt in the early part of the present century. By "hitch-hiking" rides on tramp steamers in the Atlantic, Schmidt determined that larval fish, then known under the generic designation of *Leptocephalus,* were larger as one followed the Gulf Stream away from the Sargasso Sea. They then gradually assumed the size and shape of young eels as they reached the various coasts. Schmidt also postulated that European and American stocks were specifically distinct, a suggestion that has been questioned but not yet resolved. Other species of *Anguilla* breed in an undetermined area or areas in the Pacific Ocean and ascend rivers in some large Pacific islands as well as in eastern Asia. Interestingly, they do not enter rivers of western North America.

Two species of eels mature in the rivers of New Zealand. *Anguilla dieffenbachi* may remain in fresh waters as long as 19 years and grow to two meters in length before premigratory metamorphosis. The spawning area is unknown. *Anguilla australis* occurs not only in the streams of New Zealand but on other continental fragments (Lord Howe Island, Norfolk Island, New Caledonia and the Fiji Islands, among others in the southern Pacific.[672] Other species occur in Asiatic rivers.

Migratory movements of fishes are frequently in three parts, with three goals or terminals, and may be illustrated by a triangle. The first passage is always from the spawning to a feeding area, and is more horizontal than vertical. Young Pacific salmon (*Oncorhynchus* spp.) may go thousands of miles on their feeding migration, which may last well over a year. Mackerel *(Scomber scombrus)* in the western North Atlantic move inshore in summer to feed in warm shallow water after hatching, but seek deeper water in winter as the shallow areas become cold. The first shift is a feeding migration, the second, a wintering migration. With increasing age, adult mackerel move to more southerly waters for the winter, and their passage to the north in the spring is a spawning migration.

Among the tremendous diversity of fishes are many sorts of migrations. In many species there are annual spawning migrations, but this does not occur until the individual has reached sexual maturity, at age six years or more for some Pacific salmon and eels. In both the species of Pacific salmon and eels, death follows spawning, and there is but a single spawning in the life of any given individual. Other fish, such as the Atlantic salmon *(Salmo salar)*, cod *(Gadus morhua)*, mackerel *(Scomber scombrus)* and many more, spawn for a number of years. For them birth is followed by a feeding migration, after which there are annual spawning and wintering migrations. Throughout the feeding or spawning migrations, there are sometimes vertical movements, ascending toward the surface at night and returning to deeper water in the day.

The migratory patterns of the Atlantic herring are exceedingly complex, as there are many separate stocks that breed in different seasons and on different grounds. Generally, spring-spawning stocks exhibit earlier growth than do those of summer and autumn spawners, which breed farther north. Within a week after hatching of the demersal eggs, the larvae float to the surface and drift with the current. It is not practicable to mark the larvae before they have drifted and perhaps mixed with those from other spawning grounds. It is impossible therefore to determine if herring return to their birthplace to spawn, or spawn in regions other than their parent ground. It is conceivable that a full migration does not occur within the life cycle of a given herring—that an individual from one spawning bed may drift and, when mature, spawn over a different breeding ground. The existence of discrete stocks suggests different adaptations to specific spawning areas, but historic shifts and fluctuations in hydrographic conditions in the North Atlantic suggest that the survival value of subtle adaptations must be short-lived. The characters by which the different stocks are recognized (scale and otolith features) are phenotypic and may tell nothing of the genetic background of the fish.[403] This must be one of the most enigmatic of all fish migrations.

Migrations of Amphibians. Amphibians seldom make extensive migrations, but some species make seasonal or annual journeys to and from ponds or rivers in which they spawn. Like most newts (Salamandridae), the red-spotted newt *(Notophthalmus viridescens)* leaves the water after transformation to the immature "red eft" stage. Postlarval migration of the red newts may begin in July in southern areas, but in the north gilled larvae may overwinter in ponds. In New York State there are two waves of emigrating red newts from ponds. This phenomenon results from two distinct periods of oviposition: adult females that overwinter in the ponds breed earlier than do those that hibernate on land, and these two groups of larvae move to the land at different times.[534] A Nearctic newt, *Taricha rivularis,* has a similar migration; it returns to the same site for breeding each year. Homing is remarkably precise, and a locus less than 50 feet in variation is occupied annually.[362,787,788]

Migrations of Reptiles. The green turtle *(Chelonia mydas)* nests every other year or every third year, and leads a pelagic life in the interim. The migration of various populations is regular in its schedule, but the nonreproductive home of these giant turtles is not definitely established. In its migration from Brazil to Ascension Island the green turtle apparently swims at least 1200 miles against the South Atlantic Equatorial Current. Its navigational mechanism is not known. Females tagged as they laid eggs in the sandy beaches of Ascension have been found in subsequent years not only living in the coastal waters of Brazil but breeding again on Ascension. Perhaps this turtle has nested on Ascension since an early period when South America and Africa were closer together.

Some pond turtles move from one locality to another in response to seasonal changes in water levels. The painted turtle *(Chrysemys picta)* frequently occupies a pond near its hibernation site until early spring, and then scatters to smaller temporary ponds; as the latter dry up, the turtle returns to the permanent waters in which it hibernated.[233]

Migrations of Birds. Flight, which has enabled birds to move to the very high latitudes of both hemispheres, also allows them to depart seasonally, usually well in advance of winter. Many birds of temperate and polar regions profit from the long days, which not only promote vigorous plant growth but also permit prolonged daily feeding. Although the extreme north and south offer advantages to birds in their breeding season, these same regions become uninhabitable a few months later.

Landmarks may be critical in the routes that many birds follow. Species such as hawks and storks, which frequently use warm rising air to support them, seek the shortest possible crossing over water. The white stork *(Ciconia ciconia),* migrating south from Europe, passes about the eastern and western ends of the Mediterranean (Fig. 4.8), and hawks may pass south along mountain ridges that run in a north-south direction. Logically such species always migrate in the daylight hours, for air rises over land from midmorning until late afternoon.[870,898]

Two terns make the longest migrations, and regularly move from their breeding grounds in the north to the extreme southern part of the southern seas. The arctic tern *(Sterna paradisaea)* nests on coastal areas of Eurasia and the American arctic north to the Arctic Circle, and flies across the Atlantic and Indian Oceans to the Weddell Sea, south of the Antarctic Circle. The common tern *(S. hirundo)* nests in the northern half of the Holarctic region; some individuals move no farther than the warm regions of the Northern Hemisphere, but a few fly to South Africa, Madagascar and Australia.

These great migrations are paralleled by the annual movements of many shearwaters (*Puffinus* spp.), which nest in the extreme south and move to the Northern Hemisphere in April and May. The terns, petrels and shearwaters feed on a variety of small marine organisms, including small fish, and their long annual

Fig. 4.8. Movement of the white stork *(Ciconia ciconia),* a diurnal migrant, between Europe and Africa (Schüz, 1953).

passages north and south place them at the proper season in regions of great marine productivity. Wilson's petrel *(Oceanites oceanicus),* a pelagic species, nests on islands (South Shetlands, South Orkneys and South Georgia) between Cape Horn and the Palmer Peninsula of Antarctica, mostly south of 60°, and also in the Indian Ocean on Mauritius and Kerguelen Islands. After nesting, it moves north to the North Atlantic as far as the coastal waters of Great Britain and Labrador.

Some shorebirds from the Northern Hemisphere reach New Zealand in the austral summer: the curlew sandpiper *(Erolia testacea)* nests in Siberia, the knot *(Calidris canutus)* and the golden plover *(Pluvialis dominica)* breed in the tundra of both North America and Eurasia, and all three may fly as far as New Zealand in the nonbreeding season. Some marine birds migrate over land: the emperor penguin *(Aptenodytes forsteri)* walks inland more than 100 km to its breeding site. Within Australia, avian migration is best developed along the eastern coast where seasonal climatic changes are most conspicuous. The stable marine climate of New Zealand seems not to stimulate migration, and most native birds are sedentary. The two species of cuckoos of New Zealand are exceptions:[105] the bronze cuckoo *(Chalcites lucidus)* winters in the Solomon Islands, northeast of New Guinea, and the long-tailed cuckoo *(Urodynamis taitensis)* moves north after breeding and winters in a broad expanse of Oceania, being concentrated in Samoa and the Fijis (Fig. 4.9).

Migrations of Mammals. Migratory mammals are frequently of economic value, and their annual movements have been followed by hunters, sealers and whalers throughout the world. Today there are detailed accounts of migration for many kinds of whales, pinnipeds, bats and ungulates; some species return to their birthplace with the accuracy and regularity of some migratory birds. Whales move to subtropical or tropical waters to calve, whereas many pinnipeds go to polar regions to bring forth their pups.

The sexes of many migratory mammals segregate during the nonreproductive seasons, a phenomenon uncommon in birds and fish. Winter colonies of bats fre-

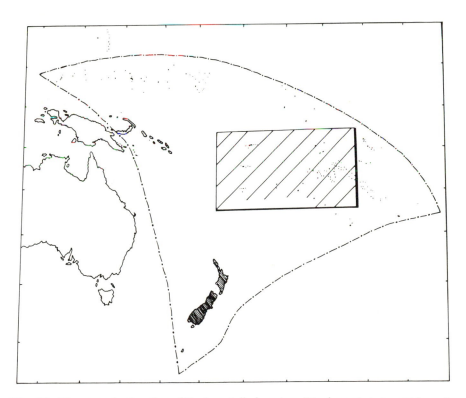

Fig. 4.9. The annual migration of the long-tailed cuckoo *(Urodynamis taitensis)* from its nesting area (New Zealand) north to many small islands. The principal winter range is the rectangle in the northern part of its winter range (Bogert, 1937).

quently consist of only males or females, and among sea lions females may migrate a thousand miles more than do males. There are sexual differences in the extent of movement of the sperm whale and perhaps for some of the whalebone whales.

The production of vast amounts of plankton in polar seas is a well-known seasonal event and probably stems from (1) the long summer days at the high latitudes, and (2) the relative abundance of oxygen, carbon dioxide and salts dissolved in the very cold water. Although the small crustacea (krill) upon which whales and many fish feed can live in the darker deep layers of the ocean, the diatoms upon which the zooplankton feed need sunlight. Thus, dense populations of marine vertebrates thrive in the polar seas and feast there seasonally.

The annual passage of whales across the latitudes is a movement to calve in the winter and to feed in the summer. Generally the krill on which the large mysticeti feed flourish in polar regions during the summer. At the end of the summer there is a simultaneous reduction in phytoplankton and an expansion of polar ice, and the mysticeti move to more temperate climates. The warmer oceans, by contrast, are usually barren of great densities of food for whales; when baleen whales move to their calving grounds, they may go four months or more without feeding. Their high intake of fat-rich food in polar seas provides energy not only for their annual migrations of up to 4000 miles but also for the production of as much as 100 or more gallons of fatty milk daily during a prolonged lactation.

The humpback whale *(Megaptera nodosa)* in the Southern Hemisphere has a conspicuous north–south migration from the Antarctic Ocean to calving grounds as far north as the equator (Fig. 4.10), and marking experiments indicate a well-developed homing ability. Different species of baleen whales have distinctive migration patterns. Although both Arctic and Antarctic whales may migrate to equatorial waters, they usually do not occur in the low latitudes at the same time, and there is very little exchange or mixing of the boreal and austral stocks of a given species. The gray whale *(Eschrichtius gibbosus)* is today confined to the North Pacific, and migrates south along the coasts of Asia and North America to calve in the waters near Korea and in the Gulf of California.[937]

The sperm whale *(Physeter catodon),* the only really large toothed whale (Odontoceti), feeds on squid, cuttlefish and bony fish, and tends to wander through tropical waters. Presumably sperm whales concentrate near large stocks of decapods, which is especially true of nursing females.

The harbor porpoise *(Phocoena phocoena)* enters the Bay of Fundy where calves are born in May and June; then herring are abundant in the shallow waters. At this time only adult females and their young are present, but in July adult males and yearlings enter the bay and there is a three-fold increase in numbers. In the autumn most of the population moves to sea and southward, remaining within eight miles of the coast.[749a]

Because all pinnipeds give birth on land or ice, there is at least a slight migratory tendency throughout the group. Some species come ashore in small groups, but breeding aggregations in others, such as the northern fur seal *(Callorhinus ursinus),* exceed 1 million individuals.

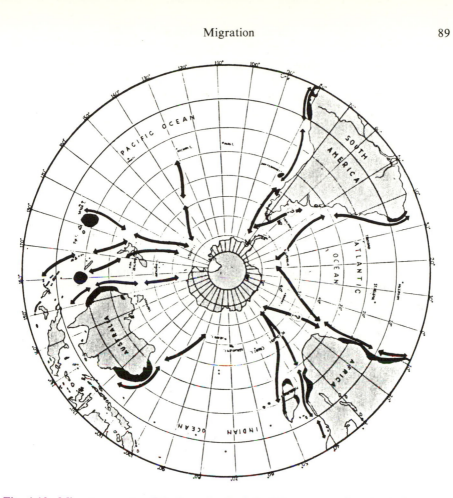

Fig. 4.10. Migratory routes of the humpback whale *(Megaptera nodosa)* in the Southern Hemisphere (Slijper, 1962, courtesy of Hutchinson & Co., London and Cornell University Press, New York).

The northern fur seal disperses from the rookeries in autumn or early winter and usually remain scattered 10 km or more from coastal waters. In autumn those from the Pribilof Islands, north of the Aleutian chain, move southeastward; males remain to the north, but females and young migrate south as far as northern Mexico. They return annually to the same site to breed, but the means by which they navigate is not known. The harp seal *(Pagophilus groenlandica)* breeds on floe-ice. Parturition occurs as early as January in the White Sea but as late as March and April in the western Atlantic, off Greenland and Newfoundland. In the summer they move to coastal regions to the north in Greenland and offshore in more southerly populations. In the autumn western contingents pass southward to the Gulf of St. Lawrence or the Grand Banks off Newfoundland, where they feed and move back to the ice floes to the north in February. Thus their annual passage depends partly on nutritional needs, and partly on the development of ice on which pups are born.[262]

The walrus in some parts of its range makes annual movements. Off the west coast of Greenland they haul out on floe-ice in late spring, and the single pup is born far from land. They move to the shore in summer and south along the coast in autumn, always south of the solid sheet of ice. The Pacific walrus makes similar annual movements, usually in groups or packs.

The caribou or native reindeer *(Rangifer tarandus)* in North America is both nomadic and migratory, i.e., although their movements may be extensive and involve entire populations, they may vary from one year to another. On the coastal plain of northern Alaska, the caribou summers in the lowlands but moves to highlands, where some ground is usually exposed by winds. The caribou of the Northwest Territories made extensive movements and sometimes covered hundreds of miles annually. Early accounts suggest that their travels varied from year to year and seemed to have been predicated on a search for food and an escape from mosquitoes. The wild reindeer *(Rangifer tarandus)* surviving today in Fennoscandia spends the summer in lichen-rich forests of pine and birch and moves to high alpine regions in the winter.[973]

Banding of several species of migrating bats (Microchiroptera) demonstrates a very strong tendency to return annually to their breeding cave or summer home. Homing is also well developed in the tropical and nonmigratory phyllostomid bat, *Phyllostomas hastatus:* removed from their roost, some individuals returned in one night from a distance of 53 km.[370]

Homing, the regularity in returning annually to the same site or cave, has been determined for many species of bats and is probably characteristic of most temperate species. The gray bat *(Myotis grisescens)* spends the summer in scattered localities in the southeastern United States. They roost and bring forth young in warm caves (about $15°-27°C$); in the winter they migrate to cold caves (with mean temperatures of about $7°-20°C$). These movements were from 17 to 437 km each way; these bats moved from 18 to 52 km nightly.[1025] *Myotis lucifugus* has been found to move 97 km in one night.[564] Once within its hibernation cave, a bat may shift its position once or several times, apparently during the periodic arousals that all hibernators experience.

In the Nearctic pipistrelle *(Pipistrellus subflavus),* there is a local differential in the sex ratio of hibernating adults. In the northern part of its range, most hibernating pipistrelles are males, but in the southeastern United States females predominate among those hibernating in caves.[219,220] In caves in southern Louisiana, most hibernating pipistrelles are females, but in the summer males and females occur in equal numbers. Apparently there is a northerly migration of the females in the spring. The hoary bat *(Lasiurus cinereus)* is highly migratory; it frequently flies over water along the coasts and is known from a number of islands far from the mainland. When they move northward in the spring, males tend to migrate northwestward and females to fly northeastward.[1000]

The free-tailed or guano bat *(Tadarida brasiliensis mexicana)* comprises at least four discrete migratory segments in southwestern United States, which migrate as separate groups (Fig. 4.11). Individuals leave Arizona in early October and move into Mexico, flying from 10 to over 25 km nightly. Populations in Cal-

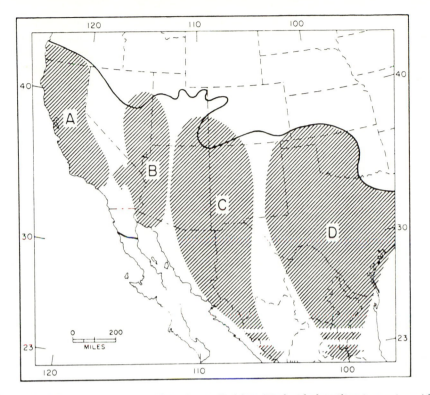

Fig. 4.11. Migratory segments of the free-tailed bat *(Tadarida brasiliensis mexicana)* in western North America. Group A makes only local movements and B moves for short distances; C and D may migrate more than 625 km (Cockrum, 1969).

ifornia seem to make only local movements, but in other regions this species moves from 250 to 625 km or more in each direction.[175]

Orientation and Navigation

The mechanisms of orientations and navigation exhibit aspects of both vertebrate behavior and vertebrate physiology, especially sensory physiology, in adaptation to the environment and geography.

Olfaction in Fish and Amphibians

Studies by Professor Arthur Hasler have revealed some means by which anadromous fish can locate the stream of their birth.[407] Both laboratory and field experiments indicate that fish learn distinctive odors of different waters. Olfaction is well developed in fish, and the blunt-nosed minnow was conditioned to distinguish between the odors of waters from two streams. This ability, determined by reward and punishment, persisted for about 6–15 weeks after conditioning; the fish, however, was unable to recognize these waters after its olfactory bulbs had been destroyed by heat.[408] Field tests with silver salmon confirmed that odor of their natal streams guided the adults to return there to spawn. Captured at the headwaters and marked and released downstream, a clear majority of the fish returned to their creek of capture. This ability to home was lost when their olfactory pits were plugged with cotton or with cotton and an anesthetic ointment.[409]

Olfactory cues guiding fish to their breeding area seem adequate to account for homing in fresh water, but there is no known mechanism for guiding fish tens of hundreds of miles through the ocean waters to the mouth of the home river. In the absence of known signals directing a fish to the river of its birth, it seems likely that random coastal movements account for the eventual arrival where home waters flow into the sea.[585] The eel (*Anguilla* spp.) can perceive magnetic fields, but it is uncertain if these influences affect their famous migrations.

Leopard frogs *(Rana pipiens)* can return home when the olfactory nerves are severed and hearing destroyed; they can orient when the sun is visible but cannot do so at night.[245,246] Olfaction is sensitive in salamanders and may function in navigation and homing. The newt *(Taricha rivularis)* becomes confused and apparently disoriented when its nostrils are plugged with vasoline.[362] Toads (*Bufo valliceps* and *Bufo boreas*) employ the position of the sun to orient, and in doing so recognize the time of day in appraising the sun's position.[356,377] The painted turtle *(Chrysemys picta)*, a North American pond turtle, orients itself relative to the position of the sun, an ability that requires an internal time-measuring mechanism.[233]

Celestial Navigation in Birds

Among the many interesting aspects of migration is the tendency to return to the same breeding area as well as the same wintering area. As with fish, birds need a mechanism for precise homing, but the means of navigation must be different. A fish moving up a river must make a decision each time a fork is reached, but a bird, flying over open skies, needs a method not only to indicate the correct route but also to relocate the route should a storm drive it off track. Considering that migration and homing have evolved in many diverse families, some of which fly only in daylight hours and others that migrate only by night, it is unnecessary to presume a single method of navigation.

Homing pigeons, which have been extensively studied, are trained by being released repeatedly over the same ground. Although skilled in returning to their home loft from unknown places, they are certainly not migratory birds. Although they may not reveal all the methods used by wild species, they suggest some of the cues used by wild birds.

Pioneering studies initially supported the concept of orientation and navigation by means other than conventional landmarks of memory. In a series of experiments, Manx shearwaters *(Puffinus puffinus),* seabirds that nest on the island of Skokholm in Wales, were taken from their nests and released far from home and in strange territory. Not only did they return quickly, traveling as much as 200 miles a day, but they flew over land, although they are naturally strictly pelagic.[810] Other seabirds have been released in alien lands, and they have usually returned promptly, even from distances exceeding 4000 miles. Difficulties in orientation seemed to occur in birds released beneath overcast skies.

Avian migration, as with migration of other vertebrates, is tied to environmental changes and is almost always annual. Thus it is convenient to separate marine migrants from those birds that breed and feed in terrestrial habitats. Some seabirds cover great distances, crossing the equator from one hemisphere to another, and many species come to land only when nesting. Such pelagic migrants navigate without landmarks, and long ago presented circumstantial evidence for the existence of celestial orientation cues. The young of some birds migrate at different times from the autumnal passage of the adults, and thus cannot benefit from the prior experience of their parents. Moreover, a spring migration does not always follow the same route as the postreproductive migration. These features all suggest that migratory birds navigate by means other than memory or "follow the leader."

Because of the regularity of daylength changes, photoperiod is a dependable *cue* to which a migrant can time its movements. Migrations are *adaptations* to seasonal changes in temperature, food, cover, growing seasons and other factors on which breeding success depends. That is, photoperiod is the timer that regulates the migratory passage, but the latter is an adjustment to the actual environmental and nutritional needs of the bird.

Among many species of passerine birds, there is a migratory urge, or *Zugunruhe*. This is a nocturnal restlessness in species that are virtually diurnal at times other than the migratory season. Zugunruhe is made apparent by a continuous nocturnal hopping and fluttering of the caged bird, and is the basis for much of the experimental work on migration. In captivity Zugunruhe accounts for activity long after sunset during the season of migration, and seasonal changes in daylight or photoperiod produce or stimulate this nocturnal migratory impulse. Photoperiodic control of Zugunruhe, hyperphagy and lipogenesis may serve as the *Zeitgeber* of endogenous cycles, i.e., photoperiod may control the *timing* of an inevitable event.[545] Several species of Old World warblers maintain annual schedules of lipogenesis and gonadal recrudescence, and molt when kept under constant light.[318] In some nonmigratory birds the increased daylength of late winter produces Zugunruhe, the significance of which is not easily explained. Autumnal migration occurs as days become shorter, but the control is not so clear as in vernal migration. Nocturnal migration may have survival value by allowing passerine birds, shorebirds and waterfowl to fly when foraging would be difficult.

Interest in position-finding by celestial bodies probably originates from human experience in navigation. Stars appear in a predictable position in the sky at a specific instant, and at a given time the elevation (angle from the horizon) is uniform when viewed from any of a number of points describing a line on the surface of the earth. By determining the lines so described by two or more stars simultaneously, one can establish that the position from which they are viewed is the point at which the lines intersect. The same information can be obtained from the sun and moon, but the exact time must still be known. The angle of the sun at its zenith indicates the latitude of the observer, but this measurement changes every day. The north star (Polaris) changes its position only slightly, so that at any hour and at any season its elevation indicates the approximate latitude from which it is viewed. For other stars, in order for an accurate position line to be determined, not only must the observed angle be exact, but the time of viewing and the distance above sea level must be accurately known. It seems unlikely that birds possess the ability to use celestial cues to determine any but their *approximate* position, but the determination of a generalized position may be sufficient when other kinds of information are available. Most studies of celestial orientation involve birds, but there are suggestions that marine turtles and even some fish use the sun in determining their positions.

A valid procedure for exploring orientation is to transport or displace migratory birds, band them, and plot the date and place of recoveries. Only when large numbers of birds are banded can one obtain sufficient recoveries to suggest the navigational procedures of migrants. Of around 11,000 starlings *(Sturnis vulgaris)* (which winter in Spain), captured at The Hague and displaced and released in Switzerland, 354 were later found. Immature birds migrating for the first time continued in the same direction (southwest) from Switzerland, as they would had they not been displaced; some immature individuals even proceeded to a new wintering ground in southern France.[807] In contrast some adults moved to the southwest, while others moved to the northwest. This suggests that adult starlings can recognize the degree of displacement and compensate for it. The absence of any compensation on the part of the immature birds does not reveal whether they were unaware of the displacement or whether they could not compensate for it.[808,809] In fact there is no evidence that an awareness of displacement and the navigational skill to correct it are different phenomena. This experiment does indicate, however, that immature individuals moved independently of adults. Displacement experiments with other species produced both types of response seen with the starling. Other displacement experiments involved transplanting young birds and rearing them in a different area. The results vary, but in general indicate that the *direction* of a migratory path is inherited, and that a displaced bird does not compensate for the displacement by changing its direction.

The experiments of Gustav Kramer illustrate the apparent ability of birds to locate their position.[559] Starlings were kept in cages under the natural sky with several perches about the sides; the perches were separately connected to counters so that the experimenter could note which perches the birds used most frequently at night. The starlings activated perches in the southwest sector in autumn and in the northeast sector in the spring. It was concluded that the birds oriented by celestial cues. An alternative possibility is that seasonal changes in photoperiod cause the birds to attempt to fly in the direction appropriate to the season. Stars can provide the observer with information about its position, but they cannot tell the direction in which to move to arrive at a new position. Only with the aid of charts can the sailor or aviator determine the direction in which he must travel to reach his goal.

Kramer's experiments were continued by Franz Sauer, who replaced the natural sky with the artificial heavens of a planetarium. Captive-reared birds oriented beneath the artificial, much as wild individuals did beneath the natural, sky: during the migratory period they fluttered in the direction of their destination. Moreover, when the artificial sky was rotated, the birds recognized the simulated displacement and changed the direction in which they fluttered. In all these experiments, when either the natural or artificial sky was obliterated, the birds became disoriented.[879]

It has long been known that migrating birds become "confused" when the sky is overcast: migrating passerines may fly low and cease to fly at all on cloudy nights, waterfowl indulge in nondirectional flights in cloudy weather, and homing pigeons perform most efficiently in clear weather. The implication is that the sky assists in avian orientation in both daylight and nocturnal migration. The sun is an obvious point of reference, even if the information provided by it is of a very general nature. At its zenith the sun indicates the latitude of the observer, and coupled with time it must provide an east–west sense of direction; both these aspects change slightly each day, and it is not obvious what information migrants obtain from the sun. Experimental observations on captive birds, reared without observing the daily course of the sun, may not indicate the orientation of wild individuals. The nature of orientation by the sun has been explored by a great variety of experiments; the results are sometimes contradictory and ambiguous, but most workers agree: (1) that the sun is of great navigational value to many diurnal migrants, and (2) when the sun is obscured by cloud cover, diurnal movements frequently become nondirectional.[658]

Daily and seasonal changes in the sun's path complicate its role as a reference point; if a bird can indeed employ the sun throughout its orbit, it must be able to predict the timing and extent of solar changes from day to day. The movement of nocturnal heavens are nearly constant throughout the year at any given point, and the night sky presents a relatively simple set of reference points. The pecten may have a role in avian orientation and navigation: it casts a sharply defined shadow on the retina from the light reflected by the image of the sun. This shadow may be pointed (depending on the position of the sun and the angle with which it is viewed), and, when viewed so as to form a shadow with the most acute angle, the tip falls exactly on the temporal fovea (Fig. 4.12). If the bird also possesses an accurate time sense, the presumed "biological clock," then the acute shadow may indicate the horizontal angle, or azimuth, of the sun.[814]

Some experiments with avian orientation suggest that direction is consistent throughout a nocturnal period. When captive mallards were kept under a schedule experimentally advanced or retarded six hours, they oriented in the natural direction under a natural sky, as did controls. Had they used any specific set of stars so as to orient with precision, their determination would have displaced them east

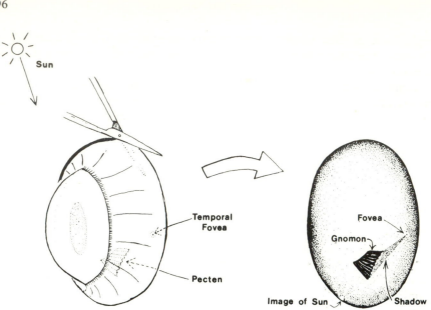

Fig. 4.12. A possible role of the pecten in orientation. On the right the image of the sun casts a shadow beyond the pecten. At one point the position of the sun's image forms a shadow with its apex on the fovea; this may indicate the horizotal angle, or azimuth, of the sun (Pettigrew, 1978).

or west, and their movement would not have been the same as that of the controls. The alternative explanation is that "stellar orientation" is fundamentally unlike the precise position-determination of human navigators.

Correct use of stars for navigation depends on exposure to the natural sky in the early premigratory period in the indigo bunting *(Passerina cyanea)*. Hand-reared individuals never exposed to the sky were unable to orient themselves properly for the autumnal migration; the birds conditioned by an "unnaturally oriented" sky in a planetarium assumed a direction correct for the experimental sky.[273,274] Thus any genetic tendency for characteristic migratory routes would seem to require early (premigratory) familiarity with movements of the sky over an individual's place of birth.

Under natural conditions some birds change their direction during magnetic storms, and in the laboratory they orient in the seasonally appropriate direction when exposed to geomagnetic but not to celestial cues.[710] The European robin *(Erithacus rubecula)* orients properly when the sky is obliterated but not when placed in a steel chamber.[1100] Similarly, indigo buntings *(Passerina cyanea)* oriented properly without a view of the sky, but altered their direction when the magnetic force was changed by activation of a Helmholtz coil about the cage. The altered direction was that predicted by the experimental change produced by the coils, but the magnitude of the response, although statistically significant, was not striking, and weakened as the migration season neared its end. The response to magnetic forces, both natural and experimental, was clearest when data from individual birds were pooled, and suggests that geomagnetic forces can assist this species in orientation.[276]

Indigo buntings were removed from their nests at 4–10 days of age and reared under two different artificial skies of a planetarium, one sky simulating the natural sky of their nesting area and the other an unnatural sky; a third group was reared in the absence of celestial cues. When the birds were tested during the migrating season beneath the artificial skies to which they had been exposed, their orientation indicated that they had been conditioned by the skies in infancy: those exposed to a simulated natural sky fluttered mostly in a true southerly direction, whereas those exposed to a sky that rotated about the star Betelgeuse instead of Polaris tried to move toward the artificial south of that unnatural sky; those exposed to no celestial cues in infancy fluttered in random directions. This type of experiment suggests that orientation by stars is learned rather than inherited. It also suggests that birds may orient with reference to the *rotation* of the heavens rather than to any specific stars, and also that geomagnetism is less important as an indicator of direction than is the sky.[274]

The experiments of Schmidt-Koenig were the first to field-test homing by nonvisual means. With frosted lenses over their eyes and nearly blinded, pigeons flew from a release point some kilometers from the home loft and could find their way to the *vicinity* of their home; when near the home loft and without visual cues, they could not pinpoint their home. This suggests that pigeons use a different means for long-distance and short-distance system homing; furthermore it indicates that long-distance homing does not require visual cues.[889] Experimentally, magnets disrupt orientation of homing pigeons, but it does not necessarily follow that the birds use natural magnetic forces in their return to home.[534]

It has long been known that electromagnetic fields disrupt orientation in pigeons; when released in the vicinity of a radio transmitter, pigeons have difficulty in direction-finding. Moreover pigeons become disoriented when small magnets are placed near their heads. In the rear of the pigeon skull are innervated magnetized crystals that may serve as a sensor for the earth's magnetic field.[1068]

Celestial cues in orientation require that a bird not only recognize the arrangement of stars over a part of that species' geographic range, but also that it be able to compensate for the constant rotation of the heavens. The case for using magnetic guideposts is statistically weak. Many migratory birds become disoriented under overcast skies, and celestial cues remain the most plausible means of navigation for nocturnal migrants. Celestial and geomagnetic cues are not mutually exclusive, and the latter may well have an ancillary role.

Some research on mammalian migration parallels studies on birds and fish. Students have banded bats for many decades, and there is a large body of data on their movements and longevity.[370] Commercial whalers have shot identifying capsules into the peripheral fat of cetaceans, but the recovery of the capsules does not approach that of numbered bands from bats and birds. There seems to be no real knowledge about the mechanism of orientation and homing in cetaceans, but patterns suggest that they sometimes swim with ocean currents. The autumnal southward passage of the gray whale in the eastern Pacific, for example, takes it offshore and into the Alaska and California Currents; in the spring it moves north in coastal waters, where the effect of the current is minimal. Similarly the bowhead whale *(Balaena mysticus)* follows the shoreline in its spring migration along the north coast of Alaska but passes well offshore on its southerly autumn journey.

Dispersal

Dispersal, the permanent exodus to a new habitat, occurs with many species, especially in the young. In birds, it is conspicuous in those individuals that move far from their natal home in the year of their birth. In North America the bald eagle regularly moves north in the year it is fledged, a habit that is also seen in the starling *(Sturnus vulgaris)* of Europe and in many herons throughout the Northern Hemisphere. Dispersal is perhaps most noteworthy in the cattle egret *(Bulbulcus ibis):* within historic times this little heron has moved from Europe to South America, from where it has rapidly spread north as far as Canada. Dispersal in immature birds can be seen in extralimital occurrences (beyond the limits of the species' normal range). Purple gallinules sometimes appear in the mid-Atlantic island of Tristan da Cunha, around 3000 km from their home, but this journey is not made by adults.

In contrast to the faithful return of many birds to their breeding area, some ducks are notoriously inconstant, nesting in regions hundreds of miles apart in successive years. This may be explained by the fact that most ducks court and form pair bonds on the wintering grounds, and that, if a pair consists of two birds from different breeding grounds, they must return to that of the individual with the stronger migrating urge.[1008]

Saturation Dispersal

Dispersal resulting from high pressure on resources in dense populations has been called *saturation dispersal.*[597] It is known in gray squirrels *(Sciurus carolinensis)* and Scandinavian lemmings.

The treks of the gray squirrel are among the most remarkable and unusual of nonmigratory movements of any mammal. These dispersals involved tens of thousands of squirrels moving dozens of miles. They crossed open fields, where they did not normally dwell, and swam wide rivers, such as the Ohio and the Niagara.[540] These dispersals occurred every so often during the mid-19th century, apparently at times of very high densities. This species was sometimes known as *"Sciurus migratorius,"* because of its famous journeys, journeys that have not taken place in modern times.

The periodic mass movements of lemmings *(Lemmus lemmus)* have been observed since ancient times, and in recent years scientists have explored details of this phenomenon. Every three to five years reproduction is accelerated, and after large populations build up they move from their natural montane home and descend to the ocean. Large numbers move more-or-less together, so that disappearances and arrivals in any given area seem to occur overnight. Actually they move about 1.6 miles daily and most activity occurs at night. Movement into the ocean is deliberate and, although lemmings swim well, most die by drowning. Several observers report that most of the individuals involved in these journeys to the sea are relatively young, nonparous animals, and that the activity should perhaps be regarded as a specialized sort of dispersal, triggered by population density.[228]

A specialized migration is the sporadic invasion of some birds from high latitudes and high elevations to warm regions. That this is a true migration is seen from the return of the survivors to their breeding areas the following spring. Seed-eating species, such as crossbills *(Loxia curvirostra),* siskins *(Spinus pinus),* red-polls *(Acanthis flammea)* and nutcrackers *(Nucifraga columbiana),* make these irregular migrations in years of scant production of forest mast (Fig. 4.13).[103] The occasional southward mass movements of goshawks *(Accipiter gentilis)* and snowy owls *(Nyctea scandiaca)* occur when populations of prey species are low.

We have seen that vertebrates perform a broad variety of movements, from the daily sallies of a small mouse foraging for seeds and insects to the extensive, round-trip migrations of salmon and marine birds. One can easily distinguish between daily movements within a home range and prolonged migrations lasting many months, or even years. They are related, however, by the similarities in (1) the goal of locating the most favorable habitat and (2) the mechanisms in navigation and orientation.

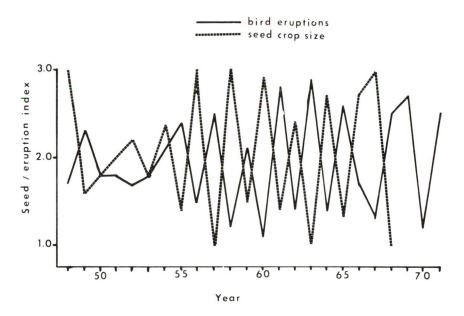

Fig. 4.13. The scarcity of seed production in boreal coniferous forests and its correlation with the southerly movement of seed-eating birds (Bock and Lepthien, 1976, courtesy of University of Chicago Press).

The differences one sees among the kinds of movements reflect (1) adaptations, (2) magnitude of environmental changes, and (3) the ability to move. Only minimal movements are undertaken, for example, by amphibians, for most species live in areas where annual environmental changes do not require extensive movements. Amphibians, however, like many birds, employ celestial cues in direction-finding. Homing by salmon primarily involves odor, a stimulus not available to most birds, but some fish apparently orient themselves relative to the position of the sun. Migrations of whales are sometimes adjusted to take advantage of ocean currents, just as are migrations of some fish.

Thus one can simultaneously point out both differences and similarities among the many kinds of movements, dispersals and migrations. There is an underlying continuity joining these activities, and it is reasonable to regard them as varying manifestations to achieve similar goals.

Suggested Readings

Dorst J (1962) The migrations of birds. Houghton Mifflin, Boston

Harden Jones FR (1968) Fish migration. Edward Arnold, London

Hasler AD (1966) Underwater guideposts; homing of salmon. University of Wisconsin Press, Madison

Matthews GVT (1968) Bird navigation, 2nd ed, Cambridge University, Cambridge, England

Mohr CO (1947) Table equivalent of populations of North American small mammals. Am Midland Natural 37:223–249

Schmidt-Koenig K (1975) Migration and homing in animals. Springer-Verlag, Berlin

Part II

Individual Environmental Responses

The first chapters of this book have dealt almost entirely with what might be termed long-term environmental responses and adaptations. The remaining three parts will each deal with matters much more related to the individual animal, namely, basic, sensory and reproductive physiology, including the resultant changes in populations. In this section, dealing with basic physiology and behavior, the topics discussed are breathing, food and feeding, thermoregulation, water balance and seasonal dormancy.

5. Breathing

All vertebrates have special surfaces for acquiring oxygen and discharging carbon dioxide. Typically, gaseous exchange occurs through gills in aquatic vertebrates and in lungs in terrestrial forms. Other respiratory surfaces are important in some fishes, amphibians and reptiles. Specializations reflect the habitats and the activities of the different classes.

Generally the concentration of oxygen in air is about 30 times as great as that of dissolved oxygen in water. Carbon dioxide is about 28 times more soluble than is oxygen in fresh water. Differences between the free gases in air and in water are marked, and seasonal—and sometimes daily—variations in gases in both marine and fresh waters are substantial. Although oxygen may occur at a relatively constant level in the open ocean, in coral reefs it may be seven times as great in the day as at night, when photosynthesis fails to replenish supply. During daylight hours, production of oxygen exceeds that consumed by respiration, but there is a decline of dissolved oxygen at night. Under conditions favoring maximal photosynthesis, dissolved oxygen in midafternoon may be nearly two-and-a-half times that present at dawn (Fig. 5.1), and much of this oxygen is released by algae. Being light-dependent, oxygen production and dissolved oxygen are greatest nearest the surface. Nocturnal increases in dissolved carbon dioxide follow the cessation of photosynthesis and increase the acidity of such waters.

At decreasing temperatures more oxygen can dissolve in water; at low temperatures not only does the need for oxygen decrease, but its availability increases. In some fishes kept at extremely low temperatures, there is a decrease in the number of red blood cells in circulation. This tendency reaches an extreme with the Antarctic icefish, *Chaenocephalus aceratus* (Chaenoichthyidae), which lacks erythrocytes altogether. The plasma of the icefish carries about 0.7 percent oxygen by volume, a very small amount, but adequate at the low temperatures at which the fish lives. Goldfish whole blood carries about 10 percent oxygen, and

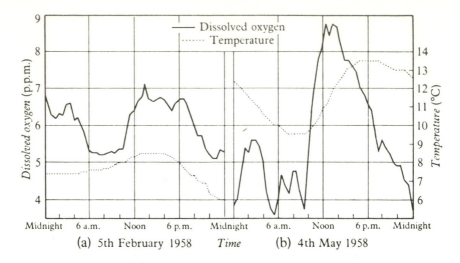

Fig. 5.1. Daily fluctuations in dissolved oxygen and temperatures in a slow-moving river (Gameson and Griffiths, 1959; British Crown copyright).

the blood of a mackerel about 15 percent oxygen, most of which is carried by erythrocytes.

Normally oxygen is carried by hemoglobin as oxyhemoglobin (HbO_2). In the lungs, as oxygen moves into the plasma, carbon dioxide moves out. The loss of carbon dioxide from the plasma, where it exists as carbonic acid (H_2CO_3), causes a rise in pH, facilitating the formation of oxyhemoglobin. Conversely, as carbon dioxide enters the plasma, the pH drops, and oxyhemoglobin tends to dissociate into hemoglobin and oxygen. This relationship is called the Bohr effect, or the Bohr shift.

Gills of Fish

Gills of fish-like vertebrates are highly vascularized membranous structures associated with gill slits. Gills are supported by gill arches, on the anterior edge of which there may be several or many gill rakers, and on the posterior edge thousands of lamellae on gill filaments arranged in two slightly divergent rows (Fig. 5.2). The blood circulates through the filaments in a direction opposite the flow of water, creating an efficient counter-current exchange.

In hagfish (Myxinidae) there is either a single pair (in *Myxine* spp.) or six pairs of gills (in *Bdellostoma* spp.). Gills of lampreys (Petromyzontidae) are enclosed in a pair of elastic branchial baskets, each with a single internal opening but seven external openings per side. Elasmobranches have five pairs of gills; exceptions are cow sharks (Hexanchidae), which have six or seven pairs, and the frilled shark (Chlamydoselachidae), which has seven. In bony fishes there are four gills covered by an operculum (or opercle).

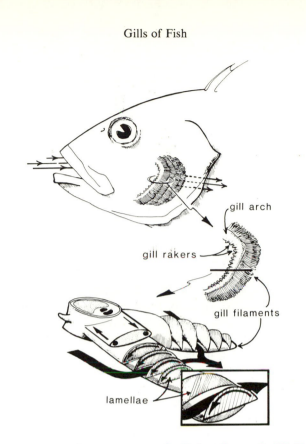

Fig. 5.2. The structure and arrangement of gill lamellae and gill filaments. Their arrangement provides for an effective counter-current exchange mechanism (Suzanne Black).

Ventilation

The passage of water over the vascularized surfaces of the gills is called *ventilation;* it varies with the structure of the gills and the habits of the fish.

Larval elasmobranches and larvae of some teleosts have external gills, extensions of posterior gill filaments. Although external gills are supplied with essentially the same circulation that serves adult gills, the ventilation is fundamentally different. External gills extend laterally above the operculum, from the side of the head, and are prominent in larvae of the bichir (*Polypterus* sp.) and lungfish (*Protopterus* spp.) of Africa and in the Neotropical lungfish *(Lepidosiren paradoxa).* They appear most conspicuously in fish from oxygen-poor waters. There is no forced stream of water passing over the external gills of larval fish, but external gills may fan out near the surface of the water and capture oxygen as it diffuses into water from the air. The yolksac of many larval fishes serves as an accessory breathing structure, as the surface is usually covered with a fine network of blood vessels.

Adult lampreys are parasitic, and attach by their sucker-like mouths to the sides of larger fishes from which they extract body fluids. When the lamprey is attached to prey, there is no water intake through the mouth. Muscles about the branchial baskets contract, forcing water out the external openings; when these muscles relax, the branchial chamber returns to its original shape, drawing water in through the gill openings. This is irregular (in–out) or *tidal* ventilation.

Ventilation in most fish is accomplished by drawing in water through the mouth and expelling it through the opercular opening. Behind the lips is a valve comprising a dorsal and a ventral flap. As the buccal cavity enlarges, water enters the mouth; closure of the mouth pushes the labial flaps together, preventing movement of water out of the mouth and forcing it over the gills. Flow of water is assisted by elevation of the opercles, which enlarges the gill chambers. This double pump moves over the gills in a continuous flow, in contrast to the tidal ventilation of a feeding lamprey. Changes in volume of the buccal chamber occur slightly before changes in the opercular chamber with the result that: (1) there is a near-constant greater pressure in the buccal chamber (Fig. 5.3) and (2) there is a continuous flow of water over the gills. The inertia and viscosity of water probably contribute to the continuous nature of the flow produced by the buccal pressure pump and the branchial suction pumps.[464] In rays (Rajiformes), which normally lie on the seafloor, water drawn in through the mouth may contain sand that can damage the gills, and in these fishes water enters through the large dorsal spiracle (a vestigial last gill slit); this short canal has a valvular flap, preventing the exit of water.

The basic patterns of gill irrigation are varied, and in such rapid swimmers as lamnid sharks and tunas irrigation is effected mainly by the passive movement of water over the gills as the fish swims.[732] The shark sucker, *Remora remora,* when free-swimming, ventilates its gills in the usual fashion; but when attached to a larger fish respiratory movements decrease as the speed of the host fish increases.

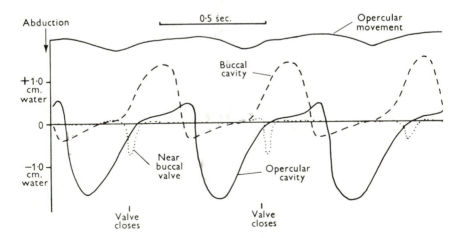

Fig. 5.3. The movements of the mouth and operculum, producing fluctuating pressures in the buccal and opercular cavities, maintain a nearly constant unidirectional flow of water over the gills (Hughes and Shelton, 1958).

At water speeds of 60 cm per second the remora needs only to adjust its mouth opening to obtain sufficient oxygen to maintain metabolism.[731] A similar example in a very dissimilar fish is seen in a sucker-like minnow (*Gyrinocheilus* sp.) of southeast Asia; it extracts nutrition from mud and at times uses its mouth for attaching itself to rocks. Water enters the branchial chamber through an upper orifice of the opercular opening, passes over the lamellae and leaves through a low exhalant opening. Some marine fishes in oxygen-poor waters at mesopelagic levels (down to 900 m) migrate to the surface waters at night. Some of them (Myctophidae, Melamphaidae and Gonostomatidae) are anaerobic in the daytime and aerobic at night. Blood oxygen capacity of these fishes is very low, from 1.5 to 3.6 percent.[252]

Fish Out of Water

Air-breathing fishes are almost entirely dwellers of fresh waters that are seasonally or sporadically low in dissolved oxygen (the mudskipper may be the only exception). Seawaters are rich in oxygen except in bathypelagic strata, and in the ocean there is little need for air-breathing. Clearly air-breathing in fishes has developed numerous times in various parts of the world, but almost always in warm, shallow ponds. Today some of these areas are seasonally dry, but very probably air-breathing began in epochs of warm humidity. A warm dry atmosphere is as dangerous to an air-breathing fish as an oxygen-deficient pond is to a conventional gill-breather. In modern times air-breathing fishes escape desiccation in rainless months by seeking high humidity in a subterranean chamber. The vascularized lungs of dipnoans provide great independence from the totally aquatic environment, but in other groups vascularized membranes line the buccal and branchial cavities and the intestine. The skin of fishes may absorb oxygen and release carbon dioxide, especially in scaleless species.

The mudskipper (*Periophthalmus* spp.), an inhabitant of tidal mud flats throughout much of the tropical Pacific, Australia and Africa, frequently flits over the mud with unbelievable speed. These little fish rest on mud with their tails in water, so that they can quickly retreat backward into their holes. By separating the gill openings from the posterior body surface in a respirometer, one can contrast the exchange of oxygen through the skin with that through the gills and the highly vascularized buccopharyngeal cavity. Although some gas exchange occurred through the skin of the head, it was calculated that about 60 percent of the oxygen passed into the body through the skin on the trunk.[998] Vascularized skin on the head and fin surfaces presumably also serves in external respiration and, while out of water, mudskippers often roll from side to side, perhaps to moisten the cutaneous respiratory areas.[971] Prior to leaving the water and moving across land, the mudskipper *(Periophthalmus sobrinus)* of East Africa gulps air, which together with water is held in the gill cavities; thus gills furnish the fish with oxygen whether in the water or on land. Feeding on land involves expulsion of air and water from the gill chambers, after which the fish returns to water.

Some Neotropical catfish of several families live in warm stagnant ponds; not only can they breathe air when in water, but several crawl over land. They breathe through the intestine; absorptive areas of the gut are well-vascularized and require minimal modification for air-breathing. Clariid catfish of tropical Asia and Africa have small gills, but vascularized suprabranchial pockets enable them to breathe air. *Heteropneustes* spp. (= *Saccobranchus*), catfish of the tropical Orient, have reduced gills and a small air bladder, but breathe through the vascularized surfaces of a pair of long caudal extensions of the gill cavities. The volume of these pockets is controlled by muscles: contraction expels air, relaxation causes air to enter. The climbing perch (*Anabas* spp.) and the related paradise fish *(Macropodus opercularis),* like the mudskippers, have vascularized chambers in the gill cavity (Fig. 5.4). These "opercular lungs" function both when the oxygen content of water is reduced and, as in the mudskipper and the climbing perch, when the fish is out of water.[465] In the snakehead (*Ophiocephalus* spp.) and the cuchia *(Amphipnous cuchia),* the gills are rudimentary, but vascularized outpockets occur on the upper pharynx. The cuchia, an eel-like inhabitant of warm ponds of the tropical Orient, readily emerges from water and, like many amphibious fishes, suffocates when deprived of air.[506]

Perhaps the best-known air-breathing fish of the Orient is the East Indian eel-like *Monopterus javanicus:* it lacks functional gills and breathes through both its skin and "pharyngeal lungs."[1110] *Monopterus* is sold for food, and vendors keep the fish alive in baskets covered by damp cloths for months as they carry them from one mountain village to another. *Monopterus albus* gulps air when in stag-

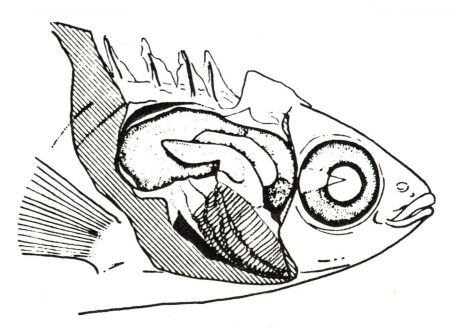

Fig. 5.4. Suprabranchial lamellae of the paradise fish *(Macropodus opercularis)* (Ebeling and Alpert, 1966).

nant, oxygen-poor waters, and can survive for extended periods out of water, when migrating over land.[503] When out of water, *Monopterus* obtains about one-fifth of its oxygen through the skin.[601]

Although the skin of fishes usually lacks conspicuous modifications, many scaleless and some scaled fishes can absorb a substantial part of their oxygen needs through epidermal capillaries. The eel (*Auguilla* spp.) is noted for its nocturnal migratory journeys about waterfalls, and its skin (with embedded scales) is the site for the entrance of oxygen as it crawls through wet grass. A loach *(Misgurnus fossilis)* may absorb more than 60 percent of its oxygen intake through the skin and up to 85 percent when oxygen is prevented from entering the gills, the remaining 15 percent being absorbed through the intestine.[216]

Lung-Breathing Fish. Placoderms possessed air sacs that may have served to hold atmospheric gases, and this air sac presumably gave rise to the air bladder of subsequent fishes. The well-known African lungfish (*Protopterus* spp.) and the Neotropical lungfish *(Lepidosiren paradoxa)* not only have reduced gills and a vascularized air bladder, but the air bladder is also deeply bilobed. The Australian lungfish *(Neoceratodus forsteri)* is an air-breather but, unlike other dipnoans, has adequate gills. Several other groups of tropical fresh-water fishes have species with vascularized swim bladders. Lungfish fill the lungs by gulping air or drawing air in through the nares and then forcing it into the lungs by a contraction of the buccopharyngeal area. Natural elasticity of the lungs and pressure of the water expel the air from the lungs. Blood leaving the heart passes through the gills before reaching the lungs. Although the lungs put about four times as much oxygen into the blood as do the gills, the latter are very important in the discharge of carbon dioxide. There is in dipnoan fishes a rudimentary double circulation: blood from the lungs is directed back through the heart and then to the systemic circulation.[588]

The Nearctic bowfin *(Amia calva)* has a vascularized swim bladder and also a cartilaginous rod connecting the distal margins of the gill filaments, so that the latter do not collapse under their own weight when in air. In the gars *(Lepisosteus* spp.) the air bladder is vascularized, and both the bowfin and the gars have reduced gills and must breathe air for much of their oxygen requirements. The gar *(Lepisosteus osseus),* a ganoid fish of the Great Lakes, is an obligate air-breather, which makes increasing use of its lungs with increasing water temperature (Fig. 5.5). At water temperatures of $25\,°C$ up to 80 percent of the total, oxygen intake is through the lungs, but less than 10 percent of the carbon dioxide is released by that path, most of it leaving through the gills and the skin.[841] The Neotropical osteoglossid, *Arapaima gigas,* has a large, cellular, vascularized air bladder, and the many alveoli give it the appearance of a true lung. It is an obligate air-breather, but most of the carbon dioxide is eliminated through the skin and gills.[697] The electric eel *(Electrophorus electricus)* of the Orinoco and Amazon drainages is another air-breather: it gulps air regularly and absorbs oxygen through its oral mucosa; as in *Arapaima gigas,* carbon dioxide leaves through the skin and gills.

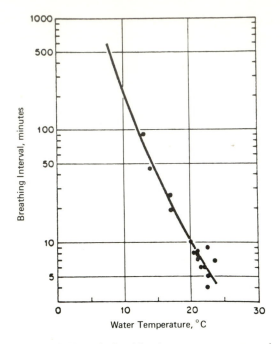

Fig. 5.5. Temperature-solubility relationship. As water temperature increases, solubility of oxygen decreases and metabolic demands for oxygen increase, all resulting in increased air-breathing by the gar *(Lepisosteus osseus),* an obligate air-breather (Rahn et al., 1971, courtesy of Elsevier/North-Holland Biomedical Press).

Amphibians

There is a diversity of breathing mechanisms in modern amphibians, reflecting their extreme responsiveness to the habitats in which they live. Amphibians can absorb oxygen through either the skin or the lungs, and some other internal surfaces, usually the lining of the mouth, are believed to exchange respiratory gases. Most modern amphibians have moist skin that is frequently very efficient in absorbing oxygen. The ability of amphibians to breathe through their skin accounts for their respiration when hibernating in deep ponds and streams. Many species of frogs and toads can spend months beneath the water, and some hibernating forms regularly do so; the cold-water temperature both increases the solubility of oxygen and reduces the metabolic needs for it.

Anurans usually have well-developed lungs; not only do these provide for the exchange of respiratory gases, but the movement of air also enables males to make their characteristic calls. Generally the skin of anurans and urodeles contains about the same degree of vascularization, but the lungs of the former provide a much greater surface for gas exchange. Some anurans apparently depend on cutaneous breathing: these are species that inhabit either deep or oxygen-poor ponds or swift streams. The so-called hairy frog develops vascularized dermal papillae in the males. The male of *Astylosternus* (= *Trichobatrachus*) *robustus,* a ranid frog of West Africa, has extensive vascularized filaments on the sides of the body and thighs and reduced lungs; it is usually a stream dweller, and the terrestrial female has normal skin and well-developed lungs, with typical anuran alveoli. The aquatic frogs, *Telmatobius* spp. (Leptodactylidae), live at high elevations in the Andes, and *T. culeus* lives in Lake Titicaca at elevations of more than 3800 m in water of about 10°C. Although the lungs are greatly reduced, the skin is furrowed and highly vascularized, and respiratory efficiency is increased by the small size and great number of erythrocytes.

Lungs are variously developed in salamanders. In pond-dwellers and terrestrial forms lungs are usually present, but in aquatic species they are important not only for breathing but also as a hydrostatic organ, and tend to be smaller in stream-dwellers than in pond species. In the Plethodontidae, found in streams and ponds and on land, lungs are lacking. Lungless salamanders breathe through the skin; none attains a large size, perhaps to preserve a high surface-to-mass ratio. Plethodontid salamanders tend to have larger capillaries than other urodeles; greater capillary diameter and slower circulation through the capillary system increase the efficiency of integumentary gas exchange in the lungless salamanders. The *Batrachoceps* spp. are lungless salamanders of the Pacific coast of North America. They are not only very small (with a high surface-to-mass ratio) but also slender and long-tailed. In addition, almost all of their red blood cells are non-nucleated; without nuclei, red blood cells have not only a smaller metabolic demand for oxygen themselves but a relatively greater surface for gas exchange.[312]

Breathing Methods

The relative importance of anatomic structures in breathing is studied by both comparing the extent of their capillary systems and measuring gas exchange at their surfaces. The skin of anurans is served by a pair of cutaneous arteries, but in salamanders the skin is supplied with capillary networks from the adjacent muscles. The relative roles of the integument and lungs in the Nearctic ambystomid salamanders are correlated with their habitats. The amphibious *Ambystoma tigrinum* and *A. talpoideum* absorb roughly 54 percent of their oxygen through the lungs; in the more aquatic *A. opacum, A. macrodatylum* and *A. maculatum,* about 34 percent of the oxygen enters through the lungs, but in *Rhyacotriton olympicus,* a dweller of mountain streams, the lungs are reduced and admit only 26 percent of the respiratory oxygen.[1093] Except for neotenic forms, the skin is the other major respiratory organ in these salamanders.

The effectiveness of the integument as a breathing organ changes with the habits of the animal; many amphibians, especially newts (Salamandridae), are terrestrial except during the mating season. When they enter the water, their skin becomes less cornified. Not only is the skin thinner in such aquatic stages, but tail fins expand and increase breathing surfaces. Respiratory surfaces of the common newt of Europe *(Triturus vulgaris)* are largely in the skin: breeding males have from 13 to 14 meters of capillaries per gram of body mass, or about 80 percent of the respiratory capillaries is in the skin and 17 percent in the lungs.[206] In a New

World tree toad *(Hyla arborea)* about 75 percent of the respiratory capillaries is in the lungs and about 25 percent in the skin.

The problems created by periods of low oxygen content and high levels of carbon dioxide are aggravated at higher temperatures, which increase respiration. In such adverse environments some fishes and aquatic amphibians become airbreathers. Experimentally the siren *(Siren lacertina)* obtains 87 percent of its oxygen needs from water at 5°C, but this drops to less than 50 percent at 25°C. The gills of the siren account for less than 5 percent of total oxygen intake, and cutaneous breathing does not increase appreciably with temperature; increased metabolic demands are met by the lungs.[1037]

At low temperatures (5°C), the gills of the mudpuppy *(Necturus maculosus)* lie motionless and appressed to the body, and the skin is then the main structure through which oxygen enters the body; as temperatures rise and oxygen demands increase, however, the gills move actively and account for a proportionately greater amount of oxygen intake. In contrast to the siren, the lungs of the mudpuppy are relatively unimportant in breathing, never accounting for more than 10 percent of the total oxygen intake.[381]

Biphasic or Bimodal Breathing

The lungs of fish and amphibians are more effective in obtaining oxygen than in releasing carbon dioxide. Thus, although the lungs of some amphibians and "amphibian" fishes provide most of the needed oxygen, carbon dioxide is lost through the skin and gills. In the dipnoan *Protopterus aethiopicus,* for example, about 90 percent of the oxygen intake is through the lungs, but they eliminate only about 30 percent of the carbon dioxide, the balance leaving through the gills and skin.[504,588] In the almost totally aquatic *Siren lacertina,* lungs account for one-half of the oxygen intake but less than one-quarter of the carbon dioxide loss.[1180] This pattern is known as *biphasic* or *bimodal gas exchange;* it characterizes not only air-breathing fishes but also aquatic amphibians. At some point in the evolution of the totally lung-breathing terrestrial vertebrate, lungs assumed the role of the skin (and gills) in carbon dioxide release, but modern amphibious breathers provide no clue as to how this was accomplished.

Because carbon dioxide is about 30 times more soluble in water than is oxygen, there is much less difficulty in releasing CO_2 by diffusion from vascularized skin than there is in obtaining O_2 by diffusion from *water*. Presumably this forms the basis of both biphasic gas exchange and the pursuit of atmospheric oxygen. The situation changes when the animal leaves the water for air, which is always richer in oxygen than is the blood, and in an aerial milieu both gases can pass through a moist skin.

Reptiles

Reptiles have achieved, with a few exceptions, total dependence on the lungs as breathing surfaces. Gills are absent, and the skin of most reptiles is dry; some reptiles can absorb oxygen through the digestive tract.[1105]

Typically reptiles are aspiration breathers: by contraction of lateral or costal muscles most saurians expel air from the lungs. In some lizards dorsolateral muscles pull the ventral skin dorsad, expelling air; because of the weight of the viscera and elasticity of the lungs, inhalation is passive. In chameleons, air may be drawn in through the nostrils by a lowering of the floor of the mouth.

Turtles

Most turtles exhale actively, by contracting muscles about the posterior openings in the shell. The mechanics of breathing varies with their movements in and out of water. In the snapping turtle *(Chelydra serpentina)*, the viscera are suspended by muscles, and the sagging gut and associated organs cause the lungs to expand; when in the water, however, pressure against the soft parts between plastron and carapace is applied to the lung. Thus the snapping turtle inhales passively on land but exhales passively in water. In crocodilians contraction of a diaphragmatic muscle causes the liver to move caudad and the lungs to expand in response. Transverse abdominal muscles cause a forward movement of the liver, expelling air from the lungs.[326]

In turtles the relative volume of the lungs varies inversely to the species' degree of adaptation to aquatic life. The turtle *Pseudemys scripta* out of water breathes in an irregular pattern characterized by long periods of apnea, but the periods of breathing activity increase with temperature (Fig. 5.6).[621] In more aquatic turtles, lungs may account for a minor part of breathing, the accessory respiratory surfaces enabling the animal to remain completely beneath the surface for long periods.

The soft-shelled turtles (Trionychidae) are almost totally aquatic: when at rest, their oxygen needs are slight, and they can remain completely submerged for days. In the Nile turtle *(Trionyx triunguis)* the floor of the mouth is covered by a dense coating of vascularized villae. The mouth is ventilated through the nostrils about 14 to 16 times a minute, and this buccal breathing accounts for about 30 percent of the total oxygen needs of the resting turtle, the balance entering through the skin. Turtles experience bradycardia when diving, with a drop from 20 to 40 beats a minute to a rate about one-tenth of normal.[343] The soft-shelled turtle *(Trionyx mutica)* can excrete as much as 64 percent of the total carbon dioxide loss through the skin, whereas in terrestrial species (*Testudo dendriculata* and *Terrapene carolina*) the skin accounts for only 3 to 6 percent of the total loss of carbon dioxide.[489]

The terrestrial snake lung consists only (or mostly) of the right lobe, the anterior part of which alone is vascularized. A snake may inhale more air than will be immediately exposed to a respiratory surface, and redistribution of air within the lung may increase the time needed between breathing pulses. Thus ventilation is a leisurely process, and many virtually terrestrial snakes can remain submerged in water for long periods.

Marine snakes, in contrast, rarely leave water, and can exchange substantial amounts of gas through their integument. The sea snake *Pelamis platurus* (Hydrophiidae) can obtain from 12 to 33 percent of its oxygen needs through the skin and excrete as much as 94 percent of its total carbon dioxide through the

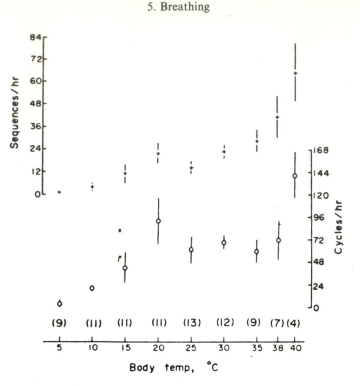

Fig. 5.6. Frequency of tidal exchange (upper) and breathing cycles (lower) in pond turtle *(Pseudemys scripta)* in relation to body temperature. Vertical lines are standard errors, and sample sizes are in parentheses (Lucey and House, 1977, courtesy of Pergamon Press).

skin.[360,417,964] Sea snakes feed near the surface, the head being submerged for long periods, and dermal breathing undoubtedly facilitates underwater ingestion of food. Some sea snakes can remain submerged as long as eight hours.

Birds

The breathing structures in birds are unlike those of other living vertebrates. Birds lack the mammalian muscular diaphragm, and their lungs, being nonelastic, do not expand and contract. They are connected to a complex system of thin-walled air sacs and hollow bones, and air is circulated by changes in volumes of air sacs. Air first enters the avian lungs through a pair of *primary bronchi*. Within the lung each primary bronchus narrows as it branches, usually into three sets of *secondary bronchi*. One set of secondary bronchi, the *medioventral* secondary (or *ventrobronchi*), forms the bulk of the anteroventral pulmonary tissue; the *mediodorsal*, or *dorso-secondary bronchi*, consist of a series of 7 to 10 dorsal branches; several *lateroventral*, or *latero-secondary bronchi*, diverge from the primary bronchi ventrally and below the *dorsobronchi* (Fig. 5.7). Each secondary bronchus produces internally a series of small *parabronchi;* these tiny tubes bear clusters of densely branched *air capillaries* into which gas diffuses. Parabronchi connect the dorso- and ventrobronchi, and parabronchi may connect with each other in the region between the secondary bronchi. This network of parabronchi and secondary bronchi is called the *paleopulmo,* and occurs in all birds.[263,264]

In addition to the paleopulmo, most families of birds have a second region of gas exchange, called the *neopulmo.* Parabronchial connections between the caudal part of the primary bronchus, the laterobronchus, and sometimes the adjacent dorsobronchus, return to the posterior part of the primary bronchus, just in front of the posterior air sacs into which the air flow is directed.

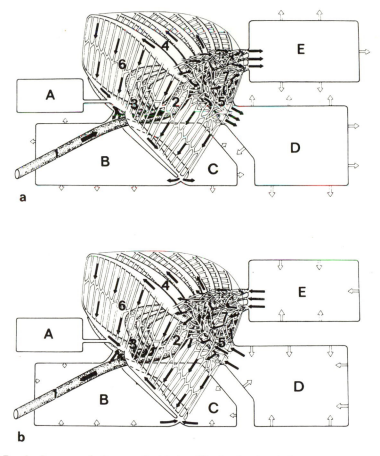

Fig. 5.7, a-b. Lungs and air sacs of a bird. **a** The inspirational phase expands the air sacs which **b** in the expirational phase results in air moving out of the air sacs. This causes unidirectional movement through the paleopulmo (7), the dorsobronchi (4) and the parabronchi (6). See text for more complete explanation (Duncker, 1974, courtesy of Elsevier/North-Holland Biomedical Press).

The air sacs are divided into two groups: the cervical, interclavicular and anterior thoracic always join the ventrobronchi; the posterior thoracic connects to the large laterobronchus, and the abdominal air sac appears as an expanded extension of the primary bronchus. Air sacs do not join the smaller laterobronchi or the dorsobronchi. Blood circulates about the parabronchi in a direction generally at right angles to the passage of air, creating a cross-current exchange. (Because in part of the parabronchi air moves in both directions, a counter-current exchange would be effective in only one direction, but a cross-current exchange operates continuously.) Parabronchi connect to form larger *saccobronchi* that in turn connect to other air sacs.[264]

Ventilation of Avian Lungs

When the respiratory muscles are relaxed, the air sac system is midway between inspiration and expiration; thus muscular effort is required to draw air into as well as to expel it from the lung–air sac system. The sternum moves relative to the vertebral column, increasing the air sac capacity by a ventrad motion and decreasing it by the opposite movement. The air sacs act as bellows and force air from region to region, through parabronchi, to another air sac and ultimately to the outside (Fig. 5.7).

As air enters the avian lung through a primary bronchus, it flows caudad to first the abdominal air sac (E) and then through the large laterobronchus (5) to the posterior thoracic air sac (D). Reversal of pressure in the posterior thoracic and abdominal air sacs forces air up and forward through the primary bronchus (2) and each dorsobronchus (4) through the paleopulmo through parabronchi (6) to the ventrobronchi (3). Simultaneously, low pressure in the anterior sacs (cervical, interclavicular and anterior thoracic) supports the flow from the ventrobronchi; prior to exhalation this air enters the anterior air sacs. Increased pressure in the anterior air sacs forces exploited air out the tracheal system. This arrangement results in a unidirectional passage of air through the paleopulmo. Because the neopulmo is arranged in series with the caudal part of the primary bronchus and posterior air sacs (the abdominal and posterior thoracic), the flow changes with inspiration and expiration. On inspiration, some air flows directly through the primary bronchus to the posterior air sacs, but some passes through the neopulmo (where gas exchange occurs) to the posterior air sacs. On expiration, the flow through the neopulmo is reversed, passing from the posterior air sacs back through the neopulmo and primary bronchus and dorsally through the dorsobronchus of the paleopulmo.[264]

The pulmonary arterioles branch out equally along the tubular parabronchi so that the movement is cross-current rather than counter-current. In the air capillaries, where most of the gas exchange occurs, blood moves opposite to the direction of air diffusion (from the parabronchi), creating a counter-current exchange.[264,885] Presumably the air sacs do not allow much transfer of gases, but, in the domestic duck, a small amount of respiratory gases is exchanged through the walls of the posterior air sacs.[636]

Mammals

In mammals the mechanical features of gas exchange are very different from those of other vertebrates and relatively uniform within the class. The thoracic cavity contains a pair of large and elastic lungs, which are provided with sac-like alveoli. Each alveolus is surrounded by a capillary bed, and gas exchange occurs to and from the bloodstream through the alveolar membrane. The thoracic cavity is closed at the rear by a broad sheet-like muscle, the diaphragm. Contraction of the latter enlarges the thoracic cavity, and the lungs draw air in through the bronchi. Relaxation of the diaphragm permits a partial collapse of the lungs, including the alveoli, expelling air. In forced breathing, intercostal muscles assist in both inspiration and expiration. Variations in mammalian breathing are modifications of this basic pattern.

Diving mammals and birds may remain underwater for 20–60 min and sustain great pressure. Clearly these animals do not suffer from the "bends" (caisson disease); it is believed that little or no air remains in their lungs at the inception of the dive, and thus there is little or no free air to be dissolved in the plasma at great depths and no bubbles to form at the end of the dive. In addition, when a diving animal submerges, its heart rate drops dramatically; blood continues to be supplied to the brain, but oxygen for the dive is that which is already in the muscles and, when that is exhausted, respiration is anaerobic.[16,272]

Both seals (Phocidae) and sea lions (Otariidae) dive well and have numerous anatomic and physiologic adaptations for remaining submerged for long periods. Various species in each family apparently differ in their performance in nature. Dives to 50 or 100 m appear to be within the ability of many species, but phocids seem to go deeper and remain submerged longer than do sea lions. The Weddell (*Leptonychotes weddelli*) of Antarctic waters is reported to descend to a maximal depth of 600 m, and several other phocids go below 200 m. Submergence for more than 30 min is within the ability of several species of phocids, but field observations are difficult and experiments in the laboratory are unnatural and always somewhat suspect.[16]

Pinnipeds typically exhale at the beginning of a dive. On surfacing from a dive, they breath rapidly. Their tidal flow is relatively greater than in terrestrial mammals but less than in cetaceans; up to 60 percent of the inspired air may be expired before submerging.[272] Some workers have suggested that the air remaining in the lungs is trapped in nonabsorptive regions by alveoli that collapse under pressure as the animal descends.[231,272] In the lungs of diving mammals (cetaceans, pinnipeds and the sea otter), the small passages are strengthened by cartilaginous rings and sometimes by muscle at the openings of the alveoli (Fig. 5.8). As a result of the pressure sustained in a dive, the alveoli collapse, driving air into the reinforced but nonabsorptive small passages, thus separating pulmonary air from pulmonary blood circulation. This displacement of remaining pulmonary gases would greatly reduce the chance of caisson disease during deep dives of these mammals. During a dive the heart of a pinniped slows (bradycardia), and a restriction of peripheral

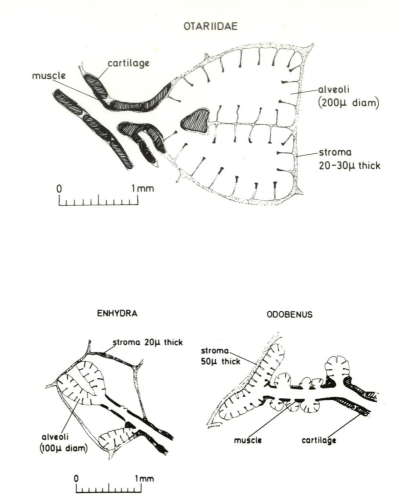

Fig. 5.8. Alveolar sacs in sea lions (Otariidae) are arranged in pairs (upper); cartilage extends to the entrance of the alveoli. Similar cartilaginous supports lead to the alveoli of the sea otter *(Enhydra)* (lower left) and the walrus *(Odobenus)* (lower right) (Denison and Kooyman, 1973, courtesy of Elsevier/North-Holland Biomedical Press).

circulation preserves blood pressure: heart beat, which is 70 to 100/min at the surface drops to 10–20 during a dive. In addition the blood is shunted to the brain and heart during a dive so that less goes to the muscles. As the animal surfaces from a prolonged submergence, its heart beat accelerates (tachycardia) to 180 beats/min, and rapid breathing replenishes the oxygen supply in the muscles.[272]

In diving, whales experience great pressures and remain submerged for long periods. The sperm whale may go to 500 m in its search for squid and cuttlefish, where the pressure is about 50 atmospheres. The sperm whale can remain submerged for up to two hours but commonly surfaces in less than one hour. The larger finback whales (Balaenopteridae) come up to breath more frequently, usually every 10–20 min. While at the surface large whales breathe repeatedly from 2 to 10 min. Dolphins and other small-toothed whales do not dive deep, nor do they remain submerged more than 5 or 10 min at most.

In contrast to pinnipeds, cetaceans inhale prior to submerging. The tidal volume, the amount of air exchanged in a single breath, is close to 90 percent versus less than 15 percent in man and most terrestrial mammals. Thus a whale can make up a deficiency of oxygen in far less time than a terrestrial mammal. Although the total lung capacity of cetaceans is not relatively greater than that of a terrestrial mammal, whales accumulate more than three times as much oxygen in myoglobin as do land-dwelling mammals. It is also probable that cetaceans accumulate a substantial oxygen debt in their muscles during a long dive. One can imagine that, when subjected to great pressure during a deep descent, a whale's lungs are so compressed that all air is forced into the pulmonary circulation.

Comparisons

Surfaces of gas exchange in vertebrates differ in their arrangement of relative movements of oxygen supply and of blood. A counter-current exchange accounts for the efficiency of the gills of fish and is possible by the constant direction of the flow of water. Gills extract approximately 80 percent of the oxygen dissolved in water passing over them; in a mammal roughly 25 percent of the oxygen is removed from air, an adequate amount in view of the greater abundance of oxygen in air.[822]

Generally in reptiles, birds and mammals, the capacity of the lungs is proportional to the body mass to the two-thirds power ($B^{0.66}$), so that in larger forms the lung volume, like the metabolic rate, is proportionately less. In amphibians lung size is highly variable, probably because (1) much of their gas exchange occurs through the integument and (2) amphibian lungs are sometimes important as a hydrostatic organ. Lungs are lacking in one large family of salamanders (Plethodontidae), and are reduced or absent in some anurans and salamanders that inhabit swift streams.

Amphibian skin is the least efficient breathing mechanism, but it is nevertheless effective by (1) the unlimited amount of oxygen surrounding it, (2) the high surface-to-mass ratio in these small animals and (3) their relatively low oxygen requirements.

Both lungs and breathing differ between birds and mammals. The avian lung has a much greater surface-to-volume ratio than has the mammalian lung, even though the total volume of the lungs plus air sacs of a bird is nearly the same as that of the lungs of a mammal of comparable size. The breathing rate is faster in mammals, perhaps to compensate for the less efficient "two-way" system in contrast to the unidirectional air flow through the avian lung. Also mammals have smaller tidal volumes than have birds of similar size. The cross-current characteristic of the exchange in the neopulmonic parabronchi of birds permits continuous gaseous movement across the membranes. In the paleopulmonic parabronchi, where the air movement is unidirectional, gas exchange is maximized through a counter-current arrangement of the air capillaries and blood capillaries. As a result of the circulatory patterns of air in birds, they remove far more (up to two times as much) oxygen from respiratory air than do mammals.[822]

Suggested Readings

Dejours P (1975) Principles of comparative respiration physiology. North-Holland, Amsterdam

Foxon GEH (1964) Blood and respiration. In: Moore JA (ed) Physiology of the amphibia. Academic, New York, pp 151–209

Hughes GM (1965) Comparative physiology of vertebrate respiration. Harvard University Press, Cambridge

Krogh A (1941) The comparative physiology of respiratory mechanisms. University of Pennsylvania Press, Philadelphia

Wood SC, Lenfant CJM (1976) Respiration: mechanics, control and gas exchange. *In:* Gans C, Dawson WR (eds) Biology of the Reptilia, Vol 5 Academic, New York, pp 225–274

6. Food and Feeding

Most vertebrates feed almost daily, and some small insectivores seem unable to survive for more than a few hours without food. There are, however, some exceptions to this generality: the great whales abstain from eating for months, most hibernators eat little or nothing during dormancy, and eels (*Anguilla* spp.) and salmon (*Oncorhynchus* spp.) do not feed on their breeding migrations. For most vertebrates, however, foraging and feeding constitute much of their daily activity.

In the long term, food intake must be adequate to meet the energy demands of the individual, but there are constant irregularities in both food intake and energy metabolism. Intake of food may exceed that consumed, in which case energy may be stored, or food may be temporarily inadequate to meet energy demands, in which case energy reserves are drawn upon. Also, energy expenditure changes markedly from hour to hour and from season to season.

Food Intake

The energy obtained from feeding depends on the time actually spent in feeding and also on the quality of food; low-energy food provides fewer calories per unit of feeding. Quality of food may vary from one season to another, generally being richer in autumn. Plants, forming the basis of all food chains, gradually accumulate carbohydrates and fats through the spring and summer to autumn; this determines the type of qualitative changes in food throughout the year, both on land and in water.

Regulation of Food Intake

The hypothalamus limits appetite, and experimental injury to the hypothalamic ventromedial region results in hyperphagia (high food intake) and increased adiposity, or fat stores. Stimulation (electrical or adrenergic hormonal) of the lateral region increases food intake and decreases drinking, and the medial region, when stimulated, suppresses hunger. The hypothalamus seems to respond to levels of circulating metabolites and other substances (e.g., blood sugar) and neural cues. In addition the hypothalamus produces a prolactin-releasing factor (PRF), which triggers the secretion of prolactin from the anterior lobe of the pituitary. Prolactin stimulates appetite in some animals and promotes lipogenesis in birds.[668,691] When prolactin is experimentally administered to crowned sparrows (*Zonotrichia leucophrys* and *Z. albicollis*), they develop *Zugunruhe* and premigratory fat.[545]

Amounts of food that individuals eat are seldom analyzed to determine possible sexual variations in eating. Nevertheless, in some domestic mammals hormonal levels and reproductive state influence appetite: high levels of testosterone retard feeding in male sheep and rabbits, whereas in female members feeding is depressed by high levels of estrogen and accelerated by increased levels of progesterone.[28] This same pattern is found in breeding populations of black-tailed deer *(Odocoileus hemionus),* in which rutting males have a low food intake and lose weight. Progesterone stimulates appetite in female rats and results in increased adiposity, and estrogens cause a depletion of fat stores. The relationship of these two hormones to food intake is known in a number of mammals, and is presumably an effect of plasma levels on the lateral regions of the hypothalamus. Observations of wild red squirrels (*Tamiasciurus* spp.) are probably typical. Males of *Tamiasciurus hudsonicus* and *T. douglasii* in southern British Columbia consumed a mean of 117 kcal/day, whereas lactating females took 323 kcal/day. During the breeding season, nursing females spent from two to three times as long in feeding as did adult males.[939] In female deer mice *(Peromyscus maniculatus),* measured amounts of food intake increased sharply after birth of the young: energy consumption rose 96, 136 and 194 percent in the three successive weeks of lactation (Fig. 6.1).[966] Similar trends are known in other small mice, and the length of such hyperphagia probably varies with the duration of lactation.

Winter dormancy with a marked decline in body temperature enables a hibernator to survive on much less food than it needs during its active period. Even a poor hibernator, such as the chipmunk, *Eutamias amoenus,* curtails its food intake in winter (Fig. 6.2).[967] Similarly the mantled ground squirrel *(Citellus lateralis),* when maintained at room temperature during normal hibernation, eats less than it does during the immediate pre- and posthibernation periods.[492]

As growth rate slows with increasing age, food intake declines. This phenomenon is well known in domestic mammals, but is not confined to captive species. Measured amounts of food eaten by young prairie dogs *(Cynomys ludovicianus)* declined as they entered their first autumn (Fig. 6.3).[397]

There is a relationship between total body fat in fishes and the depth at which they live: surface-dwelling species tend to be the fattest, fat decreasing with depth. The fat in pelagic species probably reflects (1) the fat in the plankton, which is the basis of their food chain and (2) the buoyancy required by pelagic fishes.[924]

Among wild ruminants food intake is in proportion to food quality. During

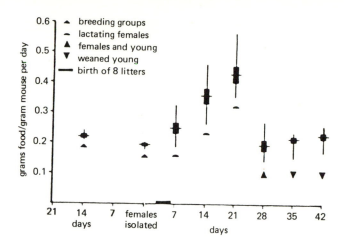

Fig. 6.1. The increase in food intake in deer mice *(Peromyscus maniculatus)* following the birth of young and the onset of lactation (Stebbins, 1977).

rainy periods, water and protein are high, and food intake is maximal; when the food deteriorates in the dry period, intake declines.[832,833]

Coprophagy, the ingestion by an animal of its own feces, is a common practice of many rabbits, rodents and insectivores. Rabbits (species of *Oryctolagus* and *Sylvilagus*), which tend to feed at night, pass soft, moist feces, which they eat during the day. This matter is passed a second time the following night, at which time it is relatively hard and dry.[38] The pocket gopher *(Geomys busarius)* reingests its feces, and cellulose-digesting bacteria in the cecum and large intestine complete digestion.[106,1096] Coprophagy not only prolongs digestion but provides the

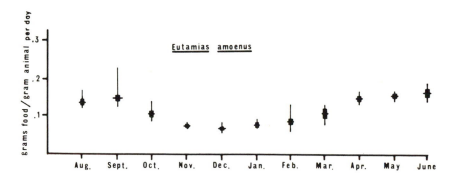

Fig. 6.2. Annual cycle of food intake in a nonhibernating chipmunk *(Eutamias amoenus)*. Data collected weekly and expressed as monthly means: blocks indicate one standard error, vertical lines indicate ranges and horizontal lines indicate means (Stebbins and Orich, 1977).

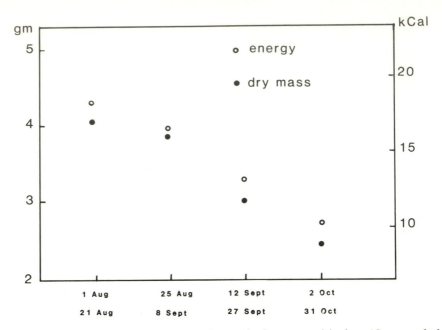

Fig. 6.3. Food intake (per 100 g live body mass) of young prairie dogs *(Cynomys ludovicianus)*, illustrating the decline with increasing age (from Hansen and Cavender, 1973).

animal with nutrition (proteins and vitamins) from the cellulose-fermenting bacteria.

Given the amount of food in any one meal (or in the stomach) and its rate of passage through the gut, one can calculate food intake of wild animals. Time between digestion and defecation can be measured by the rate of passage of dyes or other colored markers. By feeding colored glass beads, one student found that foods can pass through the digestive tract of cottontails *(Sylvilagus floridanus)* in 10 hours, and that 90 percent of the beads were recovered after three days.[38]

While foraging at night, the Japanese pipistrelle *(Pipistrellus abramus)* may emerge at 8:00 P.M. and increase its total weight 20 percent before midnight (Fig. 6.4). Mean digestion time (from ingestion to defecation) is about 45 minutes.[321] In the slightly larger *Myotis lucifugus* of North America digestion takes one hour, and in *Eptesicus fuscus* (double the size of *M. lucifugus*) digestion takes about two hours.[20]

Fat Cycles

Energy is stored most frequently as fats, but carbohydrates and occasionally proteins within the body are also available for energy. A great many vertebrates have well-marked cycles of fats and, less commonly (or to a less degree), glycogen.

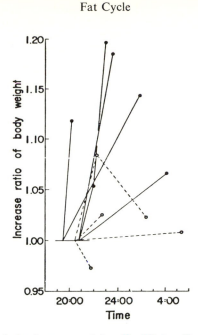

Fig. 6.4. Food intake of the Japanese pipistrelle *(Pipistrellus abramus)*, expressed as increase in body weight during evening feeding. The initial point (1.00) indicates the prefeeding weight of adult females (————●) and weaned young (- - - ○) (Funakoshi and Uchida, 1978).

These cycles are usually annual but can be longer or shorter. Fat cycles consist of an accumulation phase and a depletion phase. Generally fat cycles reflect periods of high-energy use and/or low-energy supply, on the one hand, and low-energy use and/or high-energy intake, on the other. In most vertebrates with clear-cut fat cycles, there is an annual peak, but in migratory birds there are frequently two annual fat cycles. Fat is the most concentrated energy source: for the calories it contains (9.5 kcal/g fat), fat has the least weight and volume of biologic sources of energy. In addition, fat makes low metabolic demands for its maintenance.

Annual cycles of fat characterize many groups of vertebrates and serve numerous functions; the period of fat loss may identify its role(s). Animals emerging from hibernation with their fat depleted may be assumed to have stored it for energy during dormancy. In many fishes, birds and marine mammals fat increases before migration; however, in those migrations that are followed by reproduction, fat contributes to egg-formation or, in mammals, to milk production. Some birds and mammals fast during reproduction, and they may survive solely on stored fat while caring for their young. In some fishes and reptiles there is an increase in liver fat during ovogenesis, and liver and visceral fat drop by the time of ovulation (Fig. 6.5). As in ovogenesis, increase of spermatogenic activity is marked by a depletion of (testicular) lipids.

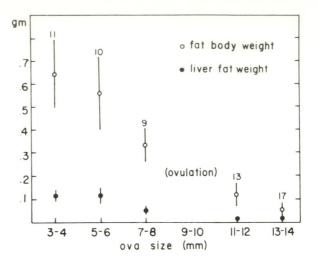

Fig. 6.5. The decrease in mass of fat bodies and liver fat during ovogenesis of the lizard *Sceloporus occidentalis* (Jameson and Allison, 1976, courtesy of the Journal of Herpetology).

Because fat is stored in various parts of the body, it is difficult to measure without separation from other tissues. Commonly fat is removed from oven-dried tissue by a solvent, such as ether, and expressed as a lipid index, namely, the ratio of fat to the fat-free dry tissue. Fisheries' biologists commonly use a "condition factor" or a ponderal index, the total mass divided by the cube of the standard length of the fish. Changes in the ponderal index do not necessarily measure differences in fat, but fat is responsible for most seasonal fluctuations in mass of fishes.

In many vertebrates, fat accumulates in clearly prescribed regions, frequently in the visceral cavity, but the muscles in some fish, reptiles and mammals are rich in fat. Fat in the Atlantic salmon *(Salmo salar)* is in the muscle, whereas in the related brown trout *(Salmo trutta)* it is stored in the coelomic cavity.[924] In birds, muscle is lean, whereas fat may accumulate in or beneath the skin or in the visceral cavity. The liver is also a site of adiposity in many vertebrates, especially certain sharks, and in whiting (Gadidae). Thus an animal with relatively lean muscle, such as a whiting or a bird, may have an abundance of fat in the coelomic cavity, beneath the skin or in the liver. Abdominal adiposity in humans tends to become concentrated above the navel in men and below the navel in women.[827]

There is a qualitative variation in fat among different groups of vertebrates. In birds and reptiles triglycerides are most frequent; these are composed of unsaturated fatty acids, and have greater mobility in the body than have saturated fatty acids, which commonly occur in mammalian fat. Also, unsaturated fatty acids are more easily converted in muscle to energy. Bird and reptilian fat has little or no cholesterol, a common component of many mammalian fat stores. Triglycerides are also common in fats of hibernating mammals.

Premigratory Fat

Amounts of premigratory fat are generally in proportion to the distance to be traveled and may in extreme cases amount to one-half of the fresh body weight. The amount of fat consumed in migration is a function not only of the mass of the individual and the distance, but the degree to which the migrant feeds en route. Small passerine birds can fly a calculated 2500 km; for shorebirds this figure may be 10,000 km, and some plovers migrate more than 20,000 km annually, but all feed en route. Annual fat cycles in fishes are associated with their migratory movements: migration is generally predicated on fat stores and a decline in water temperature.[763]

In the polar summers whales fatten on krill or euphasiid shrimp, which themselves accumulate fat from phyto- and zooplankton. There is no food shortage for whales in the summer, and they never have empty stomachs at that time. In addition to feeding on krill, whalebone whales (Mysticeti) may take herring, cod and squid, all of which accumulate fats created by polar plankton during the long summer days. As the days shorten, whales gradually move to lower latitudes with thinner populations of krill. While in the tropics and subtropics, not only do whales fast, but a large female may produce more than 100 gallons of fat-rich milk daily.

Fat cycles in seals (Phocidae) are similar to those of the whalebone whales. During lactation the female fasts and becomes lean, whereas in one month the pup may triple or quadruple its natal weight.[13,582,935]

In temperate regions insectivorous bats hibernate and some make migratory movements to suitable caves. Fat cycles are characteristic of such bats, but not of bats in the tropics and subtropics.[42]

Fat deposition is a regular preparation for migration in many families of birds and seems not to occur in nonmigratory species.[99] Such adiposity may result solely from a seasonal hyperphagia, which perhaps follows from an increase in daylight in which to feed. In the white-crowned sparrow *(Zonotrichia leucophrys gambelii)* increasing daylength stimulates appetite, and hyperphagia results in rapid increase in body fat prior to the spring migration (Fig. 6.6).[544] Autumnal increases in body fat also result from hyperphagia and are photoperiodically induced.[284,286] The great vernal increase in body fat is followed by Zugunruhe (nocturnal restlessness) in captive birds (Fig. 6.7), such restlessness paralleling the period of migration in wild populations. With daily lengthened photoperiod in the spring, gonads enlarge, Zugunruhe develops and general adiposity increases.[546] Both Zugunruhe and fat deposits appear in castrated individuals so that, although these cycles depend on photoperiod, they are independent of the reproductive cycle. Also, when kept on a maintenance diet, white-crowned sparrows *(Zonotrichia leucophrys)* develop Zugunruhe but do not get fat. Mobilization of fat for migration occurs in the liver before and during migration.

In nonmigratory passerine birds, deposition of fat is usually inversely related to ambient temperatures, but the amount stored is seldom much more than is needed during the night. Premigratory levels of fat are greater than fat stored by temperate and tropical nonmigratory species. Prior to migration, a sudden

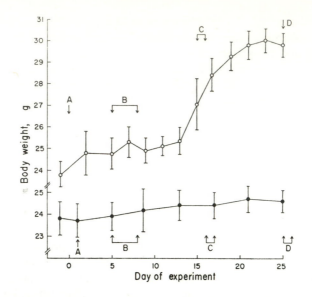

Fig. 6.6. Changes in body mass (presumably fat) in the white-crowned sparrow *(Zono-trichia leucophrys gambelii)* on a short daylight schedule (8L:16D; lower curve) and a long daylight schedule (16L:8D; upper curve). The lettered brackets indicate periods in which samples were taken to determine fat content. Vertical lines indicate one standard deviation on each side of the mean (King, 1967).

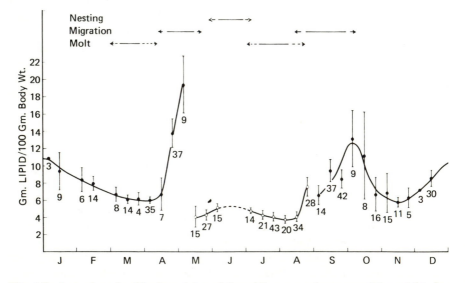

Fig. 6.7. Annual cycle of body weights of the white-crowned sparrow *(Zonotrichia leu-cophrys gambelii)* and associated behavioral activities (King and Farner, 1965).

increase in food intake accounts for both vernal and autumnal peaks of adiposity: in spring and fall many migratory birds tend to feed continuously throughout the day.

Prehibernation Fat

Prehibernation fat buildup is part of the endogenous hibernation cycle but may be timed by exogenous cues, at least in some mammals.

Small rodents that hibernate accumulate fat in late summer, and their fat stores are seldom equalled by those of nonhibernating species. Total body fat in a bat or a ground squirrel in September or October may be three times its level in July.[94,494] Among granivorous and insectivorous rodents (species of *Eutamias, Peromyscus* and *Apodemus*) in temperate regions, there is generally a peak of body fat in autumn and winter and a low in midsummer.[882] The winter peak of adiposity tends to be less pronounced at lower than at higher latitudes.[517] Voles (*Microtus* spp.), which feed predominantly on forbs and grasses, generally do not experience a clear-cut cycle in body fat.[698]

Foraging

An important element of feeding is foraging, and it is convenient to separate the search for food (or foraging) from its actual consumption (or feeding). Except for the goal of eating, foraging can be regarded as an activity that consumes energy, but in filter-feeding vertebrates, such as in whalebone whales (Mysticeti), herring (Clupeidae) and some tadpoles (Anura), one can scarcely distinguish between foraging and feeding. By reducing the energy-consuming activity of foraging, an animal can increase its efficiency; to do this, however, it must develop an ability to exploit a food that (1) requires little foraging and (2) is generally not available to other occupants of the same habitat. Such a specialization can be an advantage when it reduces the time needed for foraging; it can, however, be a disadvantage when the major food item to which it has become specialized is in short supply, for then it must spend more time in foraging. Thus a specialized feeder has an advantage that is not without risk.

A generalized feeder, or one that can subsist on a great variety of food items, sometimes spends much time and energy in foraging, but finds many items that it can eat. A generalist may profit from an abundance of food and, because of its versatility, seldom suffers a shortage. Between the extremes of a generalized feeder and a specialist lies a graded series, and some kinds of vertebrates are versatile feeders but facultative specialists, feeding on a single food item when it is abundant.

Generalized Feeders

Generalized feeders, able to exploit a broad spectrum of food items, frequently utilize both plants and animals. Because generalists usually spend more time foraging than do specialists, they are less efficient foragers. Nevertheless generalists are numerous, especially at higher latitudes, and there must be rewards for versatility. The advantage lies in their being able to survive in a region with both drastic seasonal changes and extreme fluctuations over geologic time. Survival of specialized feeders in temperate latitudes sometimes rests on either migration or hibernation to avoid food shortages in the winter, but many vertebrates are active throughout the winter, and these species are usually generalists. Some small rodents survive equally well on leaves of grasses and forbs in the spring and their seeds in autumn, and in winter they may subsist on insects and stored seeds. Small shrews, whose dentition would suggest a diet of arthropods, feed heavily on conifer seeds in years of a heavy crop. The large short-tailed shrew *(Blarina brevicauda)* of eastern North America may shift its attention from insects and snails to voles (*Microtus* spp.), when populations of the latter are abundant.[267] Undoubtedly such versatility enables these generalized feeders to remain abundant in a continuously changing environment.

Some vertebrates have survived long periods of geologic time and are little changed from what they were millions of years ago. They are among the most successful generalists and are quick to invade new, unstable environments. The opossum *(Didelphis marsupialis)* is a living fossil, but an abundant and aggressive invader of disturbed habitats. It can subsist on almost anything at all edible.

The coyote *(Canis latrans)* ranges widely over North America and thrives even close to urban areas; it persists despite intense and continuous efforts toward its eradication. It is probably one of the most opportunistic predators in the New World, adapting its diet to available plant and animal materials.[514]

Not only can generalists survive the stresses of climatic change, but their ability to subsist on many sorts of food enables them to occupy large geographic ranges. Some of the most successful terrestrial generalists are shrews *(Sorex),* bears (*Ursus* spp.), the raccoon *(Procyon lotor)* of North America and some of the dogs, such as the coyote and the gray fox *(Urocyon cineroargenteus);* whatever limits the range and abundance of these mammals, it is not their diet. There are many generalists among midlatitude birds, but the starling *(Sturnus vulgaris)* is one of the most notorious, and its feeding adaptability has no doubt contributed to its rapid spread in North America.

Finally, generalists frequently have a high reproductive capacity. There is probably a relationship between their fecundity and both their diet and geographic distribution, and very probably a more direct dependence on diet than on geography. Typically midlatitude environments have severe winters during which many plants and animals are dormant, and springs during which there is a sudden and tremendous flush of growth. Seasonal dietary changes in temperate climates may influence fecundity more than any inherent reproductive capacity.

Among some families of lizards there is a trend from an insectivorous diet in small to a herbivorous one in large species, but this is valid only in a general way. The large *Iguana iguana* at all ages feeds primarily on plants, and the very large species of *Varanus* are predatory, whereas the small (25 g) xantusiid lizard *Lepidophyma smithii* eats vegetable materials, especially figs (*Ficus* spp.).[659]

Rodents as Generalists. Rodents are, by their dental morphology, highly versatile and omnivorous. The rodent incisor can be utilized in manipulating a variety of foods, and the cheek teeth can crush seeds, seed coats, fibrous plant materials, meat and the exoskeleton of arthropods. Although some rodents are specialists in the diet, the order Rodentia with its remarkably uniform dentition feeds on a broad spectrum of items.[574]

The species of *Peromyscus* in the New World eat diverse plant and animal materials and collectively indicate the many kinds of naturally occurring substances that sustain life. The deer mouse *(Peromyscus maniculatus)* eats many sorts of plant parts but favors seeds and fruits and, in times of seed scarcity, fungi; they seem to relish insects, especially lepidopteran larvae.[490] *Peromyscus boylei* and *P. californicus* have a penchant for flowers and foliage. The oldfield mouse *(Peromyscus polionotus)* seems to prefer animals to seeds, and captives will fight over live cockroaches.[333] Lepidopteran larvae are important in the dietaries of *Mus musculus, Peromyscus leucopus* and *P. maniculatus.*[1091] Deer mice *(Peromyscus maniculatus)* eat a great variety of arthropods but seem to avoid lady bird beetles (Coccinellidae).[1079] Malaysian species of *Rattus* take land crabs as well as termites and orthopterans.[604]

The Australian squirrel glider *(Petaurus norfokcensis)*, a glissant rodent-like marsupial, normally feeds on insects and vegetative material but has also been known to attack wild birds and consume bird's eggs.[1103] The mouse-like marsupial, *Planigale,* of Australia is carnivorous: it feeds heavily on insects, and *P. ingrami* can kill the murid rodent *Leggadina,* a mouse about the same size as *P. ingrami.*[1042] The rice rat *(Oryzomys palustris)* lives in salt marshes in the southern and southeastern United States. This littoral rodent feeds largely on animal material, especially larvae of the rice borer (*Chilo* sp.) and crabs *(Uca)* in the wild, and captives also prefer meat to vegetable food.[910] Heteromyid rodents (kangaroo rats and pocket mice) consume seeds and flowers, and may at times prey heavily on caterpillars and other arthropods.[491]

Herbivores as Generalists. Among most herbivores food selection has a special effect on the efficiency of digestion. Because vertebrates that feed largely on leafy material have microbial cellulose-fermenting populations, their food must be acceptable not only to the vertebrate but to its microflora as well. Ruminants have expanded and divided segments of their digestive tract that house microbiota specialized for cellulose fermentation. Similar areas of fermentation occur in some kangaroos, some cetaceans, sloths, some reptiles and other large herbivores. In a few herbivores, such as the hyrax *(Hyrax),* fermentation of cellulose may occur throughout most of the gut. In rodents fermentation and absorption occur in a greatly enlarged cecum. The colon is the site of fermentation in elephants and horses. In gallinaceous birds there are two ceca, with a rich bacterial flora that ferments physiologic roughage.

Preference for foods by herbivores is strongly influenced by the abundance of alkaloids in certain plant tissues.[314,617] Voles *(Microtinae)* as well as ruminants avoid eating plants containing high levels of these poisons.[539] Alkaloids, such as perloline, gramine and tryptamines, not only reduce palatability to herbivores themselves but are toxic to their microflora.[777] Thus these compounds impair the digestion of cellulose. Alkaloids that inhibit or destroy gut bacteria commonly build up in such grasses as *Festuca, Lolium* and *Phalaris,* as well as in many forbs and shrubs, especially as the growing season continues. The pyrrolizidine alkaloids of *Senecio, Amsinkia* and *Heliotropium* are directly toxic to vertebrate herbivores, causing frequently fatal liver and lung damage.[678]

Palatability of some plants is also affected by the amount and type of their essential (volatile) oils. Some of these oils reduce microbial activity, causing retention of plant material in the rumen. Certain conifers, sagebrush (*Artemisia* spp.) and bay (*Umbellularia* spp.) are high in essential oils and low in palatability, and retard activity of the microbial flora of ruminants.[744,745]

When the microbial flora of a ruminant is deactivated or destroyed, the animal's hunger is not satisfied and it increases its food intake. In extreme examples there is the paradox of a deer that has starved to death with a full gut, but with its microbial flora destroyed.

One can separate herbivorous mammals into ruminants, coprophagous nonruminants and noncoprophagous nonruminants. In ruminants a bacterial flora in the rumen (or first chamber of the divided stomach) digests cellulose to sugars, which sustain the bacteria of the rumen.[469] The ruminant benefits from various by-products of sugar fermentation as well as from the cell contents (protein) of the microbes themselves, by absorption in the small intestine.[667,708] Coprophagous mammals, by reingesting soft feces, may benefit from the microbial flora and its digestive products, similar to assimilation in a ruminant. Presumably other herbivores with a bacterial flora in the cecum and large intestine derive little nutritional benefit from the cellulolytic activity of such microbes.[513]

The chuckwalla *(Sauromalus obesus),* an iguanid lizard of western North America, is strictly herbivorous. Cellulase activity is apparently absent in the stomach and small intestine but high in the large intestine. Nevertheless cellulose digestion is a minor source of energy for this lizard, as almost all assimilation of organic food occurs in the small intestine.[747]

Specialized Feeders

Many specialists are tropical and subsist on a specific food source to which they have unique access. There are many well-known examples among tropical birds and mammals living in low latitudes, especially in regions without annual droughts. Not only are annual climatic changes less severe in the tropics than at higher latitudes, but over long periods the tropics tend to be the more climatically stable regions of the world. Some animals are facultative specialists: they feed largely or exclusively on a single food because it happens to be temporarily abundant, not because there is any special ability required to obtain it. Perhaps specialized feeding developed in a situation with a single type of food predominating over others. Relative vulnerability of potential food species of the long-nosed dace *(Rhinichthys cataractae),* as an example, is determined by their degree of activity and thus exposure to the foraging dace. In the Ephemeroptera, for example, the swimming family Baetidae is far more vulnerable and more often eaten than are the creeping (i.e., nonswimming) forms, Ephemerellidae and Heltangeniidae, but this species remains a generalist.[334]

Because specialists confine their feeding to a single group of plants or animals, they have a geographic distribution not extending beyond that of their food. The everglade kite, one of the most specialized of hawks, cannot leave the marshes in which its food source, the snail, lives. The osprey is also one of the most specialized raptors, but its food, fish, occurs in suitable waters everywhere, and the osprey is cosmopolitan. Some students maintain that specialists have lower reproductive rates than generalists, but the evidence for this assertion is based on *tropical* specialists, and although most specialists seem to be tropical, most tropical vertebrates have lower reproductive rates than those from temperate regions. The relationship between specialized feeding and reproduction is not clear.

Fish. Many fishes seem to have developed adaptations for feeding, and there are many examples of species with restricted diets.

Several specialities in foraging and feeding exist among moray eels. Some of them (e.g., *Gymnothorax moringa* and *G. vicinus*) are nocturnal feeders and locate food by olfaction.[54] In contrast, *G. pictus* is a diurnal predator, emerging from the water to forage for crabs about the high tide mark. *G. pictus* has relatively short, stout teeth for the genus, apparently a modification for its departure from the piscine diet of most moray eels.[164]

The diskfishes or remoras (Echeneidae) possess a dorsal fin modified to form an adhesive disk; they attach themselves to larger fish, frequently sharks, and eat small pieces of food not ingested by the "host."[976] In addition, some species feed on host parasites. The remoras (*Remora osteochir* and R. *brachyptera*) attach near the gills of sharks and feed on copepods, which are true parasites of the sharks.[196]

The sanguinivorous catfish (Pygidiidae) of the Neotropical region are often referred to as the only parasitic vertebrates, and zoologic folklore includes stories of these little fish swimming up a stream of urine and entering the human penis. In nature they feed on the blood of many species of fishes, seeking the delicate, highly vascularized and readily accessible gill membranes.[537] Pygidiid catfish have been known to attach to the legs of bathers wading in waters where they occur.

Stargazers (Uranoscopidae) are specialized in both anatomy and behavior for an unusual way of obtaining food. These fish bury themselves in the sand and attract food items by one of several sorts of lures. The Atlantic species, *Astrocopus y-graecum,* rotates its eyes and moves its dorsal fin, both activities drawing potential prey animals, as the predator is mostly buried in sand. Electric organs discharge during capture of prey, and this activity seems to be part of the total feeding process.[1092]

Amphibians. There are few well-documented examples of extreme dietary specializations among amphibians, but specializations there undoubtedly occur. The great majority of amphibians are opportunistic in their feeding (Fig. 6.8).[392] An exception is *Bufo fergusonii* of India and Ceylon: it is somewhat fossorial and feeds almost entirely on isopterans.[549] Certain species of tropical microhylid frogs and some dendrobatid frogs subsist largely on ants.

Reptiles. Some of the semifossorial snakes have highly restricted diets, with or without obvious adaptations. The sharp-tailed snake *(Contia tenuis)* of the Pacific coast of North America feeds largely on slugs, for which purpose it has greatly elongated teeth.[1119] Although many species of the natracine genus *Thamnophis* seem to prefer anurans, Butler's garter snake *(Thamnophis butleri)* feeds largely on annelid worms. Another colubrid snake, the patch-nosed snake *(Salvadora lineata),* is a reptile feeder, and preys heavily on small lizards and their eggs; it has been observed to burrow into nests on the iguanid lizard, *Sceloporus oliva-ceous* and devour the developing eggs.[93]

The Nearctic natricine *Regina alleni* feeds almost entirely on decapod crus-taceans (*Procambarus* spp.); these it takes tail first, apparently to direct the claws so that they do not obstruct passage of the food through the esophagus.[736] In the Pacific Northwest the garter snake *(Thamnophis sirtalis)* feeds in the intertidal zone, descending at minus tides to forage for small blennoid fishes that attach themselves in patches of sea lettuce, *Ulva.*[68]

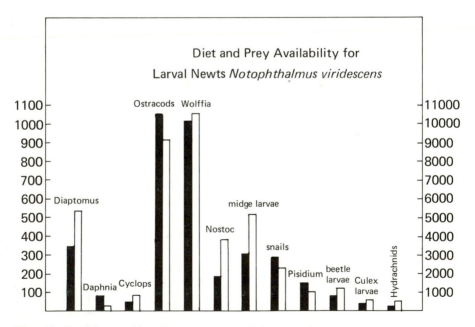

Fig. 6.8. Food items of larval newts *(Notophthalmus viridescens)* correlated with occur-rence of available food in the same habitat. Left-hand scale and black bars show occurrence of food items in 300 newt stomachs and right-hand scale and open bars show occurrence in 300 samples taken in the same habitat (from Hamilton, 1940).

Several snakes have developed a penchant for birds' eggs and are modified to such a diet. The Japanese colubrid, *Elaphe quadrivirgata,* possesses egg-crushing hypophyses on the neck vertebrae, and these pointed projections assist it in crushing eggs after they are swallowed.[352] Several other species have independently developed similar adaptations for the same purpose.

Among sea snakes (Hydrophiidae) are some rather specialized feeders: many species feed mostly on fish, but some species (such as the Australian *Emydocephalus annulatus*) have reduced dentition and feed only on fish eggs, whereas *Hydrophis melanocephalus* preys on eels of the families Congridae, Moringuidae and Ophichthiidae. Presumably such specializations permit the compatibility of several sympatric species of sea snakes.[416]

Several lizards are facultative specialists in their foraging. *Leiolopisma suteri,* a skink of New Zealand, lives among rocks just above the high tide mark; the species is nocturnal and forages for amphipods and other marine invertebrates that secrete themselves among shoreline debris.[1039] Another skink, *Cryptoblepharus boutoni cognatus* of Madagascar, feeds on arthropods and fish, especially young mudskippers *(Periophthalmus kohbreuteri),* in the intertidal zone, and moves with the tidal rhythms.[316]

Some predators are specialized in the "passive pursuit" of their prey. These are carnivores that are relatively inconspicuous and entice their quarry within range of capture. The alligator snapper *(Macrochelys temminckii)* is predatory and lies in wait for food to approach. A double-ended worm-like appendage on the tongue is a lure for small fish. This turtle clamps its mouth shut when curious fish enter. The Neotropical chelid side-necked turtle, the famed matamata *(Chelys fimbriata),* has an irregular shell and head that serve to conceal it in the quiet ponds where it dwells. It lies in wait until prey swims within range, and food is engulfed as the matamata opens its mouth. The method of lie-and-wait is much like that employed by the alligator snapper, but the matamata uses camouflage, whereas the alligator snapper entices prey with a lure within its mouth. The female of the marine anglerfish possesses a greater-modified dorsal fin, which contains luminescent bacteria. With this lure she attracts small fish, which are drawn in when she opens her mouth.

Birds. Some specialists avoid scarcities in their required food by migrating. Among the highly migratory shorebirds are species that feed on a rather restricted diet but manage to find food throughout the year. Marked selectivity is shown by many foraging shorebirds. The dunlin *(Calidris alpina),* for example, feeds heavily on polychaetes, especially species of *Nereis,* and on migration it congregates on tidal mud flats where *Nereis* abounds. During the nesting period it takes aquatic larvae of flies (Chironomidae and Tipulidae). Similarly, the knots (*Calidris canutus* and *C. tenuirostris*) are stenophagous (literally, narrow feeding), feeding heavily on minute bivalves and gathering where these mollusks occur.[528]

The gray-faced petrel *(Pterodroma macroptera)* feeds on various squids and crustaceans that during the day live at least 150 m beneath the ocean surface. These petrels feed at night, when their prey move to the surface. Approximately

90 percent of their prey animals are luminescent and presumably are detected visually.[478] The flower-feeding Hawaiian honey-creeper, *Vestiaria coccinea,* has probably fed on flowers of the lobeliad, *Clermontia arborescens,* for a very long time.[959] Many of the honey-creepers (Drepaniidae) have decurved bills, and only the Hawaiian species of lobeliads have decurved flowers (Fig. 6.9).

The everglade kite *(Rostrahamus sociabilis),* a highly specialized raptor of the New World tropics, feeds on fresh-water snails. Having captured a snail, the bird waits, perched for the mollusk to emerge from its shell; when the snail's flesh is exposed, the kite quickly jabs at an area near the operculum. The snail soon relaxes and the bird then shakes it out of its shell. Honey guides (Indicatoridae) of Africa feed on beeswax. These birds lead (or follow) predators to beehives or detect wax from its odor, and when the honey is consumed by the predator, the honey guide eats the beeswax,[21] which is digested by bacteria in the bird's gut.

The sharp-beaked ground finch *(Geospiza difficilis septentrionalis)* on Wenman Island of the Galápagos is the only bird known to feed regularly on fresh blood.

Fig. 6.9. The flower-feeding honey creeper *Vestiaria coccinea* foraging in the lobeliad *Clermontia arborescens* (Spieth, 1966; from the American Naturalist, courtesy of University of Chicago Press).

It attacks the bases of the secondary feathers of boobies (*Sula* spp.) until blood comes and picks small drops as the blood oozes out. This finch, although specialized, is a versatile feeder of insects and plant material. Bloodfeeding in this bird may have resulted from first feeding on blood-filled hippoboscid flies that parasitize the boobies.[111]

Mammals. Bats *(Chiroptera)* comprise a huge and diverse group of many families, most of which are tropical and subtropical. Among bats are numerous specialized feeders that show marked adaptations.

Phyllostomid bats have a strong predilection for feeding on nectar and pollen, but insects may be ingested as well. They feed on night-blooming plants and move seasonally when such plants are in flower. A nectar-feeding bat may protrude its tongue for as much as 100 percent of the resting length. The tongue in two glossophagine species, *Monophyllus redmani* and *Glossophaga soricina,* is forced out by muscles that not only constrict its diameter but close veins through which blood leaves the tongue, thus effecting a labial erection. Longitudinal muscle contraction withdraws the tongue to its resting position.[374]

Many bats have a very strong predilection for fruit and are well known for their attacks on cultivated stands. The Neotropical phyllostomatid *Carollia perspicillata* is greatly attracted to ripe bananas and plantains but on occasion eats substantial amounts of insects. *Artibeus jamaicensis,* a fruit-eating bat of the Nearctic region, subsists to a large extent on pulpy fruits and sometimes on nuts and leaves. Roosting individuals have been known to reach out with their wings and snatch and eat passing flies. The stomach of bats varies with their diet. In the Old World frugivorous bats *(Pteropidae)* the cardiac vestibule is quite large, perhaps to hold the relatively large amounts of food eaten at one time, similar to the gastric structure in the New World phyllostomatid bats.[353]

The sea otter *(Enhydra lutris)* seeks sea urchins and abalone; it strikes abalone, using a rock. Under continued battering the shellfish either relaxes its grip or is dislodged, at which point the sea otter carries it to the surface. The sea otter also smashes bivalves and sea urchins with small rocks. It is apparently the only carnivore to use a tool.[453]

Filter Feeders. Filter-feeding is a well-known method of foraging for many aquatic vertebrates. Many families of fishes have long, slender gill rakers that strain planktonic organisms. Filter-feeding is seen in the whale-shark, the paddlefish *(Polyodon spathula)* and herring (Clupeidae). Some water birds have grooves on the upper mandible for the same purpose; the prion (a petrel, *Pachyptela* spp.) of the austral oceans and the pantropical flamingo are both filter feeders. Except for the whalebone whales (Mysticeti), the crab-eating seal *(Lobodon carcinophagus)* is the only filter-feeding mammal. This seal is largely pelagic and feeds heavily on krill, which it strains with its deeply lobed teeth.

Among the roughly 15 species of baleen whales there are variations in their basic filter-feeding process. The right whales (Balaenidae), with very long baleen plates, swim with mouth agape at the surface, straining copepods that cluster at the water surface (Fig. 6.10). In contrast the finner or finback whales (Balaenopteridae), with much shorter baleen plates, engulf shrimp-like krill and small schooling fish below the surface, in a distensible mouth, and strains this food from water after the mouth is closed. The gray whale (Eschrichtidae) is a bottom feeder; either to scoop or lick up small benthic animals, it turns on its right side, with the result that the right side of the head and the baleen plates on the right side of the mouth are worn and abraded. Thus these three groups of baleen whales exploit different groups of small marine organisms.[824]

Black right whale (Eubalaena glacialis)

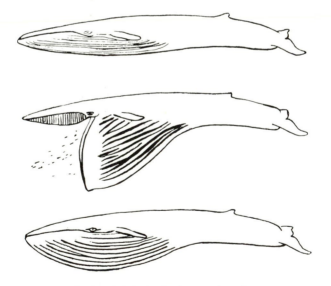

Finback whale (Balaenoptera)

Fig. 6.10. Feeding mechanisms of the right whale (Balaenidae), which filters plankton at the surface, in contrast to the finback whale (Balaenopteridae), which feeds beneath the surface (Pivorunas, 1979, courtesy of American Scientist, journal of Sigma Xi, the Scientific Research Society).

Specialization. Although the specialist is more efficient, it is also less flexible, and cannot adapt to changing conditions. Because it depends on a restricted food source, it cannot live far from that food source; if its food disappears, the specialist is doomed. Although most specialized feeders have developed their particular feeding adaptations slowly, extinction could come quickly. This fact may well account for the general scarcity of specialized feeders in the temperate world. Specializations among allied species reduce possible conflicts in food utilization and thus allow specialists to coexist where generalists would tend to compete with each other.

Although tropical rain forests are a haven for specialists, they are by no means confined to the low latitudes: some specialists are boreal species and many are migratory. Among boreal birds conspicuous obligate specialists include the cross-bills (*Loxia* spp.)—finches in which the tips of the mandibles cross so as to form a specialized beak for extracting conifer seeds from their cones.

Resource Partitioning

Resource partitioning is a useful concept and lies at the very basis of natural selection: namely, the divergence to produce differing allocations of a resource promotes speciation, and it is most clearly seen where closely allied species occupy the same area. Resource partitioning is predicated on the assumption that the resource is in limited supply. This assumption is frequently plausible but seldom proved. Many of the specializations observed among sympatric species may also represent resource partitioning.

Numerous studies of food use show that two species may eat the same types of foods without competing with each another: competition does not occur if the food supply exceeds the demand. For example, two stream-dwelling fishes, the fantail darter (*Etheostoma flabellare*) and the mottled sculpin (*Cottus bairdi*), are of similar size and eat a variety of mayflies (*Ephemerida*), stoneflies (*Plecoptera*), caddisflies (*Trichoptera*) and midges (*Chironomidae*), frequently the same species. The low temperature of the stream reduces metabolism of both fishes so that nutritional requirements are minimal, and populations of prey species are sustained.[208] Among three sympatric species of top minnows (*Fundulus* spp.) there is a broad dietary overlap and as food becomes scarce, there is a reduction (but not an elimination) in similarity of prey items, suggesting resource partitioning.[43]

In parts of the Rocky Mountain region of North America four large herbivores (two native and two introduced) occupy some of the same range. The two native species, a deer (*Odocoileus hemionus*) and elk or wapiti (*Cervus canadensis*), have diets that are almost totally exclusive.[399] The deer browses on leaves of desert shrubs and shares only 17 percent of its diet with feral horses and 4 percent with cattle. Cattle and horses, however, have a 77 percent similarity of foods and graze on the same grasses, sedges and forbes that support the elk.[400] The bush baby or galago (Galaginae) and the potto (Lorisinae), two prosimian species of tropical

Africa, are highly specialized feeders and have diets that tend to be mutually exclusive (Fig. 6.11).[163]

Sometimes related species may gather more-or-less similar food items but forage in separate areas. Among several species of heteromyid rodents (*Perognathus* spp. and *Dipodmys* spp.) in the lower Sonoran Desert of North America, there is a broad overlap of species and sizes of seeds collected, but these rodents tend to forage in mutually exclusive microhabitats.[961]

A limitation of supply is fundamental to the concepts of competition and resource partitioning, but how often and under what circumstances is food limiting? Food items typically become scarce at one season and reappear at another, even though, *as species,* they are present continuously. In many species of food plants or animals, a particular life stage may be seasonal. For example, cutworms—larvae of noctuid moths—are available as food for mice in the spring, while the adult moths are sought by bats throughout the warm seasons or, in mild climates, throughout the year. Also, many insects buried under snow become unavailable to insectivorous birds, while continuing to constitute the main fare of some mice and shrews. Similarly krill, the mainstay of whalebone whales in the

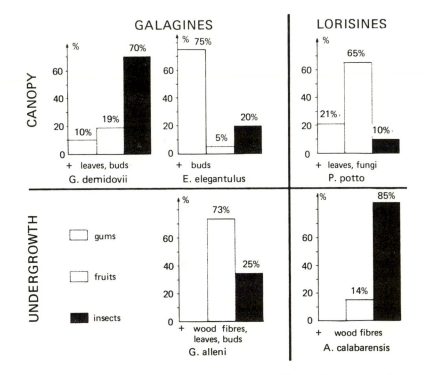

Fig. 6.11. Diets of five species of prosimians of Gabon. They reduce potential competition both by occupying different strata (canopy and undergrowth) of the forest as well as by eating different types of plant food (Charles-Dominique, 1971).

polar summer, exist in low numbers at other seasons. A less obvious case is bark-dwelling arthropods, upon which nuthatches, creepers, tits and woodpeckers feed. These birds tend to be nonmigratory and are able to forage despite probably seasonal reductions of their food source.

The question frequently asked by ecologists centers around food as a limiting factor. As often as not the answer lies, not in the abundance of food, but in its availability. The latter is determined as much by the foraging habits of the vertebrate as by the actual abundance of food. The treetop foraging of a vireo or a warbler renders it something of a specialist that must move to lower latitudes in the winter, whereas a deer mouse, frequently feeding on the same general sorts of arthropods, remains on the same ground all winter. Specialization in this sense is predicated not on the food eaten but on the manner in which it is obtained. A good example of difference in availability resulting from foraging habits is seen among squirrels: ground squirrels forage largely on the ground and, with few exceptions, hibernate; tree squirrels, on the other hand, often utilizing similar foods, are active all winter. Food storage does not explain the difference, because both groups store some seeds, but tree squirrels will forage for seeds buried in snow, while ground squirrels do not.

Special Diets of the Young

Resource partitioning is seen between age classes in many vertebrates. The simplest examples are those in which young (and small) individuals take smaller food items than do adults, but profound anatomic specializations occur in the young of many groups. The result is that different age classes do not compete with each other.

Fish. Dietary shifts with age occur in many families of fishes. Very young suckers *(Catostomus macrocheilus)* have straight, short guts and are top feeders, foraging for pelagic cladocerans and rotifers. During early growth the terminal mouth becomes ventral, the gut becomes a loop and later coiled, and the older fish become bottom feeders and ingest an increasing amount of diatoms and protozoans.[634] In the perch *(Perca fluviatilus),* young feed on zooplankton, but as they mature they prey on small fish.

Lepidophagy (scale-eating) occurs in several groups of fishes. Immature sea catfish *(Galeichthys felis)* may bite off the scales of larger fish. Scale ingestion is common among some jacks (or jack mackerel, Carangidae). Young individuals, up to 150 mm long, of the leatherjacket *(Scomberoides lysan),* a carangid of the tropical Pacific, eat scales from the silverside *(Pranesus insularum)* and mullet *(Mugil cephalus).*[637] Also, young of species of the carangid fish of the genus *Oligoplites* feed on larval crustacea and scales (and epidermal fragments) that they pick from the body of schooling fish. As these carangids attain adulthood, they themselves capture the schooling fish as well as larger crustacea. This habit of lepidophagy is reflected in a row of hooked, spatulate teeth on the outer dentary of these carangids, such teeth being lost as the predatory habit develops.[637]

Amphibians. Adults of both anurans and salamanders are predatory. Studies of their food indicate some selectivity in which size of prey is correlated with size of amphibian.[486] Larval salamanders feed in essentially the same manner and on the same type of food as do the adults; they are active predators and take a variety of small invertebrates. Tadpole larvae are usually plant feeders and sometimes filter feeders, and have the long intestine typical of herbivores. Tadpoles of the green frog, *Rana clamitans,* feed unselectively on various algae, filtering materials suspended in water ingested for breathing. Foraging continues throughout the winter in southern Illinois and is interrupted only during metamorphosis, the digestive tract being empty while being altered for a dietary of flesh.[500] Some larval anurans feed by filtering microorganisms, including flagellates and diatoms.[749] They swim slowly, in a horizontal plane, moving through concentrations of aquatic protozoa and algae. Tadpoles of the narrow-mouthed frog (*Gastrophryne* spp.) are filter feeders, but the adults feed extensively on ants.[757]

Parentally Produced Infant Food. Among the most specialized diets of young animals are those in which the young subsist on products elaborated by one or both of the parents. This manner of feeding is found in a few birds, which take secretions from the crop, a specialized region of the esophagus. In mammals the young feed on milk.

Birds: Esophageal Secretions. Pigeons and doves secrete a fat-rich material (pigeon "milk") from their crop. This substance is produced by the crop lining and is secreted under the stimulus of prolactin in both males and females. The milk-producing cells of the crop begin to develop before the end of incubation, and the young pigeons exist solely on pigeon milk for about two weeks.

The Layson albatross *(Diomedea immutabilis)* and the black-footed albatross *(D. nigripes)* range far at sea during nesting, and digestion follows ingestion. By eliminating water from their food and retaining fat, the adults can concentrate energy without a great increase in mass. On return to the colony, the adult feeds its young on the oil, which is secreted by the proventriculus. The emperor penguin, on the other hand, fasts during brooding and care of the young; the latter is fed on fluid secreted from the esophagus.[831] Similar secretions may nourish the young of other sea birds.

Mammals: Lactation. The sharpest distinction between diets of young and adults is seen in mammals. The process of nursing, or lactation, is unique to mammals and places the young in a position of extreme dependency on the female parent. Although some infant mammals [e.g., hares (*Lepus* spp.) and probably most hystrichomorph rodents] may take solid food at birth, milk from the mother is the primary means of nutrition in all mammalian young. Marked variations exist in the composition of milk and in patterns of lactation, both of which are reflected in growth rates and periods of parental care of the young.

Mammary tissue consists of either sebaceous or sweat glands. Although it occurs in all mammals of both sexes, this tissue functions only in females, and normally only after parturition. Mammary tissue develops during gestation, but occasionally virgin females produce milk. There are various arrangements of mammary glands, from the paired pectoral glands in most primates and most bats to 10 or more to a side, as in the multimammate mouse *(Mastomys)*. The number of mammae is generally correlated with the number of young per litter.

Milk consists of a protein (casein), a sugar (lactose), fats and water, with a wide variation in composition. Milk of pinnipeds has some of the highest reported fat contents, reaching 42–43 percent by weight in the Greenland seal *(Pagophilus groenlandica),* contrasting to 2.0–4.0 percent fat for human and cow milk.[935] Milk of whales varies according to the species, but fat amounts to 40–50 percent for finback whales (*Balenoptera* spp.) against only 20 percent for the beluga or white whale *(Delphinapterus leucas).* The sugar content is low (1.5 percent) in most whales, possibly lacking in milk of pinnipeds, but 6–7 percent in human milk.[84]

Lactation is prolonged among cetaceans and seems not to be correlated with size. The young sperm whale *(Physeter catodon)* nurses 10 to 15 months or more, while the very much smaller bottle-nosed dolphin *(Tursiops truncatus)* nurses as long, or even longer, than that.

There is a relatively brief nursing period in seals (Phocidae), but sea lions (Otariidae) and walruses (Odebenidae) nurse their pups for extended periods.[582] Pups of the California fur seal *(Callorhinus ursinus)* nurse as long as four months, and lactation in the austral fur seals (*Arctocephalus doriferus* and *A. tasmanicus*) persists as long as eight months. Lactation in the walrus may extend to 24 months.

The short period of lactation in phocids is correlated with the high fat content of their milk, commonly between 45 and 55 percent, and the rapid growth of the pups. The lactating harp seal *(Pagophilus groenlandica)* has 426–428 grams of fat and 105–120 grams of protein per kilogram of milk.[935] The young are lean at birth, but they nurse frequently and soon develop a subcutaneous layer of fat during their 10–12 days' nursing. For two to three weeks after weaning, seals take no food and quickly become lean again. Lactation in the gray seal *(Halichoerus grypus)* and the bearded seal *(Erignathus barbatus)* is brief and ceases about three weeks after birth. In the hooded seal *(Cystophora cristata)* pups cease nursing before they are two weeks old. Lactation in the harbor seal *(Phoca vitulina)* persists as long as six weeks.[11,26]

The amounts of fat, protein and sugar in milk change throughout lactation, at least in some mammals.[182] In captive black-tailed deer *(Odocoileus hemionus)* investigators noted an increase in percentage of protein during the initial six months of lactation, and a similar increase in fat over the same period, with a subsequent decline. In these fluctuations there is conspicuous individual variation.[729] In marsupials that nurse young of two different ages, two types of milk may be produced simultaneously, a "thinner," lower-fat milk coming from the nipple used by the younger offspring.

The milk of monotremes oozes from the bases of hairs of certain regions, and the newly hatched young suck it from the pelage. The milk of the echidna *(Tachyglossus aculeatus)* contains 15–35 percent lipids. Milk is released from the action of sucking, apparently under the stimulus of oxytocin.

Seasonal Changes in Food Supply

The annual cycle of seasons and winter changes in growth, flowering and fruiting of plants all contribute to changes in food availability. Adaptability is the hallmark of a generalist, and, if a vertebrate is a permanent, nonhibernating resident of a temperate latitude, its feeding pattern will change with the seasons.

The false vampire adjusts its foraging with seasonal changes in rainfall: during the extended dry season (May–October) it hunts close to the ground, catching ground-dwelling arthropods, but during the rains it feeds mostly on arboreal or flying insects.[1049]

The sika deer, *Cervus nippon,* grazes on the bamboo grass, *Sasa nipponica,* throughout the year and, in the spring, on buds and fresh leaves of shrubs as they appear. Leaves of woody plants are consumed until they wither in the autumn, at which time sika deer eat acorns and crabapples, and they feed only on bamboo grass in the winter.[655] Mule deer *(Odocoileus hemionus)* usually shun piñon pine *(Pinus edulis)* and juniper (*Juniperus* spp.), and most investigators rank their palatability as low. In severe winters, however, these plants may make up as much as 80 percent of the food of these deer.[398]

The red squirrel *(Tamiasciurus hudsonicus)* occupies many forest types in North America, and feeds heavily on seeds of conifers, especially species of *Pinus, Abies, Picea, Pseudotsuga* and *Tsuga.* The squirrels store cones in huge middens that may represent accumulations of several years. Some middens are 7–10 m across and sometimes 1 m deep, and may contain up to 10 bushels of cones. In years when coniferous seeds are scarce, these squirrels turn to such dicots as species of *Prunus, Ribes, Viburnum* and *Arctostaphylos.*[294]

Toxic Foods

Poisons develop in the skin, flesh and viscera of many animals, especially fishes and amphibians. The toxicity of these species provides some protection from predators, although a few of the latter are protected by immunities.

Many snakes feed on anurans, especially as tadpoles metamorphose and leave the water. In the summer, garter snakes (*Thamnophis* spp.), Nearctic natricines, eat large numbers of newly transformed toads (*Bufo* spp.), which are toxic to most frog-eating snakes. Garter snakes are immune to the steroid poisons (bufotalin and bufogin) that are secreted by the toad's parotid gland and which quickly cause death to other colubrid snakes. Larvae of *Bufo americanus* tend to be rejected by sunfish *(Lepomis macrochirus),* which prefer tadpoles of cricket frogs *(Pseudacris triseriata).*[1059] These toxins are potentially fatal to humans, and deaths have followed inadvertent ingestion of eggs of toads (*Bufo* spp.).[592]

Similarly the skin of newts (Salamandridae) is toxic, but sympatric species of grass snakes *(Natrix)* in Eurasia eat newts without ill effects. The dermal toxin of the Nearctic rough-skinned newt *(Taricha granulosa)* provides effective protection from most predators.[123] A species of garter snake *(Thamnophis sirtalis)* readily eats newts, but other snakes (*Coluber, Masticophis* and *Pituophis*) are susceptible to the toxin.[593] Toad-eating *Thamnophis sirtalis* snakes are unaffected by 3 mg of venom per gram of body weight, a dosage that causes death within four and five hours for the non-toad-eating *Salvadora* and *Masticophis*. Some Neotropical snakes prey heavily on the generally toxic dendrobatid frogs. Dermal secretions of most salamanders and anurans are offensive, if not toxic, and serve to deter predation.

Many species of tropical marine fish may contain a toxin that causes an illness known as ciguatera fish poisoning when such fish are eaten. Apparently the fish acquire the toxin in their food and may retain it in their flesh for at least 30 months even when kept on a nontoxin diet.[50]

Summary

The function of feeding is to provide energy for (1) physiologic maintenance, (2) growth and (3) eggs and general reproductive activity. Feeding is limited by the resources available to each individual and channeled by activities of other individuals (intra- and interspecific competition). Many species have restricted diets, and either live in regions where their food is continuously available or migrate so as to preserve food availability. Because there are temporal changes in the physical, floral and faunal aspects of the environment, each species makes appropriate accommodations in its foraging and feeding. These accommodations include accumulation of fat within the body as well as external food stores. Fat reserves supply energy when food intake is low (e.g., during hibernation) or when energy demands are high (e.g., during reproduction and/or migration).

Suggested Readings

de Ruiter L (1963) The physiology of vertebrate feeding behavior: towards a synthesis of the ethological and physiological approaches to problems of behavior. Z Tierpsychol 20:498–516

Freeland WJ, Janzen DH (1974) Strategies in herbivory by mammals: the role of plant secondary compounds. Am Nat 108:269–289

Morse DH (1971) The insectivorous bird as an adaptive strategy. Ann Rev Ecol Systemat 2:177–200

Paine RT (1971) The measurement and application of the calorie to ecological problems. Ann Rev Ecol Systemat 2:145–164

Schoener TW (1971) Theory of feeding strategies. Ann Rev Ecol 2:369–404

Slobodkin LB (1962) Energy in animal ecology. Adv Ecol Res 1:69–101

7. Thermoregulation and Water Balance

Temperature and humidity act in concert in their effects on the lives of vertebrates; only in fishes and aquatic amphibians can temperature be considered separately from humidity. In aerial environments humidity may have several effects. In amphibians, the skin of which is a major site of water loss, evaporation is in inverse proportion to absolute humidity of the surrounding air; the same relationship occurs in reptiles, birds and mammals, but in those animals it is less critical. Humidity in the soil, as after a rain or with melting snow, can lower soil temperatures substantially if the atmospheric humidity is low, and lower soil temperatures can retard activity not only of such vertebrates as snakes and lizards but also of the invertebrates on which they feed. Also, when humidity occurs in the form of fog or clouds, solar radiation is retarded, depressing air and ground temperatures.

Physiologic and Behavioral Relationships

Annual cycles, such as migration, hibernation (and estivation), feeding, fasting and reproduction, are related to temperature and humidity. Behavioral responses to the environment are in fact reflections of physiologic changes, such as seasonal changes in hormonal levels, sugars and fats, and many seem to be endogenous cycles. Vertebrates are also affected indirectly by the responses of invertebrates and plants to temperature and humidity. Appreciation of such relationships probably dates from the time nomadic hunters followed herds of ungulates that in turn were constantly seeking fresh pastures.

Role of the Hypothalamus

The hypothalamus, ventrad to the diencephalon, is sensitive to small variations in temperature of the blood, and initiates appropriate metabolic changes, including thermogenic response of brown fat in mammals. The hypothalamus mediates thermoregulatory responses for stimuli from various receptors. It is the controlling center for many processes that regulate water balance: it not only influences thermoregulation and, in mammals, sweating, but regulates the flow of blood through peripheral (surface) capillaries. Vasomotor changes (alterations in the diameter of the blood vessels) cause an immediate change in the rate of heat loss or gain. In birds, receptors in the spinal column appear to be more important than the hypothalamus in mediating these responses. Also, receptors themselves may cause responses: chilling of the skin of mammals, for example, may be followed by restriction in peripheral circulation before a change in hypothalamic temperature.

Temperature regulation is then a combination of responses of both superficial receptors and the hypothalamus. The hypothalamus has been compared to a thermostat: it is set for a given temperature, that which is characteristic for a given species. If the core (or deep) temperature is above or below that for which the hypothalamus is set, an appropriate thermoregulatory response occurs. In many wild animals there is obviously a seasonal change in the temperatures to which the hypothalamus initiates a response, and in some species there must also be a daily change.

Cold- and Warm-Blooded Vertebrates

Fish, amphibians and reptiles are sometimes referred to as "cold-blooded" vertebrates, a misleading expression as their body temperatures may equal those of birds and mammals (or "warm-blooded" vertebrates). Fish come close to approximating the ambient temperature but, in tropical species, this is not cold. Amphibians also usually have a body temperature close to or below the ambient temperature. Reptiles, using both physiologic and behavioral mechanisms, may attain body temperatures far above the ambient. Fish, amphibians and reptiles are also called poikilothermous, meaning "having many temperatures." This is more descriptive, but fish temperatures seldom deviate far from the ambient, and birds and mammals may have rather wide daily and seasonal variations and body temperatures.

Endotherms and Ectotherms. To avoid inaccurate implications in the term "poikilotherm," the late Professor Raymond Cowles proposed "endotherm" for those animals that derive their body heat from their oxidative metabolism, and "ectotherm" for those obtaining most of their body heat from the environment by convection, conductance or radiation.[187]

The expressions *poikilotherm* and *homoiotherm,* and *ectotherm* and *endotherm,* are not quite interchangeable. A poikilotherm has a variable body temperature, while that of a homoiotherm is constant. An ectotherm, however, derives most of its body heat from external sources, whereas an endotherm generates most of its body heat metabolically. Most ectotherms are poikilothermous, and most endotherms are homoiothermous, but the association has many exceptions. Mammals, for example, are endotherms, but the body temperature of hibernating mammals may drop to about 4°C; most fishes, although ectotherms, have a body temperature nearly identical to that of the surrounding water, with only slight daily variations. Endothermy is achieved when the rate at which an animal produces metabolic heat exceeds the rate at which such heat leaves the animal. Heat loss or thermal conductance is related not only to insulation of the surface but also to body mass, so that a very large ectotherm becomes homoiothermous or, for practical purposes, endothermous.

In endotherms there is an ambient temperature at which energy metabolism or heat production is minimal (Fig. 7.1). For practical purposes, there is a *range* of ambient temperatures, the range of *thermal neutrality,* over which energy

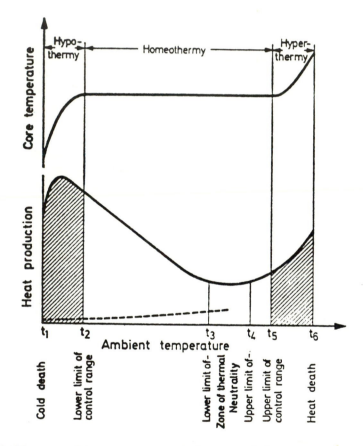

Fig. 7.1. Thermoregulatory responses and the zone of thermoneutrality relative to ambient temperature. The ambient temperature of the unshaded region (between t_2 and t_5) is the environmental range through which an endotherm can preserve a more-or-less constant core temperature. Below t_2 thermoregulatory efforts fail and hypothermia results; above t_5 temperature regulation fails and hyperthermia follows. The thermoneutral zone lies between t_3 and t_4. The dashed line indicates the relationship between ambient temperature and heat production in ectotherms (Giaja, 1928).

metabolism is virtually unchanged. At the lower point, metabolism increases and counters heat lost through thermal conductance; at the higher point, breathing, vasodilation of peripheral capillaries, and sweating promote the loss of metabolic heat. The range or zone of thermal neutrality, although subject to acclimatization, is a rather constant characteristic of a given species, and tends to be broader for species of high latitudes than for tropical endotherms. The polar bear and the arctic fox, for example, have ranges of thermal neutrality from $-25°C$ (or lower) to at least $30°C$.[100,483] The broad-range thermal neutrality of many arctic endotherms is a consequence of reduced thermal conductance, which reflects dense pelage or plumage and effective counter-current exchange in extremities. The toes of an arctic fox *(Alopex lagopus)*, for example, may be $5°C$ without a drop in body temperature.[423]

It is suggested that dinosaurs were endothermic.[1006] Three sorts of evidence support this concept. First, the bones of dinosaurs have abundant Haversian canals (housing blood vessels and nerves), characteristic of birds and mammals. Second, their bones lack "growth rings," characteristic of ectotherms that live in regions of cold winters (and minimal growth rates). Third, the apparent ratio of predatory dinosaurs to their presumed prey is closer to that among predatory birds and mammals than to that among predatory ectotherms. These ratios, among living vertebrates, reflect the greater energy utilization of an endothermic predator. This proposal can be countered by pointing out that the mass-to-surface ratio of large dinosaurs would have, in itself, effected a constant core temperature. That is, unless the Mesozoic day was much longer than 24 hours, dinosaurs would not have cooled off at night, nor would they have had time to warm up the following morning. This suggests that they had a fairly constant core temperature and lived in rather warm, temperate climates. These circumstances might explain the "endothermic nature" of the bones of large dinosaurs.

Heat Gain

Homeothermy enables endothermic animals to live in environments that might otherwise be too cold to allow minimal metabolism for feeding, digestion and reproduction. Endothermy consumes much more energy than is expended by ectotherms, and some ectotherms—especially heliotherms—are very efficient at achieving body temperatures far above the ambient. In many endotherms there is a slight circadian rhythm in body temperature: in diurnal species there is a nocturnal decline in body temperature, and in nocturnal species [such as bats, pocket mice *(Perognathus)* and flying squirrels] there is a slight drop in core temperature in the daytime. Endothermy is a costly advantage that birds and mammals enjoy over their poorly insulated and "less efficient" associates, amphibians and reptiles. For a given amount of biomass, a bird or a mammal requires far more energy than does a reptile or an amphibian for a 24-hour period, and the disparity increases if the comparison covers an annual period. Although birds and mammals can occupy regions too cold for any but the hardiest amphibians and reptiles, in temperate regions biomass of amphibians and reptiles may exceed that of endotherms.

Heat Loss

The various means of heat loss fall into one of three categories. Commonly, body heat is lost by *radiation,* as when a lizard seeks refuge from the sun beneath the cool shade of a bush and radiates heat. A number of vertebrates lose excessive heat by *conduction*, as when they leave a warm atmosphere and enter water of a lower temperature, or by *convection* of a breeze carrying away body heat (which has been lost by conduction). The circulatory system of a vertebrate can be regarded as a closed convection system, and heat is lost through the movement of warm blood to a cooled surface. In addition, amphibians and many mammals can lose heat by *evaporative cooling,* as in perspiration evaporating on the skin; this is a variant of conduction. Evaporative cooling can also be effected by panting. Radiation and conduction require an ambient temperature lower than the body temperature, and evaporative heat loss is most effective in a rather dry atmosphere.

Evaporative Cooling. Evaporative cooling lowers body temperature because large quantities of heat are required to change water from a liquid to a gas. At the interface of liquid water and air, there is a constant movement of water molecules both into the air from the liquid and from the air to the liquid. This tendency of water to evaporate from the surface is countered by the amount of water in the air. The amount of water in air is expressed as *vapor pressure,* measured in millibars. When the rate of movement of water molecules in each direction is the same, when air can accommodate no more water without an increase in temperature, *saturation vapor pressure* has been reached. With an increase in air temperature, there is an increase in saturation vapor pressure but with no increase in vapor pressure.

The difference between the saturation vapor pressure and the vapor pressure is the *saturation deficit,* the amount of additional water vapor that the air can hold at that temperature. The air at the animal's immediate surface is usually warmer than the circulating air because of continuous release of metabolic heat; this warmer air has a greater capacity to hold water than has the cooler circulating air. The rate of water loss is determined by the difference in the vapor pressures at the interface. As the water evaporates from the animal into the adjacent air, it is carried away by convection currents, or breezes. Thus capacity for evaporation increases with a rise in temperature and a decrease in water content of the air and is greatly accelerated by air movement, which brings air with a lower temperature to the animal's skin.

Mass-to-Surface Ratio and Thermal Conductance. Great size in a desert bird or mammal imposes restrictions on avenues for temperature control, but size alone tends to alleviate some of the problems encountered by small species. Great mass, by reducing the surface-to-volume ratio, slows the rate of heating so that, although an elephant or an ostrich cannot seek shade under a low bush or a burrow in the ground, the problem of excessive heat does not demand the immediate counter-action that is needed, say, for a squirrel (Fig. 7.2).

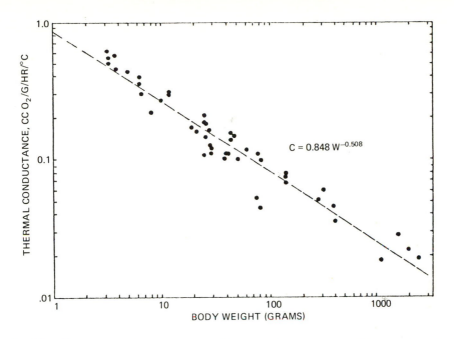

Fig. 7.2. Relationship of body weight and thermal conductance in birds from 3 to 2430 g (Lasiewski et al, 1967).

Plumage of birds and pelage of mammals become thicker and heavier with the advent of seasonally cold weather, and there is a consequent decrease in thermal conductance. Changes in feathers and hair are both qualitative and quantitative. These changes are presumably acclimatizations to lower temperatures, but this does not exclude their being at least partly genetically determined. A physical change in the nature of hair or feathers in wild populations occurs before an annual decline in ambient temperature, such a change being an adaptation of the species. In some wild species of birds, the increase in plumage (decreased thermal conductance) and an increase in metabolic rate account for an ability to withstand low temperatures in winter.

Body Temperature

Rate of metabolism is temperature-dependent, and a rise in body temperature is paralleled by an increase in energy metabolism within the normal ranges of body temperature for a given species. The speed of a simple chemical reaction increases with a rise in temperature, a relationship known as van't Hoff's Rule, but rates of change in an animal do not change in a *constant* way with temperature.

Many terrestrial vertebrates are active at body temperatures between 34° and 43°C, but for many nocturnal and fossorial ectotherms as well as for many fish and amphibians normal body temperatures are lower. Apparently different sorts of activity occur at widely varying temperatures: recrudescence of gonads in many amphibians, reptiles and mammals takes place during hibernation at from 3° to 6°C, which is much too cold for digestion or for any but the slowest muscular movement.

Fish

Some fishes exposed to seasonally cold temperatures develop physiologic means of avoiding cellular and tissue damage. This is especially pronounced in marine species, which may be subjected to temperatures below 0°C. Seawater has a freezing temperature of −1.86°C. Enzyme systems must be able to function at temperatures not generally encountered by fresh-water fishes of temperate environments. Some teleosts live in ocean depths where the temperatures are below the freezing point of their tissues, but these fish are supercooled and presumably would freeze on contact with ice.

Thermoregulation

Polar fishes have evolved various mechanisms that depress the freezing point of their blood; however, these fishes are not known to hibernate. Near the Ross Ice Shelf in Antarctica the seawater remains a nearly constant −1.85°C. Some fishes in the Antarctic, species of *Trematomus* and *Dissostichus,* contain sufficient amounts of glycoproteins in their blood serum to lower the freezing point of the latter. As water temperatures decrease, plasma protein concentrations of these fish increase, thus serving to impede ice crystal formation at water temperatures as low as −1.9°C. It is hypothesized that these glycoproteins exert their effect by being adsorbed onto the surface of ice crystals, thereby preventing water molecules from settling into the crystal's ice lattice until a much lower temperature is reached. Antifreeze compounds increase from spring to autumn in shallow-water marine fishes in the Arctic; nonelectrolytes may increase about 60 percent from summer to winter, but sodium ions may actually decrease over the same period. There may be an increase in serum osmolality with a decline in serum electrolytes in fresh-water species, such increase being in the form of glucose. Serum glucose may compensate for the loss in electrolytes and preserve a low serous freezing point.

Supercooling occurs in some other fishes exposed to very low temperatures. The killifish *(Fundulus heteroclitus),* a small cyprinodont of the Atlantic coast whose blood serum has a freezing point of −0.8°C, maintains itself in seawater at −1.5°C and yet does not freeze. When acclimated to temperatures below freezing, the blood of this species becomes very rich in glucose, which comes from liver glycogen. Such concentrations of blood glucose, 3–10 times the levels that occur above freezing, presumably account for the ability of the killifish and others to remain in the supercooled state (Fig. 7.3). Freezing in fishes in waters below 0°C is thus prevented by both supercoolants and a biologic antifreeze.[1038]

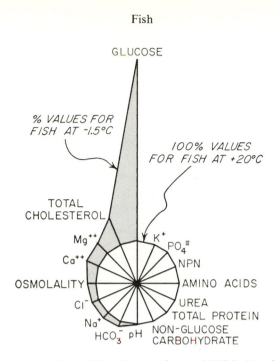

Fig. 7.3. Chemical composition of blood serum from a killifish *(Fundulus heteroclitus)* acclimated to −1.5°C, expressed as percentages from serum in a killifish at 20°C (Umminger, 1969).

Two groups of fishes are well known for their ability to maintain body temperatures appreciably above that of the water through which they swim. In the tunas and skipjacks, Thunniformes (= Plecostei), and lamnid sharks, Lamniformes (= Galeoidei), a modified circulation reduces the loss of heat through the gills and results in a greatly increased core temperature in these rapid swimmers. Unlike the arrangement in most fishes, the vascular system in these groups is largely subcutaneous, with plexuses deep within the myotomes and next to the liver. Veins and arteries running close together constitute an effective counter-current exchange system, keeping the metabolic heat produced by these predators from reaching the body surface (Fig. 7.4). In each group of fish, tunas (bony fish) and lamnid sharks (cartilaginous fish), warm muscles provide for the greater muscular activity needed for their way of life; the totality of habits, swimming activity, anatomy and thermoregulation constitutes one of the more striking examples of convergence in fishes.[150,151]

Numerous studies on fishes demonstrate that thermal preferences exist, and that fish have a delicate sense of thermal perception. Changes in thermal preferences vary with the season and may reflect seasonal differences in metabolic needs, at least more than acclimation to different temperatures.

In fishes adjustment to a change in temperature is mediated by the central nervous system. A species of fish has a characteristic metabolic rate at a given

Fig. 7.4. Counter-current circulation in the tuna; this mechanism endows these fish with a higher than ambient core temperature. A, segmental vessels; B, branches from segmentals consisting of bands of parallel arteries and veins; C, lateral artery; D, lateral vein; E, rete, passing inward from lateral artery and vein; and F, branches from dorsal aorta (Carey and Teal, 1966).

temperature within a range through which it is able to acclimate. Fish are the most difficult vertebrates to observe for any extended period, but there are well-known movements among waters of different temperatures. Depending on the species, some fish seem to seek warm, sunlit waters just as others prefer cooler, darker waters, and the various temperatures must have profound effects on such metabolic processes as gonadal development and digestion. Body temperatures of fish seldom vary much from the ambient temperatures. There is no means for fish to exchange dissolved gases through the gills without also giving up any body heat about as fast as it is made. Tunas and their allies and lamnid sharks are well-known exceptions to this generality.

Osmoregulation
Some bony fish (Osteichthyes) migrate between waters of varying salinity, but there is very little change in the osmotic pressure of their blood plasma. In marine

waters bony fish tend to lose body water and compensate by drinking seawater. Water balance is a function of the gills and kidneys, for little water passes through their skin. The neurohypophysis may secrete an antidiuretic hormone, which reduces the loss of body water. Species that move between rivers and the sea (some salmonids, osmerids, anguillids and others) are well-adapted to salt concentrations in both waters. In fresh water, fish produce abundant urine, which prevents their tissues from becoming flooded, but in the ocean urine output is curtailed, for water lost through the kidney would only augment that which passes out through the gills. Bony fishes in the ocean balance the loss of water to their salty environment by drinking seawater and excreting salts; their urine is about one-tenth that produced by a fresh-water fish. Anadromous fishes change their excretory and drinking patterns in moving between the rivers and the ocean. In moving to the sea some anadromous fishes reduce urine output by muscular constriction about the arterioles entering the kidneys, and the reverse action increases urine when they later return to fresh water.[180]

The salt content of hagfish (Myxinidae) renders its body fluids isosmotic with seawater, in contrast to all other fish-like vertebrates, which have tissue much less salty than ocean water. Cartilaginous fishes (Chondrichthyes) and *Latimeria chalumnae* (the living coelacanth) are not especially high in salt content, but urea in their tissues makes them virtually isosmotic with seawater. Hagfish are entirely marine, and Chondrichthyes or elasmobranchs nearly so: neither suffers water loss in the ocean. Those elasmobranchs that dwell in fresh water have greatly reduced urea levels, so that osmoregulation is still not a problem. Fresh-water rays (Potamotrygonidae) of South America do not build up urea when placed in marine water. In water, lungfish (*Protopterus* spp.) excrete ammonia, but, when they are dormant in dried mud when water loss must be curtailed, ammonia excretion ceases, and urea accumulates in their tissues.[619]

Elasmobranchs and the coelacanth discharge excess salts through a rectal gland. In the marine catfish, *Plotosus lineatus,* salt is apparently excreted by a "dendritic organ" near the anus. This structure is lacking in most fresh-water plotosid catfish that occur in rivers of New Guinea and Australia.[835]

Amphibians

Temperature

In amphibians it is extremely difficult to separate effects of temperature independent of humidity. The proclivity of some amphibians to remain in the immediate vicinity of water, however, may be for thermoregulatory purposes: a frog may leave a pond during the day as air temperatures rise and return to it late in the day as air temperatures fall below that of the water. The movement of some tadpoles to warm shallow waters at the margin of a pond or a creek is probably thermoregulatory, and the increased food supply in such places may also attract them. Perhaps because anurans and salamanders live in a variety of habitats and are also so thermolabile, they operate over a rather broad range of body temperatures. Salamanders resemble snakes in having a body temperature near that of their substrate, and probably these amphibians follow the substrate temperatures even more closely. Although many salamanders frequently move above ground, they rarely appear in direct sunlight but rather move at night or during the day in cloudy or rainy weather.

Aquatic Amphibians. Like small fish, aquatic amphibians have a body temperature very close to the surrounding water, and in some cases this can be very cold, or close to $0°C$. The Nearctic giant salamander, *Cryptobranchus alleganiensis,* lives in cold rivers, as does the mudpuppy, *Necturus maculosus.* Such species can change body temperatures only by moving to water of a different temperature.

Terrestrial Amphibians. Core temperatures of terrestrial amphibians frequently differ from the ambient; out of water, amphibians may lower body temperature by evaporative cooling or raise it by basking.

The response of amphibians to changes in temperature and humidity is complicated by the role of their skin in exchange of oxygen and carbon dioxide. In an atmosphere in which a frog or salamander is gradually losing body water, the surface of the skin is moist, permitting gaseous exchange. In an extremely dry atmosphere, it is conceivable that evaporation from the skin could exceed the movement of fluids to its surface and thus reduce the effectiveness of the skin as a breathing organ. This alone would be a sufficient threat to account for the disappearance of terrestrial lungless salamanders into moist soil during dry periods.

Among species of boreal frogs of the genus *Rana,* some lay eggs in a film that floats on the surface of the water, whereas others have globular egg masses. There is a tendency for globular masses to be slightly (about $1°C$) warmer than the adjacent water, whereas the surface film masses, which are characteristic of frogs that breed in warm ponds, tend to be cooler than the surrounding water.[874]

Among amphibians, basking is generally not the most common means of acquiring heat, probably because there is frequently the risk of desiccation. A number of anurans, however, bask in the spring and autumn, usually when in the immediate vicinity of water. In the Peruvian Andes at 4300 m, *Bufo spinulosus* may bask for up to one hour and maintain a body temperature more than $5°C$ above the ambient, but at a loss of 5.7 percent of its body weight (water) per hour.

Tolerance to high temperatures among desert vertebrates has been demonstrated numerous times, and this ability to tolerate heat occurs in adults as well as in eggs. Eggs of the Nearctic anuran, the desert-dwelling spadefoot toad *(Scaphiopus hammondii),* can tolerate water of $39°C$ in their later stages.[128]

Water Loss

Water loss in amphibians is related to such physical factors as temperature and the drying power of the air (or vapor-pressure deficit). Differential tolerance to desiccation in amphibians from contrasting habitats is well known: generally species that spend most or all of their lives in water can tolerate much less loss of body water than species from arid areas, such as deserts. Water loss is greater at higher temperatures and lower humidities. Because of the increased surface-to-mass ratio of small individuals, their water loss measured as a percentage of the total body water is greater than for large individuals. Mammalian oxytocin causes water absorption through the skin and a decrease in excretion in amphibians, and secretions of the amphibian neurohypophysis (or posterior lobe) have the same effect.[448,1072]

In some amphibians there is a correlation between rates of water loss and habitat. More aquatic species tend to lose water rapidly, and some are more hygroscopic than are the more terrestrial species: most desert frogs and terrestrial amphibians can tolerate a greater loss of water than can species that live most of the time in the water. Among the species of the Australian genus *Heleioporus* (Leptodactylidae), those from more arid regions seem to acquire water through the skin more rapidly than species from more humid areas, but this relationship is not yet known to be valid for amphibians in general.[1072]

Water passes readily through the skin of amphibians, and when in water they are continuously absorbing it. This is discharged in a dilute urine, and amphibians also lose water through the skin as soon as they crawl out on land. Many salamanders and anurans never enter the water, and their main problem is the retention of water. Some amphibians that lay eggs in water are aquatic only during the breeding season; their ability to absorb moisture through the skin permits them to become completely terrestrial after oviposition in water. Water loss in amphibians declines when individuals cluster together in an aggregation; under experimental conditions an increase in the number of individuals effects a decline in the rate of dehydration.

A number of anurans tolerate mild salinities and regularly occur and even breed in salinities above 3 percent. The leopard frog *(Rana pipiens)* greatly increases blood urea when placed in 0.6 and 0.9 percent NaCl for 14 days, thus preserving an internal environment osmotically above that of the external environment.[448] In southern California *Rana pipiens* is found in a saline creek that enters the Salton Sea,[865,867] and in eastern Asia not only does *Rana cancrivora* live and breed in brackish water, but adults can survive brief exposure to seawater. This frog avoids dehydration by the buildup and retention of urea in the tissues.[351] In San Francisco Bay, California, *Bufo boreas* is adapted to a life in brackish water. In Ceylon *Rana cyanophlyctis* may enter, and is said sometimes to breed in, brackish water.[549]

The clawed toad, *Xenopus laevis,* is normally totally aquatic, but parts of its native African home experience summer aridity. As its ponds dry up, the clawed toad estivates in the ground until the return of seasonal rains. While in the water this anuran excretes mostly ammonia but, as estivation approaches, nitrogenous waste is excreted as urea, with a considerably saving of water.[45,236]

Certain species of Hylidae and Rhacophoridae are uricotelic and lose little water through the skin, both adaptations to xeric environments. Species of *Phyllomedusa* (Hylidae) secrete a lipid that covers the skin, apparently retarding evaporation of body water from the skin. *Phyllomedusa sauvagei,* a Neotropical hylid, lives in relatively arid regions, has a very slow rate of water loss and is uricotelic; in this frog cutaneous water loss is about 2 percent of the body weight daily, about the same as in a desert lizard.[97,921]

Thus we see that the form in which amphibians discharge their metabolic wastes is adapted to maintain their water balance. When living in freshwater, they release ammonia and void large amounts of water in the process. When in brackish water, however, they discharge urea, some of which accumulates in their tissues, rendering them less liable to desiccation. On land it becomes necessary to conserve water, and again nitrogenous wastes are discharged as urea. Virtually all terrestrial amphibians are ureotelic. Moreover, some amphibians that release ammonia when in fresh water produce urea when they estivate. Finally, at least some desert frogs achieve maximal economy of water by the production of uric acid.

Reptiles

The reptilian integument greatly retards the passage of water: some water does leave the body through the skin, however, and the rate varies from one group of reptiles to another. Water is not known to enter through the skin of reptiles. In these vertebrates evaporative cooling is confined to panting and to evaporation of water remaining on the skin of aquatic forms as they come out on land. With very few exceptions reptilian thermogenesis is inadequate to raise the core temperatures of these organisms above the ambient temperature.

Water Balance

Loss of water in excretory waste is greatly reduced in most reptiles by the formation of uric acid, which is relatively nontoxic and which can be eliminated with little loss of water. Much of the water discharged by the kidneys of reptiles is resorbed, so that their fecal waste may be rather dry. Urine may be retained in the bladder, as is the case in chelonians, and resorption is possible either through the bladder or cloaca. Urea is excreted in the urine of some reptiles, but it is usually less than uric acid, and reptiles do not form hypertonic urine.

Turtles are versatile in the manner of forming excretion products.[724] Generally aquatic and marine species form both ammonia and urea. Pond turtles may form some ammonia, but most wastes are discharged as urea. Terrestrial and desert species (e.g., species of *Testudo* and *Gopherus*) are mostly uricotelic.[212] Crocodilians resemble water-dwelling chelonians in excreting mainly ammonia and urea.

Because the reptilian kidney does not form hypertonic urine and because some species live in arid regions where excessive water loss is a continuous threat, there is potential danger of an accumulation of salts. Marine birds and reptiles consume substantial amounts of salt in their food and water. This is not excreted in the kidneys, but salt glands remove excess salts from the blood plasma. In lizards, snakes and turtles salt glands discharge not only into the external nares but about the eyes and in the mouth. Marine reptiles feed in the water and probably ingest seawater with their food. Marine turtles of several species secrete hypertonic solutions from orbital glands, and the marine iguana *(Amblyrhynchus cristatus)* voids excess salts through nasal glands.[1073] The only totally marine reptiles occur in the family of sea snakes, Hydrophiidae; an oral salt gland in the yellow-bellied sea snake *(Pelamis platurus)* eliminates excess salts into the fleshy sheath that houses the tongue.[417]

Desert reptiles face problems of salt loading, and nasal glands occur in a number of desert-dwelling herbivorous Iguanidae. This problem is especially important, for leaves of desert plants may be very high in electrolytes in general and especially in potassium.[796] Salt in solution is secreted in the nasal gland of the desert iguana *(Dipsosaurus dorsalis)* occupying a depression near the external nares (Fig. 7.5).[735] In some desert species there is a reduction of urine to conserve water. In the desert tortoise *(Gopherus agassizii)* of the southwestern United States, the urinary bladder is permeable to water, ions and some molecules, so that certain products of the kidney can be retained. As a result urate crystals form in the urinary bladder of this species, and a highly concentrated urine is produced not only by the kidney but also by the permeable bladder. A strong sphincter of the bladder facilitates this function. Also, in the desert tortoise there is little return of water through the distal tubule, and urine entering the bladder is hypoosmotic; this allows continuation of elimination of nitrogenous wastes under some dehydration.[212]

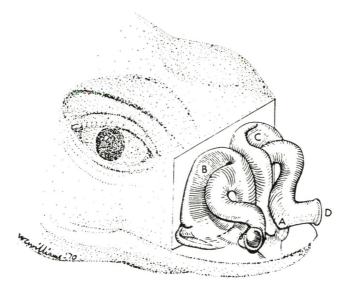

Fig. 7.5. Pattern of nasal passage of a desert iguana *(Dipsosaurus dorsalis)*. The nasal gland (B and C) secretes salt, which collects in a small ventral depression (A). As air enters the external nares (D), the water in the depression evaporates and enters the lungs with the inspired air (Murrish and Schmidt-Nielsen, 1970, courtesy of Elsevier/North-Holland Biomedical Press).

Evaporative water loss in reptiles is predicated on temperature and humidity; diurnal forms generally lose water at a slower rate than do nocturnal or fossorial species. Evaporative water loss in reptiles exceeds formation of water from oxidation. The desert-dwelling chuckwalla *(Sauromalus obesus)* loses about 2.0 mg/cm^2/day, and the highly aquatic crocodilian *(Caiman sclerops)* loses 37.7 mg/cm^2/day. In the case of two chelonians, for the rather aquatic *Pseudemys scripta* and for the terrestrial box turtle *Terrapene carolina* these data are 15.8 and 7.2, respectively. The chuckwallas (*Sauromalus* spp.), Nearctic desert-dwelling iguanid lizards, possess loose folds of tissue on either side of the trunk; these become fluid-filled during rainy periods and are later depleted during times of water scarcity.[165]

Although relatively little attention has been given to the water needs of reptilian eggs, they are usually hygroscopic. Throughout incubation there is an increase in water content and frequently a sudden increase in weight shortly after they are laid. Most geckos lay eggs with rather hard calcareous shells but, as in eggs of birds, water is absorbed during development. In low latitudes reptilian reproduction is often timed so that egg-laying occurs during the wet or rainy season; it seems reasonable that this is partly a response to a need of eggs for moisture in the soil in which they are laid.

Thermoregulation

The anatomy of reptiles does not favor high production and retention of metabolic heat: the imperfect separation of the ventricle of the heart and relatively high thermal conductance of the integument contribute to the weak thermogenic powers of most reptiles. The poor surface insulation, which allows heat to dissipate, also permits the entrance of heat from external sources.

Lethal Minimum and Maximum. Several standard expressions describe temperature relations of reptiles.[188] The *critical thermal minimum* (CTMin) is the body temperature at which locomotor ability is lost. The *voluntary thermal minimum* (VTMin) is the ambient and body temperature at which the animal begins to bask; this can vary, for basking in some reptiles depends on an endogenous rhythm. The *voluntary thermal maximum* (VTMax) is the body temperature above which the animal becomes photophobic. The *normal activity range* (NAR) extends from the voluntary thermal minimum to the voluntary thermal maximum. Heliotherms typically bask above the VTMin. The *basking range* may include most of the NAR. The *critical thermal maximum* (CTMax) is the body temperature at which the animal loses muscular coordination; death occurs within a brief period in an animal kept at the CTMax. In nature, basking (in heliothermic reptiles) may continue up to a point at which the body temperature closely approaches the lethal maximum, which is not to suggest that the NAR is a narrow range close to lethal temperature; it means rather that most species seem able to tolerate body temperatures dangerously close to that point. "Normal activity" seems to occur within a range of body temperatures, the range being much broader in some species than in others; the mean body temperature at which a given species is active may differ significantly from that of a different species in the same environment. Because some species seem to be less heliothermic than others (and therefore are active over a broader range of body temperatures), it becomes especially difficult in such cases to define basking range. The concept of a basking range is still useful when these pitfalls are remembered.

As heliotherms warm up to near their CTMax, their increased metabolism aggravates the danger. The reptile in nature can lose heat by radiation (if it is in the shade) and by panting. The relatively small size of most reptiles and the efficient thermal conductance of the reptilian integument facilitate heat loss by radiation, but a dry atmosphere is needed for panting to be effective.

Lethal temperatures for reptiles vary among species: generally lizards tolerate greater body heat than can snakes, and nocturnal and fossorial lizards are less heat-tolerant than heliothermic species. A number of heliothermic lizards survive body temperatures in excess of 45°C, but for many snakes the CTMin may be less than 30°C.

The mean body temperature (MBT) of reptiles is the mean of a series of body temperatures taken when animals are active in nature. The MBT indicates the temperatures through which a given species can operate physiologically, and should not be confused with an experimentally determined "optimum." Ecologically wide-ranging species may have a broad range of body temperatures, and environmentally restricted species may have a narrow range of body temperatures. The marine iguana *(Amblyrhynchus cristatus),* a large heliotherm of the Galápagos Islands, is active over a wide range of body and ambient temperatures. While basking in air temperatures of 50°C, they may warm up to 40°C, and while in water their body temperature approximates that of the surrounding sea, 25°–26°C.[61]

Least-tolerable temperatures have not received as much study as CTMax. One assumes that a reptile usually retreats to a site in which low temperatures are not dangerous. Some physiologic protection has also been noted. Many reptiles can withstand supercooling of their bodies to as much as −8°C and recover upon warming. *Uta stansburiana* survives −5°C for 2½ hours with a 15 percent mortality. This lizard is not a hibernator, and undoubtedly frequently experiences supercooling in nature.[388]

Natural changes, especially seasonal changes, in response to temperatures (and other environmental factors) are called *acclimatization,* and comparable changes experimentally induced are called *acclimation.* Lethal maxima and minima, for example, frequently fluctuate in nature, such annual variations being as well-marked as those involving gonadal growth.

Preferred and Eccritic Temperatures. These two expressions are commonly used interchangeably, but gradually have come to assume different meanings. The temperature a lizard or snake prefers or selects within a gradient set up in a laboratory is the preferred temperature. In the absence of solar heat, such a preferred temperature is the same as the body temperature of the animal. The temperature sought in a gradient is sometimes called the *temperature preferendum*. It is sometimes a little higher than the eccritic temperature (that obtained in the field), but generally the two are close. The eccritic temperature must be derived from a number of individuals active in the field, and thus it is expressed as a mean. The eccritic temperature is usually within a narrow range for most species, and is sometimes taken to mean the mean body or core temperature during normal activity. The MBT of a given group of lizards in the wild, at least in heliotherms, is frequently well above the ambient (Fig. 7.6).

There is generally a correlation of eccritic temperature and habits, and commonly there is a similarity among congeneric species of lizards and snakes. This is not surprising, for among congeneric species there is a much greater similarity in habits than there is within a group of species (of different genera and families) within a given area. Thus sun-loving lizards, such as *Cnemidophorus,* are commonly active in bright hot weather wherever they occur, and have much higher core temperatures than have such shade-dwelling and crepuscular species as *Gerrhonotus* spp. Snakes reflect their fossorial origin, for they are frequently nocturnal, and the diurnal species generally have body temperatures near those of crepuscular lizards.

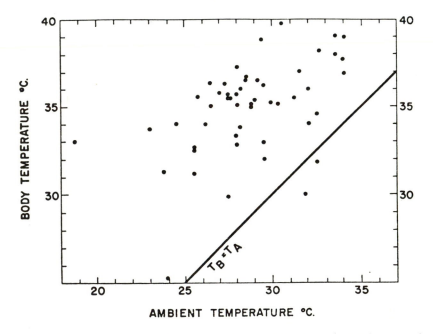

Fig. 7.6. Relationship of core body temperature to ambient temperature in a sample of 51 agamid lizards *(Amphibolurus barbatus)* measured in the field (Bartholomew and Tucker, 1963, from Physiological Zoology, courtesy of the University of Chicago Press).

The preferred or eccritic temperature changes with the season in many reptiles from temperate regions and has been shown also to change from one time of day to another, so that a relatively low temperature may be sought during the night by diurnal reptiles. *Klauberina riversiana* (a Nearctic xantusiid lizard) in the laboratory chose a warm surface in the daytime but retreated to a cool area at night and became torpid. Similarly, daily emergence in horned lizards, *Phrynosoma coronatum* and *P. cornutum,* has been observed to occur as much as 40 minutes before the "experimental sunrise" in the laboratory,[414] and this finding was temperature-independent (Fig. 7.7). Seasonal variations in both MBT and eccritic temperatures occur in several species of the iguanid lizard *Sceloporus:* in cold months not only do they seek cooler sites during the day, but their body temperatures are lower. Moreover in unusually warm winters *Sceloporus orcutti* may remain active, at which time the MBTs are significantly higher than those in cool or cold winters.[663]

Some geckos are active over a much broader range of body temperature than the ranges found in most diurnal lizards. Geckos generally not only do not tolerate the high temperatures comonly encountered by many diurnal lizards, but are poorly equipped to withstand the desiccating effects of warm dry air.[174,237,639]

In snakes body temperatures are usually close to that of the surface soil, and basking, so characteristic of heliothermic saurians, is incidental or much less common in snakes. Although some temperate and boreal snakes may bask in the sun,

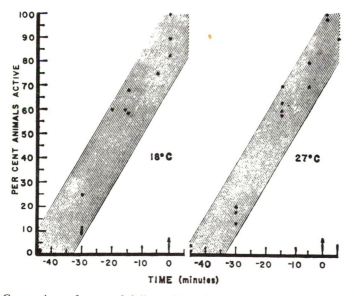

Fig. 7.7. Comparison of onset of daily activity of two groups of horned lizards (*Phrynosoma coronatum* and *P. cornutum*) kept at 18°C and 27°C. Experimental sunrise indicated by arrow at 0 (Heath, 1962, copyright 1962, by the American Association for the Advancement of Science).

especially in early morning, this may be unnecessary in the relatively mild temperatures of tropics and subtropics (Fig. 7.8).

The Warming Process. Those species that regularly expose themselves to direct rays of the sun do so in stages, beginning with the head while the rest of the body is in shade; some species remain buried in sand with only the head in the sun, whereas others hide in crevices with only the head protruding. There is frequently a temperature differential when reptiles increase their body heat: commonly the head becomes warm in advance of the postoccipital area during basking.

In reptiles a muscle surrounds the paired internal jugular veins so that virtually all blood in the head can be restricted in its passage to the heart. The internal

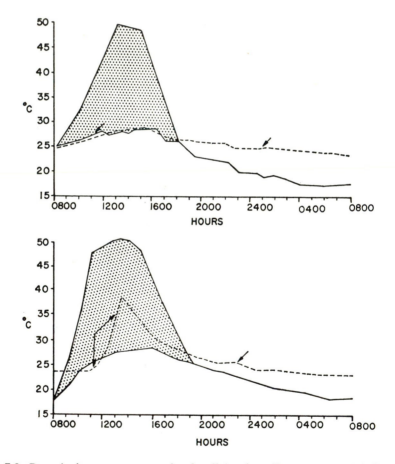

Fig. 7.8. Deep body temperatures of a free-living boa *(Boa constrictor)* indicated by dashed line; solid line is the surface temperature in the shade, and stippled area is the thermal range available to the snake above ground during the day. Arrows indicate times of movements of 10 m or more (McGinnis and Moore, 1969).

jugular veins parallel the internal carotid arteries, so that flow in both directions effects a counter-current exchange in temperature. When these veins are constricted, blood leaves the head in a shunt via the paired lateral commissures to the external jugular vein. This bypasses the counter-current exchange, allowing the reptile a means of regulating the temperature of the head.

A small reptile exposed to direct rays of the sun may quickly absorb heat; this process is called *heliothermy*. If the ambient air is cool, 10°C or less, lizards (and sometimes snakes) may bask on warm rocks; lying with its ventral or dorsal surface appressed to the rock, the lizard may also acquire heat from contact, which is called *thigmothermy* (Fig. 7.9). The various postures of basking lizards depend on body temperatures as well as on substrate; more extensive contact is made with the warmed substrate at lower body temperatures (Fig. 7.10). A reptile is not necessarily exclusively heliothermic or thigmothermic, but may easily obtain heat in both ways, either at the same time or at different times. In either case it remains an ectotherm.

As stone surfaces are quick to become warm, thermoregulation in chilly spring mornings frequently involves both heliothermy and thigmothermy. The well-known examples of *Liolaemus multiformis,* of the Andes, and *Lacerta agilis,* of the Caucasus, illustrate the remarkable efficiency of reptilian thermoregulation: *L. multiformis,* a small Neotropical iguanid, and *L. agilis,* a Palaearctic lacertid, both occur at elevations exceeding 4000 m, and both raise their body temperatures to 30°C in an ambient temperature of 0°C. In the high Andes the iguanid lizard *Liolaemus multiformis* emerges in the morning when the ambient temperature is 0°C or below. It basks first on carpet-like pads of the caryophyllaceorus plant *Pycnophyllum tetrastichum* rather than on the still cold rocks, and warms more rapidly than when tethered to a rock substrate.[798,801]

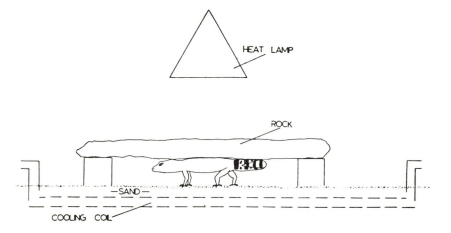

Fig. 7.9. Banded gecko *(Coleonyx brevis)* obtaining heat from contact (thigmothermy) (Dial, 1978).

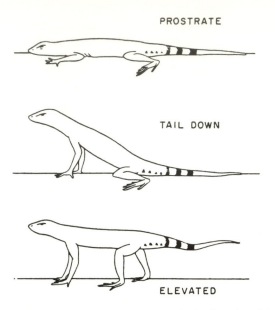

Fig. 7.10. A basking lizard adopts positions increasingly dissociated from the surface of the rock as its body temperature increases. The elevated position is assumed when both the lizard and the substrate are warmest (Muth, 1977).

Experimentally, lizards *(Sceloporus cyanogenys)* were allowed to bask under an artificial heat source (a 250-watt bulb with adjustable intensities), which they could turn on at will. The time spent basking decreased with an increase in intensity of the heat bulb. By monitoring the core (intestinal) temperature, it was found that there is a neutral range over which the animal is refractory to temperature changes within its body. That is, there appear to be high and low set points for thermoregulation responses, and between these temperatures there is no response.[330]

Rates of temperature change vary with the direction of the change: in many reptiles a rise will occur more rapidly than a decrease. As core temperature rises, acquisition of solar heat is assisted by increased production of metabolic heat; this process is accelerated by vasomotor constriction, which prevents heat loss through conduction. When the core temperature of a reptile drops, the fall is slowed by the production of metabolic heat, and this is contained again by vasomotor responses in peripheral vessels. In an Australian monitor *(Varanus varius)*, there was a rise of 0.14°C/min while the animal basked, but a decline of 0.02°/min in later afternoon. A sudden drop in body temperature occurred after drinking.[972] Also, studies of the Australian agamid lizard *Amphibolurus barbatus* showed that, under laboratory conditions, as the ambient temperature increased, there was a rise in body temperature, heart rate and metabolism. In this animal, as a rise in body temperature effected an increase in metabolism, an increase in heart beat delivered an increase in the oxygen supply. In a controlled environment this lizard increases its body temperature more rapidly than it loses it, allowing it (in nature) to preserve body heat during brief periods when clouds obscure the sun.[62]

At night diurnal heliotherms generally experience a body temperature that approximates the ambient; metabolic activities consequently drop, at a substantial saving of energy. In diurnal reptiles an insulated nocturnal retreat may effectively retard heat loss. A deep crevice in bedrock or leaf mold on the forest floor, while cool by daytime standards, may be warmer than the lowest nocturnal levels.

Some aquatic reptiles thermoregulate by moving in and out of water. Crocodilians, such as the Nile crocodile *(Crocodilus niloticus)*, which commonly sleep in the water at night, emerge and bask as air temperatures rise in the morning but return to the water at mid-day; basking may recur in the afternoon. The American alligator *(Alligator mississippiensis)* is diurnal in winter and spring, but nocturnal in summer. Heat is acquired both by thigmothermy and basking. During parts of the day when the ambient temperature exceeds the thermal maximum of 39°C, alligators maintain a body temperature of from 32° to 35°C by returning to water and sometimes reemerging to lose peripheral heat from evaporation of water on its back. This crocodilian can preserve a core temperature of 3°C above the ambient when submerged during the night. Such a differential could result from insulation triggered by vasoconstriction, from mass of the warmed animal, from metabolic heat or from any combination thereof.[943]

Water snakes (*Nerodia fasciata* and *N. taxipilota*), monitored by telemetry, revealed a daily temperature cycle that responded to changes in air and water temperatures: by basking in the morning air these snakes raised their core temperature to 26°–29°C, but when the air temperature dropped below the water temperature they returned to the water.[783]

In the hot springs area of Yellowstone Park the iguanid *Sceloporus graciosus* has an extended period of activity in the cold season. The thermal features of the ground effect a long snow-free period and permit extra seasonal warming of the ground. Using radiant heat, this lizard can preserve a body temperature above 30°C in an ambient temperature of 25°C and over a substrate of 23°C.[730]

The Cooling Process. The permeability of the skin of crocodilians and at least some snakes imparts to these reptiles a means of evaporative cooling. Certainly excessive production of saliva has a cooling effect in the snakes and chelonians in which it occurs. Crocodilians may lie with their huge mouths agape, but without the obvious panting so characteristic of agamids, iguanids and some other lizards. The more subterranean skinks (such as *Eumeces* spp.) apparently do not pant, but they may, like snakes, lose some body heat by the loss of water through the skin. The Nearctic iguanid, *Sauromalus obesus,* at high temperatures can lower its body temperature by panting, but some of the Australian agamids, which have been carefully studied, seem unable to lower their body temperatures by evaporative cooling.[62,63,1075] In the absence of the ability to reduce body temperatures by evaporative cooling, some reptiles have developed a tolerance to high body temperatures. The desert iguana *(Dipsosaurus dorsalis)* basks in the heat of the day, and may have a core temperature of 46°C.

Large reptiles tend to cool more slowly than smaller species because of the relatively low surface-to-mass ratio, and they do not always reach the ambient temperature during the night. The Komodo dragon *(Varanus komodoensis),* the largest living lizard, maintains a core temperature of 4°–5°C above the ambient of 24°C during a 12-hour night. Presumably size alone would account for a relatively stable body temperature in a reptile the size of a dinosaur.[681]

Burrowing is such a common manner for escaping extremes of temperature that the habit may partly account for the relatively small size of nontropical reptiles. Desert lizards seek refuge in cool loose soil under bushes, and desert tortoises repair to burrows in times of excessive heat and cold. In the heat of the day in seasons of maximal temperature, a retreat to cool depths in the ground lowers metabolic rates which, in turn, lower water production and loss of body water. Thus an escape from excessive heat protects desert-dwellers in two ways.

Coiling and aggregation may serve to reduce loss both of heat and moisture. In temperate regions many snakes are known to aggegate in the winter in the form of snake "balls." In captivity snakes may coil at night and so greatly retard the loss of body heat and water. The habitat of congregating at the time of hibernation may also reflect the relative scarcity of suitable hibernacula.

Although most diurnal lizards bask and are genuine heliotherms, some tropical forest-dwelling species have rather low core temperatures and manage normal activity without basking. This category includes a number of species of the Neotropical iguanids of the genus *Anolis* and the Bornean scincid, *Sphenomorphus sabanus.* These lizards, like some diurnal nonbasking snakes, manage to acquire adequate body heat from the surrounding air and are not thigmothermic.[46,866]

Although in heliothermic species MBTs of active individuals are generally well above the ambient, some shade-loving tropical anoles (*Anolis* spp.) may be active with body temperatures below the ambient. Presumably loss of water in breathing lowers the body temperature below the ambient in the absence of solar heat.

The genus *Anolis* apparently is the exception to the generalization that congeneric species tend to have similar temperature preferenda. Some species of *Anolis* normally have body temperatures substantially below the mean found in other species, and these differences are correlated to the elevations to which they ascend in mountains, the amount of sunlight in their habitats and the time of day that they are most active. This intrageneric diversity in temperature preferenda is no doubt an important factor in the abundance of species of anoles, many of which have a restricted habitat distribution. For example, *Anolis oculatus,* the only species of anole on the island of Dominica in the West Indies, has a range of 10.3°C about its MBT, equal to the range of five stenothermal species of *Anolis* on Cuba. *Anolis tropidolepis* lives in montane cloud forests in Costa Rica, and is active at a relatively low thermal preferendum characterized by rather wide limits. In air temperatures of between 14° and 20°C this anole is active; the body temperature is from 0.5° to 3° higher than the air, and the body temperature is usually less than 20°C.[299] *Anolis limifrons* lives primarily in shaded or semishaded forested areas in Panama, and has temperature preferenda of 26.4°C (in the dry season) and 27.6°C (in the wet season); these temperatures are 0.7° and 2.2° above the ambient, respectively.

Another anole *(A. cupreus)* occurs in Central America from tropical lowlands to cool uplands. It is essentially eurythermal and active at temperatures of from 22.5° to 33.5°C. At lower elevations it lives in shaded forest and does not bask, and the body temperature closely approximates the ambient; in the uplands, however, it frequently basks and has a body temperature almost above the ambient.[300] A Costa Rican forest-dwelling anole, *Anolis polylepis,* maintains a temperature slightly above the ambient even though it spends very little time basking. As with some other anoles, body temerature increases gradually throughout the day.[429]

Role of Color in Thermoregulation. Color plays an important role in thermoregulation, at least in terrestrial vertebrates. Many reptiles will be dark-skinned (due to the expansion of melanophores) when basking at lowered temperatures; darker hues allow a greater absorption of solar energy. A significant amount of infrared light is admitted through pigmented skin, resulting in a rise in internal temperature. In desert lizards exposed to very bright light, a high percentage of this light is reflected by pale skin with relatively small amounts of melanin. In such species, a light-colored skin not only reflects unwanted light (and heat) but also furnishes an effective, concealing color pattern.

Thermogenesis. During brooding, one python, *Python moluris,* is effectively endothermic. At these times oxygen consumption can be almost 10-fold the resting metabolic rate (Fig. 7.11). This snake raises its temperature as much as 5°C above the ambient by muscle contraction.[1054,1055] Females coil about the eggs during incubation. Incubation in this species is presumably required: at 30.5°C eggs develop normally, which they do not at 27.5°C.[476,1046] There is also some evidence for thermogenesis in *Python curtus* and *Chondropython viridis.*

The marine turtles seem to maintain a body temperature somewhat above that of the surrounding water. A leatherback turtle *(Dermochelys coriacea),* with a core temperature of about 25°C, taken from water at 7°C in the North Atlantic, suggests some mechanism for endothermy in these large reptiles.[313,727]

Birds

All birds are well insulated, with a layer of dead air held in their plumage. This retards the movement of heat in either direction. The four-chambered heart of birds and their very efficient lungs enable a high rate of metabolism that produces an "excess" of body heat.

Water Balance

Water intake of birds is in three forms: drinking water, water content of moist food and bound water in food. Although presumably all birds will drink when

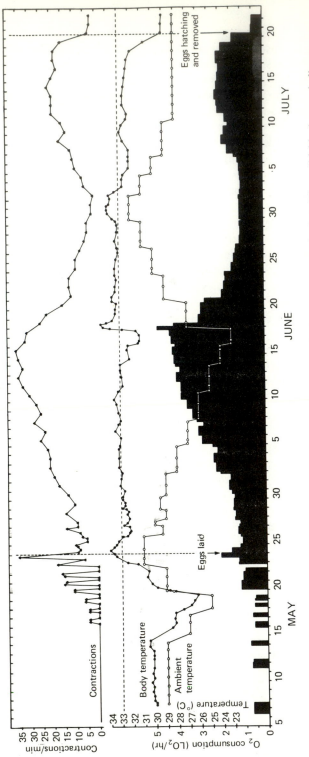

Fig. 7.11. Body temperature of a female *Python molurus bivittatus* before, during and after incubation. The solid black area indicates oxygen consumption of the snake and eggs (Van Mierop and Barnard, 1978).

water is available, insectivorous and carnivorous species can exist indefinitely without drinking water. The common parakeet *(Melopsittacus undulatus)* of Australia can survive long periods without drinking water, but one must remember that the water content of "dry" seeds may be as much as 10 percent. Raptorial birds readily drink water when it is available, but deprived of water they remain strong and healthy indefinitely on the water content of fresh meat.

During migratory flights small birds lose water from both the respiratory tract and the skin, and a part of this depletion is replenished by water produced by the metabolism of fat. One gram of fat when metabolized yields 1.07 g of water.

The avian kidney lacks Henle's loop, and birds are therefore less efficient than are mammals in concentrating materials (including salts) in their urine. (Salt marsh-inhabiting Savannah sparrows, *Passerculus sandwichensis beldingi,* are reported to have loops of Henle.) In some birds that suffer a reduction of fresh water in their diets, there are salt glands in the nasal passages that discharge salt at the external nares.[796] In petrels, or "tube-nosed swimmers," there are small horny tubes leading from the external nares to the tip of the bill, perhaps to prevent the excreted salt from entering the mouth. A single salt gland is composed of finger-like bundles of secretory tubules, each with a central canal. Blood is delivered by the internal carotid artery, and a fine network of capillaries permeates the secretory cells that discharge salt into the lumen of the secretory tubules. Like some desert reptiles, the roadrunner *(Geococcyx californianus)* has a nasal gland. Although it is larger in nestlings than in adults, the latter in the field may have salt encrusted about the external nares. These glands remove a large amount of salt from the blood serum.[778] The importance of salt glands can be observed by the inability of birds to survive when these structures have been surgically removed.

Excretion in birds involves the loss of much less water than it does for a mammal of comparable size. In birds much metabolic waste is passed as uric acid, which is virtually insoluble; therefore urine in birds contains little water. Birds and reptiles have the added advantage of deriving relatively more water from protein metabolism than can mammals, the hydrogen content of uric acid being less than in urea (in mammalian urine). In southern Africa the white-backed mousebird, or coly *(Colius colius),* from arid areas, passes dry feces, whereas two other species (*Colius indicus* and *C. striatus*), from relatively humid coastal regions, pass moist feces.[863]

Thermoregulation

Core body temperatures of birds usually vary from 39° to 42°C, which is slightly higher than those in most small mammals when active; during inactive hours, body temperatures of many small birds decline.

In terrestrial endotherms there is an increase in thermal conductance with a decrease in body weight.[424] Thus an increase in thermal conductance requires a greater expenditure in temperature regulation. By retreating beneath the ground, small mammals escape some of the problems of temperature regulation in cold climates, but small birds must seek other means of defense.

The role of color in thermoregulation is modified by the force of the wind and accompanying convective cooling. Because there is greater penetration of radiation through white plumage (or pelage) than through dark, and because dark plumage is more affected by convective cooling than is light, a white covering for polar species may be thermally advantageous.[1069]

In their behavioral aspects of thermoregulation, birds are somewhat like reptiles: both pant, both may drink water to relieve excess heat, both seek shade and both reduce muscular activity to retard production of metabolic heat. In addition most reptiles and most birds are diurnal and encounter many of the same problems of overheating. Basking in birds is much less important than in reptiles, for birds produce far more metabolic heat than do reptiles. Although amphibians and reptiles may spend long periods completely motionless, most birds (at least diurnal species) seldom rest for long. Frugivorous and insectivorous birds forage during a large part of their daily activity period, and presumably this activity is thermogenic.

Overheating. For some passerine birds, smaller broods experience greater difficulty in preserving body heat and body water. Probably a larger brood, huddling together, loses less heat and water per bird than does a smaller brood. This may explain why the female house sparrow *(Passer domesticus),* for example, spends more time brooding when the number of young is below the mean. At hatching the great reed warbler *(Acrocephalus arundinaceus orientalis)* is more or less ectothermic: when exposed to an ambient temperature of 5°C, its body temperature drops from 35° to 5°C in 30 minutes.[591] By the ninth day (at which time it leaves the nest), however, it has developed a covering of body feathers and is effectively endothermic (Fig. 7.12). Thermoregulatory huddling is also seen among penguins: the downy chicks of the king penguin *(Aptenodytes paragonicus)* may go three months without food, and during this period they cluster together in tight groups.[586] Similarly, incubating emperor penguin *(A. forsteri)* males huddle together for days at a time, when mean minimal air temperatures may be near −50°C.

The higher body temperatures of birds (when compared to mammals) perhaps assist them in coping with the threat of overheating. Many species of birds survive body temperatures of 45°C, a hyperthermia lethal to all mammals, but temperatures above 46°C are dangerous if not fatal. Although panting and gular flutter can eliminate enough water to lower body temperatures in birds by evaporative cooling, this tactic is less important than it is in mammals. An overheated bird, when not in flight, often holds its wings partly outstretched and away from its body, exposing nude skin from which heat can escape, and the membrane-lined air sacs can allow for the escape of body heat; in both these cases such heat loss would be mostly nonevaporative.

Heat Loss. Birds (rock doves, and several hummingbirds and budgerigars, for example) in flight have a metabolic rate roughly 12–14 times the basal rate, and between 10 and 15 percent of the heat dissipated is evaporative water loss; remain-

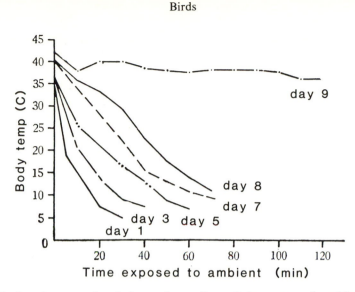

Fig. 7.12. Development of endothermy in nestlings of the great reed warbler *(Acroce-phalus arundinaceus orientalis),* from the first to the ninth day. Lines indicate the decline in body temperature when exposed to 5°C ambient temperature (Li and Liu, 1978).

ing heat loss is convective and radiative, probably mostly from the lining of the respiratory tract and the sparsely feathered underside of the wings.

During its active period many birds fluctuate 2° or 3° about a mean—not a surprising amount considering the small size of most species. Loss of heat may be concomitant with loss of water and, although this is an advantage at high ambient temperatures, it becomes deleterious in extreme cold when heat conservation is necessary. A lower nocturnal temperature (in diurnal birds) with a lowered rate of metabolism would also result in a decreased rate of evaporative cooling.

Conservation of heat is frequently a problem in birds because of (1) small size or (2) daily or seasonal cold periods. Body posture in birds can greatly alter heat conservation. Insulative qualities of plumage can undoubtedly be altered by changing orientation of the contour feathers, but measurement of the efficiency of ptilomotor activities is difficult. Frequently birds elevate their contour feathers and assume a puffy aspect, such behavior being especially characteristic when they are exposed to low ambient temperatures. Effectiveness of plumage as insulation is suggested by its generally greater weight in species from cold climates. The head is apparently a site of possible excessive heat loss; the plumage on the crown is generally thin, and the bill nude. Many birds counter possible heat loss from this region by sleeping with the head buried beneath a wing, with the bill or head covered by contour feathers of the back.[439] By placing its head beneath a wing (as in sleep), a bird reduces heat loss by more than 10 percent. Heat loss is further retarded if a bird draws up a leg close to its body or settles down on a perch so that the featherless legs are covered. Numerous wading birds reduce the degree of heat loss through their unfeathered legs by a counter-current exchange that cools the blood before it enters the foot. Shivering in birds is a frequent means of thermogenesis and is related directly to metabolism and inversely to temperature.

Daily cycles of core temperatures occur in many species, and in some cases the variation is substantial. Small birds in the Arctic with daytime body temperatures above 40°C may experience nocturnal body temperatures of 10° or more below this.[483]

Torpor as a Regulating Mechanism. Some birds experience a marked nocturnal drop in core temperature and may become torpid at night. The rate of such heat loss is in inverse relation to body weight, and also the rate of increase on arousal is slower in large species. In the Cuculiformes, nocturnal hypothermy is reported for the roadrunner *(Geococcyx californianus),* a dweller of American western deserts. At night the roadrunner becomes hypothermic where evening ambient temperatures drop below 10°C; as morning ambient temperatures rise, roadrunners sun themselves.[146] This combination of physiologic and behavioral adaptations allows this species to conserve considerable amounts of energy which would otherwise be needed to maintain a homeothermic body temperature. Such energy savings become extremely important for survival during winter, when productivity of the desert is minimal and food supply low.

Very small birds, such as hummingbirds (not only very small but mostly tropical), have the greatest thermal conductance and solve the problem of thermoregulation by becoming torpid with lowered nocturnal temperatures or low fat stores; in mountains they may find refuge in the stable environment of caves. Unlike hibernating small mammals when torpid, some species of hummingbirds apparently cannot arouse from an ambient temperature below freezing, and lower temperatures are probably an important factor in the geographic distribution of this family. Under duress of reduced foraging and lowered nocturnal temperatures, the broad-tailed hummingbird *(Selasphorus platycercus)* at 2900 m in Colorado may become hypothermic at night. A return to regular temperatures of incubation occurs before dawn.[147]

The West Indian hummingbird *(Eulampis jugularis)* exhibits daily torpor. However, unlike most other torpid small birds and mammals that reduce their body temperatures to slightly above ambient, this species drops its body temperature from 40°C to 18°–20°C, and maintains it at the latter level over an ambient temperature range of 2.5°–18°C. Oxygen consumption increases as ambient temperatures drop below 18°C.[578] The hillstar hummingbird *(Oreotrochilus estella)* regularly roosts in caves at night at elevations of 3800–4300 m in the Peruvian Andes.[154] Nocturnal torpor occurs frequently but is more common in the winter. In hummingbirds that exhibit a nocturnal drop in body temperature, the morning rise occurs well before daylight, suggesting an endogenous cycle.

Regulated body temperature during torpor may be a widespread phenomenon in hummingbirds, having now been reported in five genera (*Panterpe, Eugenes, Selasphorus, Eulampis* and *Oreotrochilus*). Moreover it is correlated with nocturnal ambient temperature, and cannot therefore be regarded only as an emergency device to counteract effects of energy reserves. It may represent a compromise between maximal metabolic savings, which occur in species that do not significantly regulate body temperature during torpor, and high metabolic expenditure, which occurs during arousal from torpor in many mammals. Several species of birds and small mammals fail to emerge from torpor if ambient temperature of the hibernaculum dips below a critical level. The ability of hummingbirds to regulate body temperature during torpor has also been suggested as a compromise adaptation, permitting them to conserve energy at night.

Some birds in cold climates regularly (or occasionally) avoid excessive heat loss by seeking some sort of shelter, a hollow tree being perhaps the most common. Although many kinds of small birds are known to seek refuge in tree holes at night, surprisingly few investigators have studied this aspect of birds' lives. Several boreal species of grouse (Tetraonidae) are known to bury themselves in snow at night, and hunters are sometimes startled by the explosive burst of a ruffed grouse *(Bonasa umbellus)* from beneath a solid cover of snow.

Panting and Gular Flutter. Overheating in birds is countered by an increase in thermal conductance in which they expose poorly feathered surfaces to evaporative cooling and air currents. The absence of sweat glands in birds does not preclude cutaneous evaporative cooling. Evaporative cooling can also occur through the complicated system of lungs and air sacs, either during normal rates of breathing or by panting, but heat loss in the respiratory system of birds occurs mainly on the breathing surfaces and not in the air sacs. Under stress of hyperthermia, panting in birds may be accompanied by an increase in breathing rate without an increase in volume of air, for breathing may (or may not) become shallow during panting. Many kinds of birds counter overheating by gular flutter, a rapid movement of the floor of the mouth (via the hyoid bones) that results in increased heat loss not only by evaporative cooling (from the lining of the mouth) but also from the circulation of blood through the highly vascularized gular skin. It is probably significant that the gular feathering in most birds is sparse. Rapid movement of wings, as occurs in flight in many birds, causes a temporary hyperthermia, but sustained flight does not result in a great loss of body water.

The roadrunner *(Geococcyx californianus)* lives in many hot arid areas of western North America. It resembles the rock dove *(Columba livia)* in its ability to dissipate heat absorbed from the environment; by evaporative cooling both species can maintain body temperatures below the ambient when the latter exceeds 42 °C. In the roadrunner heat is lost by panting and gular flutter.[146]

Other Thermoregulatory Mechanisms. Prevention of hyperthermia may be achieved in many birds by a retreat to a cool shelter; in many cases the same shelter provides aid from overheating as well as from chilling. Some small owls spend the daytime in hollow trees, but the value of such retreats as protection from excessive heat is questionable: on very hot days they may rest, panting, at the entrance of their homes. Potholes in cliffs, on the other hand, may very well provide great protection against solar radiation. Bathing may conceivably lower the core temperature of a bird (by evaporative cooling), but the fact that many birds, including small passerines, bathe in very chilly weather suggests that this may not lower body temperatures. Although there seem to be no measurements of changes in body temperatures following drinking in birds (as there are in reptiles), it must surely be an effective means of thermoregulation.

Storks (Ciconiidae) are sometimes exposed to intense sunlight and to risk of hyperthermia. Under such circumstances they commonly discharge watery excreta on their nude legs. When the ambient temperature is above that of the body temperature, evaporation of watery excreta from the skin can reduce body temperature. As this is an extremely watery release, storks are seldom far from water in hot weather.[519]

In many instances the outstretched wing posture of some birds is probably thermoregulatory. The underside of the wings is usually seminude, and often a bird will open this part of its anatomy to the morning sun. On the other hand, passerines may hold their wings protruding laterally from the body at the same time as they pant. The outstretched wings of vultures and condors in the morning, however, are more probably a means of detecting wind movements prior to soaring.

Mammals

Mammals are provided with hair that, when it forms a dense fur, lowers thermal conductance of the integument. They also have a high rate of metabolism, which usually produces an excess of body heat. Because of the great differences in size and the habitats they occupy, thermoregulation varies greatly from one group to another.

Thermoregulation

In various parts of the bodies of many mammals there are small pieces or "pads" of brown adipose tissue (BAT), which are a primary source of nonshivering thermogenesis.[988] The largest and best-studied is the BAT between the scapulas. Changes in the development and activity of BAT occur in both hibernating and nonhibernating mammals and also in cold-exposed laboratory mammals; BAT pads increase in the autumn in wild mammals and in cold-acclimated laboratory rats, such increase being in both fat and nonfat water-free components of the pad. The thermogenic activity of BAT occurs in response to norepinephrine, which response (in increased metabolism) appears to be in direct proportion to the amount of the nonfatty water-free constituents of BAT and is greater in cold-acclimated animals.[953] Species that lack BAT do not increase nonshivering thermogenesis in response to norepinephrine; this includes birds and certain mammals. Shivering increases heat production by a factor of four to five, such effort coming not only from visible muscular movement but also from muscular tension. At least three nonhibernating cricetid mice, a vole *(Microtus pennsylvanicus)*, the muskrat *(Ondatra zibethica)* and the white-footed mouse *(Peromyscus leucopus)*, experience an autumnal increase in BAT and a decrease in total body lipids. The white-footed mouse also shows an increase in nonshivering thermogenesis in early autumn.[714]

Thermoregulation in Young Mammals. The newborn offspring of many species of mammals have limited thermoregulatory ability: the young of many species are nude at birth and have very little heat production through muscular activity and a relatively high surface-to-mass ratio. The young of many birds and mammals are nude and blind at birth and require brooding by the mother. Such young are called altricial and are typical of passerine birds (and some others), rabbits (but not hares), most carnivores (not including marine forms), marsupials and some others. In some other groups, young are born covered with fur or downy feathers, the eyes are open, and they are capable of walking. This type is called precocial and is seen in gallinaceous birds, ducks, geese and in such mammals as guinea pigs, hares, most artiodactyls, pinnipeds, cetaceans and others. Thermoregulation is much better developed in precocial than in altricial young. There is not a sharp distinction between altricial and precocial young, for there is a broad spectrum of intermediate forms. In altricial young there is apparently a greater tolerance of low body temperatures. Young voles (*Microtus* spp.), for example, may be born in late winter in a nest covered by only a sparse roof of dead grass, and they may be cold to the touch, yet seem to suffer no injury.

Energy demands of altricial young that experience some hypothermia are less than those of precocial young, the metabolic rates of which are comparable to their parents. Although the body temperature of young in utero must be the same as that of the mother, the usually shorter gestation period and occasional hypothermia of altricial young result in a saving in energy over the needs of in utero and postnatal young of precocial species. This generalization would not apply to the extremely altricial young of marsupials, for they are kept warm in the pouch of the mother. With growth of the young, there is a covering of hair and a supply of fat, both components contributing to the organism's ability to regulate core temperatures; in addition most species shiver, and many young mammals possess BAT.

Heat Conservation. In the extremities of many birds and mammals a counter-current exchange arrangement of blood vessels prevents a rapid loss of heat in these structures. When subjected to low ambient temperatures, feet of arctic mammals and marine birds may become extremely cold. In these cases arterioles join venules not only by a capillary network but sometimes also by a direct connection through one or more arteriovenous anastomoses. When the muscular covering of the anastomosis constricts, blood is shunted through the capillaries and circulation slows; when the anastomosis relaxes, muscles in the adjacent venules and arterioles constrict, preventing (or retarding) flow through the capillaries and forcing blood through the anastomosis, thus allowing a relatively rapid movement of warm blood into a chilled foot or ear.

Many mammals under controlled conditions exhibit increased metabolism with decreased temperatures. In addition, under short-term exposure to lowered temperatures, there is (in experimental rats) an increased circulation to both BAT and white fat, intestine and liver, facilitating nonshivering thermogenesis.

The short pelage characteristic of small mammals does not furnish much protection from heat loss; hair of 10 mm or less in length, comprising pelage of no greater thickness, is in itself little protection against extreme cold. Pelage of arctic carnivores, such as bears, may be five or six times this thickness. Many marine mammals have sparse fur and inhabit cold regions. It is not to be expected that such thin pelts provide protection from heat loss, and the temperature of the skins of hair seals may be closer to that of the water than to their core temperature. Insulation in pinnipeds and cetaceans is apparently supplied by the scarcely vascularized subcutaneous layer of fat.

Most microchiropterans have only poorly developed thermoregulatory abilities; they experience daily and annual cycles of inactivity during which their body temperature drops to near-ambient temperature. The subtropical and tropical megachiropterans preserve relatively constant body temperatures. Thermoregulation in bats is affected by their broad, vascularized wing surfaces, which are effective in dissipating heat in flight and conserving heat at rest. At least some fruit bats (Megachiroptera) are true homeotherms, maintaining a rather constant body temperature over a range of ambient temperatures. Smaller megachiropterans encounter greater difficulty than larger species in preserving a constant body temperature at low ambient temperatures. As with other mammals, megachiropterans reduce peripheral circulation and shiver to preserve a constant core temperature; at high ambient temperatures, heat loss is accelerated by peripheral vasodilation.[626] By roosting in clusters, bats can reduce the rate of heat loss, reflecting a reduction of the surface-to-mass ratio (Fig. 7.13).

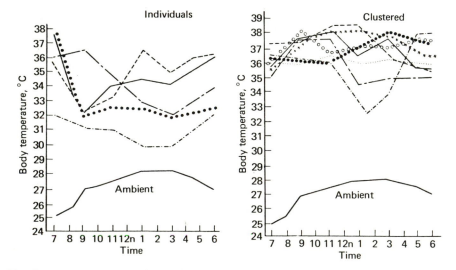

Fig. 7.13. Body temperature of roosting *Glossophaga soricina* individually (left) and clustered (right) during the roosting period (Howell, 1976).

Within a given species local populations may differ in their ability to withstand changes in temperature. The golden spiny mouse *(Acomys russatus),* a murid rodent of the Middle East, lives in deserts of the African Rift valley and at 2600 m in mountains of the Sinai desert. Individuals from the lowland desert cannot survive ambient temperatures below 18°C, whereas the montane individuals are able to maintain a normal body temperature at an ambient of 6°C; when given injections of norepinephrine, montane mice increased oxygen consumption six-fold but the desert mice showed no effect. Thus nonshivering thermogenesis accounts for the ability of a population of the golden spiny mouse to live in a cold environment from which others of the same species are physiologically excluded.[920]

The guanaco *(Lama guanicoe),* a Nearctic camelid, occurs over a great diversity of environments, from very dry to very humid and from sea level to 4500 m in the Andes. The pelage is densely matted, 30 mm on much of the dorsum, and virtually bare on the inner surfaces of the limbs and adjacent flanks (Fig. 7.14). This arrangement enables the animal to rid itself of body heat by conduction and/ or evaporation, at the bare sites or, by covering the nude parts, to utilize the insulative effect of the furred areas.[718]

Allen's Rule and Bergmann's Rule refer to geographic variations in body size that are correlated with climate in birds and mammals. Throughout its geographic range a given species is frequently characterized by a larger body size in colder regions, so that the mass-to-surface ratio increases (Bergmann's Rule) and the size of appendages (especially ears) decreases (Allen's Rule). Thermoregulation may not explain every example used to illustrate Allen's Rule. Bergmann's

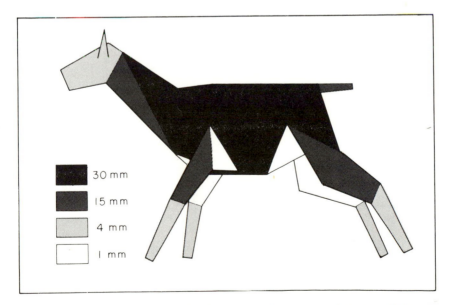

30 mm
15 mm
4 mm
1 mm

Fig. 7.14. Distribution of fur development on the guanaco (Morrison, 1966).

Rule is a useful concept when considered with such other thermoregulatory features as conductance of skin (with its pelage or plumage) and daily periods of activity.

In response to a decrease in temperature, the armadillo *(Dasypus novemcinctus)* shows a dramatic increase in oxygen consumption (Fig. 7.15) and a rise in body temperature, accompanied by a curled posture in which very little soft skin is exposed, and shivering. A vasoconstriction in the region of the dorsal plates at times of lowered temperatures may reduce heat loss in this uninsulated area. This response precludes activity during cold periods, and it is thus a response to brief cold periods (of less than one day) rather than to a cold season; it is the opposite of the lowered body temperature of hibernators.[502]

Facing Excessive Heat. Evaporative cooling in mammals is accomplished by sweating or by panting, which mechanisms cool different parts of the body: sweating reduces the temperature of the surface while the interior cools at a slower rate, while panting results in internal cooling while the surface remains warmer. Panting is probably the major means of loss of heat in mammals, although it does not occur among all groups. It is seen in many but not all marsupials and eutherians, but monotremes retard breathing under the stress of excessive temperatures. Thermoregulation by sweating occurs in many forms of mammals and

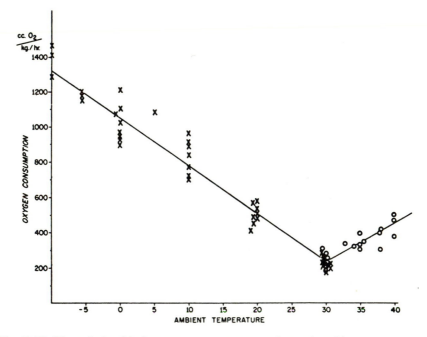

Fig. 7.15. The relationship between oxygen consumption and ambient temperature for the armadillo *(Dasypus novemcinctus)* (Johansen, 1963, from Physiological Zoology, courtesy of the University of Chicago Press).

apparently has developed independently a number of times. Theoretically, with an increase in size panting becomes more efficient as a means of lowering body temperature in mammals, and perhaps sweating then assumes a lesser importance. Sweating is the loss of body water through apocrine glands, the latter of which occur with hair follicles, and some mammals can lower surface temperature through evaporation of such water. These glands are present in most terrestrial mammals, although restricted to certain parts of the body (such as the feet) in some species and absent in many desert mammals. Mammals with few sweat glands generally distributed over the surface of the body, such as swine, may wallow in mud, the evaporation of the water so acquired lowering surface temperatures.

The jackrabbits (*Lepus alleni* and *L. californicus*) of the southwestern United States experience strong sunlight and high temperatures, and regulate their temperature through evaporative cooling and radiation of heat from their very large ears. These hares have a high lethal temperature and can recover from rectal temperatures in excess of 41°C.[891]

Cetaceans neither pant nor sweat and must lose excess heat by other means. In addition to heat loss from the blowhole at surfacing, cetaceans may use their paired fins and dorsal fin to rid their bodies of metabolic heat (Fig. 7.16). Fins of cetaceans, however, have a counter-current exchange arrangement at the base to retard heat loss when there is no excess.[395]

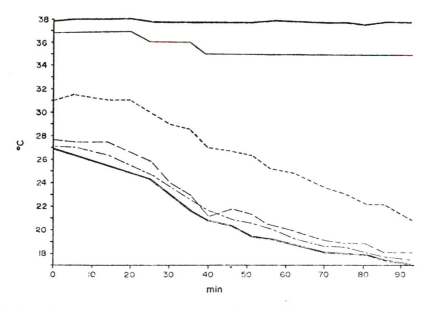

Fig. 7.16. Temperature changes in a spinner dolphin in declining water temperature: top heavy line is the deep core temperature; light solid line, flank subcutaneous; short dash, dorsal fin subcutaneous; long dash, dorsal fin surface; long–short dash, body skin surface; and double line (at bottom), water temperature (McGinnis et al., 1972).

Heat-acclimated mammals, including man, can increase their ability to tolerate high temperatures. In man, acclimation to heat stress involves the degree of exposure, duration of acclimation and amount of exercise during acclimation but seems to be independent of race. Some evidence suggests that women tend to be less able to acclimate to high ambient temperatures than are men; there are possible cultural as well as physiologic explanations.

Water Balance

In mammals the primary source of water is from drinking. However, in a few species of desert dwellers, metabolic water can be substantial and even exceed that from drinking. Most water is generally eliminated as urine, but breathing (including panting) and sweating may draw on water stores, reducing and concentrating urine; a reduction in water drunk results in a reduction of urine output. Urine production is under the control of the antidiuretic hormone and is sensitive to amounts of water in blood plasma.

Although leafy vegetation obviously contains a high percentage of free water, there is also free water in dry seeds on which animals feed. The amount varies with the kind of seed and relative humidity, but commonly it is from 5 to 15 percent. Desert-dwelling vertebrates are best able to survive on only the free and metabolic water in air-dried seeds, but most species, even in deserts, consume water-rich insects and fresh leaves. Some desert plants store large amounts of water: in the New World most cacti (Cactaceae) and in the Old World desert species of Euphorbiaceae and Chenopodiaceae are especially succulent, as are the species of *Aloe,* African plants of the family Liliaceae. Forest birds and mammals face fewer water problems, as most forest foliage has a higher free water content than have the dry and scaly parts of most desert plants. Insectivorous and carnivorous species seldom face water shortages.

In addition to free water in food and metabolic water content in very dry seeds, oxidation water is released in the metabolism of fats and glycogen stored in animal tissue. Oxidation water is frequently an important source of water in hibernating mammals. These animals may suffer some dehydration during dormancy: they invariably urinate on periodic arousal but gain some water from the metabolism of fat.[804]

Sweating can lower the temperature of an animal (especially at the surface) if the secretion and evaporation are both rapid. The secretion rate depends on the amount of water an animal can safely lose, and this amount increases if drinking water is available. Evaporation is greater if the air is hot and dry but is retarded if the air is humid; air movement accelerates evaporation. Secreted water (sweat) is cooled as it evaporates, lowering the temperature of the integument as well as the venous blood before it returns to the interior of the body. Thus, loss of heat is accompanied by loss of water, and the latter is often critical in hot, dry environments. On the other hand, the hotter and drier the environment, the greater the cooling effect of sweating. The rate of evaporation also depends on the temperature of the surface of the skin; evaporation is greater as the body temperature rises.

Water loss from breathing is an acute problem with birds and mammals, and it is remarkable that some species can survive indefinitely without any free water save that from their food. The murid *Leggadina hermannsburgensis* inhabits arid regions of interior Australia, and can subsist on dry food and without water. It produces both highly concentrated urine and partially dry feces. The kangaroo rat (*Dipodomys merriami* and perhaps other species) of Nearctic deserts and the hill kangaroo *(Macropus robustus)* of Australia seem able to maintain weight and lead a normal life without drinking, but both are nocturnal animals. The oryx and the eland, two antelopes of African deserts, survive their hot, dry environments without drinking. They derive free water from foliage, which may be one-third to one-half water by weight.[993]

Mammals feeding on relatively dry foods eat in proportion to their water intake: that is, they reduce feeding when water is scarce. The nocturnal habit prevails among small mammals in deserts, squirrels being the conspicuous exceptions. Not only do mice spend the daylight periods beneath the surface of the ground, but usually the entrance of their burrows is closed by loose sand, soil and bits of vegetation. By confining its activity to night, a mammal not only avoids the daytime heat but also restricts its periods of maximal metabolic activity to periods of cool ambient temperatures.

Bats are especially sensitive to relative humidity, and commonly seek caves with saturated atmospheres. Although metabolism of fat releases water, the high surface-to-mass ratio allows a rapid loss of water. During periodic arousals from hibernation, bats have been seen to lick water from their fur and from the cave walls.

Henle's loop concentrates solutes of urine by counter-current exchange. The longer the loop, the more complete the return of water to the blood and the concentration of solutes in the urine. The most concentrated urine is produced by desert-dwelling mammals, and these are the species with the greatest development of Henle's loop. Water is conserved also in fecal formation in many desert rodents and lagomorphs, but the mechanism is not so apparent. Although it is true that many small herbivores reingest their own feces (coprophagy) and that the redigested fecal pellets are much drier than the initial material, this habit is not especially characteristic of mammals of arid regions, and water conservation in coprophagy is probably incidental to the digestive function.

Large amounts of water are generally not lost in breathing, even in desert animals, as some condensation occurs on nasal passages during exhalation. Nasal passages tend to be cooled by air on inhalation, and the water lost enters the lungs; as air leaves the lungs, it is warmer than the nasal passages, and some of the water condenses on the nasal lining. During panting, the exchange of air is far more rapid, and water loss through breathing is much greater.[890]

Special salt-excreting glands are not known in cetaceans and pinnipeds. Whalebone whales (Mysticeti) and some sea lions possess proportionately very large kidneys and probably produce very large quantities of urine. Mysticeti feed on salt-rich crustaceans, as do many pinnipeds. Toothed whales (Odontoceti) feed on animals other than crustaceans and presumably do not ingest salt with or in their food; kidneys of toothed whales are relatively smaller than those of Mysticeti. All cetaceans and pinnipeds secrete nitrogenous wastes as urea.

A thermolabile desert animal is partly relieved of some of the stress of thermoregulation. If it can tolerate some hyperthermia, this rise in body temperature will reduce the heat flow into it from the outside and the need for evaporative cooling, thus saving water. Reptiles and birds characteristic of deserts are not conspicuously tolerant of high body temperatures. As has been pointed out, although some desert lizards do have a high CTMax, this feature characterizes taxonomic and not environmental groups. Some mammals, however, alleviate the problem of high ambient temperatures by a nonlethal rise in their own core temperatures. The camel *(Camelus dromedarius)* of Eurasia, the jackrabbit *(Lepus alleni)* and the antelope ground squirrel *(Citellus leucurus)* of the southwestern United States all[891,892] tolerate body temperatures above 40°C, and the jackrabbit may experience rises up to 44°C. As is the camel, the eland, Grant's gazelle and the oryx are thermolabile and, under duress of high temperatures, may experience a rise of 6° or 7° in body temperature. The oryx and Grant's gazelle can tolerate core temperatures as high as 45°C, thus avoiding the need to lose water in evaporative cooling; in the cool night air this heat can be lost by convection.

Although the dry air of the deserts makes evaporative cooling very effective, most desert birds and mammals cannot stray far from water holes during the dry season. A few large mammals can exist without drinking and these, such as the oryx *(Oryx beisa),* tolerate body temperatures up to 45°C; this high body temperature reduces the need for evaporative cooling and thus conserves water.[993] Small desert mammals, excepting squirrels, are nocturnal, and escape the dangers of excessive heat and desiccation. The majority of desert mammals produce a very concentrated urine.

Summary

The responses to temperature changes involve the hypothalamus and peripheral neural circuits. Hypothalamic temperature seems to govern such thermoregulatory behavior as erection of plumage in chilled birds and first daily appearance of reptiles in the morning. Humidity variations usually involve temperature changes in the environment and the organism.

Ordinarily fishes do not experience humidity variations, but they do respond to temperature changes by (1) movement to a region of a different temperature or (2) changes in the composition of blood serum.

Amphibians are generally very responsive to changes in both temperature and humidity. Their activity occurs when danger of desiccation is minimal (e.g., at night or in a rainy season), and they often retreat into the ground when the air is dry and warm. Both hibernation and estivation occur in many species.

Reptiles are very sensitive to changes in temperature but do not usually encounter a risk of excessive water loss. Body heat is acquired by basking and/or by contact with a warm surface. Daily body temperatures in some desert lizards may exceed 40°C, whereas some lizards and many snakes are normally active below 30°C, and the tuatara *(Sphenodon)* is active below 20°C. Overheating may be prevented by retreat into an underground burrow, sometimes by immersion in water, and occasionally by drinking cool water. Panting occurs in excessive heat but is not very efficient. Many reptiles hibernate, but few are definitely known to estivate. Only one reptile *(Python molurus)* is known to be effectively endothermic.

Some species of animals differ with respect to the degree of water loss they can withstand. In a variety of reptiles about 75 percent of the body weight consists of water, and some kinds (e.g., *Sauromalus obesus* and *Varanus griseus*) may store extracellular water. The lethal percentage of water loss (or the proportion of the water in a fully hydrated animal needed for life) is the vital limit; water loss below the vital limit leads to death from dehydration. The vital limit is a statistic difficult to determine, but data suggest some adaptive variability among amphibians (e.g., the Australian leptodactylid tadpoles that tolerate a high degree of desiccation). It has been suggested that reptiles differ more in rates of water loss (being less in reptiles in xeric habitats than in those from mesic areas) than in their vital limit. Certainly dermal gas exchange is more extensive among aquatic reptiles, for their skin must be moist for gas exchange to occur. Rates of water loss probably also differ among amphibians: the desert-dwelling *Phyllomedusa sauvagei* is protected against excessive evaporation by a thin film of lipids, and newts in their terrestrial stages have much drier skin than in their aquatic stages.

Birds are very much like reptiles in anatomy, but flight and endothermy enable them to use very different devices for thermoregulation. Plumage retards heat loss in cold air and flight permits escape from hot air. Panting is common in nature, and possibly the complicated system of air sacs assists heat loss. Many birds experience a daily temperature cycle, and a few species hibernate. Some birds seem to indulge in a sort of basking behavior. Generally water balance is not a problem.

Because of the relatively high body temperatures of birds (about 40°C), they can tolerate higher ambient temperatures than can most mammals. That is, at 40°C most mammals activate cooling procedures, whereas there is no temperature gradient between the surrounding air and the body of a bird. Nevertheless, when high atmospheric temperatures do occur, birds are vulnerable. On the rare occasions when ambient temperatures exceed 45°C, small passerines suffer and may perish.

Mammals, like birds, are endothermic and well-insulated. In addition, many mammals possess sweat glands. To escape overheating, small mammals can go to cool, moist underground burrows, and most mammals (especially small kinds) are nocturnal. Sweating effects heat loss as does panting, but both mechanisms entail water loss. Water conservation may be critical and result in responses in urine output. Many small species of mammals hibernate and a very few estivate. Many mammals possess brown adipose tissue (BAT), a highly thermogenic tissue used in raising body temperatures under various sorts of low temperature duress.

Suggested Readings

Chaffee RRJ, Roberts JC (1971). Temperature acclimation in birds and mammals. Ann Rev Physiol 33:155–202

Rose AH (1967) Thermobiology. Academic, New York

Swan H (1974) Thermoregulation and bioenergetics. American Elsevier, New York

Whittow GC (ed) (1970) Comparative physiology of thermoregulation, Vol. 1. Invertebrates and nonmammalian vertebrates. Academic, New York

8. Activity and Seasonal Dormancy

Activity

In natural environments animals are exposed to more-or-less regular and predictable periodic events, such as (1) daily changes in intensity of light, (2) the lunar cycle, (3) tides (reflecting the combined gravitational forces of the sun and moon) and (4) seasonal weather changes. Very few, if any, vertebrates escape the influence of these cycles.

Photoperiod

Although the seasonal cycles of animals had been well known for centuries, William Rowan performed the first controlled experimental studies that indicated an effect of *photoperiod,* or daylength, on gonadal development. In the 1920s Rowan captured juncos *(Junco hyemalis)* in the autumn and maintained them under (1) artificially long days and (2) natural (winter) daylengths. Those exposed to long days showed a rapid and unseasonal growth of gonadal tissue, whereas the control group had small gonads, typical of wild individuals in the winter.[864] Rowan's initial and subsequent studies, although significant in themselves, opened a vast area for experimental research into periodicity and light. Photoperiod as a cue has been explored in thousands of plants and animals since Rowan's first studies; although much has been added to the concept of the function of daylength in biologic rhythms, the precise mechanisms are still not well understood.

There is a regularity to many, if not most, metabolic processes. Energy metabolism, which is the sum of those processes reflecting activity, occurs in cycles, which alternate with periods of rest. This rhythm is *endogenous,* i.e., it is internally regulated and continues in the absence of external factors or cues that an animal normally encounters in nature. For example, a rat in a laboratory environment of constant light (LL), temperature and humidity will have an activity cycle that repeats itself about every 24 hours. A cycle of about one day is a *circadian* rhythm, and one repeated about once a year is a *circannian* or *circannual* rhythm. These endogenous patterns or rhythms persist in the absence of an *exogenous* cue, or *Zeitgeber* (time-giver). A circadian or circannian rhythm in a constant environment is called a *free-running rhythm* (FRR). In nature a series of events may be repeated every 24 hours or 365 days, and the ultimate Zeitgeber is the regular cycle of solar light. Some cycles are adjusted to tides or moonlight and follow the 28-day lunar cycle. An FFR of, say, 24.7 hours would soon be out of phase with the natural solar cycle if it were not reset by such regular events as sunrise and sunset. Circadian rhythms are usually measured by some sort of

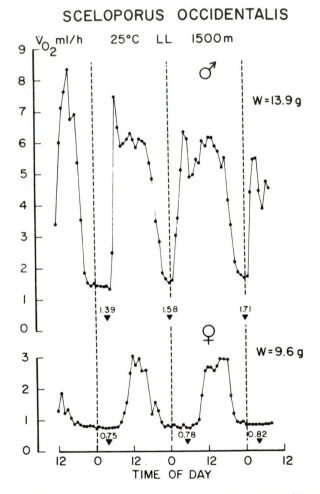

Fig. 8.1. Free-running rhythm of oxygen-consumption in the iguanid lizard *(Sceloporus occidentalis)*, showing distinct peaks and lows under constant temperature and light (Jameson et al., 1977).

activity device recording bodily movement, but rhythms are also seen in oxygen consumption (Fig. 8.1), body temperature (Fig. 8.2) and many specific physiologic events.[430]

In nature, daylength changes daily by slight increments except at the equator and, as a result, there are not only changes in the daily activity cycle but gradual physiologic changes within the animal. These changes in daylength are greater away from the equator (Fig. 8.3), and migratory animals may experience greater photoperiodic changes than do sedentary animals. Many vertebrates respond to

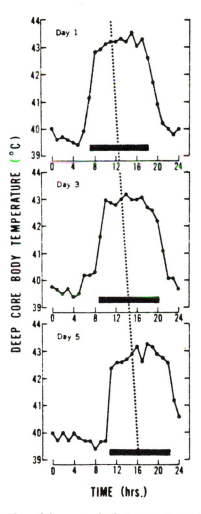

Fig. 8.2. Free-running rhythm of deep-core body temperature of a starling *(Sturnus vulgaris)*. Solid line at bottom indicates duration of perch-hopping activity, and dotted line shows the non-24-hour nature of the free-running rhythm (Rutledge, 1974; courtesy of Academic Press).

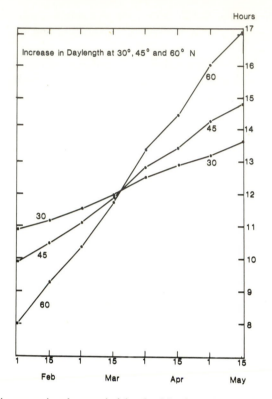

Fig. 8.3. Vernal increase in photoperiod in the Northern Hemisphere. As a bird flies north after mid-March, it is exposed to rapidly increasing photoperiod.

such changes in photoperiod by gonadal development and consequent fluctuations in hormonal levels. Such variations in the individual's internal environment may in turn cause changes in the free-running rhythm. A late winter increase in daylight, for example, may stimulate gonadal growth, and a normally diurnal species, such as a sparrow, may have an extended activity phase, including most of the night. This is called *Zugunruhe* or nocturnal restlessness, and characterizes many small passarine birds during migration (Fig. 8.4). Clearly the very slight changes in daylight (photoperiod) can have profound effects on reproductive and migratory behavior, and in this way a slight change in the light schedule may indirectly produce great alterations in activity. Nevertheless, regardless of the manner in which circadian rhythms are altered—either experimentally or in nature—they remain cycles that are repeated about every 24 hours. All that changes is the shape or pattern of the cycle within 24 hours.

Bats. Many vespertilionid bats are known to forage at dusk and again at dawn, resting in a nocturnal roost between these times. The pallid bat *(Antrozous pal-*

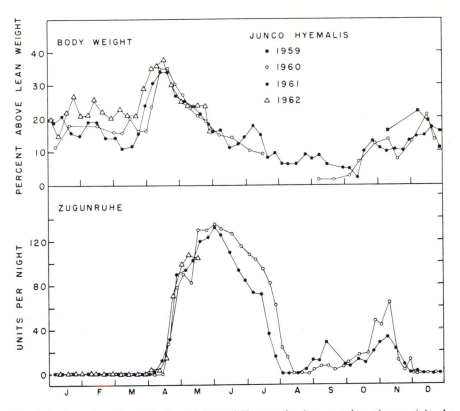

Fig. 8.4. Annual cycles of body weight and Zugunruhe (nocturnal restlessness) in the slate-colored junco *(Junco hyemalis)* (Weise, 1963).

lidus) varies its emergence with the seasonal changes in sunset, but the bimodal activity pattern persists from spring to autumn (Fig. 8.5). Nocturnal emergence of the Japanese pipistrelle *(Pipistrellus abramus)* is also closely tied to the time of sunset, but occurs earlier on cloudy evenings (Fig. 8.6) and is delayed when the temperature falls below 20°C. Lactation interrupts this pattern, but females resume their regular activity period after young are weaned. In Kyushu, Japan, the long-fingered bat *(Miniopterus schreibersi fuliginosus)* follows a daily activity cycle apparently adjusted to sunrise and sunset at any given season. At the approach of sunset it leaves the depths of the daytime retreat, moves to the twilight zone of the cave and emerges shortly after sunset (Fig. 8.7). The first movement seems to be endogenous and the second exogenous, a pattern that may characterize many cavernicolous bats.

During warm months most bats are clearly nocturnal, but during hibernation the discrete circadian rhythm breaks down and is not synchronized with the solar cycle, only to become reestablished in the spring (Fig. 8.8). The free-running cycle may exceed 24 hours in some bats in winter and be less than 24 hours in summer.

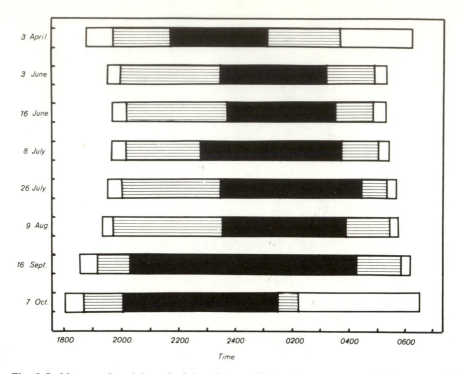

Fig. 8.5. Nocturnal activity schedules of the pallid bat *(Antrozous pallidus)*. Total span of horizontal bars indicate time from sunset to sunrise at several times of the year. Lined areas are times spent in foraging, and solid bars are night roosting (O'Shea and Vaughn, 1977).

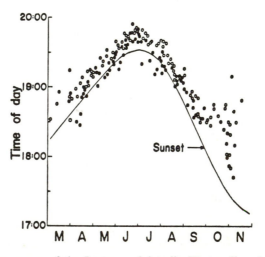

Fig. 8.6. Initial emergence of the Japanese pipistrelle *(Pipistrellus abramus)* as related to sunset (open circles indicate clear weather and solid circles, cloudy weather) for latitude 35°N, Japan (Funakoshi and Uchida, 1978).

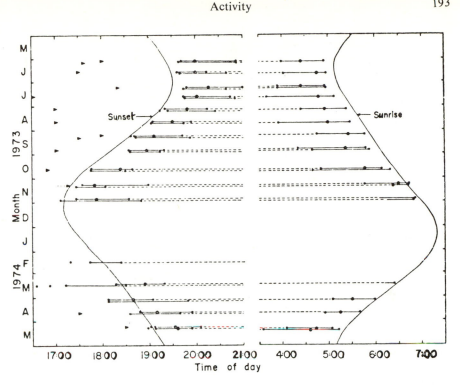

Fig. 8.7. Seasonal transitions in times of departure and return of *Miniopterus schreibersi fuliginosus* in Kyushu, Japan. With the approach of sunset the bats move from the depths of the cave to near the entrance (triangles at left), but most movement from and return to the cave occurs between sunset and sunrise (Funakoshi and Uchida, 1975).

During hibernation some bats (*Plecotus* spp.) forage and feed on occasion of periodic arousal, whereas others (*Myotis* spp.) apparently do not.

Reproductive Activities

All the critical environmental variables that regulate the timing of the reproductive cycle, hibernation or migration adjust to an endogenous schedule of events. Exogenous cues release a previously programmed routine so that courtship, brooding, care of the young and all other reproductive events occur at a logical time with respect to each other and to important environmental changes. In a constant environment with no change in photoperiod, temperature, humidity, sound or any other Zeitgeber, the endogenous rhythm persists, whether circadian or circannian.

A species in which long days stimulate gonadal growth is frequently exposed to long days after the breeding season, but the individual after the annual period of reproduction is unresponsive, or *refractory,* to such stimuli. Boreal birds that have finished breeding in the extremely long summer days of the Arctic may migrate to the Southern Hemisphere, where daylight is also long. While in their

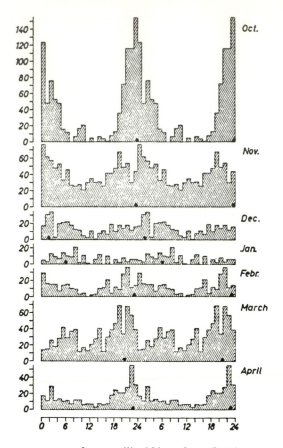

Fig. 8.8. Intracave movements of vespertilionid bats from October to April. Movements occur in the winter, but there is no discrete circadian rhythm during the period of hibernation. The bars repeat a full days record of movements (Daan, 1973).

winter home, they are refractory to long days but are responsive by the following spring or late winter. Photorefractoriness is common among many birds and mammals and sometimes resembles diapause of insects, in that a short-day regimen is necessary before the individual is again photosensitive.

In temperature and polar regions spring and summer become briefer, and photoperiod is more critical as a cue at higher latitudes. Also, annual cycles of temperature and plant growth are rather regular at high latitudes in contrast to tropical and subtropical deserts. Desert birds and mammals respond to seasonal changes in photoperiod: there may be gonadal growth, but the balance of the reproductive cycle (e.g., nesting and mating) may remain in abeyance until rain brings forth a flush of herbs and plant-feeding insects. In lower latitudes not only birds and mammals but also ectotherms respond to the effects of irregular or unseasonal rains.

One school of thought believes circadian rhythms to be essentially innate, endogenous and genetically determined; an opposing point of view, however, holds that exogenous, environmental cues or signals control circadian rhythms of various sorts. The proof of either proposal will be extremely difficult. It is most comfortable today to assume that innate patterns are modified by environmental cues. The importance of light as a cue has been demonstrated hundreds of times in a great diversity of animals, from protozoans to mammals. The duration, amplitude and wavelength of light are easily manipulated, and such experiments are logically favorite ones. The role of light is clearly operable in many vertebrates deprived of specific light-sensitive receptors, eyes and the parietal eye. Extra-optical photo-reception is a phenomenon not well understood, but it does point out the difficulty of separating light from heat produced when light hits the skin. Although students of cycles generally rank temperature as being much less important than light, temperature is less easily evaluated, and obviously affects rates of enzymatic reactions in ectotherms.

Although many students believe that the circannual rhythms of hibernators are immutable and unaffected by photoperiod, the woodchuck *(Marmota monax)*, an excellent hibernator of northeastern North America, altered its circannual rhythm two years after being moved to an external environment in Australia.[218] Such a shift suggests that light is effective as a Zeitgeber for at least some hibernators.

Timing. A daily or circadian rhythm can be adjusted or *entrained* by external cues. Both onset and duration of a cue modify phase-timing; experimentally animals accommodate quickly to such drastic events as a single 24-hour day or night, followed by a reversed 24-hour schedule or sudden changes in temperature. Generally nocturnal species, those normally active at night, are regulated by the end of light and beginning of darkness, and diurnal animals have an activity phase entrained by the onset of a light period. In nature a daily cycle in ambient temperature tends to parallel that of solar radiation, and an animal may respond to a change in temperature as well as a change in light (Fig. 8.9).

Endogenous timing, by a biological clock, fills several sorts of needs, and quite possibly the timing mechanisms or centers affecting migration may not be the same as those influencing reproduction or hibernation. In regulating such events, slight variations are tolerable, and these processes may well be modified by exogenous factors. Although the "biological clock" itself has never been identified, its presence is apparent from many daily or annual events that continue in a more-or-less predictable rhythm, even in the absence of environmental cues.

Endogenous circannual rhythms in Old World warblers (Sylviidae) involve molt, migratory restlessness (Zugunruhe), gonadal growth or recrudescence and adiposity; these cycles are probably widespread among passerine birds but undoubtedly vary among species and, intraspecifically, among geographic races.[383]

Bimodal Patterns. Some vertebrates have bimodal circadian activity patterns. Animals that have peaks of activity at dawn and dusk are called *crepuscular.*

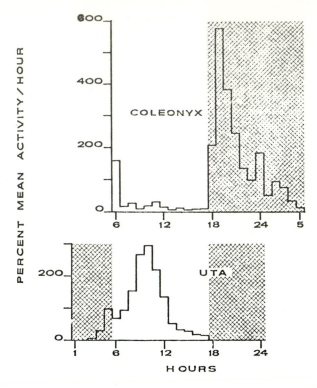

Fig. 8.9. Activity of a nocturnal gecko *(Coleonyx)* and a diurnal iguanid lizard *(Uta)* (Evans, 1966; courtesy of Pergamon Press).

Many voles (Microtinae) are more-or-less crepuscular, and the rough-legged hawk *(Buteo lagopus)* and the hen harrier *(Circus cyaneus),* two hawks that favor these rodents, hunt throughout the day but often pursue microtine rodents until well past sunset.

Some circadian patterns are seasonally bimodal. Seasonal activity shifts occur in the fresh-water sculpins, *Cottus gobio* and *C. poecilopus,* of the Swedish Arctic.[18] In the summer they are nocturnal, but with decreasing daylengths from late September through November they become totally diurnal, and during this transition they become bimodal (Fig. 8.10). Others, such as sun-loving species of lizards, have a midday peak in spring and summer, but in summer are active in the morning and late afternoon, with a midday lull (Fig. 8.11).[661,818] Such patterns may persist under constant light and temperature in the laboratory. The woodchuck *(Marmota monax),* a large ground squirrel of eastern North America, has a unimodal peak of activity in spring and autumn and a bimodal (early morning and evening) rhythm in summer.[413]

Lunar and Tidal Cycles. Lunar (or semilunar) cycles occur in several fishes that spawn at the high tide mark. The California grunion *(Leuresthes tenuis)* and a

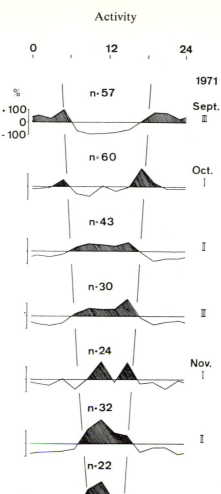

Fig. 8.10. Circadian activity of a fresh-water sculpin *(Cottus gobio)* under natural light in northern Sweden. Converging lines mark the 5-lux limits (Andraesson, 1973; courtesy of Oikos).

southern "trout" *(Galaxias attenuatus)* deposit eggs at high water; these eggs mature roughly two weeks later, and the fry are washed to sea at the next tide. The Zeitgeber in such cases is unknown, and these semilunar rhythms occur only during the reproductive seasons. Tidal rhythms of a blenny *(Blennius pholis)*, a tidepool fish, persist in constant temperature and light with no evidence of diurnality to the pattern (Fig. 8.12). Activity peaks occur with high tides, and, experimentally, hydrostatic pressure can entrain a nonphasic individual.[340] Inasmuch as these fish lack an air bladder, the mechanism for measuring pressure is unknown.

The mudskipper in Africa *(Periophthalmus sobrinus)* by choice spends most of its time out of water, and movements are correlated with tidal changes.[350] Under

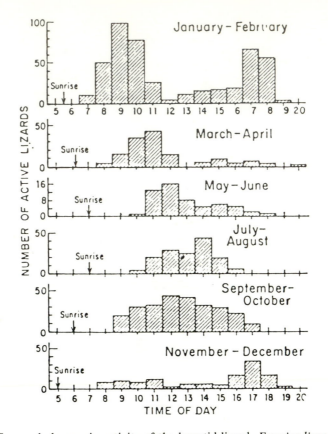

Fig. 8.11. Seasonal changes in activity of the lacertid lizard, *Eremias lineo-ocellata,* in the Kalahari Desert of southern Africa, with a bimodal activity pattern in the summer (January–February) (Pianka, 1977; courtesy of Academic Press).

experimental LD schedules, the activity patterns of some tidal fishes become a bimodal, crepuscular rhythm. It appears that some (perhaps many) tidal fishes are affected by both a tidal Zeitgeber and the day–night schedule, but in nature a Zeitgeber associated with tides is dominant.[341]

Internal Rhythms

Physiological events recur regularly every 24 hours and may be affected by external cues and in turn influence the activity of the individual. Liver glycogen and liver fat exhibit circadian cycles in some birds and mammals, such variations being partly allied to and partly independent of feeding. Many parasites in blood of vertebrates appear in the peripheral circulation at night or during the day, depending on whether the vector is nocturnal or diurnal. Cycles, such as shown by microfilaria or *Plasmodium* in vertebrates, are predicated on internal cyclical events in the host. Microfilaria in primates can have their rhythm obliterated by

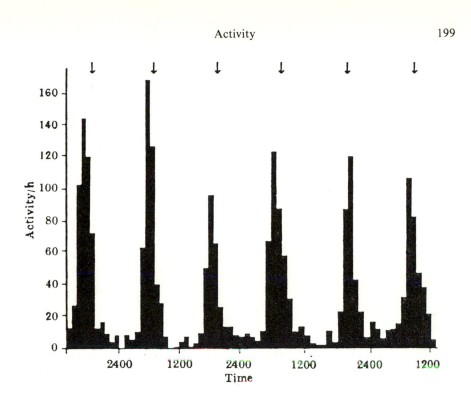

FIg. 8.12. Free-running rhythm of a blenny *(Blennius pholis),* illustrating the tidal pattern. The arrows indicate the time of high water in the area from which the blenny was taken (Gibson, 1965).

experimentally manipulating the host's body temperature. In the final analysis, bodily activity (i.e., locomotion) is related to the drive for sleep. The latter, whatever its control mechanism, is clearly an internal circadian rhythm and related to circadian rhythms in body temperature, blood pressure and oxygen consumption.

In the absence of any environmental cues, some rhythms persist. The heart beat continues, more or less regularly, according to the species and its physiologic state, and the rhythmic pulsation continues when the heart is removed and preserved in a nutrient solution. Frequently superimposed on the beating of the heart is a daily change in the rate of beating. The human menstrual cycle recurs about 28 days, varying slightly with age and race. Although numerous other cycles are known, their underlying causes are frequently not.

Sensory Receptive Structures

Most external and internal stimuli are received and mediated in several small areas in the head. On the top of some reptilian brains, just below the skull, is the parietal eye, a median photosensitive capsule with a lens and retina. A slender nerve from the parietal eye runs along the surface of the pineal body below. The latter lies on top of the diencephalon and influences metabolism via the hypothalamus and hypophysis (or pituitary gland); it also secretes a hormone called melatonin, which stimulates the expansion of melanophores, and depresses reproductive activity.

The hypothalamus is on the ventro median aspect of the diencephalon, above the optic chiasma, and it has a neural connection with the hypophysis, which lies below. Among the major known roles of the hypothalamus are: regulation of water balance, thermoregulation and control of appetite. Much circumstantial evidence points to the hypothalamus as an important organ in the control of activity of several sorts. It not only receives some innervation from the optic tracts but it is very close to the pineal body and parietal eye. In turn there are direct circulatory connections from the hypothalamus to the anterior lobe of the pituitary. Thus the hypothalamus responds to external light stimuli and mediates internal stimuli.

The hypophysis, or pituitary gland, lies below and posterior to the optic chiasma, below the hypothalamus, and projects ventrally from the floor of the brain. It consists of three parts or lobes, all of which are endocrine glands with abundant neural tissue. The pars distalis (or anterior lobe) produces six hormones that affect sexual activity, growth and metabolic rate. The pars intermedia (or intermediate lobe, absent in birds) produces the hormone intermedin. The pars nervosa (or posterior lobe) produces oxytocin and the antidiuretic hormone. In some cases, the secretions of the hypothalamus cause the hypophyseal release of hormones.

For many years there has been evidence for more than a single clock system: laboratory rats exhibit distinct rhythms for eating and exercising, but manipulated photoperiods show that these activities can occur separately. In human subjects kept under 21- and 27-hour days, "daily" cycles of body temperature became adjusted to the experimental regimens, but circadian cycles of excretion of water, chloride and potassium became dissociated, i.e., they do not occur in the same sequence as they do under a 24-hour day.

There is a role for both the parietal eye and the pineal organ in controlling rhythms in fish, amphibians and reptiles.[1039] Both structures are photosensitive and, although they are close together, they may act independently. *Sceloporus olivaceus,* a small iguanid lizard of Texas, is a sun-loving diurnal animal with both a well-developed parietal eye and a pineal body. Although parietalectomy did not affect the FRR in an LL schedule, subsequent pinealectomy resulted in either arrhythmicity or the separation of the activity periods into two separate stages. Because the nerve to the parietal eye lies on the pineal body, it is not possible to remove the latter without also destroying the parietal eye. The pineal body in this reptile seems to function as a mediator for two (or more) oscillators.[280] The two activity peaks in these pinealectomized animals may be related to the bimodal activity patterns seen in crepuscular or in diurnal animals that show a bimodal (early morning and late afternoon) rhythm in midsummer and a single activity peak in spring and autumn. Periodic changes in light schedules entrain the activity of the newt *(Notophthalmus viridescens).* Experimentally they continued to follow the light cues when blinded but, when blinded newts were pinealectomized, activity was dissociated from the light regimen.[229] Pinealectomy also eliminates Zugunruhe in birds. Experimental evidence points to the hypothalamus as the controlling or mediating structure in mammals.

Among coastal and coral-reef fishes, those that feed on small crustacea, mollusks and plankton are mostly diurnal, whereas predatory fishes are nocturnal. Primarily predatory fishes (e.g., Serranidae, Lutjanidae and Carangidae) forage at night and rest in cover during the day, whereas herbivorous and omnivorous species (e.g., Labridae, Scaridae and Chaetodontidae) are essentially diurnal. The change in activity is apparently predicated on the intensity of light. Between the time diurnal species retire and nocturnal fishes become active, around dusk, there is a quiet period. Possibly the hiatus represents the time needed for the photomechanical changes for scotopic vision of the nocturnal predators.[440]

Seasonal Dormancy

Seasonal dormancy is one of the more important endogenous physiologic adaptations. It occurs most frequently in winter, when deep snow and low temperatures severely restrict foraging, and is therefore referred to as hibernation, a modified physiologic state during which body temperature is reduced and metabolic rate slow. Many small mammals and some birds show a diurnal periodicity characterized by reduced body temperature and metabolism, and this phenomenon may be an approach to seasonal torpidity. Although hibernation can be regarded as an adaptation to climatic severity, the onset generally precedes adverse weather, i.e., most hibernators are somehow conditioned to retire early enough to avoid cold weather.

Physiologically there are some basic differences in hibernation between warm-blooded vertebrates (endotherms) and the so-called cold-blooded groups (ectotherms). Even among homeotherms there are not two discrete groups, hibernators and nonhibernators, for there are many intermediates. Many mammals do not maintain a constant internal body temperature. Fluctuations may be seasonal or daily among some monotremes, marsupials, sloths, bats, rodents and possibly carnivores. The mammalian hibernator produces its own body heat as a result of a relatively high rate of metabolism. Consequently hibernation requires some physiologic change that will retard metabolism sufficiently to lower body temperature. In contrast, the ectothermous hibernator (fish, amphibian and reptile) has a metabolic rate that is not normally sufficient to maintain its body at a constant temperature. Hibernation in these two groups must therefore be considered separately.

When seasonal torpidity starts at times of aridity or high temperatures, the process is called estivation. The distinctions between estivation and hibernation become less real when we note that many hibernators, such as adult ground squirrels (*Citellus* spp.), may become inactive and dormant in midsummer, although subadults may not begin hibernation until autumn. In still other cases, both adults and young may become inactive in early summer and subsequently "estivate" or "hibernate" until late winter of the following year. Such a cycle is seen in *Citellus beldingi, C. tereticaudus* and *C. columbianus* of western North America, where many grasses and forbs flourish with the winter rains but wither in the dry heat of summer. These species, at least in the case of old males, begin hibernation (or estivation) in early July and emerge in mid-March. Similarly early entrance into torpidity occurs in *Citellus fulvus* and *C. pygmaeus* of the southeastern Soviet Union, and is apparently an adjustment to a seasonal flourish of forbs.

Speculation on the advantages and disadvantages of dormancy is made possible by noting the regions and environments in which torpid animals occur, as well as the species that do and do not hibernate. However, one must be careful about making generalizations, since, even in a single area, closely related species may have very different annual cycles. In southeastern California, *Citellus leucurus* does not hibernate, but the sympatric *C. mohavensis* is torpid from August to March.[65]

Temperate regions, in contrast to the Arctic and Tropics, have the greatest variety of seasonally torpid vertebrates. Middle latitudes have not only alternating seasons of heat and cold not found in the Tropics but also many areas with cyclic or occasional dry periods, and even deserts, not found in the Arctic. Torpidity occurs in many temperate terrestrial and fresh-water vertebrates. Hibernating birds have been found in recent years, as well as a few species of hibernating fishes and some that endure dry periods in estivation out of water.

In their escape from a seasonally hostile environment, hibernators seek sites (hibernacula) that are somewhat warmer or more stable than ambient surface temperatures. If the temperature of the hibernaculum drops much below freezing, the hibernator may compensate with an increase in metabolism although remaining torpid, or it may die, or it may arouse and move its quarters. For this reason in regions of permafrost, in some parts of the Arctic, no hibernating mammals and reptiles exist, although there are species of amphibians and fishes that may find their winter retreat in unfrozen mud at the bottom of ponds.

In very low latitudes, with no conspicuously hostile temperature changes, seasonal dormancy is most characteristic of those fish, amphibians and mammals that must escape periodic aridity to compensate for excessive water loss. For this reason seasonal torpidity in tropical animals is often found in desert dwellers and is coupled with other adaptations to counter water loss. These include well-developed thin loops of Henle and—in fish and amphibians—hygroscopic skin, reduction of mucous glands in the skin and the seasonal formation of a cocoon.

Fishes

Seasonal torpidity occurs in both fresh-water and marine fishes. Estivation occurs in seasonally arid environments, and hibernation, insofar as is known, occurs in regions of very low temperatures.

Estivation in Fishes. Estivation in fishes probably evolved concurrently with aerial respiration, since accessory respiratory organs are needed to allow a fish to endure extended periods out of water. Air-breathing fishes characteristically inhabit shallow stagnant ponds with high temperatures and low oxygen content. These conditions most often occur in the Tropics. Estivation is unknown in marine fishes and in those that typically dwell in rivers or deep lakes. Thus estivation in fishes has primarily evolved in Africa, South America, Australia, southeast Asia and only rarely in temperate climates. This adaptive response occurs among unrelated taxa, and the respiratory modifications are of several types. In lungfish, lungs are well-developed and vascularized, and in some gobies the gill chambers are lined with capillaries. Others may breathe through the skin.

Lungfish. Lungfish (Dipnoi) are represented by three living genera: three species of *Protopterus* dwell in ponds and lakes in Africa; *Lepidosiren paradoxa* occurs in the drainages of the Amazon and North Chaco Rivers in South America; *Neoceratodus forsteri* lives in the Burnett River in Queensland, Australia, and does not estivate.[48] Torpidity (estivation) has been investigated in the African and South American species. Both genera have paired lungs and reduced gills, and require atomospheric oxygen for respiration. They inhabit swamps within river systems (not the rivers themselves) and may also occur in lakes (e.g., Lake Victoria). These swamps usually dry out annually during rainless periods, and under such conditions species of *Protopterus* and *Lepidosiren* burrow two or three feet into mud. Species of *Protopterus* become torpid, but *Lepidosiren paradoxa* apparently remains active and thus technically cannot be said to estivate.

As water in the swamps gradually disappears during rainless seasons, African lungfish *(Protopterus)* become isolated in small pools and burrow into the muddy bottom (Fig. 8.13). Species of *Protopterus* that inhabit areas that do not dry up annually, such as Lake Victoria, fail to estivate. It is apparent that *P. annectens* does not seek to avoid the approaching dry spell by entering adjacent rivers, as do other fish, but instead prepares its estivation quarters as soon as the water becomes shallow.

With hardening of the soil, a mucous film dries on the fish and creates a watertight shell or cocoon that further protects it from desiccation. Oxygen is obtained through a small opening at the anterior end of the cocoon.[345] Involution of the anterior hypophysis effects mucus secretion. Extracts from brains of torpid *P. aethiopicus* caused a 33 percent decline in metabolism when injected into rats.[983]

During estivation metabolism is retarded and no food eaten, and metabolic wastes are not discharged. Fat reserves are low and most energy comes from metabolism of protein. As protein is consumed, the estivating fish gradually becomes emaciated and its metabolic needs diminish. The remaining protein lasts longer as oxygen consumption is reduced. No great changes in concentrations of tissue glycogen or fat occur during estivation, but carbohydrate reserves increase; urea from amino acid metabolism accumulates in muscle and is excreted rapidly over the first two or three days after emergence. Temperature variations are undoubtedly slight, two to three feet below the surface of the dried swamp, yet oxygen consumption increases considerably if ambient temperatures are experimentally raised to 30°C.[497,944-946]

Additions of water reactivate the fish and terminate estivation as a result of asphyxiation. At the end of the dry period, water covers the lid and prevents entrance of air. This prevents the lungfish from breathing. Arousal and return to normal physiologic activities do not immediately follow emergence from the sleep-

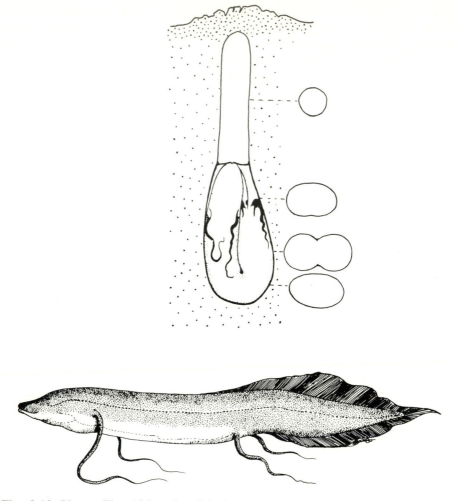

Fig. 8.13. Upper: The African lungfish *(Protoptreus annectens)* burrowing into the muddy bottom of swampy waters as they dry up (Johnels and Svensson, 1954). Lower: Lungfish in active position (Suzanne Black).

ing nest.[505] About one hour after submergence, the fish first comes to the surface for air but immediately retreats to its nest chamber. Freshly aroused lungfish breathe at 3–5 min intervals and gradually return to the normal rate of once every 10–20 min.[505] They finally abandon their nests six or seven hours after awakening.[497]

The South American lungfish *Lepidosiren paradoxa* is also a swamp dweller and fashions a retreat to which it retires during the dry winter. The sleeping nest is similar to that of *Protopterus annectens,* but some water seeps into the expanded chamber, two to three feet below the surface. The top is covered by a porous lid that admits air. A cocoon is not formed and the fish is not torpid: both vocalizations and sound made by body movements indicate the site of these lungfish to aborigines. Emergence follows spring rains, and the pattern is similar to that described for the African lungfish.[157,158] Both *Protopterus* and *Lepidosiren* respond to the same environmental conditions; the difference lies in cocoon formation and torpidity in *Protopterus*.

Estivation in Other Fishes. The eel-like *Monopterus marmoratus* inhabits the same swamps and estivates in burrows similar to those of *Lepidosiren*.[157] The meager fauna of fishes of the fresh waters of Australia has for ages been subjected to the most vigorous selective influences. This distinctive fauna has evolved from a few mostly marine progenitors under periodically arid conditions, and most naturally estivation has become a well-developed feature in the biology of some Australian fishes. These species are known from anecdotal accounts. Australian species of mudtrout (Galaxiidae) burrow into mud during dry periods. The eel mudtrout *(Saxilaga anguilliformis)* has been found buried in mud, and another species *(S. cleaveri)* was found torpid in a cavity in the base of a stump. An Australian percoid fish (*Madigania unicolor;* Theraponidae) is also known to estivate. In New Zealand the galaxiid genus *Neochanna* occurs in low-lying fresh waters that undergo seasonal drying. These fish estivate in cavities in the mud for prolonged periods. The brown mudfish *(Neochanna apoda)*, a galaxiid, is also known to survive droughts, buried in the mud.[672] A Neotropical galaxiid, *Brachygalaxias bullocki,* lives in fresh-water streams in Chile. Immature specimens appear in the spring (October), and by February and March all individuals are adult; as winter approaches, they disappear. Apparently the species overwinters in the egg stage.[886]

The bowfin, *Amia calva,* of fresh waters in North America, has a well-developed, cellular air bladder and is known to gulp air, especially in oxygen-poor waters. It is reputedly an estivator and has been found buried under four inches of hard dry soil. The eight-inch spherical chamber was moist, but the fish was a quarter of a mile from the nearest surface water, the Savannah River in Georgia. The fish was discovered by its movements and heard at the surface, and was obviously not torpid at that time. Debris on nearby trees indicated that the area had been flooded.[755] In Louisana, bowfins sometimes become stranded in low-lying fields and encase themselves in chambers in hard, dry soil; they are often plowed up during cultivation. The Nearctic mudminnow *(Umbra limi)* likewise uses its air bladder for respiration and is remarkably resistant to desiccation. During a drought it sometimes burrows into the mud and estivates, but the degree of torpidity and details of its estivation are not known.[1018]

Several species of shallow-pond fishes of India are subjected to seasonal aridity and estivate regularly. Respiratory modifications of these fishes enable them to utilize atmospheric oxygen and thus survive long droughts in the semidry mud. They characteristically pass the summer months in a torpid condition when the dry air is from 85° to 110°F.[215,216]

Gobies (Gobiidae) frequently live in tidal mud flats and construct small burrows to which they retreat at low tide. Some spend hours out of the water, either foraging over rocks and seaweed at low tide or hiding in burrows in the mud. Although gobies lack a swim bladder, their opercular chambers are highly vascularized, and some species remain out of water for several hours at low tides, or for several months during rainless periods. In San Francisco Bay, the long-jawed goby *(Gillichthys mirabilis)* may seek refuge in holes in creek banks during low tide, its retreat being low enough to catch seepage from the surrounding mud.[611] Such behavior can scarcely be considered estivation, but it is a step toward the genuine xeric torpidity seen in an Indian goby *(Pseudapocryptes lanceolatus)* of the Ganges delta. This goby, like the unrelated lungfish *(Protopterus annectens),* burrows head downward in soft mud to a maximum depth of six feet, where a small amount of ground water prevents desiccation. This soil, being high in colloidal materials, retains moisture so that estivating fish remains moist while the ground surface is hot and dry.[452] Such burrows provide protection not only from desiccation but from excessive heat. At a depth of 12 inches pond water was found to be 35°C, but 12 inches in the burrow the temperature was 26.5°C. There the fish remains for many weeks, until the rainy season restores the aquatic environment.[215,452] Thus within the family Gobiidae there is a transition from species that burrow into mud during low tide to others that burrow deeply into firm clay and withstand seasonal droughts in a torpid state.

Other fishes have evolved different mechanisms that allow them to invade habitats which are only seasonally habitable. Pearlfish or toothcarp (Cyprinodontidae), of the genera *Cynolebias* and *Simpsonichthys,* live in temporary ponds in southern South America. As the surface water disappears, the adults perish and the entire population is maintained by the dormant embryonated eggs, which are buried about half an inch in moist soil. Drying is usually essential to development of the embryo, and hatching rarely occurs in eggs not subjected to desiccation. Eggs of many species of *Cynolebias* normally pass from three to six months in moist soil without benefit of rains. The eggs mature during the rainless season, during which temperatures range from near 0° to above 30°C, with a wide pH range. Young fish emerge in small numbers following flooding of the land. Time of hatching varies, some eggs of *C. ladigesi* hatch three to seven months after spawning, whereas others remain dormant for more than a year. The staggered hatching is apparently genetic, and prevents all the young from being stranded in temporary ponds created by light or unseasonal rains.[71,435]

In East Africa, species of the cyprinodont genera *Nothobranchius* and *Aphyosemion* inhabit temporary ponds. The adults die as the rains cease and standing water disappears, leaving fertilized eggs to develop in surface mud during the following dry summer. Eggs of *Nothobranchius guentheri* require desiccation for development and will not hatch if constantly submerged. This phenomenon is unique, in that eggs must lie dormant in dry soil before hatching, and thus represents another form of embryonic diapause. During this interval, metabolism is slow and little of the remaining yolk is consumed.[435]

Hibernation of Fishes. Winter torpidity in fishes is known in both fresh and marine waters. Hibernating fishes are usually buried in the muddy bottoms of ponds, estuaries and slow-moving rivers; they are seldom encountered and difficult to study. Undoubtedly, as underwater exploration develops, we will find that some fishes regularly pass the winter in a torpid state. Winter torpidity in fresh-water fish may occur in response to lowered oxygen content of water under ice or seasonal disappearance of vegetation for herbivorous species.

Several marine teleosts, such as the flatfish, anchovies and mackerel, accumulate fat in late autumn. In these fish, "hibernation" may consist of nothing more than a retreat to deep water and cessation of feeding, but torpidity is not known.[763]

Carp *(Cyprinus carpio)* and other cyprinoid fishes seek deep still waters beneath the ice. When the water is at 9°C, they no longer eat, and at 4°C they are torpid.[886] During this time they may be buried in soft mud. Sport fishermen report that very large striped bass *(Morone saxatalis)*, when caught in late winter and early spring, often have mud packed beneath their scales, suggesting that they have been buried in the river bottoms and perhaps were torpid. The killifish *(Fundulus heteroclitus)*, a cyprinodont of the Atlantic coast of North America, exhibits what might be an early stage in the development of hibernation. When ice forms over shallow ponds, this fish burrows six to eight inches in the mud, which is about 7°–8°C. On bright days the sun may warm the water beneath the ice and a few fish become active, but almost no feeding occurs.[168]

The Arctic fresh-water *Dallia pectoralis* is well-known for its ability to survive in solid ice, and must be a hibernator. The assumption is that, although it is encased in ice or frozen mud, hibernating dallias are not frozen. Death from freezing occurs only if ice crystals form in the muscle and blood cells. It lies encased in a cocoon formed of mucus and surrounding silt, and this capsule provides some protection from the cold.

Amphibians

The three living orders of Amphibia are historically associated with humid environments, and today achieve their greatest development and abundance in areas where water loss is not usually a major problem. Amphibian skin permits rapid loss of water, and their kidney likewise discharges copious amounts of dilute urine. On the other hand, the integument of amphibians is hygroscopic, so that in a humid atmosphere a frog or salamander experiences no serious deficiency of body water. Most terrestrial amphibians manage to survive in semiarid to arid habitats by adjusting their behavior to minimize water loss rather than by employing physiologic adaptations related to water, electrolyte and nitrogen metabolism. Such adaptive behavior and anatomic modifications have enabled them to expand their geographic and ecologic ranges. It should also be noted that a dry atmosphere is not always hostile to amphibians, for evaporation form the body surface lowers body temperature, a valuable if costly protection against overheating.

Estivation. Estivation represents a physiologic adaptation that permits amphibians to escape desiccation during rainless periods, which usually occur in summer. Amphibians may become torpid during any rainless period, but the physiologic processes are the same regardless of whether seasonal aridity occurs in winter or in summer. Amphibian estivation almost always involves descent into soil.

The siren *(Siren intermedia)*, a salamander, is capable of estivation, but this species seems not to have been carefully studied. Several hundred sirens *(Siren intermedia)* were plowed up in Texas, having been buried about 10 inches in the ground for at least five rainless months. These estivating specimens were torpid and seemingly lifeless when found and were covered with "a crust or sheath of dried black slime." They became active five minutes after being returned to water. No discrete burrows were discernible. In Georgia, the Congo "eel" *(Amphiuma means),* another salamander, has been found in midwinter 3–3½ feet below the surface in firm soil in an area where the water had been drained away.[554]

Several frogs are well-known estivators. In areas of extreme aridity, amphibians move above ground only during the brief rainy season or at night. Throughout the dry part of the year they burrow to escape desiccation and extreme heat. The Neotropical anuran *Hemipipa carvalhoi* burrows two feet into the mud when ponds dry out,[515] and the highly aquatic ranids, *Ooeidozyga laevis* and *Rana cancrivora,* of the Philippines burrow deep into soil during dry periods. Most species have no special modification to withstand desiccation but seek cool humid tunnels of fossorial mammals. However, spadefoot toads (*Scaphiopus* spp.) of the southern United States are armed with a sharp, cutting metatarsal protuberance used for digging. As it digs backward into soft earth, the loose soil falls in front so that it is eventually buried in an earth-filled tunnel. Spadefoot toads disappear in the ground after the rainy season in July and August, and may descend to 24 cm by September and October; by midwinter they are 54 cm below the surface. Additional protection is developed by a hard, gelatinous coat that reduces water loss.[662]

Cocoons are formed from epithelial tissue in several desert-dwelling frogs of Australia and function in retarding water loss.[584] The Australian leptodactylid frogs *(Cyclorana alboguttatus, C. platycephalus, C. australis, Lymnodynastes spenceri* and *Neobatrachus pictus)* remain underground in dry clay for 3–10 months. The cocoon is derived directly from the stratum corneum and surrounds the estivating frog except for tiny tubular invaginations projecting into the internal nares, which permit respiration. Water loss from estivating *C. alboguttatus* and *N. pictus* was 0.65 and 0.51 mg of $H_2O/cm^2/h$, respectively, whereas the rate increased to 4.9 and 9.0 mg $H_2O/cm^2/h$ for active frogs lacking cocoons. These data were not adjusted to compensate for water loss through the lungs; however, the cocoons obviously provide substantial protection against water loss for estivating frogs.

Other Australian frogs, such as *Heleioporus albopunctatus,* estivate in plugged burrows up to 33 inches beneath the surface, but no cocoon has been reported.[85] In estivation, *Lepidobatrachus llanensis,* a Neotropical leptodactylid, forms a cocoon from epidermal cells, and this covering greatly retards water loss.[515] The Indian bullfrog *(Rana tigrina crassa)* digs into soft earth at the close of the rainy season and surfaces at the onset of the subsequent monsoon.[549]

Hibernation   Hibernation in amphibians is generally an escape from effects of low temperature. When amphibians are chilled, circulation, respiration, digestion and other physiologic processes are retarded. Low temperature by itself does not necessarily induce torpidity in amphibians, for physiologic cycles prepare them for winter.

Maturation of gonads during hibernation occurs in *Rana esculenta, R. fusca* and *R. pipiens,* and these species prepare to mate immediately after emergence; this is probably a widespread phenomenon.[60] Gonads of *R. pipiens* failed to mature when the frogs were kept at room temperature and not permitted to hibernate.[447] Many species of *Rana* spend the cold winter in an inactive state at the bottom of ponds. *Rana pipiens* has been observed unburied in small pits in the bottom of a pond at depths of from 1 to 3 m by skin divers. There was slight and slow circular movement. Such movement was sufficient to clear the frogs of debris and prevent anoxia. Among Nearctic species of *Rana,* the bullfrog *(R. catesbeiana)* and the closely allied green frog *(R. clamitans)* are almost exclusively aquatic. In winter they burrow into soft mud where they remain torpid.[1098] The green frog may also hibernate in surface duff away from water and survive subfreezing temperature. Green frog tadpoles remain active, and probably feed beneath the ice.

The Manitoba toad *(Bufo hemiophrys)* undergoes hibernation for 8–9 months of each year. These toads burrow into Mima mounds (small hills) located about 75–115 feet from margins of ponds in late August or early September. As soil temperatures decrease, the toads dig progressively deeper (to a depth of 140 cm), burrowing just enough to stay ahead of the frost line (Fig. 8.14). As soil temper-

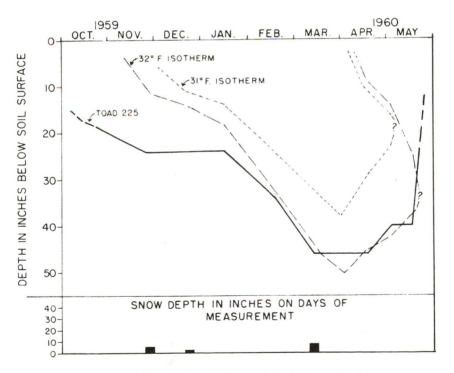

Fig. 8.14. Burrowning activity of a hibernating toad *(Bufo hemiophrys)* in relation to soil temperature (Tester and Breckenridge, 1964).

atures rise in spring, the toads gradually go toward the surface and eventually emerge in May or early June. Emergence appears to be associated with periods of elevated ambient temperature and precipitation. Their ability to respond to small changes in soil temperatures suggests hibernation and not hypothermia.[1003]

The red-backed salamander *(Plethodon cinereus)* may descend as much as three feet beneath the surface, well-insulated from winter frost. During winter, body water of this salamander drops from about 81 to 75 percent, a physiologic adaptation that would lower the freezing point of the tissues, and thus provide protection from freezing. Larval red-spotted newts *(Notophthalmus viridescens)* leave ponds in summer and in their first winter hibernate on land. Subsequently adults hibernate either in ponds or on land. This is a common pattern in most newts (Salamandridae) in temperate regions of the Northern Hemisphere.

Many species of amphibians remain active throughout winter in warmer regions in the temperate zone: in Georgia ambystomid salamanders remain active throughout the winter, and frogs remain active throughout winter in Louisiana but hibernate for increasing periods farther north, not appearing until April or May in New York. In a given region small bullfrogs emerge before large individuals in the spring and remain active later in autumn, perhaps because their small bodies can absorb heat more rapidly or perhaps because smaller individuals have shallower hibernacula.

Reptiles

Many reptiles become inactive in winter in temperate regions and escape freezing temperatures in deep soil, among rocks or in caves, and some even hibernate in water. On the other hand, estivation is rare. Many desert reptiles escape heat and aridity by foraging at night.

Many pond turtles (Emydidae) occupy small ponds in the winter and become terrestrial after egg-laying in the spring. The spotted turtle (*Clemmys guttata*) hibernates in the shallow margins of temporary ponds; in the summer heat it seeks cover under fallen vegetation and returns to the water after autumnal rains. Blood glucose levels increase in red-eared turtles *(Pseudemys scripta)* subjected to low temperatures. This response is most pronounced when the temperature drop is sudden. Increased glucose levels may result from a sudden drop in utilization of glucose following a decline in metabolic activity during torpidity, or it may be associated with increased hepatic glycogen catabolism. Glucose and lipids accumulate in autumn in many reptiles. The box turtle, *Terrapene carolina,* becomes dormant when air temperatures are below 18°C, and hibernation is reportedly more complete for females.[153,806]

Mortality of Reptiles in Hibernation. Loss of weight and mortality are features of winter torpidity, for hibernators are not always sufficiently prepared to survive an extended period without food. This is especially true for ectothermous vertebrates. Individual snakes, marked and weighed as they entered hibernation, showed a weight loss the following spring. The striped whipsnake *(Masticophis taeniatus)* exhibited a mean weight loss of 9.4 percent, six young rattlesnakes

(Crotalus v. viridis) had declined by 22.5 percent of their original body weight as compared with 6.3 percent (males) and 8.8 percent (females) for adults. Sixty-one percent of the whipsnakes of this marked population and 66 percent of the rattlesnakes were found the following spring at emergence, i.e., mortality approached 39 and 34 percent during hibernation for whipsnakes and rattlesnakes, respectively.[438]

Birds

Temperature fluctuations in birds have been known for a long time. Some of the temperature declines and consequent torpidity are nocturnal, and in other cases a seasonal torpidity resembles mammalian hibernation. Although seasonal torpidity is still known for only a few kinds of birds, more examples may await discovery. To date, hypothermia (periodic or facultative) and hibernation are known for several families of birds in the somewhat related Cuculiformes (cuckoos), Caprimulgiformes (poorwills) and Apodiformes (swifts and hummingbirds) but are not definitely known among any of the many families of passerine birds.

The insectivorous and generally nocturnal nighthawks and poorwills become extremely fat as cold weater approaches. Poorwills *(Phalaenoptilus nuttallii)* were found torpid in the winter, in crevices in rocks, and the torpor lasted many weeks. This bird had a body temperature of from 18° to 20°C when the ambient varied from 19° to 24°C, and one individual returned to the same crevice three years running.[488] Captive poorwills become torpid with denial of food during cold weather, and nighthawks *(Chordeiles acutipennis)* in captivity accumulated fat in the autumn and underwent torpor in cold weather.[64]

Torpor in birds is apparently achieved by a passive drop of body temperature to below 10°C, at least for some hummingbirds (Trochilidae) and goatsuckers (Caprimulgidae) and subsequently remains about the same as the ambient. At such temperatures these birds remain torpid for some hours without physiologic damage but lose the ability to arouse spontaneously. They seem unable to arouse when the ambient drops below 15°C.[154] Above 15°C arousal occurs, probably in response to a rise in temperature due to heat generated by shivering, and this manifestation is accompanied by a greatly increased heart rate. The inability of birds to arouse when ambient temperatures fall below 15°C may well stem from their lack of brown fat.

Although early naturalists believed that swallows hibernated, such suggestions in modern times have been humorously dismissed. Under circumstances suggesting hibernation, however, the Australian white-backed swallow *(Chermoeca nuttallii)* has been found torpid in an aggregation of about 20 individuals in an enlarged nesting cavity of an embankment.[905]

Mammals

Hibernators are found in the monotremes, marsupials and placental mammals. Hibernation is not related to taxonomic position, and hibernating species are frequently closely related to nonhibernating species. Although mammals are homothermous, hibernators tend to be imperfectly so and are seasonally thermolabile. Insectivorous bats are sometimes said to be heterothermous. Daily cycles or rhythms in body temperature are well-known for some Microchiroptera, and also occur in a number of other small mammals.[220,461] Such brief torpor seems to differ from that of hibernation only in degree. In fact daily cycles apparently differ with the season, with some mammals showing greater thermolability as the season of hibernation approaches. Greater daily changes in body temperature occur with approach of hibernation and are reported for *Myotis lucifugus* and for *M. myotis.*[693,825] Several species of pocket mice *(Perognathus)* are thermolabile in the laboratory, having daily periods of torpor, especially when food is withheld.[1021,1104] At least some pocket mice must be presumed to hibernate in the field, for they are not generally trapped in winter.

In temperate regions of North America, hibernation and sometimes estivation occur in verspertilionid bats, jumping mice (Zapodidae), many species of ground squirrels, chipmunks (Sciuridae) and pocket mice (Heteromyidae). Estivation occurs in the cactus mouse *(Peromyscus eremicus)* (Cricetidae), both in the laboratory and in nature. These little mice do not hibernate and cannot survive cold that is induced experimentally.[633] Estivation in other mammals is generally followed without interruption by hibernation. Insectivores, such as the Eurasian hedgehog *(Erinaceus europaeus)* and the tenrec *(Centetes ecaudatus)* of Madagascar, hibernate in captivity and in nature.

There are also hibernators among marsupials and primates (lemurs). In the Monotremata, the echidna *(Tachyglossus aculeatus)* is known to become torpid.[30] Most hibernators are rodents, from the small jumping mice *(Zapus* spp. and *Napeozapus insignis),* birch mice *(Sicista subtilis* and *S. betulina)* and pocket mice (*Perognathus* spp.) to such large sciurids as woodchucks and marmots (*Marmota* spp.), and they inhabit a variety of temperate environments, including forests, grasslands and deserts.

Conspicuously excluded from the above list of hibernators are large mammals of temperate regions and most, if not all, genuinely tropical mammals. Ungulates do not hibernate, even though a number of species occur in temperate regions, and a few enter the Arctic.

Despite many reports that certain carnivores regularly become torpid, body temperature, metabolism and respiration are only moderately reduced. Some carnivores, such as bears, badgers, and skunks, become fat in autumn and are sometimes inactive in winter, but none enters torpidity. Carnivores, such as bears, experience a $4°–5°C$ drop in body temperature, in contrast to a $25° +$ decline in dormant rodents. The rectal temperatures of black bears *(Ursus americanus),* grizzly bears *(Ursus arctos)* and polar bears *(Ursus maritimus)* are near $38°C$ when active and close to $34°C$ when lethargic.[443] The great mass of a bear would require a tremendous amount of energy for the periodic arousals characteristic of hibernating rodents. Bears neither forage nor feed during the winter but survive on body fat. There may be a sexual difference in the degree of hibernation of bears, for females give birth to young in midwinter, and the infants take milk until they emerge in the spring.

Hibernation can be divided into four phases: (1) *preparation,* (2) *entrance into torpidity,* (3) *torpidity* and (4) *emergence.* There are four physiologic states of hibernators: (1) a *homothermic* period in spring, during which reproduction occurs and the animal maintains an active thyroid gland and a high body temperature whatever the ambient temperature; (2) a *preparatory* period, with fat accumulation, gradually reduced thyroid activity, initial gametogenesis and increased responsiveness to low temperatures; (3) a *hibernation* period, with periodic arousals and gradually increased endocrine (and gonadal) activity; and (4) *terminal-arousal,* the end of winter sleep. These phases are not the same in all hibernators.

Preparation for Hibernation. Mammalian hibernators usually prepare by either storing substantial amounts of food for sustenance in winter or by accumulating large deposits of fat as material for metabolism until spring.[624] Intermediates between adipose nonhoarders and relatively lean hoarders exist, and even the latter accumulate some fat in late summer and autumn. The famous hoarders in North America are the chipmunks (*Tamias striatus* and *Eutamias* spp.), whose feverish autumnal activities are described in accounts by a host of early naturalists. Adipose hibernators in the Old World include many species of ground squirrels *(Citellus),* marmots *(Marmota),* birch mice *(Sicista),* many insectivorous bats (Microchiroptera) and dormice *(Glis, Glirulus* and *Muscardinus).*[916,917]

Some hibernators, such as the dormouse *(Muscardinus avellanarius),* increase conspicuously in weight, varying from 23 g in spring to 43 g just before hibernation. Autumnal fat deposition in *Citellus lateralis* increases three-fold from August to September.[494,677] Fat storage occurs in mesenteries, between muscles and especially in subcutaneous tissues. The Arctic ground squirrel *(Citellus undulatus)* has fewer phospholipids and fatty acids in its liver and kidney in July than in September, indicating a substantial fat metabolism in midsummer.[1095] Some individuals that fail to become fat seem unable to hibernate, thus suggesting fat storage as a prerequisite to successful hibernation in some species.[817]

Disappearance may be associated with summer drying of grasses and forbs or with lower temperatures in the fall. In either case, hibernation is associated with an environmental change. Nevertheless, when mantled ground squirrels *(Citellus lateralis)* were kept at a constant low temperature for as long as three years, they preserved their annual cycle of summer activity and winter torpidity.[805] Most female *Citellus* spp. enter hibernation after males, perhaps because of their extended reproductive responsibilities.[494,512,912] This is in fact a common pattern, but is not true of captive individuals that have not bred.

Shortage of food possibly lowers body temperature and metabolic activity in some hibernators but certainly not in all cases. The eastern woodchuck *(Marmota monax)* enters its winter quarters when days are warm and food is succulent and plentiful.[79,218] The birch mouse *(Sicista* spp.) also enters torpidity when food is present,[507] as do most species of *Citellus* and some *Eutamias.* It may be more accurate to say that deprivation of food may initiate torpidity in individuals already physiologically prepared for hibernation. A decrease in the water content of food can accelerate entrance into torpidity (estivation) in *Citellus fulvus*, and may well be a cause of estivation in nature in regions where estivation is concomitant with the end of winter rains.[485] It seems then that both hibernation and estivation occur in response to a variety of different stimuli and that the similarity lies in the response rather than in the stimulus.

Hibernacula. The hibernating cell of Columbian ground squirrels is fairly typical of most hibernacula of rodents. It is usually about 2½ feet below the surface, consisting of a spherical grass-lined chamber about 8½ inches in diameter with a drain leading down from both the entrance and hibernation cell. In general it resembles the summer burrow system, which indeed it usually seems to be, but it may also have chambers for food stores as well as a drain and plugged entrances. The plugged entrances probably protect the hibernator from predators and, in estivation, from water loss and excessive heat. The hibernaculum then protects the inhabitant from extreme winter cold, excessive summer heat and heavy rains.[913,914] Chiropterans may seek winter quarters close to their summer caves, or they may migrate many miles, the distance probably being determined by the availability of appropriate caves. In western North America, dark and cold abandoned mines provide suitable hibernation quarters for species of *Myotis* and *Plecotus*, with both dry and humid shafts being used.[209,220]

Bats typically seek retreats with a constant temperature and humidity, although different species have their own preferences. Unlike the hibernacula of ground squirrels, some bat caves may fall below 0°C in winter, but some measure of insulation from freezing temperatures may be provided by the clustering of hibernating bats.[457] Near the entrance of caves conditions tend to fluctuate with the seasons, and bats roost in sites appropriate to their seasonal needs. In summer, when they feed at night, they remain where the temperatures are high (about 18°–24°C) and where digestion can proceed, but in winter a hibernating bat seeks a cool part of a cave, usually from 3° to 9°C.[320]

Tree-roosting species seem adapted to withstand lower ambient temperatures during hibernation. In northern Japan, *Nyctalus lasiopterus* has been found hibernating in hollow trees when the ambient temperature was −6°C.[635] Similarly the noctule *(Nyctalus noctula)* hibernates in hollow trees, where the ambient temperature may drop to −7°C. These bats assemble in clusters of from 150 to 250 individuals and maintain a temperature above the outside ambient.[220] The red bat *(Lasiurus borealis)* of the Nearctic region is also a tree dweller and is one of the most cold-tolerant bats. In the laboratory hibernating individuals remain dormant at temperatures down to −5°C but respond to ambient temperatures below that by increased breathing and heart beat.[220]

When hibernating in caves, dormant bats have a body temperature slightly above the ambient. *Rhinolophus ferrumequinum* has been found with a body temperature of 11.5°C in an ambient temperature of 11.0°C,[220] and *Miniopterus schreibersi* when dormant maintains a core temperature of 7.9°C in an air temperature of 7.6°C.[321]

Endocrine Glands. Endocrine glands may aid in synchronizing the overall hibernation cycle and in regulating metabolism during torpidity, but their exact roles are generally not clear. A general involution of endocrine glands and a decrease in energy metabolism precede hibernation. Such changes occur before the autumnal decline in mean daily ambient temperature: that is, low ambient temperatures may trigger torpidity, but endocrine preparation occurs prior to temperature drops.

Increased secretion of thyroxin raises metabolic rate, increasing heat production. Both the adenohypophysis and the thyroid gland exhibit a seasonal cycle of activity, including autumnal involution and a gradual resumption of activity in midwinter.[485] In the course of hibernation, substances accumulate in the blood, and, when the blood (serum, cells or whole) is transfused into the circulatory system of an active individual, dormancy follows.[222] The longer the donor has been in hibernation before the transfusion, the shorter the time for the recipient to become dormant (Fig. 8.15).

The adrenal cortex helps the hibernator to preserve its normal body temperature and, like other endocrine glands, varies seasonally in size and activity, becoming involuted in the winter. The adrenal medulla of hibernators presumably secretes large quantities of epinephrine during periodic arousals, which undoubt-

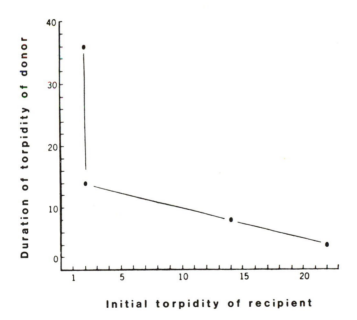

Fig. 8.15. Dormancy following the transfusion of blood (serum, cells or whole blood) from a hibernating ground squirrel *(Citellus tridecemlineatus)* into active ground squirrels (Dawe et al., 1970; copyright 1970 by the American Association for the Advancement of Science).

edly plays a role in cardiac activity and increasing the body temperature above that of torpor.[531]

Gonadal growth and recrudescence take place during hibernation. Mating almost always occurs immediately after emergence from hibernation and in some vespertilionid bats, there may be sexual activity during periodic arousals. In other words, the seasonal cycle of gonadal activity is not interrupted by hibernation, although it is not certain that the gonads affect the time of arousal.[677] Gonadal recrudescence occurs during winter, regardless of whether the individual hibernates, i.e., torpidity and gonadal growth occur simultaneously but independently.[531] The final vernal arousal from hibernation, however, may result from intense gonadal activity and increased sex drive.

Body temperatures of most hibernating rodents, like those of bats, are usually slightly above ambient temperatures and fluctuate with the latter. Mammalian hibernators differ from ectotherms in their ability to increase their energy metabolism and body temperature during hibernation. Hibernating rodents usually respond to lethally low ambient temperatures by increasing their metabolic rate.

Brown Fat. Brown fat or brown adipose tissue (BAT) is deposited about the neck, interscapular region and major blood vessels of many mammals. It was once called the hibernating gland and thought to govern metabolism during torpidity, if not the entire process of hibernation itself. Its role has been reexamined, and the function is clearly associated with the rapid production of heat, as during arousal from torpidity.[953]

Brown fat occurs in many mammals, including all forms known to hibernate, in which it is more abundant. White and brown fat differ cytologically in that cells of the latter contain relatively more cytoplasm and less fat than do those of white fat; the nucleus is more or less central in cells of brown fat but in white fat is flattened against the cell membrane. This tissue has a rich autonomic nervous and vascular supply.

Brown fat increases not only throughout the summer in hibernators but also in cold-acclimated rats. During the first six weeks of hibernation, the reserves in brown fat tissue decrease rapidly, but the white fat subsequently decreases much more rapidly, so that by spring a substantial amount of brown fat remains, although the white fat may be nearly exhausted. Brown fat is thermogenic, producing heat during arousal. This heat production is induced by the release of norepinephrine from the autonomic nerves that so abundantly innervate that tissue. Norepinephrine in turn "activates" enzymes in brown fat, which oxidize the stored lipids to produce heat, and at the same time produces large amounts of oxidizable substrates for metabolism in other organs of the body.

Physiologic Processes and Hibernation. Most mammalian hibernators gradually become dormant over several days when subjected to lowered ambient temperatures. The Beechey ground squirrel *(Citellus beecheyi)* experiences a decline in body temperature at 48-hour intervals at the onset of hibernation accompanied by morning resumption of normal activity.[978] Periods of torpidity at the beginning of the season are relatively brief in most rodents, gradually increasing to about 10 days to two weeks in such species as *Citellus lateralis* and *Eutamias speciosus*.[492] Initially the slowing of heart rate is gradual and irregular and, as it slows, body temperature, metabolic rate, respiration rate, oxygen consumption and velocity of nerve impulse conduction decrease and the animal becomes torpid. Slight disturbances to a hibernating mammal result in increased heart rate, the latter almost always being faster at higher ambient temperatures. Heart rates range from two to eight beats per minute for most mammals (sciurids, hedgehogs, echidnas) in deep hibernation. Respiration rate decreases during hibernation, often reaching rates as low as two per minute. Periods of apnea (absence of respiratory movements) may grow longer with increasing depth of hibernation. Some chipmunks (*Eutamias* spp.) breathe slowly but regularly, whereas the mantled squirrel *(Citellus lateralis)* exhibits typical Cheyne–Stokes respiration or highly irregular breathing.[492]

The nervous system continues to function, but at a reduced rate, to coordinate body functions during hibernation, and at least some neurons can function at body temperatures as low as 4°C. Irritability to tactile stimuli is obvious when one touches a torpid animal. It usually responds by immediate (though slow) movement of the entire body, and a moderate amount of handling will initiate arousal. Nervous activity is also indicated when a torpid animal arouses spontaneously after the ambient temperature drops below freezing. However, hypothalamic thermoregulating neurons appear to be inactive during hibernation, although capable of being readily "activated" by sensory stimuli or neurotransmitter substances. Response to auditory stimuli appears to be minimal during hibernation.

Periodic Arousal. Throughout hibernation torpidity is interrupted by brief periods of activity during which the animal may eat, defecate and urinate. Periodic arousal of the manteled ground squirrel has been discussed at length by Professors Pengelley and Fisher[804] and, although not emphasized by previous workers, seems to be characteristic of all hibernators in which it has been investigated.

The stimulus for periodic arousal during natural hibernation is still unknown, but several causes are proposed. One suggestion is that metabolic wastes accumulate in the blood during torpidity and, when such wastes reach a threshold, they trigger arousal and their elimination. During torpidity kidney function is retarded. In torpid ground squirrels *(Citellus columbianus)* urea accumulates in blood serum and is rapidly removed by the kidney upon arousal, whereas in active animals it is excreted continually. Moreover periodic arousal is more frequent at higher temperatures, suggesting that it is a function of the accumulation of metabolic wastes.[1026,1030]

Thermoregulating hypothalamic neurons play an early and important role in arousal. Intrahypothalamic injections of norepinephrine or serotonin (naturally occurring neurotransmitter substance of the brain) produce hyperthermia in active mantled ground squirrels *(Citellus lateralis)*. Both agents initiated complete arousal when injected into the anterior hypothalamus of hibernating squirrels, and hypothalamic temperatures rose from 8° to 39°C within two hours of injection.

Moreover both compounds failed to initiate arousal when injected into other parts of the brain, as did intrahypothalamic injections of control solutions.[159]

Nervous activity may also function to receive internal and external stimuli and coordinate such sensory input with the arousal process, especially for the final arousal in late winter or spring. Many workers have shown that in nature a higher temperature (outside the hibernaculum) is associated with a relatively early emergence, although temperature perception and response may have occurred after arousal and not during torpidity.[1210,913,1063]

Normally arousal occurs with little or no change in ambient temperature and in a sense is the reversal of entrance into hibernation. The animal must begin in a very cold state and raise its body temperature far above the ambient temperature. In most cases arousal is characterized by shivering, especially in the anterior part of the body. Circulation is confined to the anterior part of the body at the beginning of arousal. This restricted circulation effects a warming of the lungs, heart and brain, thus increasing their efficiency. Renal circulation does not occur until the animal is almost normothermic. Arousal in bats seems to involve much less shivering than in rodents, and differential significance of brown fat in arousal is suggested. Bats may experience a very rapid arousal, lasting about 30 minutes, and in such cases the temperature of the brown fat may exceed even that of the heart.

The marked loss of appetite (anorexia) during most of the hibernation period reduces the problem of excreting metabolic wastes and permits extended bouts of dormancy (Fig. 8.16). From the onset of hibernation in autumn the hibernator loses weight continuously until the return of the active state in the spring.[726] It has been suggested that there is a series of continuously decreasing set points in body weight, and that the set point of any particular state in hibernation may determine the frequency of periodic arousal and appetite (Fig. 8.17).

Vernal Arousal and Breeding. Most hibernators mate shortly after the beginning of activity in the spring, which requires spermatogenesis during hibernation. The Arctic ground squirrel *(Citellus undulatus)* mates promptly after emergence; testicular growth begins before hibernation and gametogenesis occurs during torpidity (Fig. 8.18). This phenomenon also occurs in bats and in the Eurasian hedgehog.

In eastern Washington, *Citellus columbianus* has an extended period of estivation and hibernation, lasting about 190–240 days. Females arouse later than do males, and in mid-March almost all active Columbian ground squirrels are males. Males are in breeding condition at the time of emergence in early March. Initiation of breeding varies with the orientation of the den site: squirrels emerge and mate 10 days later from dens on northeastern-facing slopes, where snow lies late in the spring, than from dens on sunny southwestern slopes.[913] Time of arousal of the Japanese dormouse *(Glirulus japonicus)* varies with local temperatures; the hibernation period is from roughly December 6 to March 22 at temperate Shimoda (Japan) but extends from November 2 to April 13 in Subashiri, where temperatures are relatively low.[916,917]

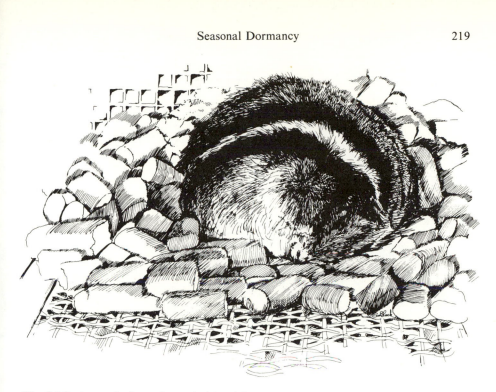

Fig. 8.16. Anorexia (loss of appetite) in a hibernating ground squirrel *(Citellus lateralis)*. Although this species frequently hoards food for the winter, it eats very little during the season of dormancy (Suzanne Black, after Mrosovsky and Barnes, 1974).

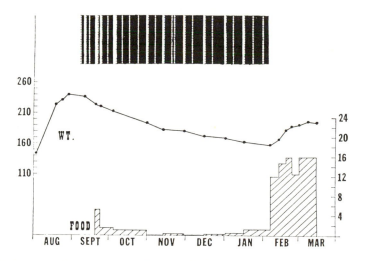

Fig. 8.17. Frequencies of arousal and eating and changes in body weights of a thirteen-lined ground squirrel *(Citellus tridecemlineatus)*. The solid vertical bars (upper) indicate periods of torpor, interrupted by arousals, the white bars. Food eaten, shown at bottom, indicates mean grams eaten per day (scale at right) (Mrosovsky and Fisher, 1970; courtesy of The Canadian Journal of Zoology).

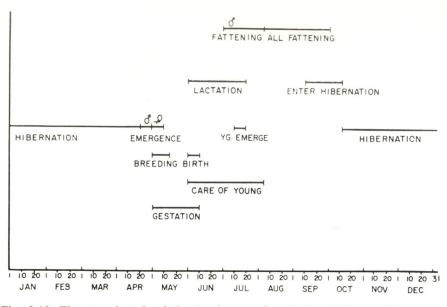

Fig. 8.18. The annual cycle of the Arctic ground squirrel, *Citellus undulatus,* near Anchorage, Alaska (Hock, 1960).

The causes of the beginning and end of hibernation have been studied intensively but remain largely undetermined. The difficulty is two-fold: (1) hibernation is a heterogeneous process and (2) the various theories that presumably explain hibernation are themselves both difficult to prove or disprove. To date, no environmental or other factors in nature have been shown to cause or stimulate hibernation. Scholars have explored temperature, photoperiod and food as possible stimuli for winter torpidity, and, although these factors influence natural hibernation under certain conditions, some unidentified natural factor remains that prepares for a drop in body temperature following exposure to low temperature. In the absence of such preparation for hibernation, a decrease in ambient temperature does not result in torpidity.

Some investigators suggest that hibernation is part of an endogenous rhythm: that is, hibernation is a cyclic phenomenon with timing unaffected by environmental features such as temperature, food, light and relative humidity. The mantled ground squirrel *(Citellus lateralis)* has a genetically determined endogenous rhythm that controls the timing of hibernation—allowing, however, for some adjustment by external factors, especially temperature. An endogenous rhythm has been suggested for *C. mohavenesis, C. tereticaudus, C. variegatus* and *C. beecheyi* from studies made on animals caught in the wild.[803] Animals in the laboratory hibernate more readily during their hibernating period in nature (usually the winter) than at other times, a situation that argues in favor of an endogenous cycle. One should bear in mind, however, that virtually all laboratory research with *Citellus* spp. has been done with specimens captured in the wild and hence subjected to external conditions for months and sometimes a year or more before being brought into the laboratory.

The endogenous rhythm may thus be physiologic movements or patterns established in the wild. If an endogenous rhythm is determined genetically and is more-or-less independent of changes in temperature, photoperiod and rainfall, one must account for the regularity of torpidity and spring arousal in species that have two litters or produce young over an extended period. In northern Texas, *Citellus tridecemlineatus* may produce young over a period of 9½ weeks, and entrance into hibernation occupies an extended period, but emergence is confined to 3 weeks from mid-March to early April.[669] Thus the duration of hibernation must vary greatly for individuals, and the date of arousal cannot be predicted from the date of entrance of hibernation.

Absence of food has been suggested as a cause of torpidity. Although a reduction of food in a low temperature milieu may induce hypothermia in both hibernators and nonhibernators, it is probably not a factor in natural hibernation, as most hibernators enter torpidity before food shortages occur.

Photoperiod has a role in fat deposition and migration in many birds, and has been studied for its possible role in hibernation. In general there is little evidence that photoperiod directly affects the timing of hibernation, but fat accumulation by the Arctic ground squirrel *(Citellus undulatus)* and the mantled squirrel *(C. lateralis)* may be accelerated by a shortened light period in the summer.[442] Fat accumulation may be prerequisite for survival of hibernation in such species as ground squirrels, some chipmunks, dormice, jumping mice, pocket mice and at least some vespertilionid bats. Deprived of fat, *C. lateralis* fails to become torpid.[817]

Causes for arousal are also uncertain. Because the hibernaculum is usually so well-insulated from seasonal changes of temperature and photoperiod, these factors would not seem to trigger vernal activity. Arousal may very possibly be predicated on hormonal levels, especially by gonadal activity. Such is suggested for *Citellus lateralis*[1028] and for *C. tridecemlineatus*.[512] Increasing sex drive throughout the winter may ultimately be reflected in emergence.

Relative humidity changes seasonally, both in subterranean hibernacula and in caves, and may possibly announce seasonal sequences to torpid occupants. In regions of seasonal extremes in rainfall, precipitation is reflected in increases in soil moisture, and a proper atmospheric humidity may promote both torpidity and final arousal. Temperature may possibly affect the time of emergence. Frozen soil is reported to be a deterrent to the emergence of *Citellus franklini* and *C. tridecemlineatus*.[1063] Cold air temperatures, however, do not retard the emergence of *C. tridecemlineatus*.[669]

Relation of Hibernation to Phylogeny. Attention has been given to phylogenetic development of an organism's ability to become torpid and to recover naturally.[144] The phenomenon of hibernation has appeared several times among unrelated groups of mammals and other vertebrates. Hibernators occur in advanced groups (rodents) as well as in rather primitive mammals, such as bats, insectivores, marsupials, and monotremes. The majority of modern students believe that the ability of an animal to become torpid and later to assume a more-or-less homeothermic condition depends on a highly specialized set of adaptations, and has developed independently among many groups of mammals. It is tempting but untenable to argue that hibernation is evidence of direct relationship.

Summary

There is a pulsation of most biologic events, in a broad variety of intervals: heart beats and breathing are frequent except during dormancy; other events, such as eating and urination, occur several times a day but are influenced by conscious choice. Regardless of the timing of the cycle, research shows many rhythms to be endogenous, i.e., they are generated by stimuli within the animal but are usually adjusted by exogenous Zeitgebers.

Many rhythms are timed or entrained by photoperiod, but temperature and rainfall are Zeitgebers for some animals. Gradual changes in photoperiod cue many seasonal events, such as reproduction and, possibly, hibernation.

Hibernation and estivation are two major modifications in energy metabolism that are adaptations to annual weather cycles. In mammals hibernation is clearly an endogenous circannian rhythm, the Zeitgeber of which is unknown. Hibernation is preceded by an accumulation of body fat which, totally or in part, provides energy for the hibernator during dormancy. Hibernation is interrupted by periodical arousals in which the animal warms and awakens, urinates, and once again becomes torpid. The energy for periodic arousals is provided by brown fat.

During hibernation gonads ripen (recrudesce) and the periodic arousals become more frequent. Mating usually occurs soon after hibernation.

Suggested Readings

Cloudsley-Thompson JL (1961) Rhythmic activity in animal physiology and behavior. Academic, New York

Fisher KC, Dawe AR, Lyman CP, et al (eds) (1967) Mammalian hibernation III. American Elsevier, New York

Harker JE (1964) The physiology of diurnal rhythms. Cambridge University, London

Kayser C (1961) The physiology of natural hibernation. Pergamon Press, Oxford, England

Lofts B (1970) Animal photoperiodism. Arnold, London

Lyman CP, Dawe AR (eds) (1960) Mammalian hibernation. Bulletin of the Museum of Comparative Zoology. Cambridge, England

Palmer JD (1976) An introduction to biological rhythms. Academic, New York

Pengelley E (1974) Circannual clocks. Academic, New York

Saunders DS (1977) An introduction to biological rhythms. Wiley, New York

Suomalainen P (ed) (1964) Mammalian hibernation II. Annales Academiae Scientiarum Fennicae, Ser. A. IV Biologica 71, Helsinki

South FE, Hannon JP, Willis JR et al (eds) (1972) Hibernation-hypothermia, prospectives and challenges. Elsevier, Amsterdam

Part III

Communication

Vertebrates have a broad variety of means for transmitting feelings and physiologic states both within species and among different species. Types of signals are characteristic of a given species and are appropriate to their habitats and activity patterns. Some kinds of communication are similar among unrelated taxa and serve for both intra- and interspecific communication. These signals constitute the means for the cohesiveness of family and larger groups as well as for hunting, prey avoidance and territoriality.

Communication between individuals may involve only a single medium, or it may combine several sorts of signals and receptors. Communication in the broadest sense involves: (1) all the stimuli, including all the colors, color patterns, odors and sounds that an animal produces as well as (2) all the stimuli that an animal can perceive. Most of this communication is intraspecific, and most students of behavior have concentrated on communication within a given taxon.

Systems of intra- and interspecific communication and dissemination of information vary with the habitat, the animal's activity pattern, the nature of its enemies and its ability to generate signals. Animals must have mechanisms for emitting signals to, and for receiving such signals from, one another. Logically a given species has sensory receptors for the kind of stimulus it reproduces. There are a few exceptions to this rule: rattlesnakes produce a loud buzzing sound but are deaf; some elasmobranchs are sensitive to electrical impulses and yet most of them do not emit electrical signals, and many snakes are colorful and yet are color-blind. Thus in some cases stimuli produced by one species are important signals for another.

Among the kinds of information provided by exteroceptors is that pertaining to a specific object. Visual and auditory receptors, because they receive a given stimulus at slightly different moments (for sound) or from slightly different directions (for a visual or infrared impulse), contribute different information. An animal with binocular vision, such as a hawk or a primate, is far more efficient in using its eyes (by triangulation) than is a rabbit or a tuna, which has a very small field of binocular vision.

9. Sensory Receptors and Perception

Vertebrate sensory systems are mostly very old and can usually be traced to structures in the earliest forms. Most of the oldest fossils exhibit some sort of olfactory system, lateral lines and eyes. As needs develop, these systems become modified to serve changing environments and evolving social demands. The evolution and adaptations of these systems are the topics of this chapter.

Olfaction and Taste

Chemoreception in fishes can be separated into *olfaction* (innervated by cranial nerve I), *taste* (innervated by cranial nerves VII, IX and X) and *general chemoreception* (innervated by spinal nerves). The olfactory tissue lies in pits or sacs on the surface of the body above the mouth.

Fishes

The nares in fishes consist of a pair of canals, open at each end, through which water flows as the fish moves. In eels (Anguillidae) and some eel-like fishes, which tend to be nocturnal and guided strongly by their sense of smell, the nasal passage is long, running from near the upper lip to the eye. Chemoreceptors in the oral region are classified as taste receptors, and taste buds occur on fins and barbels of some species. Other chemoreceptors are in the fins and skin of many fishes. Quite likely, general chemoreceptors provide information about the presence of potential prey, while mates' and enemies' olfactory and taste receptors evaluate palatability of food.

The olfactory part of the brain of fishes develops as two extended grooves, concave dorsally. The apex of each forms the floor of the short tube through which water passes as it enters the anterior, and leaves the posterior, nares. The grooves are lined with frequently folded olfactory epithelium, the resulting lamellae increasing the odor-sensitive layer and retarding the speed through which water passes over them. The olfactory tissue in elasmobranchs is covered by a flap, not completely roofed over as in teleosts. In lungfish (Dipnoi) and ancient Crossopterygians the posterior nares open in the roof of the mouth and become internal nares, and in some bony fishes the anterior or the posterior nares are closed. As a fish swims, its movement accounts for some passage of water through the olfacgory chamber, but the epithelial lining is frequently ciliated, and there are often nasal sacs whose pulsations create currents across the lamellae.

The pelvic fins of some bottom-dwelling fishes are modified and supplied with chemoreceptors that are sensitive to food. The gadid fish, the squirrel hake *(Urophycis chuss)*, has greatly specialized and separated pelvic fin rays that possess taste buds enervated from fibers of cranial nerve VII and from spinal nerves IV and V. Swimming, hakes extend their filamentary pelvic fins when approaching bottom and detect food by both chemical and physical stimuli.[55]

Amphibians

Olfactory organs began to develop for aerial function in the crossopterygian fishes; the olfactory passage in some species contained lateral or ventral pockets that could well have been the start of Jacobson's organ. Jacobson's organ, or the vomeronasal organ, is served by a segment of the olfactory nerve and had from its inception an olfactory function. In both urodele and anuran amphibians it is a ventral diverticulum of the nasal passage, and in reptiles it is a pair of outpockets in the roof of the mouth. In reptiles with the characteristic long forked tongue, the tip of the resting tongue lies in the concavity of Jacobson's organ. When the tongue is thrust out in the air, apparently volatile materials adhere to its surface and these are detected by the epithelial lining of the organ.

The nasolabial groove of plethodontid salamanders is especially well-developed in males, suggesting a sexual function, and this structure is best developed in species that conduct courtship on land (Fig. 9.1). Fluids experimentally placed in the grooves are carried internally to Jacobson's organ. By tapping the substrate with the upper lip and specifically the nasolabial groove, lungless salamanders pick up scents that are then carried to the internal nares. The red-backed salamander *(Plethodon cinereus)* can distinguish between odors left by familiar and unfamiliar individuals.[673] Aquatic salamanders also have an excellent sense of smell. Males of the newt *(Taricha rivularis)* can detect odor released by a sponge soaked in water in which females have been kept (Fig. 9.2).

Although frogs and toads rely heavily on sight in feeding, experimentally leopard frogs *(Rana pipiens)* and toads *(Bufo boreas)* can recognize food by odor in the absence of visual stimuli.[654,919]

Reptiles

Odor is of major importance in courtship in snakes. Prior to mating, male brown snakes *(Storeria dekayi)* and garter snakes *(Thamnophis sirtalis),* both New World natricines, follow females, frequently touching various regions of their

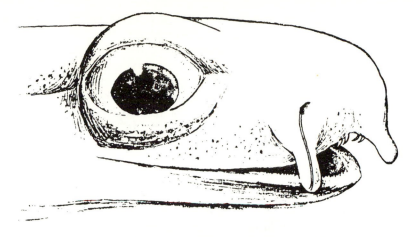

Fig. 9.1. Nasolabial groove of male *Eurycea quadradigitata,* a plethodontid salamander. The projection from the upper lip is a secondary sexual character and is usually less well-developed (Dunn, 1926).

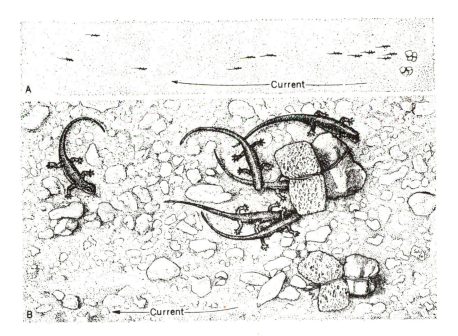

Fig. 9.2. Male newts *(Taricha rivularis)* attracted to a sponge soaked with water from a container in which females had been held. The males were drawn from many yards downstream but ignored a sponge soaked with water not containing odors of the female (Twitty, 1955).

bodies with their tongues and rubbing their chins on the females' backs. Moreover males can follow trails made by rubbing glass with the skin of females but show no interest in similar trails made by either skin of males or cloacal secretions of females. The males clearly followed odor trails of females, but, when the nostrils of the males were plugged or covered or when the tips of their tongues were severed, they ceased pursuit of the females. Thus, although odor proved to be critical, it was detected by either the nostrils or Jacobson's organ.[765]

Birds

Taste and smell have traditionally been assigned minor roles in total sensory perception of birds. In several seed-eating birds small numbers of taste buds are at the rear of the tongue. Most commonly salts and acids cause responses in olfactory nerves, and a bitter response (to quinine hydrochloride, for example) occurs in chicken nerves. Raptorial birds have a great aversion to bile and skillfully pick off the liver from around the gall bladder of their prey. Experimentally chickens can distinguish certain tastes and have aversions to salty and sour substances and the ability to distinguish between sugar and saccharine; some birds seems to prefer sugar.

Olfaction, likewise, is not ranked high in avian sensory perception, but olfactory epithelium occurs in the nasal passages, and the olfactory bulbs in some species are noticeably enlarged. Odors elicit nervous activity in the olfactory endings of some birds. Behavioral tests point out a fine sense of smell for the pigeon, and some behavioral responses to odors in a broad taxonomic spectrum of birds.[960,1085]

The choana of birds is best developed, and contains an enlargment or tubercle in the kiwi (Apteryx spp.)[849] Both experiments and observations on their behavior clearly demonstrate that smell is important in the lives of the kiwi and the turkey vulture. Anatomical evidence suggests that some rails (Rallidae) and the button quail (Coturnix suscitator) possess an ability to detect odors: the olfactory membranes and olfactory bulbs are larger than in most birds.[49,855] When exposed to odors in the laboratory, a spectrum of birds, but especially shearwaters (Procellariidae), responded by varying impulses recorded by EEG patterns; separate odors produced characteristic patterns. Olfaction in birds seems best developed in species living close to the ground, where odors are strongest.

Some of the tube-nosed swimmers (Procellariiformes) have well-developed olfactory tracts. Sponges soaked with cod liver oil and suspended one meter above water in the Bay of Fundy were approached significantly more frequently by Wilson's petrel (Oceanites oceanicus), Leach's petrel (Oceanodroma leucorrhoa) and the great shearwater (Puffinus gravis) than were water-soaked sponges. In view of the well-developed olfaction of petrels, it seems likely that these birds locate their nesting colony, and perhaps even their own nests, by its powerful odor.[378]

Kiwis (Apteryx spp.) have long been known to have an efficient perception of odors. Their olfactory bulbs are very large, and their external nares are placed at the top of the bill, a location unique in birds. Experimentally they are seen to locate food solely by odor. Honey guides (Indicatoridae) feed on wax from beehives and locate and feed on the wax from burning household candles; some observations suggest that they locate the hives from the smell of the beeswax or from some other material associated with the hive.[21] Homing pigeons may use olfactory cues in locating their home loft: birds with their nasal passages plugged or olfactory nerves severed returned home late if at all, whereas sham-operated birds returned earlier.

Mammals

Olfaction is highly developed in most mammals and serves several important social functions. Odors seem to convey messages among individuals in a population, but most certainly do not operate as simple stimuli. More likely odors fall into a category of "reminders," which are modified or conditioned by experience. Scent marks—whether left by urine or by products from glands or legs, head or other parts of the body—are recognized by others of the same species. Scent marking by males serves to indicate territorial boundaries to other males. The effect on conspecific females has been less well studied. Female house mice respond by showing increased activity when exposed to scent from male urine, although females show no such response when the urine is from a castrated male.

Chemoreceptors of mammals are concentrated in (1) the inner reaches of the nasal passages, (2) the vomeronasal organ (or Jacobson's organ) and (3) the tongue. In each case the chemoreceptive surface is moist and the scent, when dissolved on the surface, initiates a response in the receptor neurons.

Mammals use odors not only to locate prey but also to detect the presence of predators, and olfactory input can give specific information. The ground squirrel *(Citellus beecheyi)* can distinguish between the (venomous) rattlesnake *(Crotalus v. viridus)* and the predatory (but nonvenomous) gopher snake *(Pituophis melanoleucus)* when provided only with olfactory cues.[422]

The vomeronasal organ, poorly developed in birds, is present in mammals and is usually well-supplied with olfactory epithelium. In some rodents there are distinct neural pathways from the olfactory epithelium and the vomeronasal organ, and the latter structure provides important olfactory information to the medial hypothalamic region of the brain.[883] In the laboratory rat, destruction of the nerves from the vomeronasal organ is followed by a conspicuous weakening in the male's mating effort. In the laboratory mouse and hamster, severing the vomeronasal nerves causes cessation of copulatory attempts. In these rodents males sniff and lick the female genital region prior to mating and it seems that signals from both the olfactory tract and the vomeronasal input mediate copulatory efforts in males.[72]

The female laboratory rat becomes anovulatory when maintained under a schedule of constant light (LL): there is continuous estrus but ovulation is irregular. When such anovulatory females were exposed to bedding soiled with urine from adult males rats for as briefly as 30 minutes, ovulation occurred within 22 hours. When the ducts leading to the vomeronasal organ were occluded by electrocautery (but with the nasal passages intact), there was no ovulatory response to the bedding soiled by male rat urine.[125]

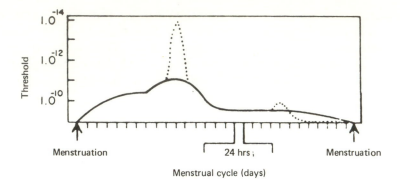

Fig. 9.3. Fluctuation in sensitivity to exaltolide (a synthetic musk) during the menstrual cycle, showing increased sensitivity close to the probably time of ovulation; solid line a mean of 10 women, dotted line exceptional sensitivity (Doty, 1976, based on data from Le Magnen, 1952; courtesy of Academic Press).

For many years there have been statements arguing for olfactory superiority of men over women, and sometimes studies assert the reverse dichotomy. There are many difficulties in detecting significant sexual differences in olfaction: there can be variations with the odor tested, some persons unknowingly suffer from anosmia (loss of olfactory sense), and there is a possible variation in olfactory perception in different parts of the human ovulatory cycle. In human beings the psychologic response to odors may be difficult to separate from the perception of the odor itself, and the response may thus be conditioned by experience.[250]

There is an apparent increased sensitivity to the artificial musk exaltolide after menstruation and before ovulation (Fig. 9.3), and minimal olfactory sensitivity in general is found just before and during menses. Olfaction sensitivities evidently vary directly with estrogen levels within the ovulatory cycle. This relationship is neither definitely known to exist in all women nor to recur regularly in every cycle. There are also marked variations in olfaction and taste through the duration of pregnancy, but trends are not always clear and individual variation is substantial. In some women, moreover, ovariectomy is followed by a decline in olfactory sensititivity, and this tendency is reversed after estrogens are administered. In a similar way, administration of androgens in women lowers olfactory sensitivity. Interesting though these trends are, they are suggestions, and patterns are far from discrete.[250]

Vision

Among the major groups of vertebrates, eye structure and function vary not only with taxa but also with habitat (Fig. 9.4). In water, the original home of vertebrates, the eye is faced with problems of vision different from the aerial type.

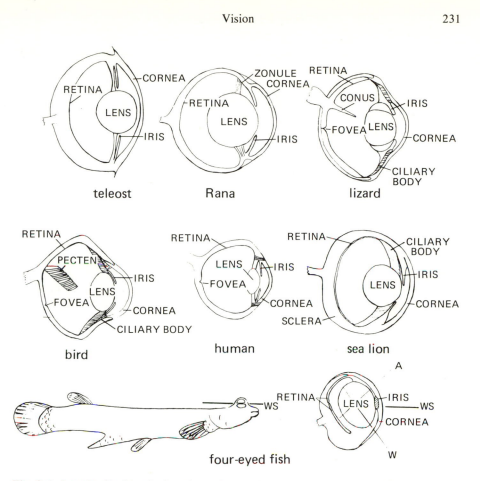

Fig. 9.4. Longitudinal vertical sections of several representative vertebrate eyes. Aquatic forms (fish and amphibians) generally have nearly globular lenses to compensate for the relatively great refractive index of water. The four-eyed fish *(Anableps)* views an aerial image (line A) through the short axis of the lens, while viewing an aquatic image (line W) through the long image; WS indicates the water surface. The bird, lizard and human lenses are soft but flattened, and the sea lion has a typical spherical eye of an aquatic animal. The sclera is the outer supporting tissue present in all eyes but thicker in those subjected to pressure. The fovea is characteristic in eyes of species with especially good vision, and some animals (e.g., raptors and chameleons) have two fovea.

As vertebrates became increasingly terrestrial, the eye became more complex. As light enters the eye, it is refracted by everything through which it passes until it hits the retina: the cornea, aqueous humor, lens and vitreous humor all have refractive indices greater than 1.0. That is, the speed of light in clear air is greater than in other transparent materials, and this difference is indicated by the refractive index.

In water the eye faces the refractive index of water, which is always greater than 1.0. As vertebrates become terrestrial, the eye must be kept clean and moist and adjust for viewing objects at varying distances in air (with a refractive index of 1.0) and in light of varying segments of the visible spectrum.

Structure of the Eye

Fish lack eyelids and lachrymal glands; fleshy or fatty tissue about eyes of some fishes usually serves to reduce turbulence. Light is reduced in water; the refractive index of clear seawater is 1.33, and the pupil and entire orbit of fishes are adjusted to the amount of light received and the daily activity of the fish. With few exceptions (e.g., anguillid eels and some elasmobranchs) the iris does not move and the pupil diameter is fixed. There is a crude correlation between the depth at which fish live (or the amount of light in which they move) and the diameter of their eyes: at and near the surface eyes tend to be small, and they become larger at greater depths as light decreases but smaller again at depths not reached by light. Bathypelagic fishes have a modified iris (it may even be lacking), so that some light can fall directly on the retina, by passing through an opening in the iris that does not overlie the lens. The eyes in these fishes are usually large: the height of the eye may be one-half the total height of the head. Cave-dwelling species may lack functional eyes.

This dichotomy in visual acuity relates to the difference in communication between deep-sea fishes and cave-dwellers. Bathypelagic fishes lack an air bladder (or, if it is present, it is grease-filled), and cannot produce or receive sound in the manner of most fishes; many of them produce light in patterns characteristic of their own species. Most cave-dwellers possess an air bladder and presumably communicate by sound or olfaction. Two species of catfish (Ictaluridae) from drilled artesian wells of Texas have obsolete air bladders, probably reflecting great pressure of their habitat.[460]

The amphibian eye itself is rather like that of teleosts, but in metamorphosis eyelids, lachrymal glands and their ducts develop. The reptilian eye is markedly different from the eyes of fishes and amphibians. Lachrymal glands and ducts are usually well-developed and eyelids are always present, although sometimes fused. The lachrymal duct is reduced in turtles, although in marine chelonians the gland is large and serves to excrete salt. In most fossorial lizards and in snakes the eyelid is fused, and a transparent lower lid allows undisturbed passage of light while providing physical protection to the cornea. Turtles and crocodilians possess a translucent nictitating membrane, between the cornea and the lower lid.

The eyes of birds and reptiles are similar in many respects. There is a slender rod, the *conus papillaris,* projecting through the retina at the rear of the eye of lizards and some snakes, and the conus is apparently homologous to the *pecten* of birds; indeed the pecten of the kiwi (*Apteryx* spp.) is more reptilian than avian. The pecten (or conus) has been known for more than 300 years, and many functions have been proposed for it. A gradient of decreased oxygen has been detected away from the pecten in some birds; when the pecten is deprived of its vascular supply, it withers and the retina degenerates. The pecten is highly vascularized and provides oxygen to the retina.

The eye of a bird is unusual in its relatively large size and complexity. Visual acuity is possibly increased by the somewhat tubular shape of the eye of most raptors; eyes of such birds are deeper (as measured along the axis to the retina) than they are wide, and the image is presumably magnified proportionately. The shape of the eye is reinforced by a series of from 10 to 18 scleral ossicles in birds, much as in the typical reptilian eye; there are usually 14 in reptiles and 15 in birds. The cornea is responsible for most of the refraction, and the lens, together with the cornea, accommodates for changes in distance. That is, in birds striated muscles alter the shape of both cornea and lens. The avian lens is rather soft, and in most species capable of rapid accommodation over a great distance. In terrestrial birds and mammals the lens is flattened, but it is nearly globular in aquatic forms.

Truly amphibious mammals, such as pinnipeds, must adjust to aerial vision when leaving water. The refractive index of the cornea and the nearly spherical lens account for sharp vision in water, and a slit-like pupil provides for visual acuity in air.[821] The eyes of whales resemble in some respects those of fishes: their immobile eyes lack eyelashes and are directed slightly downward. The lens is spherical in toothed whales, much as it is in fishes.

In a great many vertebrates there is some overlap of the fields of vision of the eyes, and in this area the animal has binocular vision. Depth perception is possible with binocular vision, because each eye is trained at a slightly different angle from a line between the viewer and the object viewed. With experience the brain learns to interpret the degree of tension of eye muscles as distance. Generally predators have binocular vision as have arboreal animals, such as squirrels and monkeys. In predators, such as dogs, cats, owls, hawks or rattlesnakes, binocular vision is of great value. Although the advantages of binocular vision are clear it reduces the total field of vision, and in prey species, such as rabbits, the value of a wide field of vision apparently outweighs that of binocular vision.

Accommodation

All vertebrates face the problem of focusing, either on objects that are moving, or when approaching objects. This is called accommodation (or refraction), and can be achieved by changing either the curvature of the lens or the distance between lens and retina.

In accommodation of lampreys the lens is moved by changing the shape of the eyeball of external muscles, but in sharks and rays internal smooth muscle moves the lens. The teleost lens is firm and nearly spherical, with a refractive index of from 1.65 to 1.72, and teleosts accommodate to distant objects by movement of the lens. The cornea of a fish eye has a refractive index nearly equal to that of water and is thus optically nonfunctional. In some bony fishes the center of the lens is denser (and with a greater refractive index) than the periphery; this characteristic presumably enables a fish to view a close object through the middle of the lens and a second object closer to the edge of the field of vision, and yet keep both in focus simultaneously.

Fish are adapted for aquatic vision and are myopic in air, but some amphibious forms have acute aerial vision. The Neotropical four-eyed fish *(Anableps anebleps)* has each eye divided horizontally, the dorsal pupil functioning for aerial vision and the lower one for seeing objects in water; a horizontal sclera separates the two. The ventral part of the lens is more sharply curved for aquatic vision, and the dorsal window is less curved for aerial images. Thus the eye forms two images simultaneously (Fig. 9.4). The retina is also modified: the dorsal part, which receives the image from water, has mostly rods, whereas the ventral part, which receives the aerial image through the dorsal window, has both rods and cones.

The mudskipper (*Periophthalmus* spp.) is an abundant inhabitant of tidal flats from Japan to South Africa, and spends much of its life out of water. Its eyes are very large and can be protruded from their sockets for greater aerial vision; they can also be withdrawn and rotated to prevent desiccation and to cleanse their surface. In the retina of the mudskipper the dorsal rim, which receives images from below, consists of rods only, but the ventral half, viewing objects from above the horizon, contains rods and many cones. The archer fish (*Toxotes* spp.) a pond-dweller of Southeast Asia, feeds on flying insects that it shoots from the air by "spitting" a stream of water. The brain apparently corrects for the bending of light as the image enters the water, and accuracy improves with age. When discharging its stream of water into the air, the fish is close to the surface, and error due to refraction is small.

Accommodation in amphibians is achieved by movement of the lens, but by a slightly different means. Because most fish are near-sighted, accommodation requires that the lens be moved closer to the retina; amphibians, however, are far-sighted, and accommodation is effected by moving the lens *away* from the retina. The amphibian eye, adapted for aerial vision, has a thick cornea that functions optically, together with the lens.

Accommodation in snakes is achieved by internal pressure moving the lens away from the retina, but the lens changes shape in several colubrid snakes. In lizards the lens is flexible, and its shape is altered by contraction of the ciliary body. The refractive function of the cornea varies among different reptiles according to their habitat. It is frequently convex but, inasmuch as its refractive index is close to that of water, the cornea in aquatic reptiles is rather flat and mostly protective.

Accommodation in lizards and turtles also involves a slight movement of the lens toward the snout, increasing the binocular field while focusing on a nearby object. Crocodilians and marine turtles have poor powers of accommodation, and the latter probably do not see well when in air. Accommodation in sea snakes (Hydrophiidae) is a function of pupil size, which can be greatly reduced. Birds and reptiles have sets of muscles that change both the shape of the lens and the cornea; birds often have a need for rapid accommodation, and perhaps for this reason the muscles involved are striated.

Retinal Function

The role of the eye is to translate electromagnetic light waves into nervous impulses in an orderly fashion so that the brain can perceive the image on the retina. Light is translated to nervous impulses as a result of a reaction involving one of several substances called visual pigments.

Pigments that function in dim light or at night are called scotopic in contrast to those that function in bright light, which are photopic. Characteristically rods produce scotopic, and cones photopic, pigments. Visual pigments of fishes possess absorptive maxima from about 473 to 551 nm, a broad range, reflecting the different sorts of light to which they are exposed. Fishes, pinnipeds and whales may have more than one scotopic visual pigment, and these may vary either seasonally or during the life of the individual, apparently as the latter changes its environment and is exposed to different segments of the visual spectrum. Visual pigments in rods are carotenoids, easily dissolved and extracted and well known. Technical difficulties in the removal of pigments from cones require that the theory of cone function rest partly on what is known of rod activity. Visual pigments in the retina break down (or "bleach") in light; a given pigment is maximally sensitive to a specific wave length. Energy released by this reaction stimulates the retinal receptors.

In the early period of retinal research, certain visual pigments were identified with certain segments of the visual spectrum. As more pigments were recognized, it became apparent that a system of specific names for visual pigments was cumbersome. Nevertheless some of the names persist in literature, and it is convenient and helpful to retain them.

Fresh-water fishes have in the rods a visual pigment, rhodopsin (based on vitamin A_1) and another common pigment, porphyropsin (based on vitamin A_2). Porphyropsin is maximally sensitive to light between 520 and 540 nm, and rhodopsin breaks down at wavelengths between 500 and 545 nm. The occurrence of these two pigments varies in ways that may not necessarily be adaptive. Under increasing photoperiod, rhodopsin increases in proportion to porphyropsin, with a corresponding increase in sensitivity to blue light. In deep-sea fishes there may be a type of rhodopsin especially sensitive to blue light, with a maximal absorption of 480 nm (formerly designated as chrysopsin), an adaptation to the deeper penetration of water by light in the blue end of the spectrum.[9]

Variation in visual pigments frequently occurs in migration of fishes. In some anadromous fishes, species of smelt *(Osmerus)* and Pacific salmon *(Oncorhynchus),* for example, porphyropsin occurs when they dwell in fresh water and rhodopsin when they dwell in the sea. When in fresh water, the eel *(Anguilla anguilla)* has porphryopsin in its retina, but the retinal pigment is chrysopsin when it enters the sea in its spawning migration. These two pigments vary seasonally in some fishes, rhodopsin and porphyropsin prevailing in summer and winter, respectively. The two pigments also vary with temperature, porphyropsin being more common at lower temperatures in the golden shiner *(Notemigonus chrysoleucas).* To some extent the spectral sensitivity (e.g., shifts between rhodopsin and porphyropsin) is labile, responding to changes in light intensity, photoperiod, temperature and migration between environments of contrasting light.[9]

Retinas of many ectotherms, birds and primates contain both rods and cones and are called *duplex retinas,* but cones are greatly reduced in nocturnal forms. The retina of birds has one or two regions in which there is a concentration of cones, and this latter is associated with a pit or fovea, sites of increased visual acuity. Most vertebrates have a single fovea, but hawks and falcons and some other birds are bifoveal; the second fovea is temporal in position, occurs in birds in which the eyes are directed forward, and is believed to serve in binocular vision. The structure is usually associated with binocular vision and in fishes, if present, lies near the rear of the retina. The lizard *Anolis* is bifoveal and certainly possesses binocular vision.

In the eye of many nocturnal vertebrates is a reflective layer, the tapetum lucidum, which lies in the pigment epithelium and directs light outward, toward the retina. The tapetum lucidum is well developed in many families of nocturnal fishes, which either inhabit murky waters or are deep-water dwellers. The light that the tapetum lucidum reflects augments the light absorption in the retina.[22,762] The reflective material in most species in guanine, but small spheres of lipids occur in the tapeta of some species. In the daytime (or in bright light) melanosomes move inward to cover the tapetum but move outward, exposing the tapetum, when light is reduced (Fig. 9.5).

In some fishes there is a reflective area, the argentea, in the iris. Possibly it shields the eye from glare, or it may serve to camouflage the eye, reflecting light as do the silvery sides of the head and body in many fishes.

Both rods and cones (in fishes and amphibians, at least) have myofibrils that can extend or withdraw their distal ends: in bright light rods extend into the melanin of the tapetum, and light hitting the cones activates their pigments; in dim light the positions and roles of the rods and cones are reversed (Fig. 9.6). In teleosts melanin granules of the pigment epithelium retreat from the rods in reduced light. In bright light the melanin moves forward, covering the rods. The movement of melanin coincides with that of the rods and cones. These two photomechanical phenomena constitute a circadian rhythm that maximizes the roles

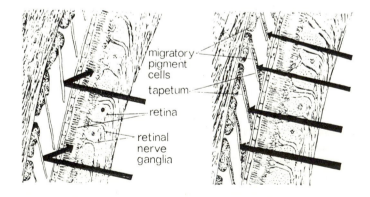

Fig. 9.5. The retina and tapetum of the blue shark *(Prionace glauca)*. In the darkened condition (left), the tapetum reflects light back through the retina, increasing the stimulation to the retina. In bright light (right), pigment covers the reflective tapetum (Budker, 1971).

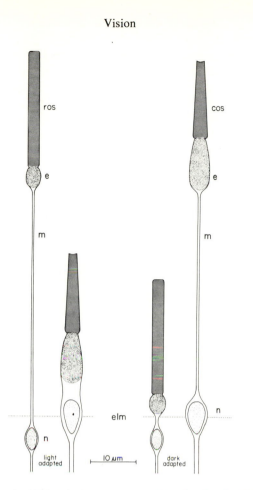

Fig. 9.6. Photomechanical light–dark movements of the visual cells of the goldfish retina. In the light-adapted state, the outer segment of the rod (ros) and ellipsoid (e) are moved from the nucleus (n) by extension of the myoid (m). In the dark-adapted state, the cone outer segment (cos) makes a similar movement away from the external limiting membrane (elm) (O'Day and Young, 1978; courtesy of The Journal of Cell Biology).

or rods and cones. Dark-adapted retinas, which register light through rods, tend to be more sensitive to shorter wavelengths: they are most responsive to wavelengths around 500–510 nm, and light-adapted retinas are maximally effective at wavelengths of 550 nm or more. The difference in maximal sensitivity to the light-adapted and dark-adapted retinas is called the Purkinje shift.

Color Vision. Color vision occurs in many diurnal vertebrates and is generally lacking in those that are not regularly abroad in sunlight. Color perception, a function of cones, depends on the combined responses of cones that are sensitive to one of the three primary colors. Different types of cones have pigments sensitive to separate (but overlapping) parts of the spectrum but, because the cones are crowded together, physical isolation of their different pigments is difficult. In color vision the perception of greens, oranges and other secondary colors is a function of the brain. The retina of most fish eyes possesses both rods and cones; sharks and rays possess retinas of both rods and cones, and probably all elasmobranchs possess color vision.[379,380] Color vision is well developed in most bony fishes, especially those species exposed to bright sunlight and those living near the surface. In deeper waters fish are most sensitive to blue light. Color vision is a property of all diurnal birds and at least some nocturnal species. Some mammals seem to lack color vision, but it occurs in cats, squirrels, primates and probably other animals too.

Cones have different sizes and forms. In the goldfish *(Carassius auritus)* some long cones are sensitive to red rays (max = 625 nm), and very short cones are sensitive to blue light (max = 455 nm). The cones may be in pairs or even threes or fours in different species, individual cones in a group differing in their visual pigment.[974] Underlying the layer of cones and rods is a layer of several horizontal cells that seems to mediate responses between nearby rods and cones (Fig. 9.7) and bipolar cells that are proximal to the horizontal cells. Most amphibians possess both rods (of two types) and cones, and presumably see color. Some variation and specialization occurs in the reptilian retina. In diurnal lizards only cones, both single and double, occur in the visual layer, and some species possess a fine ability to distinguish different colors. Even the nocturnal Gila monster *(Heloderma suspectum)* and the fossorial *Anniella pulchra* possess only cones.

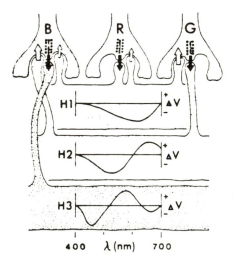

Fig. 9.7. Three cones with underlying horizontal cells. The pigment of each type of cone is sensitive to maximally its own segment of the spectrum (B, blue; R, red; G, green) (Stell et al., 1975; copyright 1975 by the American Association for the Advancement of Science).

Geckos and snakes possess cones and "rods" that seem to have been derived from cones. These visual cells in many lizards and snakes contain a colorless oil of unknown function; in turtles and some birds these oil droplets are pigmented from red to yellow. Crocodilians and most turtles possess single and double cones and true rods.

Behavioral experiments illustrate an animal's ability to discriminate between colors. This ability was demonstrated by conditioning lizards *(Cnemidophorus tigris)* to eat mealworms displayed before one color and to reject them when displayed before a contrasting color, which was offered together with a mild electrical shock. In time this species learned to distinguish between hues only one Munsell standardized color step apart.[80]

Most avian cones contain pigments (yellow, orange and red) dissolved in oil, such droplets filtering light before it reaches the photosensitive visual pigment. Theoretically this arrangement could establish distinctive spectral sensitivities in cones, with no variation in visual pigment among different cones.[109] Some turtles use pigmented oil droplets, together with three different visual pigments in cones, thus refining the selectivity of the cone pigments.[600] Similarly in pigeons, combinations of three visual pigments and oil droplets of several colors provide possible total broad spectral sensitivities, from less than 400 (near-ultraviolet) to almost 700 nm (Fig. 9.8).

Several kinds of terrestrial vertebrates respond to ultraviolet (UV) light (320–400 nm), and it appears to be an important requirement not only for health but also for certain types of behavioral activity. Whereas the lenses of some mammals (including man) are yellowish and prevent the transmission of UV wavelengths, UV light does reach the retina of some birds, reptiles and amphibians. These vertebrates respond behaviorally to UV.[704] Persons who have had their natural lenses replaced with plastic lenses may see UV light.

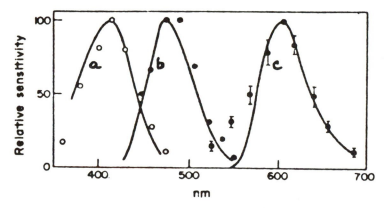

Fig. 9.8. A possible range of spectral sensitivities resulting from combinations of pigmented oil droplets and three visual pigments in the pigeon retina. The colored oil filters the light before it reaches the cones, thus increasing their selectivity (Govardovskii and Zueva, 1977; courtesy of Pergamon Press).

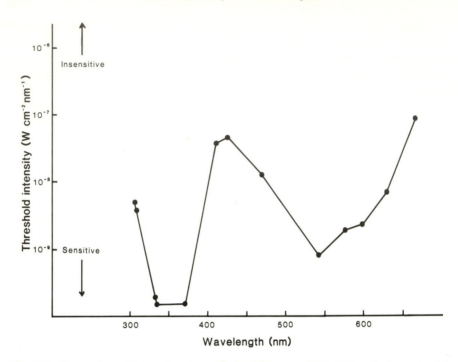

Fig. 9.9. Spectral sensitivity of a pigeon in the UV range (325–360 nm) determined by conditioning to a mild electrical shock. Detection of UV light is not definitely known to be retinal (Kreithen and Eisner, 1978, courtesy of Nature).

By exposure of pigeons *(Columba livia)* to UV light just before a weak electric shock, an increased heart beat occurs with the UV light, in anticipation of the shock. The experimental pigeons were sensitive to 325–360 nm (Fig. 9.9). The mechanism of reception is unknown, but retinal response seems to be a possibility.[652] Ultraviolet discrimination has been demonstrated for a toad *(Bufo bufo)* and for a hummingbird *(Colibri serrirostris)*. Some captive agamid and iguanid lizards also displayed agonistic behavior in UV light; this behavior (courtship, threats and so on) usually began within two minutes of the UV light being turned on, approximately the time for UV radiation to commence.

Hearing and Vocalization

Hearing involves the sympathetic vibration or other response to external compression waves. In air these waves cause movement of some structure, such as an eardrum, which bounds a sealed volume of air. As the fluid in the inner ear vibrates, nerve endings translate the pitch into nervous impulses. Among various taxa there is a great variation in sensitivity to extremes of wavelengths. Human hearing is restricted generally to sounds between 50 and 20,000 cycles/s but is much narrower in some individuals. Frequency is indicated in cycles per second, or hertz (H_z) Vibrations above human hearing (about 20 kH_z) are called *ultra-sonic,* and echolocation frequently employs ultrasonic frequencies. Although water is virtually incompressible, sound is produced and perceived by aquatic vertebrates. Sound passes through water at 1500–1540 m/s, whereas in air its speed is only 330 m/s. Thus for a given frequency, wavelength (λ) is five times as long in water as in air. Water surface is a barrier to sound: aerial sound enters water at a greatly reduced volume, and sound produced in water scarcely enters air.

The spiracle, so conspicuous in some elasmobranchs, arises from the second pharyngeal pouch, as does the eustachian tube of terrestrial vertebrates. The outer opening of the external end is usually modified to collect sound waves and direct them toward the eardrum. In amphibians and reptiles, when there is a functional eardrum, it lies near the surface, but in birds and mammals it may be in a recess. The pinna of mammals is absent in whales, reduced in pinnipeds and large in such mammals as rabbits and hares and insectivorous bats. In birds some specialized feathers both protect the eardrum and allow the passage of sound. The tube leading to the eardrum is funnel-shaped in some birds, and the converging tube may concentrate sound waves. In most owls the soft parts and bony elements of the middle and inner ear are markedly asymmetrical. Many owls possess large ear tufts of feathers borne on fleshy flaps. Such tufts are absent in owls that are essentially diurnal; apparently they aid in determining direction of sound.

The eardrum, or tympanum, marks the external end of the middle ear, and is joined by a bone or ossicle. In anurans, most reptiles and birds, a columella abuts the tympanum, but in mammals sound waves are transmitted by a series of three bones, the malleus connecting to the tympanum. The columella contacts the center of the tympanum in most birds but is eccentric in owls.

The inner ear effects the translation of sound waves to nervous impulses. The base of the columella or the stapes (in mammals) abuts the oval window, which continues the auditory wave into the inner ear. The nature of the cochlea of birds and mammals is similar, but substantial differences occur at the family level. The cochlea is fluid-filled; the fluid vibrates in response to movement of air against the tympanum. The cochlea of birds is not coiled as it is in mammals, and the basilar membrane is shorter but with broader sensory tissue than in mammals. In both birds and mammals the vibration of the oval window causes the waves to move through the perilymph down the cochlea; for each frequency a specific area of the basilar membrane moves in response. A given area of the basilar membrane vibrates in maximal response to a specific frequency.

On the surface of the basilar membrane and beneath the tectorial membrane are numerous clusters of cells bearing fine hairs, and their vibration (relative to the overlying tectorial membrane) initiates a nervous impulse. The round window vibrates sympathetically with the oval window, because of the incompressibility of the cochlear fluid. The avian ear detects sounds within the auditory range of most human beings, but many mammals detect frequencies far above the usual upper limit (20 kHz) of human sensitivity.

Fishes

Most fishes perceive sound, and sound-sensitive species greatly outnumber the sound-producers.[991] The sound-analyzer of most fish is the sacculus–lagena structure of the labyrinth (of the inner ear). In the herrings (Clupeidae) the air bladder abuts the area of the utriculus, which appears to be sound-sensitive. Sensitivity is greatest in those fish that possess Weberian ossicles (Ostariophysi), a series of bones that extend from the air bladder to the labyrinth; destruction of the Weberian bones or the air bladder impairs hearing in species of catfish. The air bladder is clearly associated with hearing in teleosts, but elasmobranchs, lacking an air bladder, respond to frequencies between 100 and 1500 Hz up to 200 m.

The basis for both hearing and frequency discrimination in fishes is unknown. Because they lack the cochlea so important in birds and mammals, fish obviously employ a different mechanism. Determination of the sound source is poorly developed in fishes and seems to operate best for the near-field. This may rest on the bilateral development of the lateral line system (the near-field receiver) in contrast to the single, median air bladder (the far-field receiver). Near-field vibrations are those of low frequency arising close to the receptor, and far-field sounds are of a higher frequency, coming from some distance away.

Near-field reception involves the lateral line system, quite independent of the sacculus–lagena complex. The lateral line system is old, and is complex in primitive fishes living today. Ancient aquatic vertebrates apparently had elaborate lateral line developments, and in more advanced modern spiny fishes the system is sometimes simplified and reduced: the phylogenetic trend is toward simplification. Neuromasts occur over the integument of fishes and aquatic amphibians in diverse situations; they lie not only in canals but also in pits and on surface elevations.

The pressure-sensitive structure of the lateral line is the *cupula,* which lies either exposed or in the *lateral line canal,* which communicates with the surrounding water. The cupula covers a saucer-like depression in which lie many hair cells, and fits snugly, filling the lumen of the canal (Fig. 9.10). On the heads of many fishes and amphibians the cupula is not within a canal but projects rather from the contours of the animal, and in an exposed neuromast the cupula may be columnar (Fig. 9.11). Beneath the cupula are several or many *hair cells* from which a kinocilium and many sterocilia project into the lumen of the cupula (Fig. 9.12). Movements in the surrounding water cause slight displacements of the cupula, and the sensory hairs bend in response to its movement. Adjacent hair cells are arranged so that the kinocilia are usually adjacent to or opposite each other, perhaps a provision for equal sensitivity to shearing displacement of the cupula in either direction.

The distinction between near-field and far-field results from the virtual incompressibility of water. Movement of a nearby object in water causes a displacement of the surrounding water, but the effect dissipates with the cube of the distance. A swimming fish moves water as it progresses, and such an aqueous displacement can cause a slight shift in the neuromast of a nearby, but not in a distant, fish. This hydrodynamic effect can be called sound, and the rapid passage of a fish can set up low-frequency vibrations in the adjacent waters, the lateral line moving

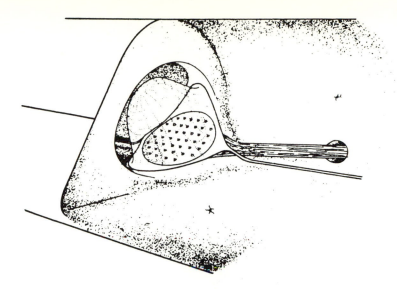

Fig. 9.10. Bell-shaped cupula lying in a constriction of the lateral line canal; hair cells are scattered on the surface of the sensory area (Flock, 1967; courtesy of Indiana State University Press).

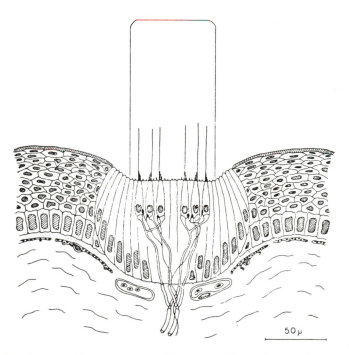

50 μ

Fig. 9.11. The free-standing neuromast of the clawed frog *(Xenopus laevis)*, with supporting epithelium. The hair cells are arranged so that the long kinocilium is either adjacent to or farthest from the kinocilium of the neighboring hair cell (Flock, 1967; courtesy of Indiana State University Press).

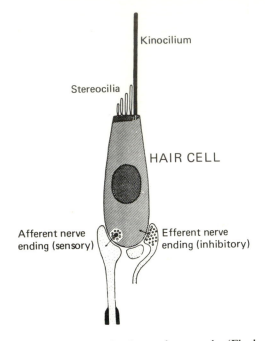

Fig. 9.12. A single hair cell from the base of a cupula (Flock, 1971; courtesy of Academic Press).

sympathetically to these vibrations. The detection of movement of a distant source (distant in the sense of not being in contact) justifies calling such displacement sound. The real distinction between the near-field and the far-field is not the distance from the origin of the sound to the receptor but rather the frequency of the sound. The lateral line receptors are sensitive to low-frequency sounds (with maxima from 140 to 500 Hz, depending on the species). Many low-frequency sounds, which arise from the displacement of water and to which the neuromasts are especially responsive, have very limited range.[241,303] Information from the lateral line seems to relate largely to *direction* and little to distance.

The cave-dwelling poeciliid, *Poecilia sphenops,* known from near Tabasco, Mexico, varies behaviorally and morphologically, according to the habitat. The cavernicolous form is almost devoid of pigment, fleshy circumorbital tissue obscures the small eye, and such fish are quite blind; the cephalic lateral line system is extremely well developed with exposed neuromasts in individuals from deep within the cave.[1071] Representatives from the cave entrance are pigmented and almost always have good vision. The blind form feeds on the bottom; epigean individuals forage in the middle and upper strata and obviously feed by sight. Courtship in the eyed epigean form follows the patterns of other poeciliids, but in both the epigean and the cavernicolous forms courtship in the dark is notable in the great amount of contact involved. Some surface-feeding fishes, such as the top-minnow *(Fundulus notatus),* can detect the direction of slight surface movements, and this ability must help in finding food.

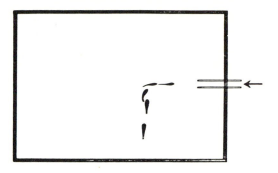

Fig. 9.13. Detection of a local water flow, by means of lateral line organs, in a blinded minnow, showing successive positions of a single fish (Dijkgraaf, 1967; courtesy of the Indiana State University Press).

In detecting objects in water, the lateral line system enables fish to detect both moving and stationary objects. Greatest sensitivity seems to lie between 50 and 150 Hz. Schooling fishes seem to employ both vision and signals received by the lateral line system. When separated by plexiglass (a barrier to faint low-frequency vibrations), minnows continue to swim and turn in unison; when vision is destroyed, however, and the fish are not separated by plexiglass, they do not lose their schooling ability. Blinded minnows can slso determine the direction of water flow (Fig. 9.13).[241]

About the head of elasmobranchs, ampullary lateral line organs, or the ampullae of Lorenzini, are distinctive both in lacking the cupula and in being sunken in pits (Fig. 9.14). Ampullary lateral line organs are very sensitive to weak electric fields—even to electricity emitted from nonelectric prey species—an ability not shared by ordinary lateral line organs. In weakly electric fishes, specialized lateral line units called tuberous and ampullary organs line units contain electrosensitive

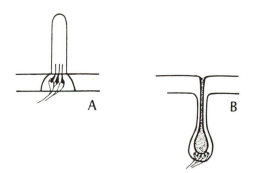

Fig. 9.14A–B. Lateral line system. **A,** A typical superficial neuromast, as found on the heads of some fishes and amphibians, with the cupula projecting above the surface; **B,** ampullary lateral line organ, lacking a cupula, but with a duct filled with a gelatinous substance (Dijkgraaf, 1967; courtesy of the Indiana State University Press).

neuromasts. They occur in addition to the usual lateral line organs. Ampullary organs also occur in some catfish, which may be electrosensitive.

Although the lateral line system is the primary receptor for near-field low-frequency signals, this does not preclude detection of such vibrations by the air bladder, especially in Ostariophysi. Inasmuch as both signals go to the labyrinth, there is no clear distinction in signals received by either the air bladder or the lateral line. In groups lacking an air bladder, such as gobies (Gobiidae), the lateral line system is not the sole sensor for acoustic signals; the lagena of the inner ear may also register near-field movements.

In a group of fishes called the Ostariophysi, or Cypriniformes (minnows, some electric fishes, suckers and the many families of catfish), a series of three (sometimes four) bones (the Weberian ossicles) connects the air bladder and the endolymphatic space extending to the lagena. The Weberian ossicles are apparently derived from the first three cervical vertebrae, and these latter in the Ostariophysi are fused to each other and to the occipital area of the skull, forming a firm support against which the ossicles can transmit delicate vibrations. Thus movement of the air bladder is passed along the paired Weberian ossicles to the endolymph (of the semicircular canals); displacement of the endolymph in turn moves an otolith, which is attached by fine hairs, and movement of the otolith initiates nervous impulses. Incompressibility of the endolymph allows the otolith to move because of the sympathetic movement of a very thin bony area near the lagena. This area functions much as does the round window of birds and mammals.

Not surprisingly, species of Ostariophysi have excellent hearing, and sound plays an important role in their lives. Their auditory thresholds are rather low, and the range of frequencies detected is from 100 to 7000 Hz. In the goldfish *(Carassius auratus),* a member of the Ostariophysi, sensitivity ranges from 100 to 3000 Hz or more. Sensitivity is greatest between 500 and 1000 Hz. In a catfish *(Ictalurus nebulosus),* another ostariophysine, maximal sensitivity is near 800 Hz. Some of the characid fish (tetras and their allies) hear sound up to 7 kHz.

In the herring and their allies (Clupeidae and Engraulidae), the air bladder actually penetrates the inner ear as a pair of small tubes or ducts that are surrounded by the semicircular canals and which abut the endolymphatic fluid. Movement of the fluid induces nervous impulses much as in the ostariophysine fishes. In both the Clupeidae and Engraulidae, as well as in the Ostariophysi, sound is passed directly to the inner ear, presumably with little loss of amplitude. In the elephant-nose fish (Mormyridae) of Africa and the knifefish (Notopteridae) of South America, extensions of the air bladder penetrate the skull. In the elephant-nose, the air bladder ends as a bladder-like sphere surrounded by a semicircular canal, much as in the herring, but with maturity the connection to the main body of the bladder is lost. These sacs are separated from the endolymph by a delicate membrane and seem to function as miniature air bladders. In the knifefish the extension lies near the inner ear. The function must be to preserve the strength of the sound impulse received by the air bladder.[854] In other groups that hear by air bladder, vibrations must pass through the flesh and bone of the fish, probably with some loss of amplitude.

Amphibians

Male anurans call in patterns distinctive and characteristic of each species. Sonagram oscilloscope recordings, graphic representations or recorded calls, permit visual contrast of intra- and interspecific variations. At least some typical features of the call of given species are determined by central nervous system control of intrinsic laryngeal muscles.

The greatest auditory sensitivity lies in the spectral region of greatest energy produced by the call. In the Nearctic cricket frog *(Acris crepitans)* the calls vary from 3550 Hz in New Jersey to 2900 Hz in South Dakota, and this geographic difference in the call is paralleled by variation in sensitivity to these two spectral differences. An additional restriction to this auditory response is the high threshold to the call, with even higher thresholds above and below the mean frequency of the call. Presumably this auditory selectivity forms the basis for the well-known role of the anuran call as an isolating mechanism.[149]

As sound is produced, it hits both sides of the eardrum, reducing sensitivity to the animal's own voice. The selectivity of acoustic sensitivity is determined by not only the structure of the eardrum but also the nature (size and shape) of the mouth cavity. Anurans have a continuously open connection from the mouth to the inner side of the eardrum, through the eustachian tube. Because the eustachian tube conducts sound from the middle ear to the mouth, the latter becomes a resonator; and, when the eustachian tubes are blocked (as with cotton or wax) or if the mouth is held open, as with a toothpick, the acoustic selectivity disappears.[170]

Reptiles

The reptilian ear is relatively simple and variable among the several orders. The eardrum, or tympanum, is present in most lizards except in fossorial forms and absent in snakes. The typanum is obscure in most turtles and crocodilians. The membrane, when present, is rather thick, and this may account for the lack of sensitivity of the reptilian ear when contrasted with that of mammals and birds. Not only does the columella extend to the oval window of the inner ear, but a supporting strut attaches to the quadrate. Sound waves are transmitted through the oval window through perilymph. Vibrations are converted to nervous impulses in a papilla between the small basilar and the tectorial membranes, less developed than in birds or mammals.

The hearing of reptiles, when known, varies with the degree of complexity of the ear structure, and most reptiles lack the sensitive ears of birds and mammals. Apparently snakes are deaf to most sounds produced by animals, although they are capable of hearing some loud, low-frequency sounds. Rattlesnakes are unable to hear the sound that they produce. Reptilian sensitivities range from about 100 Hz to 10 kHz, and some geckos detect frequencies as high as 20 kHz.[813]

Birds

Small birds (with small heads, and ears relatively close together) probably have increased directional acuity with increased frequency, a direction source of higher pitches having shorter wavelengths. Also, if the song comes from a point to the side of the bird's head, that side will receive the sound with a greater intensity than will the ear on the opposite side.

Owls have very sensitive hearing, and the asymmetry of their external ears may account for their acoustical skill in prey location. The basilar membrane of the barn owl *(Tyto alba)* (Fig. 9.15) is much longer than in some diurnal birds, presumably reflecting a better ability for discriminating frequencies and perhaps intensities. Owls are distinguished not only by asymmetry in otic construction but also by a very broad head; the breadth of the head results in the external ears being much farther apart than in most birds (Fig. 9.16). In *Aegolius funereus,* an owl of Europe, frequencies greater than 12,500 Hz provide directional cues so that, if the owl is not facing the prey when first heard, components of the sound reach the two ears in different intensities and out of phase. By noting the angle between the source of high-pitched sounds from prey and the horizontal, an owl can judge the necessary angle of descent to capture the prey. When the owl faces the prey, its sound (12,500 Hz or greater) is heard with equal intensity and in phase by both ears. Thus the extreme accuracy of prey-location by owls depends on both ears and, if one ear is plugged, an owl cannot find prey by sound alone. Asymmetry in the ears of owls has apparently developed several times, and provides the basis for determining vertical discrimination.[767,768] Auditory sensitivity of owls is increased by their large meatus: their external openings have an area of

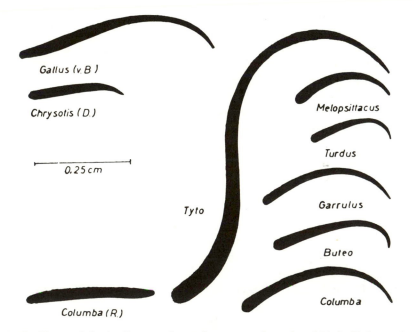

Fig. 9.15. Shape of the basilar membrane from several unrelated birds (Schwartzkopff and Winter, 1960).

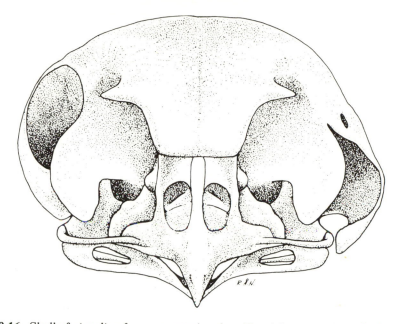

Fig. 9.16. Skull of *Aegolius funereus,* anterior view. The right otic aperture is about 6.3 mm above the level of that on the left. The eardrums are symmetrical (Norberg, 1967).

25 or 50 times that of the tympanum in nocturnal owls, in contrast to 3–6 times that in passerine birds.

Sensory Pits of Snakes

Perception of infrared radiation is important in certain snakes, especially some nocturnal species that feed on warm-blooded prey. Pit vipers (Viperidae, Crotalinae) have a deep pit in the face, between the nostril and the eye. This pit is directed forward and is deep, entering a cavity in the maxillary bone. In the pit is suspended a thin membrane richly supplied with free endings from the ophthalmic and maxillary branches of the trigeminal nerve and also a fine capillary network. These nerves consist almost entirely of warm fibers. These heat-sensitive nerve endings are rich in mitochondria and finely divided so that the entire membrane constitutes a single, broad, sensitive plate (Fig. 9.17).[696,766,1001] Behind the pit membrane is an inner chamber, so that the membrane faces a gaseous space on either side. The nerves in the pit membrane fire when stimulated by transient temperature changes as small as 0.003°C, a degree of sensitivity unknown for any other heat receptor. The nerves respond to changes produced by small prey, such as a mouse, even when it is far beyond the striking distance of the snake.[353] The openings are directed forward so that the fields have a 25°C overlap at only 1.5

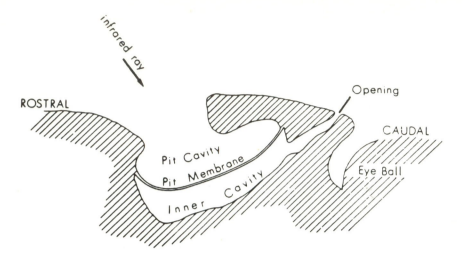

Fig. 9.17. Sagittal section through the pit organ of a pit viper (Crotalinae) (Goris and Terashima, 1976, courtesy of Elsevier/North-Holland Biomedical Press).

cm from the snout (Fig. 9.18). Even when blinded, a pit viper can strike accurately at both still and moving prey.

The labial pits of pythons (Boidae) are also specialized heat receptors. The pits resemble the facial pits of crotaline vipers except that the pits of pythons lack an inner chamber. In species that possess them, the pits lie in a series between the scales of the lower or upper jaw, or both. Interestingly some boids that lack labial pits nevertheless have an unusual sensitivity to heat. In some genera of boid snakes (*Boa, Eunectes* and *Lichanura*) the labial region is infrared-sensitive, but there are no pits.

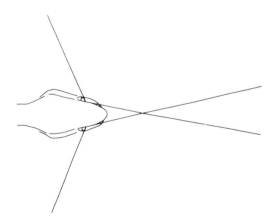

Fig. 9.18. Receptive field of *Agkistrodon blomhoffi brevicaudus*. The two fields have about a 25° overlap about 1.5 cm in front of the snout (Terashima and Goris, 1974, reprinted with permission, Progress in Neurobiology, Pergamon Press).

Early studies by Karl Schmidt and G. K. Noble established that these pits are heat-sensitive. Experimenting with blindfolded boids, they demonstrated sensitivity to heat given off by a lightbulb lit and wrapped with cloth.[766] Similar experiments with pit vipers showed that their sensitivity to heat was lost when the facial pits were plugged with wax, grease or other insulator. It is now known that facial and labial pits respond to radiant (infrared) heat, and can locate the source against a background of "neutral" temperature.

Such heat-sensitive pits are not definitely known for other snakes, but the Indian colubrid *Elachistodon westermanni,* a rear-fanged snake, has a pair of large pits similar to those of pit vipers.

Extraoptical Photoreceptors

Many vertebrates, especially amphibians and reptiles, can perceive changes in intensity in light in the absence of eyes. Extraoptical photoreceptors (EOPs) receive light in sufficient amounts for entrainment of activity schedules, orientation and chromatophore changes. The parietal eye of some reptiles contains a retina with cells resembling the cones of lateral eyes. The pineal body develops in all vertebrates and is also light-sensitive. In anurans the pineal end organ *(Stirnorgan)* is an established photoreceptor; it is homologous to the partial eye of *Sphenodon* and lizards. The pineal end organ is an important receptor in frogs and toads; it acts as a light receptor in blinded animals, a function that is lost when the end organ is also destroyed or shielded with an opaque cover.

Extraocular photoreception occurs in many vertebrates and is a property of the pineal body. The pineal body of amphibians can detect not only intensity but also direction of light from overhead; it thus serves not only for orientation but also for reception of light cues for photoperiodic cycles. Blinded amphibians can orient movements with reference to the position of the sun and synchronize their internal endogenous rhythm to the natural environment schedule.

The simplest light receptor is the middle or third eye, which is composed of the parietal eye and pineal body, and lies above the midbrain and below a translucent window at the top of the head. This composite structure is found in jawless fishes, and many amphibians and reptiles. In cyclostomes it consists of two distinct elements: an anterior structure is homologous to part of the pineal body of reptiles, and the posterior one corresponds to the parietal eye. The structure (or pair of structures) occurs in fishes but without an orifice in the cranium or unpigmented skin covering. Ancient reptiles had well-developed parietal eyes, but they occur in the tuatara *(Sphenodon punctatus)* and most lizards.[268] The third eye also occurs in a broad spectrum of endothermic vertebrates.

The parietal eye characterizes many modern lizards, especially heliothermic species. Covering the structure is a pigmentless, translucent scale. A thick cornea overlies a lens composed of columnar cells. Among retinal cells are pigment-containing cells, and a light–dark movement shift (Purkinje shift) in pigment is seen in the retina of some lizards. Although a visual pigment is not known in the parietal eye of reptiles, when the eye is exposed to light, the nerve leading from the retina responds with a change in electrical potential. As a photoreceptor the parietal eye in lizards seems to prevent overexposure to solar radiation, and, when this structure is covered or destroyed experimentally (e.g., as in *Sceloporus virgatus*), such individuals spend more time under light than do whole individuals.[968,970]

Light responses can occur in blinded animals in all classes of vertebrates: individuals in which the eyes are permanently destroyed respond to photoperiod in gonadal development, activity phasing and orientation. Light perception in blinded animals is called extraretinal or extraoptical photoreception (EOP). In several frogs and amphibians activity peaks can be shifted by a shift in the light–dark (LD) schedules in the absence of eyes. In the slimy salamander *(Plethodon glutinosus)* of eastern North America, the free-running rhythm (FRR) is slightly more than 25 hours, but both eyeless and intact animals are entrained by an experimental LD schedule. Green frogs *(Rana clamitans)* adjust their activity to a given LD regimen whether eyed or eyeless, but cannot do so when the pineal end organ (homologous to the reptilian parietal eye) is destroyed. The cricket frog *(Acris gryllus),* a small hylid of central United States, orients its movements toward light even when blinded, showing that it perceives both the presence and the direction of light. The cricket frog loses this ability, however, when opaque tape is placed beneath the skin on the top of the head.[994]

There is evidence of light perception in the skin of some amphibians, but there is no photoreceptor known to account for this circumstance. Bright flashes of light on the skin stimulate nervous impulses, but the mechanism of the stimulation remains unexplained.[74,975] Despite the importance of the parietal eye in lizards, the iguanid *Sceloporus olivaceus* can be entrained to a light cycle after destruction of the paired eyes, the parietal eye and the pineal body.[1039] In the house sparrow *(Passer domesticus)* entrainment to light occurs in blinded and pinealectomized individuals. In both *Sceloporus olivaceus* and *Passer domesticus* the EOP has not been identified. Light elicits electric responses from the skin of several vertebrates, including fish, amphibians and a mammal, but the function of such unidentified dermal photoreceptors is not established.

Electroreception

In those fishes producing low-voltage impulses, the lateral line system becomes an electroreceptor system: the canal walls are resistant to electrical transmission and act as an insulator, whereas the contents of the lumen have a low resistance and transmit the impulse to receptors within the canals. Electroreceptors are modified neuromasts of the lateral line system and are specialized for detection of low-frequency signals (ampullary organs) and high-frequency signals (tuberous organs). These receptors fire in response to the animal's own electric organ discharge (EOD), and receptors of each species are sensitive or tuned to the frequencies that it produces.[136] Some types of receptors code discharges in a yes–no manner, responding to one spike above a certain threshold, whereas others encode intensity and are suitable to monitor modulations in discharge amplitude.

In electric fishes there are great modifications in the development of the lateral line nerves: in *Malapterurus* a branch of cranial nerve VII dorsally extends caudad to the adipose fin and sends branches to the skin, and the lateral line nerve runs along the lateral (outer) margin of the horizontal myoseptum, as in most teleosts, but it is insulated from the electric organ by fatty and loose connective tissue. In the electric eel *(Electrophorus)* the lateral line nerve runs caudad deep in the body, producing lateral branches to the surface; its electricity is both of high voltage, for obtaining prey, and low voltage, for communication. In fishes that produce a pulsating field (e.g., the mormyrids of Africa), the lateral line nerve runs along the medial (inner) part of the horizontal myoseptum, adjacent to the vertebral column.[81]

Electroreceptors occur in some nonelectric fishes; they are modified lateral line organs and serve to detect very weak currents from potential prey species that are not generally considered to be "electric" fishes. Some elasmobranchs can detect small amounts of d.c. electrical currents produced by normal metabolism of prey species. Small pits, ampullae of Lorenzini, on the head of elasmobranchs are the sense receptors in this case. Electrical detection of prey may occur in fishes other than elasmobranchs, for similar pits are widespread among teleosts. Experimentally the Nearctic catfish, *Ictalurus nebulosus,* can perceive d.c. currents through the pit organs on their heads. Such electrolocation is based on currents only from the potential prey species and is thus a different phenomenon from the electrolocation of gymotids, mormyrids and their allies.[812]

Electrolocation is predicated on the ability of the fish to detect irregularities in the electric field about its body. The flow of current is from the electric organ, near the tail, forward to the anterior part of the body, and outward, about the fish and perpendicular to its surface, returning to the tail. (Some species possess additional electric organs in the chin and behind the gill openings; these may differ in the frequency of their discharges.) In the surface of the forward part of the body receptors detect transepidermal flow, which is part of the electrical discharge. A nonconductor in this electric field distorts the flow of current about it, changing the transepidermal current pattern to form an electric image (Fig. 9.19). The electric image thus formed increases with an increase in size of the object and with a decrease in its distance from the fish. Generally, if the threshold distance is doubled, the diameter of the object must be multiplied by four. This limits the effective distance of electrolocation, and probably the image is less distinct than a visual image on a retina. Clarity of the image is presumably enhanced by the bending of the tail filament (Fig. 9.20).

Not only are fish unique among vertebrates to have evolved such a variety of behavioral patterns based on electrical discharges: electric species are unknown among invertebrates. This is a superb example of a totally novel sense and means

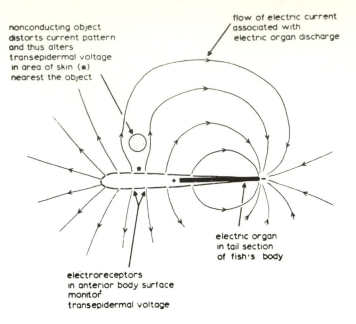

nonconducting object
distorts current pattern
and thus alters
transepidermal voltage
in area of skin (∗)
nearest the object

flow of electric current
associated with
electric organ discharge

electric organ
in tail section
of fish's body

electroreceptors
in anterior body surface
monitor
transepidermal voltage

Fig. 9.19. Mechanism of electrolocation. Longitudinal horizontal section through an electric fish. The black bar indicates the typical position of the electric organ (Heiligenberg, 1977).

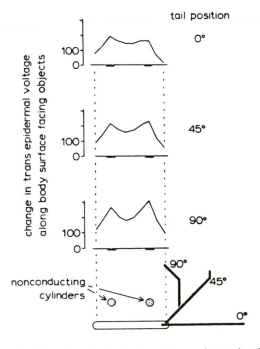

tail position

0°

45°

90°

change in trans epidermal voltage
along body surface facing objects

100

0

100

0

100

0

90°

45°

0°

nonconducting
cylinders

Fig. 9.20. Function of tail-bending in electrolocation: as the angle of the tail from the axis of the fish moves from 0° to 90°, the distortion of the electric field is increased (Heiligenberg, 1975).

of communication derived from preexisting materials that are widespread among the vertebrates.

Summary

The emphasis on one sensory system or another reflects the habitat of a group. Apparently sense receptors are very plastic, for they become modified to suit the changing life of a taxon as it diverges into various types. Receptors for odor, sound waves and light waves all become changed from one class to another; some, such as the infrared receptors of some snakes, develop de novo.

Chemoreception is perhaps the most ancient sensory system, and olfactory lobes are very well developed in the earliest fishes. Taste buds in aquatic vertebrates occur on various parts of the skin, sometimes concentrated on barbels, as in catfish, and sometimes scattered. In water, where light is reduced, odor is of paramount importance for some major activities, such as food-getting, homing and many aspects of reproduction. Much of the evidence for efficient odor perception in amphibians is circumstantial, but there is no doubting its importance.

Although olfaction in birds is secondary to vision and hearing, some avian groups can detect odors, and olfactory signals assist in food-finding in such unrelated groups as petrels, indicator birds and kiwis. Olfaction in mammals not only serves in the detection of prey and predator but is also vital in ordering mammalian social life. Olfaction is very well developed in all mammals except primates and marine mammals.

Vision and paired eyes occur in almost all vertebrates, but sight varies greatly from one group to another, generally reflecting the intensity and quality of light available to the individual during its active period. Among diurnal fishes bright colors and sexual dichromatism are frequent, and color vision prevails. The colors that we see in a reef fish in an aquarium, brightly illuminated, are not those seen by the same species in 20 or 30 feet of water. Red penetrates water to a very limited degree, and colors are modified accordingly; even in rather shallow waters, a reddish fish may appear a shade of gray to other fish. Nevertheless the cones, which form the basis for color perception, are well differentiated in many fishes. Vision in fishes is generally much less important than is olfaction.

Adult amphibians are predatory and have good vision. Prey is usually located visually by its movement when the frog or salamander is out of water. Sexual variation in color is slight in amphibians, and sight seems to be less important than color, sound or touch. Reptiles see well and, except for snakes, some perceive color as well as we do. Experimental work verifies color vision in lizards and turtles. Sexual coloration is not only conspicuous in many lizards (which possess only cones), but also accentuated during the season of courtship. The role of color vision in some lizards probably approaches that of birds. In the latter, vision is the primary sense for receiving environmental signals and is associated not only with courtship but with flight and the accommodation required for rapid flight.

Mammalian vision is apparently much inferior to that of birds, and many mammals may lack color vision. Circumstantial evidence suggests that most mammals, primates excepted, do not perceive distant images as well as do birds. Many mammals seem unable to detect an observer (downwind) until the latter moves, but this particular aspect of vision is difficult to explore experimentally.

Sound is produced by the majority of all classes of vertebrates, and most sound-producers have efficient hearing. Although vertebrates have evolved many kinds of noise-making structures, the organ of hearing is the inner ear, or the lateral line system which is derived from it. In fishes the receptor of higher-frequency sound waves is the air bladder, an unpaired structure, and several means have been developed to deliver impulses from the air bladder to the lagena (the forerunner of the cochlea). In other vertebrates sound is received by a tympanum and transmitted directly to the inner ear.

At the extreme red end of the visual spectrum, electromagnetic waves become infrared. Infrared rays are not perceived by photoreceptors, but at least two families of snakes have independently developed sensors that respond to infrared radiation emanating from the warm-blooded quarry on which they feed. Infrared perception is not known in any other vertebrates.

In a number of vertebrates certain "nonoptical" structures register varying amounts of light. The partial eye, together with the adjacent pineal body, is especially important to the tuatara *(Sphenodon)* as well as to a number of families of diurnal lizards, and is a receptor involved in entrainment to photoperiod. Homologous structures in amphibians also assist in orienting movement toward a light source, which in nature could be a body of water.

Reception of very weak electrical currents depends on water as a transmitting medium and is known for certain fishes. Their sensitivity is adequate to detect impulses by "nonelectric" species, and some sharks and perhaps catfish can locate their live food in this manner. It is not known how many fish can perceive very weak currents, but several diverse families not only produce and communicate by low-voltage emissions but can electrolocate nearby objects in turbid water by a disruption of their own electric fields.

Suggested Readings

Disler NN (1971) Lateral line sense organs and their importance in fish behavior. (Izdatel'stvo Akademii Nauk SSSR, Moskva, 1960). Israel Program for Scientific Translations. Jerusalem

Doty RL (ed.) (1976) Mammalian olfaction, reproductive process and behavior. Academic, New York

Eakin RM (1973) The third eye. University of California Press, Berkeley

Fox HM, Vevers G (1960) The nature of animal colours. Sidgwick and Jackson, London

Protasov VR (1970) Vision and near orientation of fishes. Israel Program for Scientific Translations, Jerusalem

10. Signals

Vertebrates use the sorts of communication suitable to their habitats and habits. Thus birds, mostly both diurnal and endowed with the ability to fly, are largely vocal and visual in their communication. On the other hand, mammals, mostly nocturnal and moving more slowly than birds, are vocal and olfactory. Mammals live largely in an olfactory world and produce many scents for various purposes. Few mammals, however, are as colorful as birds.

Most signals among vertebrates are intraspecific and frequently sexual in significance. Some signals relate to nonsexual functions, such as scents or sounds to indicate alarm and sounds for echolocation. Because of our own limited ability in hearing and olfaction, it is difficult to appreciate the diversity of sounds and odors used by other vertebrates. Other signals, such as low-voltage electricity and bioluminescence, are limited to only a few vertebrates.

Fishes

Odor plays a major role in the breeding patterns of many fishes.[815] Some fish, which spawn in nests of other species that defend their nests, receive protection for their own eggs. The odor of the milt and eggs of spawning green sunfish *(Lepomis cyanellus)* stimulates the redfin shiner *(Notropis umbratilis)* to oviposit in the nest of the former. The minnow spawns only in occupied nests and only when the sunfish are themselves spawning.[471] *Hemichromis bimaculatus,* an African cyclid, cares for its young, which it recognizes by their odor, and the behavior is probably common in this large group of fishes.[656]

The sturgeon chub *(Hybopsis gelida)* is adapted for life in rapid, turbid streams; ventral rugosities are richly provided with taste buds and probably compensate for poor vision in the water they occupy. In clear waters this minnow lacks this extensive system of ventral pores.[115] Taste buds occur on many species of minnows (Cyprinidae) and are concentrated on the head region, pectoral fins and barbels.[711]

Experiments demonstrate the extreme sensitivity of fish to odors and their importance in social behavior, food-finding and migration. Such experiments usually involve (1) concealing food and (2) destroying part of the olfactory receptor or anesthetizing the olfactory epithelium, sometimes with obliteration of the eyes. Most fish can find hidden food, even when blinded, unless the olfactory structure is not functioning.

The well-known alarm behavior in fishes is a response to rupture of the skin. It is olfactory, for it does not occur when the olfactory tract is anesthetized or destroyed, and it is apparently confined to species of Ostariophysi (minnows, catfish and characins). Injured skin in cyprinid fishes and their allies (Ostariophysi) causes other individuals to become agitated, and the "alarm substance" seems to alert nearby fishes to possible danger.[19,1050]

Amphibians, Reptiles and Birds

Among *amphibians* chemoreception seems to be most important in salamanders. Many salamanders develop hedonic glands during the season of courtship, which produce substances that are stimulatory on contact, suggesting that chemoreceptors are involved. Also Jacobson's organ, well developed in many reptiles, is first seen clearly in amphibians. In the salamander family Plethodontidae, the nasolabial groove is lined with ciliated epithelium that conducts a current of fluid to the mouth and ultimately to Jacobson's organ.

The sensitivity of *reptiles* to odors suggests that they produce special smells. Anecdotal accounts indicate that this is so, and there is some evidence for the release of sexual odors in snakes. The subject has not been explored to the depth that it has been in mammals, and very probably students of behavior will make important discoveries in the realm of reptilian scents.

The role of special scents is unknown in *birds*. Olfaction is poor and, where present, seems to function mostly in food-finding. There is no evidence that birds produce scents that carry messages to others of the same species.

Mammals

Odors are appropriate for communication in air because of the rapid diffusion of small amounts of scent and the almost constant movement of air. Not surprisingly, terrestrial mammals have developed numerous scent glands and a high degree of olfactory sensitivity. Among arthropods there are hundreds of species in which one sex, usually the female, releases an odor that produces a specific response in the other sex. These odors are called *pheromones*. Some workers prefer not to use this term for mammalian scents because: (1) the responses are less precise than they are in arthropods and (2) much olfactory communication in mammals is intra- as well as intersexual. In contrast to vocal and visual communication, scents can be placed on an object where they persist for hours or even days; such *marking behavior* is a major means of mammalian communication.

In many species of mammals both sexes produce scents that serve different social functions, and some of these odors apparently attract and stimulate the opposite sex during the reproductive period. Although it may be an oversimplification to equate these substances with pheromones of insects, some mammalian scents are quite specific in their effects. Female rhesus monkeys in estrus, for example, release a vaginal secretion that stimulates males to mating behavior. On the other hand, human males declared no discrete impression, either pleasant or unpleasant, from a sample of unknown odors that came from tampons used by a series of women during their ovulatory cycles.[251]

A constant production of scents serves to identify urine of mammals not only as to species but also as to individuals, and in females, and probably males, it carries information about the individual's reproductive state. There seem to be identifying odors in fecal matter as well. These odors are the products of glands associated with the urinary tract or the anal area. In addition there are many glands on various parts of the body of many mammals, exudates of which may be placed either on the individual's body or parts of its environment.

Scent Production

A variety of structures are specialized to produce scents. In carnivores they usually secrete into urine, and artiodactyls have scent glands not only on the feet but sometimes on the head. Rabbits have chin glands, side glands characterize most shrews, and kangaroo rats (Heteromyidae) have dorsal glands. Not only do these scents mark the ground where the mammal urinates, but scents may be placed above the ground on rocks and on vegetation.

Lipids from the preputial gland are probably released with urine and are associated with social dominance. Within a given population, preputial size increases with dominance, and preputial lipids stimulate activity of sexually experienced but nonpregnant females. Does the sexual experience enhance olfactory perception or does memory affect the message of the odor? Furthermore odors of laboratory mice affect the frequency of estrus: an odor in female urine retards or suppresses estrus, and male odor accelerates the cycle.[722]

Characteristic scents distinguish both species and individuals, and most mammals generously spread their own odors throughout their territories. Even with our own poor olfactory ability, humans can detect scents of some species of mammals. The peccaries (*Tayassu* spp.), when alarmed, release an odor well known to hunters. The short-tailed shrew *(Blarina brevicauda)* can fill the woodlands with its heavy odor on a damp spring evening. Most mammalian scents, however, go unnoticed by us. A stroll with one's dog, however, suggests the existence of an olfactory world: the dog recognizes not only other dogs but cats and birds, and is sure to identify a women in menses.

Some of the prosimians have discrete skin glands that seem to produce scents used in marking, but monkeys and apes apparently do not communicate by olfactory signals. Monkeys detect the odor of a female in estrus, but they lack well-developed marking behavior. When the response by an individual is to a conspecific individual of the opposite sex, then the scent is not only sexually stimulating but also a reproductive isolating mechanism.

Anal glands release characteristic odors in many groups of mammals, and these glands are often larger in males than in females. Many kinds of hystricomorph rodents mark their mates with urine. Generally the male urinates on a prospective mate, but in some the female ejects a strong stream of urine posteriorly at the male.

Coloration

Color comes or is reflected from animal surfaces, and the type of color depends on the spectrum of light waves emanating from the integument. Colors result from pigments, or zoochromes, and from structural colors, or schemochromes. A combination of zoochromes and schemochromes can produce colors.

Zoochromes

Because of the widespread and diverse use of the word pigment, zoochrome (or biochrome) is specifically applied to nonstructural colors in plants and animals. Zoochromes absorb some band of the visible spectrum and reflect other wavelengths to give the animal a color. The color of a zoochrome depends on its molecular structure. Zoochromes are usually carotenoids, pterins or melanins.

Most carotenoids are orange or red; they occur widely in cells of plants and animals but are not synthesized by animals. A common carotenoid is carotene, the orange pigment in the carrot, but related carotenoids occur widely in vertebrate tissues. Carotenoids are fat-soluble and frequently account for the color of the scales in birds' feet and the gold-red color in many fishes. Carotenoids in birds are derived entirely from fats contained in their food, but these pigments are sometimes modified before final deposition in bird tissue. Some yellow carotenoids result in a green color when deposited over a guanine layer which is reflecting blue, but green carotenoids seem not to occur in birds. Certain carotenoids in animals can be split to yield vitamin A, which is a component of some visual pigments. Pterins account for many of the colors in fishes and amphibians. Frequently they resemble carotenoids in color, but pterins are insoluble in oils and fats and usually insoluble in water.

Zoochromes are contained in special cells, *Chromatophores,* which can expand, covering a large area and at the same time intensifying the color within or contracting, thus minimizing the effect of the contained color. Zoochromes also occur in *chromatocytes,* which do not expand or contract and which change color slowly, if at all. Chromatophores often take the name of the pigment or zoochrome within. Erythrophores are mostly reddish and xanthophores yellow, from their pigments erythrin and xanthin, both carotenoids.

Melanins, universal products in animal tissue, are usually black or brown, and melanophores are common in the skin of fish, amphibians and reptiles. Melanins are extrememly stable materials formed from tyrosine (an amino acid). The sepia of squid is a melanin and is sometimes preserved in fossil squid many millions of years old. Guanine also occurs widely as a waste product of metabolism, and is the essential component in guanophores.

Schemochromes

Structural colors, in contrast to pigments, derive their color(s) from the configuration of their surface, and the colors produced may change with the direction of the light they reflect or the direction from which they are viewed. Structural colors are produced by one or a combination of three phenomena: interference, diffraction and scattering.

Interference is produced by reflection of light from two surfaces, which are usually close together and parallel, or nearly so. Light passes through different substances at different speeds, variation being caused by the refractive index of the material. The refractive index of air (or a vacuum) is 1, and anything denser has a refractive index greater than 1. When light strikes a film of oil, for example, part of it is reflected from the outer or upper surface of the film and part of it passes into the film and is reflected from its inner or lower surface (Fig. 10.1). Thus the reflected light one sees comes from two surfaces; as the light penetrates the first surface, its speed is slowed. When it is reflected from the lower surface and reenters the air, the specific wavelengths may be out of phase with those that were reflected from the upper surface of the film. Thus when beams of light are parallel and superimposed on one another *and out of phase,* destructive interfer-

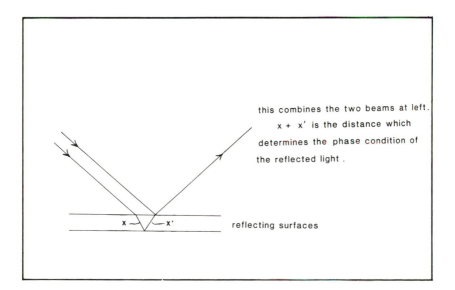

this combines the two beams at left.
x + x' is the distance which
determines the phase condition of
the reflected light .

reflecting surfaces

Fig. 10.1. Reflection of light from both upper and lower surfaces of a transparent substance, with a refractive index greater than 1.0 producing interference.

ence results (Fig. 10.2). Color is produced from white light when one part of the spectrum is reinforced at the expense of another. The degree of interference is the result of the angle at which light hits the outer surface, the refractive index of the material and its thickness. Interference is a source of iridescence in animals.

As light passes through a small orifice or slit, white light separates into its several parts, blue light bending most and red light least. This is *diffraction*. As light is diffracted in passing through a fine hole, the color seen depends on the angle from which it is viewed (Fig. 10.3). Viewed from position A light passes through the slit or orifice unbent and unaltered but, when light is viewed from position B, only diffracted light is seen. The degree to which light is bent increases inversely with its wavelength, so that as the position of viewing moves from points A to B, the light seen changes from red to blue. And, as one views an iridescent bird or fish that has a diffraction grating in its feathers or scales, its color changes with the viewing angle. Inasmuch as diffraction is an "edge effect," it can sometimes be observed from an angular or a pitted surface. Reflection from an angle is a common type of diffraction in vertebrate coverings. Reflection from a pitted surface causes diffraction of "white light" into different monochromatic or "pure color" light. Diffraction is a common source of interference, producing iridescence in such birds as hummingbirds (Trochilidae). Diffraction can emit a single color when light passes through successive layers or reflecting points, spaced so as to allow only one color to pass through, and this only when viewed from the proper angle. This is called a space lattice, and occurs in certain iridescent feathers and fish scales.

Scattering of light occurs when small reflective particles redirect light. When these particles are extremely fine, too small indeed to reflect any but the shortest wavelengths in the visible spectrum, they reflect bluish light and allow the longer wavelengths to pass about them. This is called *Rayleigh* (or *Tyndall*) *scattering*.

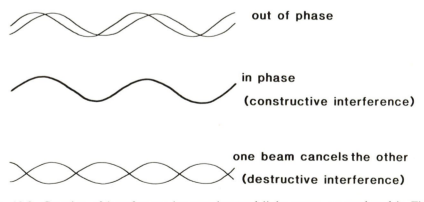

Fig. 10.2. Creation of interference by superimposed light waves, as produced in Fig. 10.1. Superimposed rays, out of phase, cause various spectral changes. Light is unchanged but intensified by constructive interference. Destructive interference is caused when the light rays are out of phase.

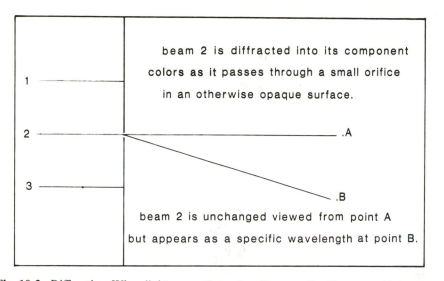

Fig. 10.3. Diffraction. When light passes through a slit or small orifice, part of it becomes bent, or diffracted. The light seen varies with the angle from which it is viewed. Shorter wavelengths are bent at greater angles.

In vertebrates guanine crystals are smaller than the wavelength of red light, and thus selectively reflect blue light. Tissue with layers of guanine tends to vary from whitish blue to deep blue. Tyndall scattering accounts also for the blue color of the sky when the sun is high; at dawn and dusk, when sunlight nearly parallels the earth's surface, it passes through a relatively long layer of dust particles large enough to reflect red rays, a different effect of scattering. Tyndall scattering, or Tyndall blue, accounts for the blue on the undersides of many fish, amphibians and reptiles, as well as for the blue buttocks of mandrills. Blue in feathers is always a result of scattering. When a guanine layer lies beneath a yellowish carotenoid, the blue passing through yellow endows the animal with a green color. The green of many amphibians is derived in this manner. Green may also reflect the pigment biliverdin; this accounts for the color in some wrasses (Labridae) and some anurans. With somewhat larger particles, too large for scattering, the effect varies and depends on both the size of the particle and the relationships of the scattered light with that which is refracted and reflected. The result is called *Mie scattering,* and can range from bluish to white. Most animal color effects from scattering are due to Rayleigh or Tyndall scattering.

The predominant color of an animal is usually a combination of several colors. The blue of mackerel and many other fish is a product of bluish light reflected through a thin layer of melanin. The same blue reflected through a yellow appears green, and guanine alone accounts for silver in such fish as silversides and jacksmelt (Atherinidae). Iridescence in vertebrates is sometimes produced by guanine crystals, such as in the scales of many snakes.

Color Changes

Color changes involve the skin and its products (feathers and fur). The ability to change rapidly belongs to the dermis and epidermis, and color changes in plumage and pelage require their physical replacement or wear. Color changes with development or season are *chromogenic*. With a few exceptions, color changes in birds and mammals are chromogenic, as are, for example, the different color patterns of young and adult racers (*Coluber* spp). The winter development of white fur of weasels *(Mustela erminea, M. frenata)* and hares *(Lepus americanus)* results from molt as does the development of the white plumage of ptarmigan (*Lagopus* spp.). The vernal darkening of the snow bunting *(Plectrophenax nivalis)* is caused by the differential wear of the white in its feathers during the winter; by spring this whitish sparrow has become blackish. This is a chromogenic change and occurs in the absence of molt or feather growth. Not all chromogenic changes are seasonal; in some shrews (e.g., *Sorex fumeus, S. arcticus* and *S. trowbridgei*) the adult pelage is unlike the color of the juvenile and, because of their short life span, these mammals rarely experience a second molt.

Color changes affecting both sexes equally are usually protective. Many arboreal anurans and lizards can become green when on a green background or brown when on a bark or a rock of a brown or gray color. Some changes result in changes of intensity of a gray or brown, such as are commonly seen in terrestrial lizards; in these species a darker color is often seen on a cool day and may be a thermoregulatory device. Color changes that affect only one sex are most commonly associated with courtship or territoriality, or both.

Rapid color changes result from movement of pigment within the chromatophore, as in some anurans, some lizards and many bony fishes. These changes are called *chromomotor*. Although chromogenic changes usually serve a seasonal purpose, such as courtship or concealment in the snow, chromomotor changes are responses to an immediate and unanticipated need. As a flounder swims from one color of substrate to another, its "eyed-side" changes color accordingly, and certain tree frogs (Hylidae) change from brown to gray or green, depending on the surrounding color.

The distinction between chromogenic and chromomotor changes is at the cellular level: chromomotor changes result in movement of pigment within a chromatophore, and the slower chromogenic changes occur with gradual changes of pigment within a chromatocyte.

The direct response of a chromatophore is to light and reflects also the cell's genetic capability. This is called *primary* response: it is not modified by neurohormonal stimuli, not is it dependent on ocular function. A microbeam directed on a single anuran chromatophore can result in movement of pigment, and probably chromatophores vary in their responsiveness to different wavelengths. Thus chromatophores are light receptors in themselves. In contrast, *secondary* responses are triggered by neurohumoral stimuli initiated in the eye, parietal eye, pineal body or hypothalamus. Color changes in the chromatophores may be coordinated with behavior, as in courtship of a fish or a lizard.

In most ectotherms most changes are of intensity and are effected by changes in size of melanophores: contraction of melanophores results in a lightening of the animal's general color. In a few fishes, frogs and lizards one color can be suppressed and replaced by another. Color change in bony fishes is under nervous control, but under hormonal control in amphibians. In amphibians, intermedin (from the intermediate lobe of the pituitary) causes expansion of the melanophore. Species of the iguanid lizard *Anolis* effect extreme color changes from bright green to dark brown; as in amphibians, color changes in anoles are caused by changes in intermedin levels in the blood.[8,840,996]

The chromatophores are arranged with melanophores underlying the color-bearing chromatophores and iridophores containing guanine. The anole *(Anolis carolinensis)* is greenish when the melanophores contract but expanded, they change to brown by allowing melanin to shield reflection from the iridophores; in this lizard only the melanophores cause color variations. True chameleons (Chamaeleontidae), on the other hand, respond solely to nervous stimulation of the chromatophores. Color change occurs also in some geckos (species of *Sphaerodactylus*), agamids and true chameleons (Chamaeleontidae), but not in snakes or turtles and only rarely in crocodilians. At least one crocodilian, the caiman *(Caiman sclerops)*, can change color. When the animal is chilled, its color darkens, obliterating the distinctive banded pattern.

Color Patterns

Many intra- and interspecific messages are conveyed by visual means. Although some visual signals consist of specific sorts of body movements, many others consist of colors and distribution of colors on the animal's exterior. Other types of coloration seem to relate to environmental adaptations without reference to behavior between conspecific individuals. It might be well to point out that, although hundreds of biologists have written reams on animal coloration, there is still active controversy on many interpretations of color patterns.

Conspicuous color patterns in animals sometimes indicate a dangerous or distasteful element in the bearer. It is well established that predators avoid insects with *warning* or *aposematic* coloration, but such coloration in vertebrates is less common. Black and white is the most common warning pattern, and the New World skunks (Mephitinae) are the best-known examples. Like other aposematic animals skunks are fearless, and their odor is unforgettable.

Sexual Coloration

Bright colors in one sex are usually associated with courtship. In many fish and lizards courtship colors develop during the breeding period in the presence of circulating androgens; nuptial plumage in many birds is molted when nesting ends. Many male lizards develop bright colors as reproduction approaches, such colors intensifying with gonadal development. Distinctive sexual coloration is absent in snakes, most mammals and other nocturnal vertebrates that are fossorial or color blind, or both.

In a large number of fishes and lizards males develop distinctive nuptial colors to which females of the same species respond. The red of courting sticklebacks (Gasterosteidae) is produced by an androgen, as are the various nuptial hues of many male agamid and iguanid lizards. Sexual hormones produce colors in females of some species: postovulatory levels of progesterone result in pink sides, the color disappearing after the eggs are laid. The function of such colors in females is not known. In snakes, turtles and crocodilians both sexes are usually identically colored. Sexual differences in the flesh of several primates is apparent but is scarce in the pelage of all mammals.

Many avian males have distinctive coloration whereas females of the same species tend to be drab. This difference is a specific feature and occurs in some species of a given family, whereas in other species male and female are alike. Among species of the puddle duck genus *Anas,* for example, males of most species have a colorful plumage for the greater part of the year, but in several species of *Anas* both sexes have the same coloration. In most seabirds (loons, cormorants, shearwaters, alcids and others) the plumage is alike in both males and females. Among the crows, jays and magpies (Corvidae) both sexes have the same colors, but in the birds of paradise (Paradiseidae) the males are gaudy and distinctive and the females dully colored. In these examples sexual differentiation in plumage is associated with visual displays in courtship and territoriality; in the absence of sexually distinctive plumage, cues of color and form are replaced by behavioral cues.

Concealing Color Patterns

In many vertebrates zoochromes and schemochromes are distributed so as to hide the creature within its environment. *Concealing coloration* effects the blending of the animal with its background; in most cases the apparent function of such concealment is protection from predators.

Pale colors in desert animals occur in many lizards, birds and mammals. Most workers believe such paleness to be cryptic and adaptive to pale desert substrate. Concealment by cryptic coloration must be especially important in deserts, where protective vegetation is frequently sparse. The pale coloration is apparently adaptive to some extent and genetically fixed, and some small mammals on dark substrates in deserts have dark pelage. According to Gloger's Rule, however, animals in warm dry climates tend to be pale, implying that a reduction in melanin somehow occurs in warm dry air.

Coloration in water is affected by selected penetration of light, longer wavelengths at the red end of the spectrum penetrating least (Fig. 10.4). Thus red pigmentation in the absence of red light appears blackish and tends to be concealing. In the bright illumination of a shallow-water, coral-reef habitat, most fishes are various shades of yellow, green and blue.

A number of pelagic larvae of fishes have almost no pigmentation except for the melanin in the epithelial lining of the eyes, and they are more-or-less transparent as they drift passively at the surface. The larvae of some eels (Anguillidae)

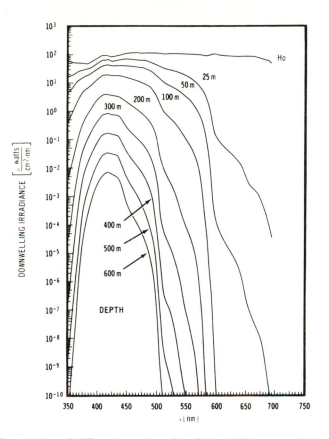

Fig. 10.4. Penetration of different wavelengths of the visible range of the spectrum to various depths of clear fresh water (Crater Lake, Oregon) measured as intensity (watts/cm^2) of light (Smith and Tyler, 1967).

and herring are at first nearly transparent. The protective value of such concealing coloration is questionable, for larval mortality among such species is high.

Light penetrating a forest is modified by chlorophyll, which absorbs light at both ends of the visible spectrum and reflects and transmits greens; this phenomenon causes a maximum of ambient light near 550 nm (Fig. 10.5). In such an environment unpatterned greenish animals may be less conspicuous than reddish or bluish species.

Many ground-nesting vertebrates have their venter concealed from view, and the dorsal pattern may resemble the substrate. This resemblance is frequently achieved by a bold pattern of contrasting lines and patches that follow irregularities and their shadows on the ground. *Ruptive coloration* is most familiar in birds: the woodcock *(Philohela minor)* nests on the ground in open woodlands or thickets and is an uneven design of earth-browns and short black stripes; the various spe-

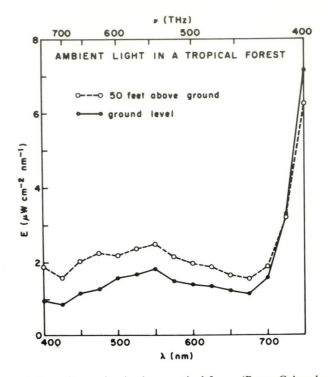

Fig. 10.5. Spectral irradiance-density in a tropical forest (Barro Colorado Island, Panama). The maximum near 550 nm is mostly caused by the absorption of chlorophyll above and below the greens, giving ambient light a greenish hue (Hailman, 1977).

cies of goatsuckers (Caprimulgidae) nest on gravelly ground and dorsally are a mottled gray. Chipmunks (*Eutamias* spp. and *Tamias striatus*) are conspicuously striped as are many species of ground squirrels *(Citellus);* they are diurnal and spend most of the time on the ground. The disruptively patterned *Tamiops* spp. of tropical Asia closely resembles the chipmunks, but it forages high in the tree-tops. Disruptive coloration does not always require a terrestrial background.

Many snakes that are slender and move rapidly are conspicuously marked with longitudinal stripes. As they crawl through grass, they seem immobile until they suddenly disappear. Similarly many slender and rapidly moving lizards tend to have longitudinal stripes.

A universal principle in the distribution of melanin is that the rate of distribution increases with exposure to sunlight. The Nile catfish, which swims "belly-up," has the morphologic venter darker than the dorsum, and flatfish, which have the eyeless side without melanin, develop pigmentation on that side when kept in glass-bottomed aquaria with strong illumination from below. Under most conditions the animal becomes darker dorsally than ventrally; the result is that its shaded (ventral) part appears lighter than it would if the melanin were evenly distributed. The "dark above, light below" pattern is called *countershading,* because it tends to reduce the conspicuous effect of a shadow on the venter. The difference in the melanin on the upper and lower sides of an animal increases with the intensity of dorsal light. Thus countershading is most obvious in diurnal animals that stand in sunlit areas and least apparent in species that are nocturnal or dwellers of the forest floor.

Most shorebirds are clearly countershaded, but the phenomenon is scarcely apparent in bats or moles or forest-dwelling thrushes. There are exceptions among shorebirds: the golden plover, black-bellied plover and dunlin all have *black* venters, which render them conspicuous on the beaches that they frequent. Countershading is also less apparent in an animal standing high above the ground, such as a giraffe or a crane, and is more obvious in a mouse, ground squirrel or lizard. In pelagic fishes countershading is doubly protective: the blue-gray dorsum of a mackerel or a herring is concealing when viewed from above, and the silvery-white belly tends to resemble waves and foam of the ocean surface when viewed from below.

Although countershading is best developed in invertebrates living near the ground, the venters of many tree-dwelling birds are pale; such birds may achieve some concealment by the reflection of ambient light: viewed from the ground, the paler underparts of a bird in branches overhead may assume a greenish aspect, reflecting the color filtered by and reflected from the leaves. The advantage of such concealment, although perhaps slight, varies with the habitat.

Protection from UV radiation would also lend survival value to countershading. In a group of anoles (*Anolis* spp.), the degree of melanin in the peritoneum is positively correlated with the amount of time spend in the sunlight, forest-dwelling species with this membrane being but lightly pigmented.[830] The reduction of melanin in the skin of lightly pigmented desert species is compensated by melanin in the muscles and especially the peritoneum; these internal layers of melanin protect the viscera, and the gonads in particular, from the damaging effects of UV light.[830]

Mimicry

Many unrelated taxa are colored similarly. Among arthropods are numerous examples of one species' resembling other phylogenetically distant species. Sometimes predators have the form and color of an innocuous species, and the predator is said to be a mimic. Mimicry is proposed to explain many examples of similar color patterns in vertebrates, and there is a large body of literature on the subject. A commonly cited example consists of a pair of species, one of which is dangerous, harmful or distasteful, designated the *model,* and another, the *mimic,* which lacks the distinctive characteristics of the model but which resembles or mimics it in color. This color pattern, when distinctive, is a *warning coloration;* presumably it alerts a potential predator to the dangerous or undesirable nature of its wearer. Such a predator–prey pair of model and mimic constitutes a *Batesian mimicry.* Students of animal coloration believe that the mimic benefits from the protection derived from the color pattern, which resembles that of the model. Not all examples of similarity constitute mimicry; before deciding for himself, the student should consider the advantage to be gained from mimicry and the likelihood of alternative possibilities.

In some animals interspecific similarities in color patterns include an array of species that embraces a broad diversity of offenseiveness or danger, so that the species reinforce each other's similarity. In such examples no species is either clearly the model or the mimic, and this is called *Müllerian mimicry*. As in Batesian mimicry the basic color pattern is presumed to be a warning coloration. The coral snakes (Elapidae) are strikingly banded with black, red and a light cream color, and several unrelated colubrid snakes have similar color patterns. The colubrid species include both harmless and mildly poisonous snakes, and perhaps the harmless colubrids and the highly venomous coral snakes mimic the mildly toxic colubrids. This being the case, these species would constitute Müllerian mimicry. One objection to this supposition is that most of these snakes are largely nocturnal, fossorial, or both, and that their mammalian predators may be color-blind. In addition, Müllerian mimicry depends on the members of the mimicry ring being more-or-less sympatric, but in reality this is not always so.

In a number of snakes the underside of the tail is distinctively colored; when raised, the tail resembles a head. Such deception is based on both coloration and behavior, and seems to divert a predator's attention from the snake's real head to the conspicuous tail. In one such species, the rubber boa *(Charina bottae)* of western North American, museum specimens frequently have bruised tails.

Numerous species of birds in several families, mostly passerines, normally lay their eggs in nests of other species, and these social parasites produce eggs that are pigmented like those of the host species. Presumably this similarity prevents the host from recognizing the parasitic nature of the alien egg; the latter having been reared by the same host species makes the association at nesting time a year later natural. Among Eurasian cuckoos *(Cuculus canorus)* there is frequently one host in any given geographic range; thus the eggshell pigmentation evolves to resemble that of the host's eggs. Not surprisingly, when an egg is laid in the nest of a species other than the normal host, survival is low. Similarity between the egg of the host and its parasite occurs among many parasitic species.

Sound Production

Animal sounds differ not only by species but by age, sex and season. Certain sounds of course may convey specific meansings, and some vertebrates produce a large number of distinctive sounds. The song of a male bird or sea lion in the breeding season announces his presence not only to potential mates but also to other males that might invade his territory; the call of a frog may serve not only to attract females to his pond, but also to define his territory.

Not all animal sounds are produced vocally. Fish are nonvocal but produce sounds from the air bladder and sometimes by moving one bone over another. Many birds produce sounds from the passage of air through feathers and the woodpecker drums on trees or drainpipes of buildings. The "drumming" of the Nearctic ruffed grouse *(Bonasa umbellus)* is made with its wings. Some mammals make their typical sounds with foot-stamping.

Sound in Fishes

Many diverse fishes produce sounds, but the air bladder is the most exhaustively sonic organ in these vertebrates. Many species can produce characteristic sounds by moving one bone against another, but this action may occur in some only when the bones (such as a pectoral spine) are manipulated by the human hand. A silurid catfish *(Clarias batrachus)* of India makes a loud clear sound by moving the pectoral spine. The base of the spine as well as the socket of the cleithrum into which it fits are corrugated; such rugose surfaces are absent in silurids unable to produce sounds. Pharyngeal teeth stridulation accounts for acoustics produced by many families of bony fishes; these sounds are characteristic for each species and sometimes sex, and must be important for courtship and other types of intraspecific identification.

The swim bladder produces sounds by vibrations. Drums and croakers (Sciaenidae) expand the air bladder by muscles extending from ribs on the sides of the visceral cavity to the air bladder, and sudden relaxation causes the sound from which these fish derive their names.[37,758] In other families there are different muscles extending to the air bladder but, in each, sound is produced by air bladder vibrations, which then enter the water. In some families the muscles attach only to the air bladder. In either case, muscular contraction causes tension on the walls of the air bladder, and the elasticity effects a return to the "normal" shape on muscular relaxation. The frequency emitted results from both the size of the air bladder and the frequency of the muscular contractions. Air bladder sounds seem not to exceed 500 Hz, but stridulatory sounds may exceed 8000 Hz. Nonostariophysine fishes generally have high thresholds with low ranges of frequencies, mostly less than 2000 Hz. Several of the electric fishes (e.g., Electrophoridae, Mormyridae, Gymnotidae and Gymarchidae) are Ostariophysi, and as such possess Weberian ossicles: they are producers not only of electricity but also of sound, and probably communicate in both media.[979]

In some families of fishes, such as cods (Gadidae) and toadfish (Batrachoididae), the males emit a distinctive sound to which the females respond. The sexual significance of these sounds seems confirmed by their occurrence during the breeding season. Several kinds of priacanthid fishes produce sounds by vibration of the air bladder. Sounds produced by the cusk-eel *(Rissola marginata)* differ between the sexes and seem to serve for sex recognition. In drums (Sciaenidae) sound is an important aspect of courtship.

Meso- and bathypelagic fishes generally lack drumming muscles on the swim bladder, and in many of these species the swim bladder is filled with grease. Most deep-sea fishes are silent and communicate by light and odor. Those with efficient sound-making swim bladders dwell on the bottoms of deep-sea slopes. Males of brotulid fishes, for example, are provided with well-developed drumming muscles running from the anterior part of the swim bladder or ribs of the first three vertebrae to the large otic capsule (Fig. 10.6).

Sound in Amphibians

Among modern amphibians the anurans alone are conspicuously vocal, and they alone have a middle ear designed to capture air-borne sound. The middle ear of early amphibians (and anurans, today) may have developed from a pharyngeal diverticulum of rhipidistian fishes. Opening to the outside and leading to the inner ear, this passage needed only to be closed by a membrane to transmit air-borne vibrations to the inner ear. The inner part of this passage became the middle ear of anurans and ultimately of other terrestrial vertebrates.

Anurans produce sounds by forcing air out of the lungs through the larynx to the air sacs. This air is held in the sacs until either (1) they are maximally distended or (2) no more air can be forced from the lungs. As the same air is returned to the lungs, no sound is made. Thus this sound is made in a closed system, and a frog can call from beneath the surface of the water.[887]

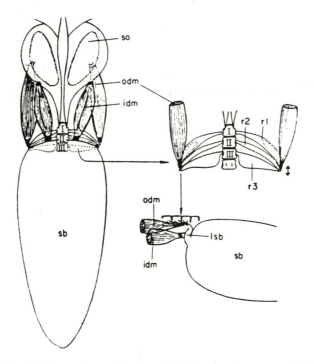

Fig. 10.6. Swim bladder and associated sound-producing structures of the brotulid fish, *Monomitopus metriostoma.* At left, the swim bladder (sb), with outer (odm) and inner (idm) drumming muscles extending from the anterior end of the swim bladder or the first three vertebrae (I, II and III) to the saccular otolith (so). Upper right, modified ribs I–III. Lower right, ligament (lsb) attaching swim bladder to ribs I–III (Marshall, 1967, reprinted with permission, Marine Bio-Acoustics, Pergamon Press).

The air sac is either median and gular or lateral and paired (or somewhat intermediate). Air sacs are thin and elastic and may contain striated muscle, so that they play an active role in returning air to the lungs. The nature of the call is determined not only by the air sacs but also by the larynx, the frequency of the call note being higher in smaller species.

In almost all frogs and toads, males are vocal and females silent. Both sexes of the bell toad *(Ascaphus truei)*, *Bufo boreas* and *B. bocourti* are silent and also lack both a typanum and a columella.[888] The two auditory receptors of the anuran inner ear, the pars basilaris and the pars amphibiorum, are nevertheless present in both species. Females of several species are known to call. Among the terrestrial breeding leptodactylids, there are females that call and attract males to their subterranean nests. Females of at least some species of the leptodactylid frog genus *Tomodactylus* call, but they lack vocal sacs.[243]

Virtually all male anurans produce specifically distinctive sounds, which attract conspecific females during egg-laying. Because fertilization is external, there is a possibility for interspecific mating, and vocalization is the most important isolating mechanism for anurans. Calls differ in frequency and duration and may vary intraspecifically according to the species composition; the calls of *Hyla ewingi* and *H. verreauxi* of Australia, for example, are similar when the two species are remote from each other, but the calls diverge when they are sympatric. Among all anurans studied, call distinctiveness increases with sympatry.

Although most salamanders are apparently silent, the spotted salamander *(Ambystoma maculatum)* produces low sounds of 1500 Hz; both sexes are vocal, but the role of their voice is unknown.[1111]

Sound in Reptiles

Although vocal lizards occur in virtually all families, communication by sound is conspicuous in only two, the geckos (Gekkonidae) and the snake-lizards (Pygopodidae).[1080] Geckos produce a variety of "clicks and rattles," which appear to be specific and may be territorial signals.

Although snakes are generally silent creatures, many make hissing sounds. Rattlesnakes (Crotalinae) are an exception: at each molt a fragment of the skin remains loosely attached to the tail and, with successive molts, a rattle develops. Some turtles are vocal, and crocodilians can be heard for miles; in each group voice is associated with reproduction, but the exact function is unknown.

Sound in Birds

The avian voice box is the syrinx, an expansion in the breathing passage where the trachea branches into the two bronchi. Within the syrinx two tympanic membranes vibrate with the passage of air; muscles about the tracheal and bronchial rings change the size of the passages, tension on the membranes and type of sound. The two tympanic membranes vibrate independently of each another and may produce different song elements simultaneously. Sound is amplified by the vibration of air in the interclavicular air sac about the syrinx and further modified by the trachea and the mouth. Sexual differences in vocalizations reflect differences in the sound-producing structures. Among ducks the female utters the familiar quack, whereas males (or drakes) have a soft, mellow whistle; the latter have a conspicuous swelling or enlargement in the upper part of the trachea.

A song consist of a succession of notes given in characteristic sequence and interval. Although songs are usually typical of a given species, conspecific males may differ among themselves, and one male may utter a song that differs slightly from that of a neighboring bird. Evidence indicates that such minor variations are recognized by both other males and their mates.

Parts of bird songs are innate—parts are learned; experimenters have developed meticulous procedures for determining the extent to which vocalization is learned. When kept in soundproof rooms, birds utter only inherited notes. Students of bird songs have placed eggs of one species in nests of another species, outside its own breeding area, and the infant learns songs and call-notes of the foster parent. Typical songs have innate elements to some extent, no doubt dictated by the morphology of the syrinx; young males, however, learn parts of their song by listening to adult males. In a few species, males may have a broad repertoire, which is acquired by listening to males of other species. Call-notes are apparently innate and may be uttered before they are heard. The *location call,* a short call given by a fledged bird to announce its position to its parents, is genetically determined.

Songs of most birds fall well within the range of human hearing, not exceeding 8 kHz for most species, and the upper limits of the auditory range are not known to exceed 20 kHz for most birds, although some species emit weak ultrasonic components as part of their normal sonic songs.[1009]

Imitation may account for geographic variation in bird song. These variations sometimes involve the speed of the songs along the Atlantic Coast of North America. The yellow-throat *(Geothlypis trichas),* a warbler, sings its characteristic song more rapidly in the north than in the south. Commonly such local dialects include changes in pitch and separation of notes, as in the chaffinch *(Fringilla coelebs)* (Fig. 10.7). Some birds sing only on their nesting grounds, and young males hear only the adults in the region of their birth. Sedentary species may also develop dialects. Thus vocal variations may have a geographic basis, although they are probably learned and not genetic.[575] From birth, females become accustomed to the dialects sung by males; this may serve to strengthen geographic ties of the female.

Most songs are associated with courtship and territoriality and are heard most frequently during the breeding season when androgens are at high levels. In the tropics, where populations are sparse and territories generally widely spaced, song serves primarily to attract a mate.

Although the song of the male may serve both to identify his territory to other males and to remind the female of his presence, in some species males use a distinct vocalization for each purpose. Careful observations on captive budgerigars show that ovulation occurs after a specific precopulatory call by the male, recalling the visual stimulation of a courting ring dove needed before the female lays her egg.[122] There may be stimulatory effects of a song on other males and con-

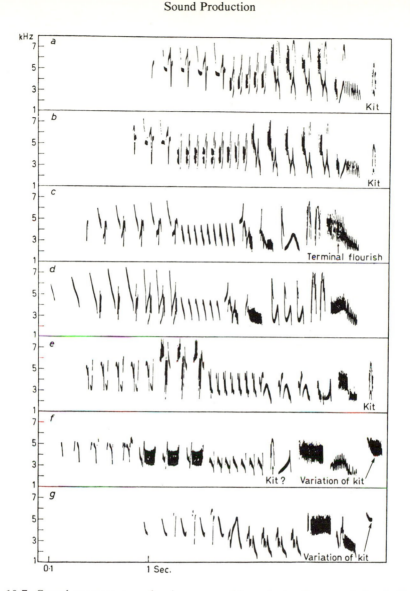

Fig. 10.7. Sound spectrograms showing geographic variants of songs of the chaffinch *(Fringilla coelebs)* (Thielcke, 1969; courtesy of Cambridge University Press).

ceivably other females; apparently songs intended for a mate are of low volume and do not carry far. Some birds have two or more courtship songs; a few, such as the mimic thrushes (Mimidae), not only have several distinctive songs, but some of them, such as the mockingbird, may imitate songs of other species.

Several types of brief utterances are classified as calls. They seem not to have any sexual significance but are used to communicate among a group that may comprise one or several species. Some calls increase the cohesiveness of a flock, others seem to prepare birds for flight to a feeding or resting ground (especially in the nonbreeding season), and still others are alarm calls that announce danger. These calls are remarkably similar in frequency and duration among species in different families (Figs. 10.8 and 10.9). A few bird calls are sexually distinctive

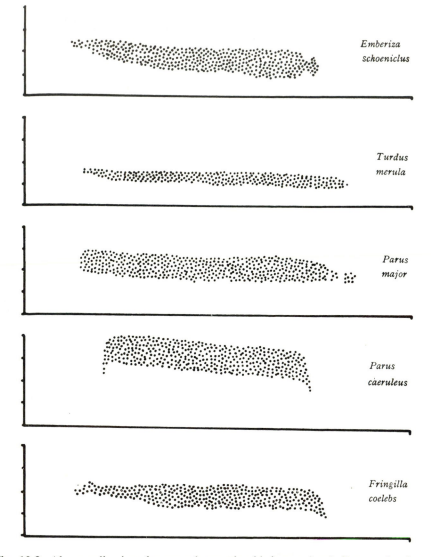

Emberiza schoeniclus

Turdus merula

Parus major

Parus caeruleus

Fringilla coelebs

Fig. 10.8. Alarm calls given by several passerine birds as a hawk flies overhead: reed bunting *(Emberiza schoeniclus)*, a blackbird *(Turdus merula)*, great tit *(Parus major)*, blue tit *(Parus caeruleus)* and chaffinch *(Fringilla coelebs)* (Marler, 1957; reprinted from *Behaviour,* courtesy of E. S. Brill).

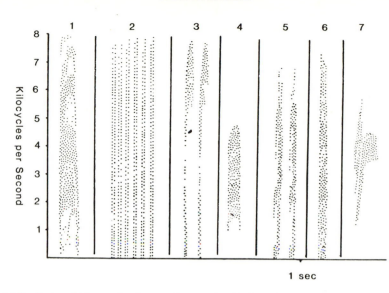

Fig. 10.9. Owl mobbing calls of several passerine birds: 1, blackbird *(Turdus merula)*; 2, mistle thrush *(Turdus viscivorus)*; 3, European robin *(Erithacus rubecula)*; 4, garden warbler *(Sylvia borin)*; 5, wren *(Troglodytes troglodytes)*; 6, stonechat *(Saxicola torquata)*; 7, chaffinch *(Fringilla coelebs)* (Marler, 1957; reprinted from *Behaviour*, courtesy of E. S. Brill).

and may serve for sexual recognition in species in which both male and female are similarly colored.

Sound in Mammals

Vocal communication among mammals is more complex than the roar of a large cat or the higher-pitched bellow of an elephant. Although the sounds of mammals are usually brief in contrast to bird song, some are long and complex. Vocalizations of most mammals fall within the range of human hearing, but some rodents and insectivorous bats are ultrasonic (with frequencies above 20 kHz).

Understanding of mammalian sounds suffers from the fact that most mammals are far more visual and olfactory than vocal in their communication. Also, when we hear mammals in nature, they are frequently not visible, and we cannot associate the sound with behavior. Nevertheless some mammals, especially carnivores and primates, make a great variety of sounds, and it is reasonable to assume that these vocalizations convey distinctly different kinds of information.

Prolonged vocalizations may be pulsations of varying pitch but, in contrast to bird songs, they are usually not complex. They may consist of rapid repetitions of monosyllabic barks. Some long utterances seem to occur spontaneously or in response to a physiologic state, such as rut, and not necessarily in response to another individual. These "songs," while lacking the complexity of many bird songs, may serve to advertise the presence of the caller. Very frequently the caller elevates its head or stretches out its neck; either action tends to straighten the path of sound and enhance its intensity.[720]

Young mammals of many species give characteristic calls that elicit a response from the mother: among colonial groups, such as sea lions and bats, a mother can distinguish the voice of her own young from among the squealing horde. Infant mice of several species of Cricetidae and Muridae produce distress calls, both within sonic (8–20 kHz) as well as ultrasonic (60–140 kHz) ranges. These calls are given when the young are chilled or touched, and the mother is quickly drawn to played-back recordings of these calls. Experimentally the maternal response is best developed in lactating females, which are even attracted to calls from young of other species, whereas males and virgin females show a lower level of response.[676]

Many mammals produce nonvocal sounds, most commonly by foot-stamping or tail movement. Rabbits, kangaroos and some rodents frequently make thumping sounds with their feet when under duress, and some rodents create a tapping sound with their tails. The beaver is famous for the loud, sharp slapping noise it makes with its tail in water.

Many short utterances, called barks, are given not only by various species of dogs but also by deer, especially the species of muntjac (*Muntiacus* spp.), which are frequently called barking deer. Behavior associated with barks suggests that they may correspond to the alarm calls in birds.

Carnivores are also extremely vocal and communicate by a variety of sounds. Studies on domestic and captive dogs and cats reveal some meanings of different sounds, but meanings may also vary with the circumstances. Generally low pitches indicate aggressiveness and high pitches, alarm or fright. Their voices do not approach the complexity of vocal communication of the primates.

In migratory pinnipeds, sound has the same roles as with many migratory birds. In the Steller sea lion *(Eumetopias jubatus)* the bulls go to a rookery about two to four weeks before the arrival of the cows and vocally establish territories. Their voices may not seem musical, but they are sexually distinctive, and the grunts and howls of a bull sea lion serve as well as the song of a thrush in announcing his presence to unattached females. Although bull sea lions (Otariidae) produce sounds of moderate frequency when in the rookery, the vocalizations of sea lions in water may reach 8000 Hz. In contrast, seals (Phocidae) are relatively silent.

Among nonprimate mammals, whales produce the most complex sounds, and some of the gigantic mysticeti have extensive variations in song. Such vocalizations were known to early whalers from New Bedford, but hydrophones and recorders permit modern students to document whale music. The humpback *(Megaptera nodosa)* produces a prolonged song varying from 0.05 to 10 kHz, and a song may extend from 7 to more than 30 minutes. Individuals seem to produce their own songs; a given whale is prone to repeat his own pattern, but another humpback will sing on a very different theme. The songs of the humpback whale have been heard at various seasons in both the Atlantic and the Pacific, but their function and the sex of the singers remain a mystery.[795]

Some rodents communicate by sound, and a few are on occasion as musical as the oscine birds. The house mouse *(Mus musculus)* and the grasshopper mouse *(Onychomys leucogaster)* are known to sing. The grasshopper mouse (Fig. 10.10) has a vocal repertoire including both sonic (>16 kHz) and ultrasonic (20–64 kHz) elements.[384] The function of these songs is unclear. Precoital communication between rats *(Rattus norvegicus)* involves ultrasonic vocalization by males. Sounds around 22 kHz by males stimulate receptivity of females.[676] Most squirrels are loquacious and perhaps produce a greater variety of sounds than do most rodents. There is a dichotomy in sounds made by lagomorphs: rabbits and hares (Leporidae) are foot-stampers, but pikas (Ochotonidae) utter clear whistles.

Primates are definitely the most vocal of all mammals. Not only do monkeys, gibbons and their allies communicate in a large measure by voice, but a given individual may have a repertoire of 30 or more distinct sounds. Higher primates have a varied series of vocal signals, and sound and sight replace scent as the primary means of communication. These sounds vary from 10 to perhaps as many as 25 for any one species, but they grade from one to another, usually without boundaries. Vocalizations of monkeys and apes are generally low-pitched barks and grunts, but some higher-frequency whistles and howls approach 5 kHz. A

Fig. 10.10. Adult grasshopper mice *(Onychomys leucogaster)* in typical position (on haunches, propped by its tail) when giving a clear call of from 9.5 to 13.5 kHz (Suzanne Black, after a photograph by Hafner and Hafner, 1979).

variety of sounds suggests a variety of meanings, but observers believe that the messages of primate vocalization are more emotional than informational. In addition many primate vocalizations occur, together with specific facial expressions, which themselves are rich in meaning.[721]

Students of primate behavior have correlated sounds with actions and responses and found some similarity of vocalizations for attention-getting, alarm, aggression, affection, hunger and so on. There is a pattern of similarity of sounds between the first vocal repertoire of a human infant and some nonhuman primates. Circumstances seem to alter the meaning of some sounds, such as a soft, low sound given by a female to her infant in contrast to a potential mate. The Kloss's gibbon *(Hylobates klossii)* uses loud vocalizations in intergroup signaling for both alarm and spacing in dense vegetation, over distances as long as 500 m.[721]

In a general way calls of birds and mammals are correlated with the message. Low frequencies are associated with hostility or aggressiveness; gutteral sounds or growls, sounds that are low and harsh, suggest a willingness to attack or to fight in defense. In contrast, highly pitched whines suggest submissiveness or friendliness, and these emotions may be emphasized by tonal purity of the utterance.

Echolocation

In the early part of this century ships at sea determined the depth of the ocean floor by the time required for an impulse (the sound) produced on the ship to reach the bottom and return. Such a sounding machine was crude and accurate only in relatively shallow water. The concept ultimately resulted in the detection of echos of radio waves or radar. The high frequency of radio waves accounts in part for the accuracy and range of radar. As sound waves increase in frequency, they decrease in size as do the objects from which they can be reflected. Ultrasonic sound then permits greater definition of the object from which they echo comes. Also a narrower wavelength can be aimed in a more restricted direction.

Echolocation in vertebrates usually involves sounds of very high frequency, above the range of human perception. Echolocation is known for several groups of mammals (most notably bats and cetaceans but also shrews) and some birds. As with other types of environmental information, paired receptors greatly enhance the quality of perception; the apparent scarcity of echolocation in fishes may reflect the fact that the primary receiver, the air bladder, is unpaired.

Echolocation in Bats

In the late 18th century an Italian scholar, Lazaro Spallanzani, noted that bats not only maneuver well in a dark room but avoid objects and feed efficiently even when blinded. He learned moreover that bats were unable to maneuver in flight when their ears were plugged. Spallanzani had no notion of how they navigated within a dark room or how blinded bats found and captured flying insects. More than 200 years later Professor Donald Griffin determined that small bats emit sounds far above the 20 kHz that our ears detect and that they maneuver by locating objects by echos of their ultrasonic pulses.

Their echolocation is effective for short distances, perhaps 5 m or less, and observations on captives the voices of which were recorded indicate that detection of small insects (such as mosquitoes) occurs at about 1 m.[372] The *distance* of the source of the echo can be measured by a single ear, but the *directional aspect* of echolocation is the same as with sound emitted by the prey itself, i.e., both ears are usually needed to determine the direction of the source of the echo and, when one ear is blocked, echolocation may be seriously impaired. Presumably, foraging bats might suffer from a plethora of high-frequency sounds, those being received together with echoes of their own signals. Sounds of higher frequencies, however, are rapidly absorbed.

Knowledge of some salient features of sound helps in understanding the nature of echolocation. Sound travels through any substance and, within any uniform medium, at a constant velocity (v). Sound consists of variations in pressure, called sound waves; a wavelength, designated by lambda (λ), is the linear distance between crests of sound waves. The number of waves produced within a unit of time is the frequency (f) or pitch. Because velocity is constant, wavelength varies inversely with frequency, and this relationship is indicated by $v = \lambda f$. In air at 20°C sound travels 344 m/s, and in water the speed of sound is approximately five times as fast, or 1543 m/s. Therefore, at any given frequency a sound wave in water is about five times as long as in air. Velocity also varies with temperature and, in water, with salinity.

Although some insectivorous bats (e.g., Vespertilionidae) experimentally register cochlear potentials for frequencies as low as 30 Hz, sensitivity drops off markedly below 10 kHz. There is in some species a bimodal pattern of auditory thresholds: above 55 kHz the threshold is reduced from that which obtains at 40 kHz.[322]

Bats may obtain an additional advantage by varying the frequency of a brief emission, so that a pulse consists of a series of gradually changing length. Thus, because each successive wave is slightly different from that before and after it, the echo from a given pulse reflects this change. This enables the bat (1) to distinguish the times required for the return of each wave and thereby (2) to determine if the object reflecting the echo is moving or stationary. This frequency modulation (FM) of the ultrasonic impulses of bats is an important element in echolocation, especially in finding and capturing flying insects. In some bats, sound emissions begin at a high frequency and drop very rapidly, and these FM pulses are especially typical of the insectivorous Microchiroptera (e.g., Vespertilionidae). The higher frequencies may be near 100 kHz, dropping to a minimum of about 20 kHz.

Vespertilionid bats increase the number of impulses when an object is first identified (Fig. 10.11), and the effective range of echolocation can be determined by measuring the distance at which this increase occurs. The pulse must be very brief so that it does not meet its echo from an object at close range. Hunting at close range, bats emit a pulse of from 0.2 to 0.5 msec, and these brief signals

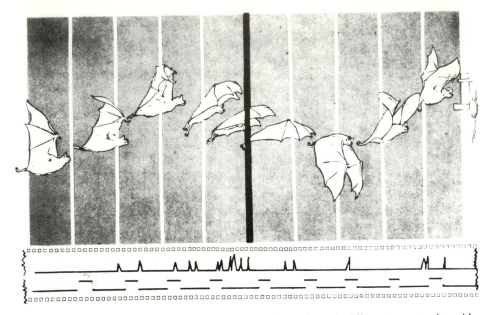

Fig. 10.11. Path of bat flying through a wire grid. Dark vertical line represents the grid, and the sound emissions are correlated with the strobe flashes (below) on which the sketch is based (Howell, 1974).

return as distinct echos from objects less than 10 cm away. Experimentally *Myotis oxygnathus* (a Eurasian species) can detect wires 4 mm in diameter at 400 cm but wires of 0.21 mm at 110 cm.[2]

The horseshoe bats (Rhinolophidae) have a large fleshy development above the nostrils (Fig. 10.12). In contrast to vespertilionid bats, they produce an unmodulated frequency or constant frequency (CF) of from 80 to 100 kHz, varying with the species. Horseshoe bats are especially sensitive to the range of their own impulses. When in flight, they keep their mouths closed and emit sounds through their nostrils. Rhinolophid bats have an unusually well-developed larynx, and their sound emissions are synchronized with their breathing. Plugging their nostrils prevents echolocation, although the bats can breathe through the mouth. Because they emit impulses through their nostrils, they can eat and yet continue to send out signals. Unlike the short signals of vespertilionid bats, rhinolophid bats have rather long emissions, and their echos return before the end of their signals. As they approach the returning sound, the Doppler effect may account for an increased frequency of the echo, and the difference in frequency between echo and emission indicates distance. When a rhinolophid bat detects an object, the signals are shortened in duration and increased in number. As a flying insect is approached, rhinolophid bats cease sound emissions and follow the sound produced by the prey. This is in contrast to vespertilionid bats, which increase the frequency of their signals as prey is approached.[706,707]

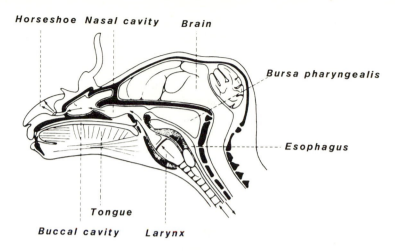

Fig. 10.12. Longitudinal section of a horseshoe bat (Rhinolophidae) showing anterofrontal projection and enlarged, muscular larynx (Möhres, 1953).

Rhinolophus ferrum-equinum emits impulses of 80 kHz, a frequency with a wavelength of 4.25 mm, or twice the distance between the nostrils from which these sounds come. This spacing allows the two sound waves to reinforce each other directly in front of the nostrils (Fig. 10.13). The strength of the sound emitted by a horseshoe bat is much greater than that for a vespertilionid bat, and their range is correspondingly greater, up to 8 m or more. Unlike vespertilionid bats, rhinolophid species can echolocate with one ear plugged, and they apparently use different mechanisms for determining direction. Rhinolophid bats move their ear

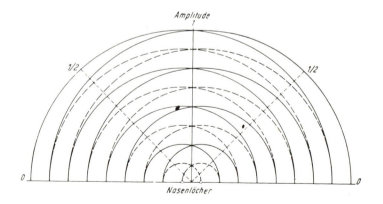

Fig. 10.13. Sound impulses from the nostrils *(Nasenlöcher)* of a horseshoe bat. The impulses have a wavelength about twice the distance between the nostrils, so that the sound is reinforced directly in front of the bat, and the waves tend to cancel each other out on either side (Möhres, 1953).

pinnae independently of each another; this ability may provide the directional aspect of echolocation much as does a radio-direction finder, by determining the angle from which the strongest signal is received.

Leaf-nosed bats (Phyllostomatidae) of the New World tropics feed on pollen and nectar of night-blooming flowers. They take their name from a peculiar foliate appendage on the snout. *Carollia perspicillata* issues very brief (1.4 ms) pulses of from 60 to 130 kHz, but of such a weak intensity that they have been called "whispering bats."[369] These very weak emissions lack the FM aspect of pulses emitted by some insectivorous bats, yet the nectar-eating phyllostomatid bats maneuver very well by echolocation. They emit their signals through either nose or mouth, and also move their ear pinnae independently of each another. Phyllostomatid bats avoid objects with great skill and, as they are mostly frugivorous, they presumably use this ability to maneuver through the forest canopy. The vampire bat *(Desmodus rotundus),* a Neotropical phyllostomatid, feeds on blood, mostly of cattle and other ungulates, and is also adept at flying through wooded areas.

The well-developed and complex ears of insectivorous bats seem to be correlated with their ability to receive ultrasonic emissions. Certainly the large ear pinnae and unusually pronounced tragus, together with the small eyes of insecteating bats, contrast with the relatively large eyes and small ears of the Old World fruit-eating Megachiroptera. Most of the fruit-eating bats do not maneuver by echolocation but rather use their eyes to avoid objects in dim light. An exception is the tomb bat *(Rousettus aegypticus),* which navigates visually when there is light but in darkness emits sounds varying from 6.5 to more than 90 kHz. This bat flies with its mouth almost closed and produces sounds with its tongue.[770]

Perhaps the most interesting type of echolocation is the detection of small fish by the fish-eating bat *(Noctilio leporinus)* of the Neotropical region. This bat reaches into water with its hind feet and snatches small fish from the water. It locates its prey by irregularities or ripples on the surface, and these minor irregularities created by movements of the fish are detected by echolocation.[981,982] This bat emits brief pulses (200 s^{-1}), starting at about 60 kHz and dropping to about 30 kHz (Fig. 10.14).

Although there have been many studies of echolocation in bats, there are many species in many families, and overall our knowledge is fragmentary. The rather different sorts of echolocation in bats indicate that they have developed this ability independently several times.

Echolocation in Birds

Species in two families of birds are known to maneuver in the dark by echolocation. The oilbird *(Steatornis caripensis)* is a frugivorous bird of South America, and nests in dark recesses of caves. In the interior of such caves it is very noisy, producing loud calls of brief pulses (1–2 ms) of sonic frequencies (6–10 kHz) but, when flying in the forest, it is quiet. Apparently, sonic pulses are adequate to locate perches and nests in the caverns they inhabit.

In southeast Asia live several species of small swifts *(Collocalia)* called swiftlets. These include the species that make nests of dried saliva, delicate struc-

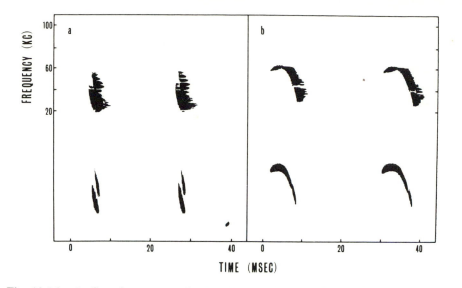

Fig. 10.14 a–b. Sound spectrographs of **a** *Pizonyx vivesi* and **b** *Noctilio leporinus,* fish-eating bats; upper tracings are made with a narrow-band filter setting, showing the frequency spectrum, and the lower tracings, made with a wide filter setting, show best the pulse duration (Suthers, 1967).

tures used in the creation of bird's nest soup. These birds also nest in caves, frequently beyond the twilight zone where there is no light. In caves of Sarawak on the island of Borneo, the swiftlet *Collocalia maxima* flies in total darkness, emitting sounds of 1–4 kHz. In New Guinea, *Collocalia vanikorensis* maneuvers in the dark, using pulses of 4.5–7.5 kHz.[371,686,687]

Many birds are theoretically able to determine their altitude by echolocation. Many nocturnal migrants emit calls, and geese, when flying in the fog, call almost constantly. Variations in the surface of the ground below affect the nature of the echo; such acoustic reflectivity determines not only the strength of the echo but the degree of alternation of the original impulse. Thus an echo can reveal not only the altitude of a calling bird but also the nature of the substrate below.[373]

Echolocation in Water

Echolocation was developed in World War II as a means of underwater detection of submerged objects. The process, SONAR, took its name from so(und) n(avigation) a(nd) r(anging), and employed high-pitched pulses of short duration. When SONAR equipment was used in naval searches of enemy submarines, many animal sounds were heard. Postwar studies on small cetaceans, porpoises and dolphins revealed their ability to navigate and locate food by echolocation. Transmission of light in water is poor, and cetaceans have little if any binocular vision; moreover, in turbid water vision is useless in navigation.

Sound-producing and -receiving equipment enabled biologists to explore animal sounds in water. They learned that some animals have highly refined means of echolocation, and this ability has been studied extensively in cetaceans. Porpoises produce short pulses or clicks of extremely high pitch, well above the usual upper limit of 20,000 cycles of the human range of hearing. Very high frequency, or ultrasonic sound, offers some marked advantages in echolocation. A sound with a frequency of 5 kHz, traveling through water at a velocity of 1543 m/s, has a wavelenth of 154.3 cm. But if the frequency is increased by 10 to 50 kHz, the wavelength is decreased by the same factor, to 15.43 cm. Porpoises have long been known to produce a variety of sounds, even though they lack vocal cords. Many porpoise sounds are from 7 to 15 kHz, well within the range of human hearing, but they also emit short pulses, sometimes to less than 1 ms, ranging from 20 to 120 kHz. They emit pulses that change frequency, using frequency modulation as do insectivorous bats.[538,769]

The sea catfish *(Arius felis)* uses low-frequency sound for echolocation. The lateral line responds differentially to sounds within the 50–150 Hz range, and the swim bladder has directional ability in production as well as in reception of sound. Both systems may reinforce each other to produce an effective auditory orientation system in detecting reflections and reverberations of sounds from the air bladder.[992] Catfish are ostariophysine fish, and possibly the paired aspect of the Weberian ossicles contributes to their ability in directional hearing.

Those vertebrates that use frequencies below 20 kHz do so only for position-finding. The tomb bat, oilbird and swiftlet use sonic signals for finding their way about their caves. Those that echolocate by ultrasonic frequencies use their signals not only for navigation but also for prey location.

Bioluminescence

The production of light by animals, bioluminescence, occurs in protozoa, coelenterates, ctenophores, nematodes, annelids, crustacea, insects and many marine fishes. Among vertebrates there are some luminescent elasmobranchs and many luminescent bony fishes; in many fishes the light is extrinsic, produced by bacteria, but there are numerous intrinsic species that produce light themselves.[610] In most cases the light is greenish or bluish.

Luminescent organs of fishes may be (1) open, in which there is a pore through which bacteria enter and (2) photophores with a lens, gland and a reflector. These fishes producing intrinsic light have photophores, from which light is discharged. The photophore is composed of an internal reflector, a vascularized glandular tissue that produces an enzyme, luciferase, and a lens (a modified scale). Glandular tissue releases luciferase, in the presence of which luciferin and oxygen join to form oxyluciferin, a reaction accompanied by the release of a yellowish light. Light is emitted under both hormonal and neural stimuli. The emission can be constant or variable in intensity. Photophores are more active at night than during the day.

Luminescent bacteria occur in a number of other unrelated bony fishes but in no deep-sea groups. Various species of rattails (Macrouridae) have an open gland anterior to the cloaca from which a luminescent fluid produced by resident bacteria can be forced. In two species of Monocentridae of the marine waters of eastern Asia to Australia, bacteria provide luminescence. *Monocentris japonicus,* the knightfish of Japan, as a pair of organs near the top of the lower jaw and in the Australian pine-cone fish, *Cleidopus gloria-maris,* the light-producing organs lie near the eyes. In both species the bacteria emit a bluish-green light, but a reddish filter in the light organ of the pine-cone fish allows only red-orange light to escape. The Anomalopidae are small tropical fish known as flashlight fish because of their ability to turn their lights on and off quickly. *Kryptophaneron alfredi* is found in the Caribbean Sea, and two Old World species, *Photoblepharon palpebratus* and *Anomalops katoptron,* occur in shallow waters from the Indo-Pacific area to the Red Sea. They all have large subocular pockets that contain cultures of symbiotic luminescent bacteria that emit a pale green light. *Photoblepharon palpebratus* raises a small opaque movable membrane or curtain to obscure or "turn off" its light, and *Anomalops katoptron* rotates the light organ so that the opaque inside lining is on the outside. In life the subocular organs emit light continuously except for the brief periods (200–800 ms), producing an occulting light in contrast to the flashing light of most bioluminescent fishes.[396] These fish in darkness conceivably use their lights to confuse a predator by covering their light and moving away.

In some species specialized luminescent organs are lures for attracting potential prey species. Perhaps the best known are the various kinds of anglerfish (Ceratioidea): a greatly modified dorsal fin ray has a branched lure that the angler can suspend before its mouth. Bacteria in the lure create a glow that attracts small fish. As the angler opens her huge gape, the prey is engulfed; male anglers are minute and parasitic on the body of the female. In the lure or esca of anglers is a pore through which bacteria enter or escape. When cultured in seawater broth, these bacteria do not luminesce; presumably the angler contributes material that enables them to glow when in the lure.[466,773]

At least two species of fishes of the family Pempheridae—shallow-water marine fishes of eastern Asia, New Zealand and Australia—have luminescent parts of the lower gut. These ducts and pyloric ceca, which are embedded in thin, translucent ventral muscles, are rich in luciferin, which is the same as that in the small crustaceans *(Vargula* [= *Cyrpidinia*] *hiligendorfi)* on which thse fish feed. The geographic distribution of these luminescent species of Pempheridae is the same as that of species of *Vargula.* One or two kinds of codfish (Gadidae) also contain pockets of luminous bacteria. Luminescent bacteria occur also in intestinal diverticula embedded in the translucent muscles of *Apogon elllioti,* a cardinal fish (Apogonidae).

Photophores occur in many species of fishes, especially those dwelling in deeper water. The structure has evolved independently many times and yet retains its fundamental structure in unrelated families.

Species of lanternfish (Myctophidae) are marine fishes that live at depths down to about 1000 m; they have a series of photophores that glow with a steady intensity and also luminous patches, near the head or tail, that emit variably rapid flashes. The arrangement of photophores is sexually uniform and constant for a given species, but the luminous patches may vary sexually. The photophores and luminous patches are under neural control, and the production of light in both is by the luciferin reaction, but the function in light in these fishes is not understood.

Hatchetfish, expecially species of *Argyopelecus* (Sternoptychiidae), have rows of large reflectors associated with photophores, and the light produced resembles irregular flashes of natural light which reach the depths at which these fish live. Luminescene is apparently protective in the hatchetfish.

Numerous species of dogfish (Squalidae) produce light on the ventral surface of the body and paired fins. They produce a greenish light by tiny photophores that are directed downward, parallel to the fish's vertical axis. This light may serve to illuminate possible food directly below the swimmer. A deep-water electric ray, *Benthobatis moresbyi* (Torpedinidae), has photophores about the edge of its disk, perhaps to lure prey.[427]

The midshipman *(Porichthys notatus)* is known to flash its photophores for one or two seconds at varying intervals in the presence of a female; luminescence in this fish possibly functions in courtship, but photophores in both sexes are the same.[193,977] The photophore of the midshipman has both a nervous and a vascular supply, and can be stimulated both through the spinal cord and by injection of epinephrine. The reflector is rich in guanine, as are the reflectors of other luminescent fishes, including those species with light-producing bacteria. It has been suggested that species of *Porichthys* obtain luciferin from luminescent euphasiid crustaceans *(Vargula)* upon which they feed; the light-producing materials from the two animals are chemically similar and in the Puget Sound area, where the euphasiids do not occur, *Porichthys* does not produce light.[1020]

Although most deep-sea fishes are sensitive only to the blue end of the spectrum (the colors that penetrate deepest in water), at least two predatory fishes that produce red luminescence are visually sensitive to red light. One such fish, *Aristostomias scintillans,* produces and is sensitive to both red and green light.[774] Another bathypelagic predator, *Pachystomias atlanticus,* has large luminescent organs that produce bright red flashes. *Pachystomias,* unlike most deep-sea fishes, is also sensitive to red light and thus can illuminate prey species without being seen itself. The deep-sea eel, *Saccopharynx harrisoni* (Saccopharyngidae), is a predator with a huge mouth and distensible gut; on its tail it bears an expanded organ with scarlet luminescence.

The function of luminescence in fishes is mostly unknown. When both sexes produce similar amounts and patterns of light, a sexual significance seems unlikely. In the lantern fish photophores, which are essentially alike in both sexes, may be a specific recognition character, but the luminous patches of lanternfish, which vary between the sexes and which are flashed in a conspicuous manner, may well have a sexual role. The lure of luminescent anglers is to attract prey, and the concentration of photophores about the head and mouth (or even within the mouth) of bathypelagic species suggests a prey-attracting function. The red flashes projected by the bathypelagic *Pachystomias atlanticus* and others almost certainly serve to illuminate their prey. Most luminescent fishes are mesopelagic, living in the twilight zone, and not bathypelagic. Those luminescent fish that are bathypelagic have large and elaborate photophores, with large lenses and reflectors, and they are predatory in habit. The ventral photophores may, when viewed from below, simulate overhead irradiance or light transmitted through waves.

Electric Fishes

A variety of fish produce and discharge appreciable amounts of electricity. This ability has developed independently many times, and perhaps as many as 500 species in 10 families (Table 10.1) can be included among the "electric fishes." The electric properties of some of the more powerful species have been known since ancient times, but recent studies have revealed the role of low-voltage currents in the behavior of these fishes.

Electric fish fall into two groups, depending on the voltage of their impulse: predatory species that use electricity to stun or kill prey, and species that use low-voltage emissions for electrolocation and electrocommunication (Fig. 10.15). In most electric fish the current is generated from a series of modified end-plates (called electroplaques or electrocytes). In one family (Apteronotidae) the organ is derived from modified nervous tissue. In either case, potentials are established in the same manner as in unmodified nervous or muscle cells. Large numbers of electroplaques oriented in series account for the 500 or more volts (of low amperage) reported for large electric eels (*Electrophorus* spp.) of the Orinoco and Amazon drainages. The electric ray (*Torpedo* spp.) has fewer electroplaques in a column but more columns than are in the electric eel, and the ray produces 50 V of roughly 50 amp. Both fishes are efficient predators of large prey species. The electric catfish of the Nile drainage produces charges of several hundred volts. Some skates (Rajidae) are weakly electric; they produce impulses sporadically, and the function of electricity in this family is unknown.[523]

The stargazers (Uranoscopidae) are the only marine bony fishes with electric organs; other electricity-producing teleosts are fresh-water specimens, and the other electricity producing fishes in marine waters are elasmobranchs. Stargazers discharge electricity when feeding. The electric organs are derived from ocular muscle tissue and, together with the eyes, are exposed when the rest of the fish is buried in sand or in mud. In each feeding there is an initial discharge of 800 mV with a frequency of 200 impulses/s occurring when the fish opens its mouth. The burst lasts from 20 to longer than 200 ms, longer bursts being typical of larger prey items. This finding suggests that the duration of the burst may depend on the

Table 10.1 Taxonomic and geographic occurrence of the several families of electric fishes.

Marine	South America	Africa
Uranoscopidae	Gymnotidae	Gymnarchidae
Rajidae	Electrophoridae	Mormyridae
Torpedinidae	Rhamphichthyidae	Malapteruridae
	Apteronotidae (= Stenarchidae)	

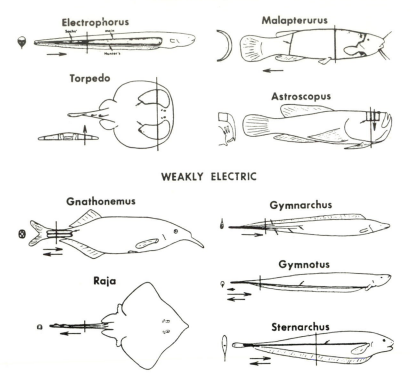

STRONGLY ELECTRIC

Electrophorus

Malapterurus

Torpedo

Astroscopus

WEAKLY ELECTRIC

Gnathonemus

Gymnarchus

Gymnotus

Raja

Sternarchus

Fig. 10.15. Some representative electric fishes. Electric organs are indicated by small arrows, and large arrows show directions of flow through electric organs (Bennett, 1968; courtesy of The University of Chicago Press).

degree to which the stargazer opens its mouth. After the initial feeding burst, there is a series of separate pulses of 10–25/s that may last up to 19 s.[819]

The producers of low-voltage impulses inhabit murky waters of South America and Africa. Their electricity is too weak to serve any offensive purpose but provides a method of communication and electrolocation. Electrical impulses are characteristic of a given species and may differ between the sexes.[450,607] In these species the fish generates a small electrical field around itself, and any interference in the field (e.g., a nonconducting object) is apparent to the fish.[420] Some electric fish, which emit continuous low-voltage impulses, have very long unpaired fins, either dorsal or anal (e.g., *Gymnarchus, Gymnotus, Sternarchus*) (cf. Fig. 10.15), and these fins propel the fish while its body is held straight. Apparently a straight, rather rigid body is required to minimize distortions of the electrical field about it.

Different species of weakly electric fishes produce distinctive electric signals. Some discharges are long, with little or no variation within a rather constant frequency: these are called *tone* or *wave* discharges in distinction to *pulse* discharges, in which there is a long pause between signals (Fig. 10.16). Tone or wave pro-

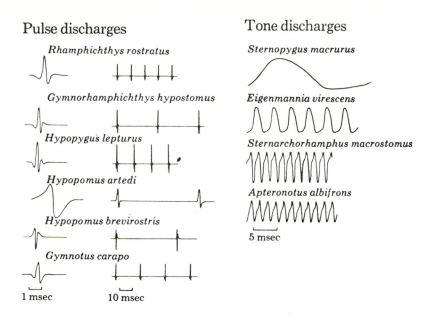

Fig. 10.16. Forms of electrical impulses from 10 species of gymnotoid fishes from Moco-
moco Creek, Guyana (Hopkins, 1974, courtesy of American Scientist).

ducers tend to emit prolonged discharges of varying frequencies. Pulse producers
increase their frequencies at night; both types can modulate their frequencies,
perhaps forming signals with rather specific meanings. Within these variations
electrical impulses retain features serving to identify the species and sex. In these
ways electric signals are remarkably similar to vocalizations. Weak electric signals serve also to preserve the schooling pattern in the mormyrid fish, *Marcusenius cyprinoides,* a nocturnal species that lives in turbid waters.[709]

Environmental "electric noise" is a potential interference in electrolocation of
fishes, especially if the frequency is close to the species' own electric organ discharge (EOD). To avoid such jamming, a fish emitting wave signals can shift its
frequency; this jamming avoidance response (JAR) allows individuals of a given
species to continue recognition of their own EODs, even if the competing signal
is considerably weak. For those species producing a pulse signal, the JAR consists
of a change in rate of pulse emission (instead of a pure change in frequency).
Thus the JAR of a wave species is a gradual shift in frequency, whereas pulse
species alter the timing by acceleration or deceleration of the EODs.[421]

Summary

In all these sensory systems emphasis is on those stimuli most appropriate to
a given activity pattern and environment. A watery home is ideal for the transmission of odors, especially where light is poor and any water-soluble substance

is dispersed slowly. Odor moves at a variable speed, quite dependent on movement of water, and olfaction tells little of the direction and distance from which the odor enamates. Odor is unique among stimuli produced by vertebrates: odors last for hours, days or even weeks, whereas sound, light and electrical signals die at birth. Learned odors remain long in the memory and are very important in the homing of fishes. Odors are also of paramount significance in snakes; they tend to live in regions of limited light, and they lack the ability to hear, so emphasis on olfaction was inevitable. Most birds are diurnal, and their poor sense of smell is compensated for by sharp hearing and excellent vision. Even nocturnal birds, with the apparent exception of the oilbird, have virtually no sense of olfaction: excellent hearing compensates for diminished visual signals in nocturnal birds. In contrast to most birds, most mammals are nocturnal; mammalian communication is one first of odors and second of sound, with vision third. Odor in mammals assumes a dimension not available to fishes. Odors in water float about the medium through which fishes themselves move: neither is static. A mammal can place its special scent on a stationary object, and the odor carries the additional aspect of location. The exploitation of odor in marking territories is unique to mammals.

For all these sensory signals, including electricity and infrared radiation, the receptors (or receptor systems) are paired. The sole exception is the air bladder. The arrangement of receptors in pairs provides the receiver of a stimulus to estimate not only its direction but also its distance. (External nares are so close together that their paired condition provides no information about the source of an odor but, because odors persist, the proximity of the two nares imposes no real limitation on their direction-finding ability.) This ability to triangulate reaches the ultimate sophistication in the asymmetry of ears of some owls.

Thus it is that each group of vertebrates lives in its own sensory world. The stimulus conveys not only the kind and usually the sex of its maker but frequently its psychologic and hormonal state, and much of this information is excluded from or unintelligible to individuals of other species.

Suggested Readings

Airapetyants ESh, Konstantinov AI (1970) Echolocation in animals. Israel Program of Scientific Translations, Jerusalem

Bennet MVL (1971) Electric organs. In: Hoar WS, Randall BS (eds) Fish physiology. New York, Vol V: 347–491

Eisenberg JF Kleiman DG (1972) Olfactory communication in mammals. Ann Rev Ecol Systema 3: 1–32

Griffin DR (1958) Listening in the dark. Yale University Press, New Haven

Hailman JP (1977) Optical signals. Indiana University Press, Bloomington

Harvey EN (1952) Bioluminescence. Academic, New York

Hinde RA (1969) Bird vocalizations. Cambridge University Press, Cambridge, England

Johnson FH, Haneda Y (1966) Bioluminescence in Progress. Princeton University Press, Princeton

Needham AE (1974) The significance of zoochromes. Springer-Verlag, New York

Sales, G (1974) Ultrasonic communication by animals. Chapman and Hall, London

Thielcke GA (1976) Bird sounds. University of Michigan Press, Ann Arbor

Part IV

Population Phenomena

Certain activities collectively constitute the means for the preservation of populations. These processes are usually confined to intraspecific events and to responses of species to environmental changes. Reproduction is the means for not only maintaining the species but effecting increases in population. These increases are commonly temporary and eventually to be followed by population declines. In some species population densities fluctuate cyclically. These aspects of reproduction are tied to individual growth, and the two topics account for variations in structure and size of populations.

11. Reproduction

Because ovulatory cycles are best known for several species of laboratory mammals, it is helpful to become familiar with the phenomenon first in mammals. Except for data on the domestic fowl, information for the other classes is generally not extensive. There seems to be a similarity in gonadotropic and sex hormones throughout the several classes, but their specific functions and the target organs are not always the same, and there are huge areas in which there is little or no information.

Vertebrate Ovulatory Cycle

Many species ovulate only once a year; such a species is *monestrous*. Others may experience a succession of ovulatory cycles in which one of more ova are released each time; these species are *polyestrous*. In some polestrous species a postpartum estrus occurs and then a female may be both lactating and gravid; in such cases several broods may follow in rapid succession. Species in which the population may have some breeding members at any season but in which individuals have a long anestrous period are not necessarily polyestrous.

Ova must be released while sperm are present if a zygote is to develop. In many species the female is responsive to courtship at the time of ovulation, and ovulation occurs even in the absence of coitus, such species being called *spontaneous ovulators*. Most mammals seem to be spontaneous ovulators, and a balance of estrogen and progesterone stimulates rupture of the follicle and release of an ovum. In other species copulation is needed for ovulation; these species are *induced ovulators*. Numerous mammals are known to be induced ovulators. The domestic cat, rabbit and ferret have long been recognized as requiring the stimulus of coitus before they ovulate. At least some voles (Microtinae), the mink *(Mustela vison),* the short-tailed shrew *(Blarina brevicauda),* the mole *(Scalopus aquaticus)* and the camelids are induced ovulators.

In spontaneous ovulations the regular level of circulating luteinizing hormone (LH) increases just prior to ovulation, as part of the negative feedback pattern (in polyestrous species). The physical stimulation of copulation in induced ovulators causes secretion of luteinizing hormone-releasing factor (LH-RF) from the hypothalamus (via a nervous impulse from the cervix to the hypothalamus). As an adjustment to this difference, induced ovulators are receptive at any time during their reproductive season, and such species do not experience regularly repeated estrous cycles characterized by ovulation. There is instead an estrous phase, during which receptivity increases, and, if mating does occur, this stage is followed by anestrus.

There appears to be a longer period of receptivity in an induced, compared to a spontaneous, ovulator. It is perhaps significant that species with great reproductive potential (e.g., rabbits and voles), which regularly experience cyclical changes in density, seem to be induced ovulators. This suggestion is speculative, for it is not known that *all* species of rabbits and voles are induced ovulators.

The distinction between spontaneous and induced ovulation may not always be sharp. Although the laboratory rat is considered to be a spontaneous ovulator, vaginal and cervical stimulation may be followed by a reflex ovulation, the pelvic nerve carrying impulses to the central nervous system (cns). It is clearly established that conception may occur at any time of the ovulatory cycle in women, even during menstruation. In cases with known dates of the last menstruation, the intense coitus of rape has been known to stimulate ovulation and to be followed by conception at any point in the cycle. Variations in the length of the human ovulatory cycle might be caused by coitus-induced ovulation, inasmuch as the luteal phase is a relatively constant 14 days.[172] This variability is well known among people who use the rhythm method of birth control.

Among some aboriginal societies in which nutrition is minimal, conception is spaced by the term of lactation, sometimes four or more years. With the acculturation of some peoples, exercise is reduced and diet upgraded; infant diets are supplemented by cow's milk and grains, and a shortened period of nursing and an increased birth rate result. The increased caloric intake seems to be followed by more regular ovulation.[556]

Hormonal Aspects

The ovulatory cycle is adjusted to climatic changes, which may very well be registered in the hypothalamus. The cycle begins with the secretion of the *releasing factor* of the *follicle-stimulating hormone* (FSH-RF), which passes from the hypothalamus through a hypophyseal portal system to the anterior pituitary (Fig. 11.1). When it reaches the latter, FSH-RF stimulates the release of the follicle-stimulating hormone (FSH). Although it circulates throughout the body, FSH affects the ovaries (specifically the follicles); there is promotes growth of the follicle(s) and the ovum developing within. As the latter matures, the follicle produces *estrogen,* which prepares the female for mating, both physically and psy-

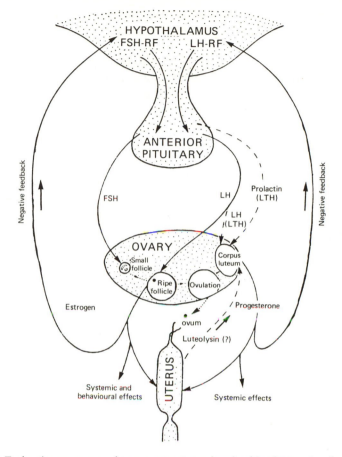

Fig. 11.1. Endocrine sources and target structures involved in the ovarian function in a mammal (Perry, 1971; courtesy of Oliver and Boyd).

chologically. In oviparous reptiles and birds the follicle deposits fat in the ovum, and in mammals its role is the production of estrogens and some progesterone.

Another effect of estrogen is the *negative feedback,* by which it retards the release of FSH-RF. At this time a second product of the hypothalamus, the *luteinizing hormone-releasing factor* (LH-RF), moves through the hypophyseal portal system, effecting the release of the *luteinizing hormone* LH from the anterior pituitary. The accumulation of LH is believed to trigger ovulation in spontaneous ovulators. The postovulatory follicle under the stimulation of LH becomes the *corpus luteum* (pl., *corpora lutea*), which produces progesterone. In mammals progesterone effects endometrial development of the uterus in expectation of the latter's role in housing the blastocyst. Progesterone also has a negative feedback to the hypothalamus, suppressing the release of LH-RF. The preservation of the corpus luteum depends in part at least on one or another substance from the pituitary gland, and the necessary hormone seems to vary among species: the role is filled by LH in some mammals; in others the pituitary gland releases a postovulatory *luteotrophic hormone* (LTH).[811]

Thus there is a rise in estrogen in the preovulatory phase of the estrous cycle, and at the same time there is a drop in the progesterone level. The accumulation of both steroids prior to ovulation suggests that each one plays a role in the release of the ovum from the follicle, and that the estrous cycle cannot be simply divided into preovulatory estrogen and postovulatory progesterone phases.

Fluctuations in estrogen and progesterone affect many aspects of reproductive behavior. In addition to responding to male courtship activity, the estrous cycle has ovulation and sexual receptivity as its central themes. Nest-building in rabbits, for example, is a hormonal response, and disappears after ovariectomy.

The *distinct* gonatropin-releasing factors of the mammalian hypothalamus do not occur in the avian hypothalamus; some birds, however, do have hypothalamic secretions that stimulate LH and FSH production, and avian LH and FSH are similar to those of mammals.[779]

Surgical destruction of the portal vessels from the hypothalamus to the anterior pituitary is followed by regression of gonadotropic tissue of the anterior pituitary and the gonads, and by the failure of the usual responses to changes in photoperiod. Ovulation in birds is apparently caused by high levels of LH, the release of which is under neural control. The latter in turn is stimulated by progesterone. Sequentially, circulating progesterone attains a high level before release of LH, which peaks four to six hours before ovulation to release LH; when injected, progesterone is followed by ovulation.[692] It is not clear whether the bird pituitary secretes one or two gonadotropic hormones resembling LH and FSH. It has been suggested that the presence of an egg in the oviduct depresses the release of gonadotropin(s), and that there is no negative feedback mechanism in birds. The avian follicle may produce small amounts of androgens.

After the eggs are laid and incubation begins, LH and progesterone levels decline and the *prolactin* level rises. If the eggs are removed, LH and progesterone once again increase and prolactin levels fall, suggesting that prolactin may suppress gonadotropin output.[692]

Ovulation in birds is sequential, usually with one egg being laid each day until a clutch is complete. In reptiles, on the other hand, the entire clutch is laid at once, placed in a nest, covered with sand or soil and usually given no parental care or protection.

In addition to these hormones, one other, prolactin has various roles and occurs in a number of vertebrates. It is secreted by the anterior pituitary and is traditionally associated with activation of mammary tissue. It also seems, at least in laboratory mice and rats, to stimulate and maintain the corpus luteum, a role that may be widespread in mammals. In pigeons, prolactin, together with estrogen, stimulates proliferation of the crop lining to produce "milk," a source of nourishment for nestlings. Not only does prolactin stimulate the crop of produce "milk," but its release induces the feather loss and vascularization in brood-patch formation. In some birds (e.g., pigeons) prolactin is antigonadotropic and suppresses ovulation. But in other species (some ducks and gallinaceous birds), high prolactin levels parallel gonadal development.[692]

In males of some fishes reproduction may be controlled by hypophyseal gonadotropins rather than by testicular hormones. Testicular tissue seems not to produce androgens in such species, and castration does not stop courtship. Prolactin occurs in fishes and is known to stimulate deposition of melanin in melanophores in the skin.

Variations in levels of these hormones cause changes in reproductive structures, not only in the ovaries but in the oviducts, uterus, genitalia and mammary tissue as well as in behavior. Collectively these changes constitute the estrous cycle, and can be separated into a preparatory *proestrus, estrus, metestrus,* and *diestrus.* The quiescent period of the estrous cycle is *anestrus.* Ovulation normally occurs during estrus.

In humans, if the ovum is not fertilized, the vascularized uterine tissue, which was built up under the influence of progesterone, sloughs off and passes out of the body. This phase may be brief or take several days, and is known as menstruation. This phenomenon is characteristic of humans and Old World primates, but some post-ovulatory bleeding and loss of endometrial tissue occurs in the elephant shrew *(Elephantulus)* of Africa and perhaps in some bats. In the former, endometrial growths develop as ova mature, and serve for attachment of the zygote. In a sterile ovulatory cycle, these growths are lost, such discharge being likened to menstrual flow of the Old World primates. In the domestic dog a small amount of uterine bleeding occurs prior to ovulation, but his does not involve any sloughing off of the endometrium and is therefore not comparable to menstrual bleeding.[811]

The corpus luteum was first studied in mammals and is best known in these vertebrates. It produces substantial amounts of progesterone, at least during the early phase of pregnancy, and at a reduced level it produces estrogen. In addition to its effect on the uterus, progesterone also suppresses development of ova. In the Monotremata the corpus luteum is large but, during the prolonged passage of the egg, it shrinks and virtually disappears by the time the egg is laid. Corpora lutea also occur in fishes, amphibians, reptiles and briefly in birds (possibly a similar structure develops in some ascidians and echinoderms). Although researchers do not agree on its function in nonmammalian vertebrates, the corpus luteum may stimulate the oviduct to provide a jelly-like covering in fish and amphibians or shell in reptiles and birds. It begins to form after ovulation by hypertrophy of the cells lining the follicle, and extension vascularization occurs in all classes (although it is short-lived in birds). The progesterone secreted by these structures is similar in all classes.

The postovulatory follicles in the hagfish (Myxine) enlarge and become granular, and may well be comparable to mammalian corpora lutea. These jawless vertebrates also form "preovulatory corpora lutea" (or corpora atretica), which seem to be secretory. In elasmobranchs there is a corpus luteum that apparently produces substances which cause the uterus either to secrete material for the egg-case (in oviparous species) or to prepare the uterine wall for its role in pregnancy. In the eel-pout *(Zoarces viviparus)* the preovulatory corpora lutea are enlarged

throughout pregnancy and are also conspicuous during pregnancy in the Poeciliidae (guppies). Preovulatory corpora lutea develop in amphibians and secrete sex
hormones. They develop from the granulosa layer and apparently are part of the
regular sequence of unovulated eggs. Postovulatory corpora lutea also may
develop in amphibians, but their function is not clearly established.

In mammals progesterone induces vascularization of the uterine endometrium
prior to implantation. Following implantation, the placenta in some species produces large amounts of progesterone. In other species, however, such as the nutria
or coypu *(Myocastor coypu)*, the placental attachment is preserved with the influence of progesterone from the corpora lutea. In pregnant nutria, ovariectomy is
followed by abortion, in contrast to the guinea pig, in which the progesterone of
pregnancy appears to be placental, and ovariectomy does not trigger abortion. As
in the nutria, in the domestic rabbit as well as in the ferret, the pig and the cow,
the corpora lutea are the sole source of progesterone.[811]

In amphibians the corpus luteum is generally short-lived, there being little
evidence of hormonal production in oviparous species. A corpus luteum persists in
viviparous amphibians, such as live-bearing caecilians, live-bearing salamanders
(Salamander atra) and anurans (*Nectophrynoides* spp).[613,1113] Also, in anurans
that carry the eggs in dermal pockets (e.g., *Pipa pipa*), the corpus luteum persists
until birth or release of the young, but its function is not established. A corpus
luteum is produced in the postovulatory follicle of all reptiles. It regresses quickly
in oviparous species, but in live-bearers the corpora lutea remain throughout most
of gestation. There are conflicting reports and opinions on its function in pregnancy: it may be necessary in the early phase but not necessary closer to parturition. The corpus luteum in *Thamnophis elegans* contains and presumably
secretes progesterone during pregnancy (Fig. 11.2). Luteectomy prolongs pregnancy and reduces the number of normal births, but does not otherwise alter normal embryonic development.[433] The corpus luteum of the snapping turtle *(Chelydra serpentina)* synthesizes progesterone, the function of which is not known.[205]

Accessory corpora lutea form in the ovaries of several mammals that release
many more ova than are fertilized, and in some mammals corpora lutea develop
from unruptured follicles. In either case they become luteinized and presumably
secrete progesterone. In the domestic horse the original corpus luteum (of pregnancy) regresses about four weeks after ovulation and is replaced by a succession
of several corpora lutea of later ovulations. In some other mammals the released
ova greatly exceed those fertilized, and all follicles become functional corpora
lutea. This phenomenon is known in some hystricomorph rodents (e.g., the porcupine, *Erethizon dorsatum*) and the elephant shrew *(Elephantulus myurus)*. The
African elephant *(Loxodonta africana)* frequently has several corpora lutea but
only a single embryo. In the elephant a corpus luteum forms even in the absence
of fertilization and may persist after a corpus luteum forms from a subsequent
(fertile) ovulation.[951] In addition, some follicles may become luteinized after pregnancy, and corpora lutea may persist from one pregnancy to the next. Presumably
all these corpora lutea produce progesterone.[811]

As sexual maturity occurs, the cycle of sexual hormones from the hypothalamus, pituitary and gonads begins. In man and in some domestic animals repro-

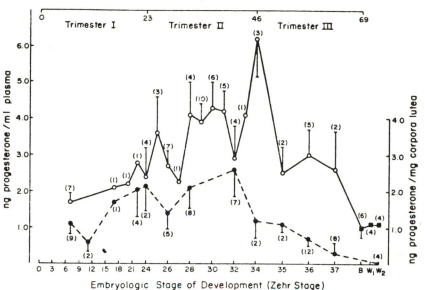

Fig. 11.2. Luteal concentrations of progesterone (●) and peripheral plasma levels of progesterone (O) in *Thamnophis elegans*. Means are in parentheses, vertical lines indicate the standard error of the means, B is birth and W_1 and W_2 are week 1 and week 2 following birth (Highfill and Mead, 1975a; courtesy of General and Comparative Endocrinology).

ductive activity is more-or-less continuous throughout the year. In most species, however, there is an annual period during which males are fecund and females are responsive and receptive. In the nonbreeding period, gonads are generally refractory to exogenous stimuli. Gonads undergo a cycle, usually annual, and produce gametes and hormones. Generally the annual cycles of males and females are synchronized or at least broadly overlapping, but in some groups ovulation may occur at a time other than the mating period, and in some vertebrates mating may occur before the peak of sperm and androgen production.

Although gonadal cycles of males and females are asynchronous in some species, the two sexes must be prepared for mating at the same time. While this seems obvious, some workers refer to supposedly different breeding seasons for males and females of a given species, when what is meant is that the two sexes have different gonadal cycles. If the two sexes of any species actually had different reproductive seasons, it would quickly become extinct.

Ovulatory cycles are affected by the presence of or contact with other females or males of the same species. Some presumably polyestrous species, such as voles (*Microtus* spp.), ovulate repeatedly in the breeding season in the wild; when kept

Behavioral Aspects

isolated in the laboratory, however, females do not ovulate. Thus ovulation is induced by mating or by some other form of sexual excitation.

The role of odors in ovulation is known for some mammals and may be more common than the literature suggests. Groups of virgin female laboratory mice in an experimental wind tunnel were placed upwind to, downwind from and below a group of males. Estrus, determined by vaginal smears 48, 72 and 96 hours after the placement of the mice, was significantly greater in females downwind, whereas those upwind had a frequency characteristic of females kept remote from males; estrus was most frequent in those females kept directly beneath the males.[1094] Presumably odor from male urine stimulates estrus in female mice—a phenomenon called the Whitten effect.

Synchrony of menstrual cycles has been noticed for groups of women living together: female students, for example, in an all-girl college dormitory were much closer in their spring menstrual onset than they were the previous autumn, when they began to live together. The synchronizing agent is unknown, but odor is a possibility.[670]

In the Neotropical cichlid fish, *Aequidens portalegrensis,* the sight and contact of the pair stimulate both maturation and deposition of ova. The pair bond persists after egg-laying, with both parents guarding the eggs and fry. When separated from the male, the female does not spawn but, if a mirror is placed by a freshly isolated female, she continues the cycle of spawning and care of the eggs, the reflection providing the needed stimulation.[826]

Egg-laying in the budgerigar, a hole-nester in the wild, is stimulated by darkness; it does not lay in extended light or in the absence of a nestbox. A particular low warbling call of the male also stimulates ovarian development and egg-laying.[121,122]

Both internal and external factors affect the ovulatory cycle and are mediated through the hypothalamus (Fig. 11.3). It seems that behavioral studies have just begun to reveal the extent to which intraspecific contacts influence the female sexual cycle.

Clutch or Brood Size

The ova released at one time (or within a restricted period) constitute a major factor in determining the capacity of a species to reproduce. The mean number of ova released at one time is characteristic of a given species and may vary from one to millions. Many mammals (most primates, many ungulates, proboscidians, most bats and so on) bear but a single young at a time, and many oviparous vertebrates lay but a single egg. The kiwi, petrels and shearwaters lay but a single egg each season; the frog *Sminthillus* of the West Indies and many lizards lay solitary eggs, one or more times a year. Clutch size (in birds) or brood size (in mammals) varies with certain environmental features. Clutch and litter sizes tend to increase with the latitude and sometimes with elevation at a given latitude; the eggs are sometimes fewer in insular populations than in mainland ones. In many vertebrates there is a tendency for larger (and usually older) individuals to lay more eggs each season, but the relationship is not necessarily linear.

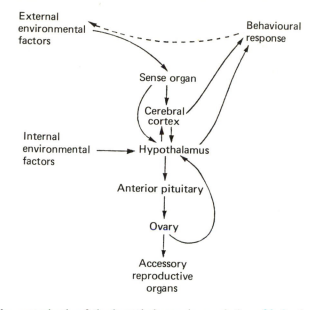

Fig. 11.3. The central role of the hypothalamus in regulation of behavior and the ovulatory cycle (Perry, 1971; courtesy of Oliver and Boyd).

The young in a brood or eggs in a clutch that mature at one time is predicated on the amount of FSH, and the actual number ovulated is proportional to the level of LH. Thus higher levels of both hormones promote increased brood size.

Clutch or brood size varies about a mean in many species of vertebrates; as the mean is exceeded, survival of young may decline. Density in this situation refers to the number of eggs or young in the nest; as eggs or offspring increase, so does the problem of brooding and food-getting. In other words, there is indeed a limit to the eggs and young that females can care for at any one time.[570,572]

This simple generality is confirmed by comparing egg-production among mating types. Behavioral mechanisms controlling brood or clutch size, and fecundity in general, affect the female more critically in monogamous or polygynous species. In polyandrous birds, in which a female may lay a series of clutches, each brooded by a different male, the female may lay many more eggs than a single parent can brood. The same limitation is placed on mammalian litter size, especially for a placental mammal. In a reptile, amphibian or fish in which no parental care is provided, other factors must determine the upper limit of egg production.

Role of Food. Food and feeding may underlie the relationship between fecundity and latitude. At higher latitude the longer days in spring coupled with gradually increasing temperatures greatly accelerate growth of plants and abundance of plant-feeding insects. This provides not only a sudden flush of food for young birds, but also food that the adult females can convert into eggs. In addition to the vernal increase in food at higher latitudes, the increase in daylength allows the parents more time to forage for food and to feed the nestlings. Variations in clutch size in birds and litter size in mammals indicate that fecundity is determined not only by endogenous factors but also by exogenous elements, among which food available to preovulatory females must be paramount.

In oviparous fishes, production of eggs tends to increase with the size (and age) of the female. The number of eggs does not increase in proportion to ovary weight, because there is a gradual increase of ovarian nonfollicular material in older females; egg number, however, reflects the amount of fat available for vitellogenesis. Senility does occur in some fishes, and in the Poeciliidae fecundity in smaller, younger females exceeds that in larger, older ones.

Latitude. Among passerine birds, clutch size tends to increase with latitude. In tropical regions, for some species [e.g., some sunbirds (Nectariniidae) and some birds of paradise (Paradisaeidae)] a single egg is the rule, and in the boreal regions some birds may lay 10 or more eggs in a single clutch. The increase in clutch or brood size in the high latitudes varies inversely with length of breeding season and with number of clutches annually. At lower latitudes clutches are smaller and nesting is prolonged.

In mammals tropical forms have smaller litters than related taxa farther from the equatorial regions. Perhaps because mammals are nocturnal and commonly have hidden nests, most relevant data relate to birds. The trend toward smaller litters has exceptions, however. Reared under the same laboratory condtions of light, temperature and food, the white-footed mouse *(Peromyscus leucopus)* from Michigan has mean litters of 4.28 (SE 0.11), whereas the same species from Campeche (southeastern Mexico) has a mean litter size of 4.67 (SE 0.11).[573]

Adequate food is unquestionably necessary for reproduction; abundance is frequently followed by larger-than-usual broods and, in years of scarcity, reproduction may not occur. This phenomenon has been observed repeatedly in boreal birds. Many workers have commented on the gonatropic effects of food on wild vertebrates. Sprouted wheat (five days old) added to the diet of *Microtus montanus* was followed by and presumably caused a 41 percent increase in pituitary delta cells; this area of the anterior pituitary secretes gonadotropins. Also, when fed sprouted wheat, this vole had more postpartum matings and a higher survival of young.[750,752] The Levant vole *(M. guentheri)* of the eastern Mediterranean produces large broods when fed an abundance of legumes.[104]

The relationship of quality and quantity of food to ovulation, brood size and reproductive success is undoubtedly not simple and direct. The significance of food in ovulation does not exclude a density-dependent role, such as will be discussed (in Chapter 12, on Growth). Adequate food is relative to the demands made on it, so that brood or litter size is affected not only by the abundance of food in the environment, but also by population density.

Mortality. It is tempting to assume that the number of eggs or young is adjusted to the mortality rate of the species in any given region. This concept fits comfortably with the tenets of natural selection but is extremely difficult to demonstrate.

Most evidence supports the opposite concept, namely, that mortality more frequently varies with productivity, and that local and annual variations in productivity are possibly produced exogenously and are possibly nonadaptive. Mortality is frequently variable and, in nature, litter or clutch size is known to vary, sometimes widely, about a mean.[618]

Mating

Meeting potential mates is commonly assured by a preliminary set sequence of events. In migratory species males arrive at the breeding site before females and advertise their presence by characteristic displays, sounds or odors. This is typical of many anurans, pinnipeds and passerine birds. Some migrants form aggregations, and mating may occur in autumn (as in many insectivorous bats) or pairs may form in winter (as in many species of ducks). Also, many hibernating species aggregate in the winter and emerge more-or-less as a group in the spring. In nonmigratory and nonhibernating vertebrates, such as shrews and mice, populations tend to cluster so that individuals are usually not separated by distance. In all these situations there is very little need for a vertebrate to expend time and energy in search of a mate.

Mating patterns fall into several types, and in nature the distinctions may not always be clear.[781] In monogamy there is a pair bond involving one individual of each sex; in a monogamous species each sex contributes equally to the genetic constitution of the next generation. In polygynous mating one male is associated with several or many females, and more females than males contribute gametes. Polyandry in these respects is the opposite of polygyny. Not only are the distinctions not readily apparent in nature but, because most mammals are very difficult to observe, basic mating patterns can be inferred only from observations and a comparison of home ranges of the two sexes. Moreover careful observations tend to refute the long-held concept that in polygynous species the dominant male is the exclusive, or even predominant, mate of the females in his group. There is increasing evidence of mating by subordinate males.

Considering the apparent variation in mating patterns in different classes, it is difficult to see clear-cut characteristics for any one class. Among fishes are numerous variations in mating groups, but no single pattern prevails. Spawning schools of fish may have a unique pattern in which simultaneous release of milt and roe by many males and females results in random fertilization. This pattern would occur in pelagic spawners, such as most species of herring (Clupeidae). Among amphibians, detailed records do not suggest monogamy for most groups. The ease with which mating occurs in many pond-breeding anurans indicates a random mating affair in which polygyny may be common. Also, the complicated courtship of some salamanders does not preclude multiple matings for either sex. Most reptiles are not clearly known to have discrete or discernible mating patterns. Some lizards (Iguanidae and Agamidae), however, exhibit marked sexual differences in size and color and are polygynous. Crocodilians are apparently monogamous.

The prevailing pattern of monogamy in birds may be more imagined than real, for evidence points to a greater association between each member of a monogamous pair to their specific nesting site than to each other as mates. Female American robins and redwings tend to return to their original nesting locality but at a date later than the return of the males. When one member of a breeding pair of birds is removed experimentally, it is often promptly replaced by a new mate. The net effect of such "monogamy" may be the same as if it were based on a strong pair bond. If the real bond is to the nest site rather than to the mate, however, polygamy may not constitute a difficult departure, behaviorally speaking, from typical avian monogamy. A variety of polygamous patterns is noted among birds, but polygyny is more common than polyandry.

In most mammals that have altricial young, the female has a strong attachment to the nest, and in all mammals only the female can nourish the infant. The role of the male seems ambiguous, but clear-cut monogamous pairs are rare. There are conspicuous examples of polygyny among pinnipeds and artiodactyls, and some other ungulates form family groups consisting mostly of females. Because the great majority of mammals are both small and nocturnal, mating patterns must usually be inferred from sex ratios and home ranges; these sorts of data do not support a concept of monogamy. In many species of small mammals, promiscuous mating is apparently the rule: males mate with any female in estrus and, when in estrus, females sometimes accept several mates.

Several concepts have been proposed to explain variations in monogamous mating. Any deviation from monogamy involves an increased expenditure on the part of the "minority" sex (i.e., the male in polygyny): this cost may be reflected in defending territory, in supplying food to the majority sex and the young or in defending the individuals themselves (and the young) against intrusion by an outside member of the minority sex. A polygamous pattern does not reflect a marked disparity from a 1:1 sex ratio, but instead results in a population of sexually unemployed members of the minority sex. A polygamous mating arrangement— assuming a nearly equal number of males and females in the population—must benefit the genetic constitution and survival of the young. This suggests a behavioral or ecological situation differing from that found in a monogamous system.

In a polygynous system one male has two or more mates simultaneously, and his role in rearing the young varies among the several classes. He may take an active role to the extent of brooding the eggs (as in some fish, some amphibians and many birds) and of feeding the young (in many birds); or he may lie at the opposite end of the behavioral spectrum and kill his own or any other young. If he shares the care of the young, his effort will be diluted with an increase in the number of broods.

Several students have pointed out a relationship between an irregularity in the occurrence of a critical natural resource and polygamy; this suggestion requires that a dominant male defend a favorable nesting or feeding area, and that the females entering the area remain under the control of the local male. This would seem to describe the harem formation in some sea lions (Otariidae): a male defends a rookery (a rocky island for such an animal), and females later come to the rookery to calve. Presumably the females are attracted to the rookery and not to the defending bull, for courtship in such species is minimal.[365] In many seals (Phocidae) the family is a less rigid organization: in the gray seal *(Halicoerus grypus)* a harem of 20 cows may be under the control of a succession of two or three bulls in one season. In defending his cows against intruding males, the resident bull also defends the young. There is no obvious advantage to the females in a harem, but a dominant male mates with all the females of his rookery, and such vigor is a theoretical advantage to the young.

If in a monagamous pair a male contributed nothing to the care of the young, it would seem to make no difference to the female in question whether or not he remained after mating *unless* he defended her territory. If the nesting area contains a concentration of such resources as food or nest sites that are scarce elsewhere, and the male neither broods the eggs nor feeds the young, polygyny could easily result. Perhaps polygyny in the ring-necked pheasant *(Phasianus colchicus)* developed in such a situation. Where paternal care is needed for the proper growth of the young, as with many gulls, cormorants and pelicans, a concentration of nest sites does not result in polygyny. Several species of falcons (*Falco naumanni* and *F. vespertinus*) nest in colonies, but males share in incubating the eggs and hunt for food for the young, and pairing prevails.

In certain gallinaceous birds, males assemble and display to each other, establishing a hierarchy, and attract females to their display grounds. The aggregation of displaying males is called a lek, and this word is used also to designate their area of display. Both subordinate and dominant males display to females that enter a lek, but only the dominant males mate. After mating, males defend neither the females nor the females's territories.

Polyandry, the mating of one female with more than one male, seems to evolve if the male assumes complete care of the young. In some ground-nesting birds the male may incubate a first clutch, although the female may not lay a second clutch. In many fishes (anabantids, ariid catfish, some osteoglossids, some cichlids and others) males brood the eggs in their mouth, and in many other fishes and some amphibians males guard the eggs. In either case the female's parental role ceases at oviposition, and she does not usually lay a second clutch the same season. In a few ground-nesting birds, some plovers (Charadriidae), some pheasants, quail (Gallidae) and others, the male may incubate the first clutch, and the female lays a second clutch, for which she assumes total responsibility. Polyandry does not necessarily follow in this situation, but it is found in a jacana *(Jacana spinosa),* the northern phalarope *(Phalaropus lobatus)* and some other birds. The cause of polyandry in birds is not clear, but if polyandry is to be effective (1) the males must incubate the eggs (or at least the initial clutch) and care for the young and (2) the female must be able to develop successive clutches.[277] This possibility does not occur in mammals because of the male's inability to nurse the young.

Mating arrangements in vertebrates appear to be relatively responsive to both population levels and availability of necessary parts of the environment. Thus they may vary both among groups of species and within a given species. Monogamy, however prevalent it may seem, may itself be facultative and certainly is not a genetically fixed social pattern.

Parthenogenesis and Intersexual Conditions

Parthenogenesis, reproduction without fertilization of ova, occurs in a few vertebrates—mostly amphibians and reptiles. In fish there are very few known examples of parthenogenesis, which perhaps rather reflects the limits of our knowledge than the biology of fish. In maturation of ova in a parthenogenetic species there is no reduction division in meiosis, so that the maternal diploid condition is continued. In a few parthenogenetic species, however, occasional fertile matings (with males of related species) do occur, and the offspring are then triploid. Diploid parthenogenetic daughters are genetically identical to the mother.

The absence or scarcity of males in nature is evidence of possible parthenogenesis. Two types of experiments show whether the population is actually unisexually reproductive. By rearing isolated females through two generations, one can eliminate sperm storage as a factor and establish that viable eggs have been laid by the second-generation virgin females. Also, successful skin grafts between two individuals (between mother and daughter or between siblings) constitute evidence that they are genetically the same; the rejection of a skin graft suggests that the two animals are genotypically dissimilar.

In fishes parthenogenesis is best known in some of the Neotropical poeciliid fishes, which include the mollies, guppies and swordtails.[522] *Poecilia formosa* of northeastern Mexico and southern Texas is an all-female species that resulted from hybridization between *P. sphenops* and *P. latipinna;*[459] *P. formosa* reproduces by a variation of parthenogenesis called *gynogenesis*.[47] The female mates with a male of one of the parent species, and the sperm—presumably by rupturing the fertilization membrane—stimulates cleavage without fertilization (i.e., fusion of the two pronuclei). Fertilization may actually occur in *P. formosa,* in which case such offspring are triploid.

In a related genus, *Poeciliopsis,* there are parthenogenetic species that are also apparently the result of a hybrid mating, and which mate with males of one of the parent species. A sperm does fertilize the egg, and the result is a self-sustaining, all-female triploid species. Such triploid species show features of both parents, but the paternal genome is deleted in ova development, so that these triploid females produce diploid eggs that are fertilized by normal haploid sperm.[712] Parthenogenesis resulting from hybridization exhibits typical hybrid vigor and preserves the F_1 condition. In the all-female species of *Poecilia,* the condition is clearly parthenogenesis from the genetic if not from the behavioral point of view. In *Poeciliopsis* there is a return from the parthenogenetic condition to a specialized bisexual type of reproduction called *hybridogenesis.* The original hybrid genome is perpetuated in each case.

In amphibians and reptiles all-female parthenogenetic populations are apparently more common than in fishes. In naturally occurring populations of the Nearctic salamander, *Ambystoma jeffersonianum* "complex," there are both diploid and triploid parthenogenetic species. Oviposition occurs after the female picks up a spermatophore of a normal diploid male, recalling the sterile mating in *Poecilia formosa.* The parthenogenetic salamander, *Ambystoma tremblayi,* for example, is triploid, but cleavage of its eggs is stimulated by sperm of the sympatric *A. laterale.*[1040]

Parthenogenetic species occur in at least six families of lizards and in one family of snakes. The first of unisexual lizards was in the Eurasian genus *Lacerta.* The scarcity of males in some populations of *Lacerta* suggested parthenogenesis. Several diploid species of these lizards are apparently the products of hybridization, the two genomes being heteromorphic. Occasional mating with males of a sympatric species produces sterile triploid females.[213] There are also several parthenogenetic lizards of the New World teiid genus *Cnemidophorus.*[656] Diploid, triploid and tetraploid individuals occur in nature; mating of triploid females with normal males of a sympatric species produces a tetraploid offspring. Parthenogenesis in these species of *Cnemidophorus* is verified by the production of fertile eggs by virgin females and by skin grafts.[620] The widespread gekkonid lizard *Hemidactylus garnotii* consists of triploid females, males being unknown;[551] reciprocal skin grafts verified suspected parthenogenesis in this gecko.[269] In the xantusiid lizard, *Lepidophyma flavimaculatum* of Panama, males are unknown; in view of the established existence of parthenogenesis in the Lacertidae, Teiidae and Gekkonidae, an all-female population in the Xantusiidae is strongly suggestive of unisexual reproduction.[999] The gecko *Lepidodactylus lugubris* is diploid, and males are all but unknown. As in *H. garnotii,* histocompatible skin grafts confirm parthenogenesis. The Australian gecko *Gehyra variegata ogasawarisimae* consists of triploid females and must be parthenogenetic. Also the agamid lizard *Leiolepis triploida* is an all-female triploid species, and the chameleon *Brooksia spectrum affinis* is all-female; probably both these lizards are parthenogenetic.[657]

The worm snake *Typhlops braminus* is the only parthenogenetic snake known. It is a burrowing creature, and conceivably a bisexual fossorial snake might experience difficulty in finding mates.

Not all reptilian parthenogenetic species are hybrids; homomorphic pairs of chromosomes characterize at least some individuals of *Lepidodactylus lugubris* and *Lepidophyma flavimaculatum.*

Natural parthenogenesis does not occur in wild birds and mammals. Parthenogenesis in birds was established by incubating large numbers of eggs from virgin domestic hens. Hatching of a few eggs provided the beginning of a partly parthenogenetic strain. Because the avian female is heterogametic and because YY individuals do not survive, all young birds produced parthenogenetically are males. Such avian parthenogenesis is known in the domestic chicken and domestic turkey.

Experimentally, virgin births have been induced in some domestic animals, such as rabbits: by rupturing the zona pellucida with a needle, cleavage is initiated.

The origin of parthenogenesis in some natural populations of poeciliid fishes, salamanders and lizards can usually be traced to hybridization. Its persistence testifies to the existence of some desirable features of unisexual species. Inasmuch as productivity is a characteristic of females, an all-female population will be potentially twice as productive as a bisexual one, in which one-half is female. For a given biomass, a unisexual species should presumably make the same demands on its environment as does a bisexual species. Thus a parthenogenetic species can theoretically be twice as efficient in converting environmental energy to its own reproduction. Traditionally, unisexual species are considered to be evolutionary dead-ends because of their preservation of homozygosity. If a parthenogenetic species has once developed from a hybrid mating, it could do so again, probably most commonly along the area of sympatry of the two parent species. Each time a hybrid mating results in a parthenogenetic female, an element of heterozygosity is introduced into the parthenogenetic population, and the all-female population can adapt in step with its two hybrid parents.

The presence of functional ovaries and testes in the same individual constitutes *hermaphroditism* from Hermaphroditus, son of Hermes and Aphrodite. In hermaphroditism the production of gametes is usually sequential, the individual being first one sex and later changing to the other; simultaneous occurrence of active gonads of both sexes is rare. In species in which testes mature prior to ovaries, the condition is *protandrous;* in *protogynous* species the female sex develops first. Such individuals have a unique chance to function as both male and female parent in passing on vigor and desirable adaptations. Many families of fishes have bisexual members; this has developed independently and there is a variety of gonadal arrangements.

Gonadal plasticity or lability in fishes is apparent from experiments on some domesticated cyprinodonts. In these fish, males are heterogametic (XY) but, when exposed to estrogen solutions, they develop into functional females. Normal females are homogametic (XX). Not only are such hormonally induced females— still heterogametic—fertile when mated with normal (XY) males, but some off-spring from such matings are YY, confirming that these sex-reversed females are genotypically males. Similarly, young females become functional males when exposed to critical levels of testosterone; when mated with normal females, the offspring are all females (XX). The guppy *(Lebistes reticulatus)* of Central America and the Asiatic ricefish *(Oryzias latipes)* have been extensively investigated in such studies.[1114]

Monopterus albus, an eel-like fresh-water teleost of the tropical Orient, is a well-known hermaphrodite: young individuals develop ovaries as long as 30 cm (about 24 months of age), and at lengths of 42 cm or more (at about 36 months) most individuals are males. In this species sperm and ova are not produced simultaneously in the same individual. As an individual changes from female to male, production of estrogens gradually decreases and output of androgens increases.

Protogynous hermaphroditism is found in various species of sea basses, or Serranidae (species of *Epinephelus, Polyprion, Myctoperca, Petrometropon* and *Cephalopholis*). In some species of wrasses (Labridae) and parrotfishes (Scaridae), some genotypic males develop directly into functional males, while others function first as females and later as males, a condition known as diandry. Genotypic females do not change into males. In *Thalassoma bifasciatum,* a wrasse of the West Indies, there is a series of color phases predicated on both sexual condition and social dominance. Those in a dull color phase can be either male or female. When they mate in schools, the dull-colored males participate in spawning, but in a pairing situation brightly colored males are dominant. In the Australian labrid fish. *Labroides dimidiatus,* a given group comprises a single dominant male, which displays aggressively toward the others, all of which perform as females. When the male dies (or is removed experimentally), there may be (1) a takeover of the harem by a neighboring male or (2) aggression by the dominant female. Maleness is displayed by the dominant female in a matter of hours and is completed and functional in from two to four days.[29]

Hermaphroditism occurs in several deep-sea fishes,[682,941] and has been reported for some species of the marine fishes of the family Platycephalidae; an ovotestis occurs, but individuals function as males early in life and later become females. Among the aulopiform fishes, bathypelagic families are all hermaphroditic, whereas the only two shallow-water families (Synodontidae and Harpadontidae) are bisexual. In the deep-water species the gonads are separate but apparently function at the same time. The bisexual condition in the shallow-water species may have evolved from the condition in the bathypelagic forms simply by the loss of one set of gonads. In bathypelagic fishes populations are low and scattered, and hermaphroditic or monoecious (literally, "one house") species ensures reproduction. The hermaphroditic condition is common in the essentially benthic families Alepisauridae, Paralepididae, Chlorophalmidae, Ipnopidae and Evermannellidae, but it is not known if these species are self-fertile. Most of these species have a dorsal bilobed testis and a ventral bilobed hollow ovary.[682]

The best-documented hermaphroditic fish is the sometimes self-fertilizing cyprinodont of the Caribbean, *Rivulus marmoratus.* The testicular element in the ovotestes increases with age, and eventually egg-laying ceases, secondary male features develop and the individual becomes a functional male. Maleness develops in response to short-day seasons, the time required on temperature and the individual's genetic background. Reared at moderate or low temperatures, this species rarely changes to males.[404,405]

The occurrence of many sorts of hermaphroditic patterns among different taxa of fishes points out the absence of any single evolutionary trend. Those forms, such as the bathypelagic aulopiform fishes, which can produce both types of gametes simultaneously, may or may not be self-fertile. One can imagine that a deep-water fish population must be scattered, as is its food supply, and encounters must be infrequent. If this is actually so, it is easy to appreciate the importance of any meetings' being bisexual. In a bisexual species with a 1:1 sex ratio, only one-half of the encounters would involve two individuals of the opposite sex.

There is early gonadal plasticity in amphibians and reptiles as there is in fishes, and there are both natural and experimentally produced sex reversals as described for cyprinodont fishes. There is a difference in response of immature gonadal tissue to temperature: in amphibians and reptiles low temperatures retard or even repress the development of testicular tissue. In northern and alpine populations of the Eurasian grass frog, *Rana temporaria,* all newly transformed individuals possess well-developed gonadal cortices with premeiotic ovogonia, the testicular medullary part of the gonad remaining at first undeveloped. In those individuals that are genotypically male, the cortical (female) part of the gonad is gradually replaced by medullary cells that become functional testicular tissue. At low elevations and in central and southern Europe the gonadal tissue is clearly differentiated by the end of metamorphosis. Some populations of the bullfrog *(Rana catesbeiana)* in Taiwan are all female at metamorphosis, and males develop testes about six months later.[458] Alternation of gonadal tissue can also be experimentally achieved in amphibians, by either manipulating temperature or rearing the larvae in male or female hormones. Generally testicular development occurs at higher temperatures (25°–30°C), whereas at lower temperatures (10°C) gonadal tissue becomes ovarian. The same effect of temperature on gonadal development has been reproduced experimentally in lizards (Agamidae) and turtles (Testudinidae). Functional males emerge from eggs incubated at low temperatures and females from those at higher temperatures, regardless of genotype.[135]

Conceivably, sequential hermaphroditism endows an individual with a greater reproductive potential, for every individual could function as a female—fecundity or reproductive potential resting on the performance of females. It would also reduce sibling mating.

Parental Care in Fishes

If there is a hypothetical primitive breeding pattern for bony fishes, it must include the production of a large number of small eggs shed and fertilized by a spawning school. This arrangement characterizes many pelagic species, such as cods (Gadidae), herrings (Clupeidae), the ocean sunfish (Molidae), flying fish (Exocoetidae) and others. The eggs usually float at or near the surface where there is abundant oxygen, sunlight, food for the fry and predators. The parents do not guard the eggs. In some other species living in shallow waters, the eggs are demersal and scattered or placed in a nest (a pile of small rocks), where they remain unguarded. Typically many minnows (Cyprinidae), trout (Salmonidae) and perch (Percidae) spawn in this manner. Many other fishes, however, make some provision for one or both parents to maintain a close relationship with the developing eggs and in some cases with the young.

The mere construction of a nest may be regarded as incipient parental care. Many species of minnows build elaborate nests of small stones that hide or cover the eggs left unattended. The usual sequence is for one parent, usually the male, to remain until the young emerge. Whereas the larger percid fishes (Percinae) attach their eggs to submerged vegetation and abandon them, their small relatives, the darters (Etheostominae), place their eggs on plants partially hidden by rocks, and the male guards the site. The nest of the Pacific salmon (*Oncorhynchus* spp.) protects not only the eggs but also the young fish (or alevins) there from predators, until about four months of age.

A few examples illustrate the broad taxonomic spectrum through which parental care is well developed and even elaborate. The ancient and relict osteoglossid fishes exhibit a highly developed routine for protecting their eggs and small young. The African *Heterotis niloticus* constructs a wall of aquatic vegetation about a shallow basin some one and a half m across. One parent guards the nest and later the small young. The Neotropical *Arapaima gigas* builds a nest in deeper water and both parents protect the fry. The oriental osteoglossid *Scleropages formosus* carries the eggs in the mouth, thus obviating the need for nest building. The bowfin *Amia calva,* a holostean fish, builds a large nest, and the male guards first the eggs and later the brood of fry.

Among the common families in which males protect the nests are most of the sculpins (Cottidae), fresh-water sunfish (Centrarchidae), tidal blennies (Blenniidae), gunnels (Pholidae), midshipman (Batrachoididae) gobies (Gobiidae) and sleepers (Eleotridae). In the fresh-water leaffishes (Nandidae) and catfish (Ictaluridae), the parents not only guard the nest but remain and protect the small young.

Nest-building, courtship and spawning in sticklebacks (Gasterosteidae) have been extensively studied and are well known. As spawning approaches, the male makes a spherical nest of small bits of vegetation attached to a rooted plant. He is very bright red at this season and, stationed at the entrance of his nest, he emerges at the approach of a dull-colored, egg-distended female. Spawning occurs within the nest, after which the male drives the female away and waits for another. He guards the fry and herds them together, even to the point of engulfing a stray individual in his mouth and putting it back into the nest.

Oral incubation, the care and protection of eggs in the buccal cavity of one of the parents, occurs in a number of unrelated families. During this period, the incubating parent apparently fasts. Among the best-known oral incubators are the marine catfish of the family Ariidae, all of which except for *Hexanematichthys australis,* care for the eggs in this fashion. The eggs are large, from 15 to 30 filling the mouth of the male, and he cares for them until they hatch, sometimes as long as a month. In the cosmopolitan cardinal fish (Apagonidae) the eggs are incubated in the mouth of one of the parents, usually the male. The jawfishes (Opisognathidae), secretive bottom-dwellers of warm oceans, are also mouth-breeders.

Oral brooding is not confined to marine fishes but occurs in some fresh-water families. The fresh-water anabantid fishes are widespread in the Old World tropics and include the gouramis, the fighting fish (*Betta* spp.) and many other favorites of tropical fish culturists. Parental care is well developed in these species: in some of them the eggs are carried in the mouth of the male, and in others the male places them in a bubble nest, where he remains until the fry are past the larval stage. The Neotropical armored catfish (Callichthyidae) have several ways of protecting their eggs. A number of species of *Callichthys* build froth-like nests of bubbles, which house the eggs and young. The male forms the floating nursery of mucus-covered bubbles from the gill clefts and maintains the nest throughout the development of the eggs, much as do the Old World anabantids. In another species, *Hoplosternum thoracatum,* after the male completes the bubble nest the female places her mouth against the vent of the male and withdraws milt. She places the sperm and then the eggs in the froth nest, after which the male assumes his role, caring for the nest and its contents and driving away intruders, even the female whose eggs he defends.

Among the African cichlid fishes there is a transition from those forms that are substrate spawners to those in which the eggs and also the young are brooded in the mouth. In *Tilapia galilaea,* which extends from Jordan to West Africa, both parents carry the eggs and fry in the mouth, and biparental mouth-brooding also occurs in *Tilapia lohbergen, Pelmatochromis guentheri* and species of *Tristarnella.* In most species of mouth-brooding Cichliae, however, only the female cares for the eggs.

There are some other very specialized forms of parental care. The male pipefish (*Syngnathus* spp.) and seahorse (*Hippocampus* spp.), both of the Syngnathidae, is provided with a marsupium, which in courtship he places in front of the vent of the female; the transfer of eggs to the male is quickly done, and fertilization occurs at this time. The eggs are large with abundant yolk. During brooding, the marsupium becomes highly vascularized, and through this mechanism the eggs receive oxygen; a circulatory supply of oxygen may also bring nutrients. In both pipefish and seahorses the young are not cared for after they are discharged from the brood pouch. In the Neotropical banjo catfish, *Aspredo laevis* (Bunocephalidae), the eggs are held in cup-shaped receptacles on the ventor, and the small young also remain in these cups for a short time. In the East Indian humphead (Kurtidae) the eggs are attached to a hook-like outgrowth from the head of the male.

The development of parental care in fishes has occurred many times in diverse families and in many habitats of both marine and fresh waters. There is clearly no single trend in this behavior, except perhaps within a single family (e.g., Cichlidae). In almost all cases (except some of the Old World cichlids) the male assumes the responsibility for care and protection of the eggs and young; presumably the female is in more urgent need of nourishment than he immediately after spawning. The environments provided by parental care assure not only physical protection but also an adequate supply of oxygen. In the oceans, nest-building is almost confined to littoral areas, and pelagic spawners occupy the surface strata of the open seas. Thus, if one of the parents carries the eggs orally, deeper waters can be occupied, and the oral environment insures a continuous flow of water (and oxygen) over the eggs. In fresh waters, either oral incubation or the construction of a bubble nest provides maximal oxygen, and it may be significant that all these cases are in the tropics, where lentic waters are frequently low in oxygen. It is suggested that the male in guarding the nest on a substrate not only provides protection against intruders but also, by constantly fanning the nest with his pelvic fins, maintains a constant supply of fresh water over the eggs.

Phoretic Spawning in Fishes

Many fishes use a surrogate parent to house the developing eggs, and the complexity of some of these relationships attests to their antiquity. Although these foster parents neither supply nutrition to the eggs nor actively defend them, they do give them physical support, and it seems appropriate to call the relationship phoretic.

The finspot *(Paraclinus marmortus)*, a littoral clinid fish, lays its eggs in the lumen of a sponge *(Verongia fistularis)*, and the male then guards the eggs. A small liparid *(Careproctus melanurus)* of the North Pacific lays its eggs on the gills of a box crab *(Lopholithodes foramindtus)*, and the small larvae remain there until their yolk sac is absorbed. Similarly the Japanese tubesnout *(Aulichthys japonicus)* lays its eggs in the peribranchial cavity of an ascidion *(Cynthia roretzi)*.

The most elaborate case of phoretic spawning is that of the bitterling *(Rhodeus* spp.), a widespread minnow of Eurasia. In the breeding season the female develops a long tubular ovipositor with which it can determine the exact site of egg deposition; the odor from a mussel (a species of *Unio* or *Anodonta*) stimulates her to place the tip of the ovipositor within the excurrent siphon of the bivalve, and pressure within the oviduct forces the tightly fitting edge down its lumen and onto the mussel's gills. The male also responds to odors from the mollusk and releases milt into the incurrent siphon. The young emerge about four weeks later.

In all these cases the surrogate provides physical protection for the eggs and also ensures that they have an abundant and constant supply of oxygen.

Parental Care in Amphibians

Among amphibians several groups of frogs and salamanders provide some protection for eggs and young, but it is not always clear whether nourishment is involved. If there is a maternal contribution to the nutritional needs of the developing young and the embryos are in the uterus, this is called viviparity, but there are some borderline cases that are here considered to be parental care. Although most amphibians lay eggs in water, most examples of parental care in this group occur in species that breed on land.

Most of the species of salamanders in the world occur in eastern North America, and in this area are represented all families except the Hynobiidae. Much of our knowledge about their reproduction is the product of the lifelong studies of Professor Sherman C. Bishop.[92]

Salamanders that deposit their eggs in streams usually attach them to the underside of rocks. In these cases one parent is almost always with the eggs, the assumption being that the parent is in fact providing them with some measure of care and protection. Adult males regularly occur at nests of the hellbender *(Cryptobranchus alleganiensis;* Cryptobranchidae) and are aggressive in defending the eggs. The mudpuppy *(Necturus maculosus;* Proteidae) also guards its eggs, the protective role in this case being assumed by the female. In both the hellbender and the mudpuppy, the guarded nests are attached to the lower surface of some submerged object. The Congo eel (*Amphiuma means;* Amphiumidae) lays its eggs, concealed at or near the margins of ponds, and the female coils about them during incubation.[554] In the families Hynobiidae, Ambystomidae, Sirenidae and Salamandridae, the eggs are usually laid exposed in shallow water and attached to vegetation. In these four families neither parent remains with the eggs in water. In *Ambystoma opacum* and *A. annulatum* eggs are laid on land, in depressions beneath forest litter, and in these two species the female guards the eggs. In some species of these four families the eggs are scattered, attached singly or in small clusters, and parental care then is impossible. In the Plethodontidae, eggs are laid either in water or on land but always attached to the lower surface of some protective cover, such as a stone or rotting log. In all plethodontid salamanders whose breeding habits are well known, the female remains with the eggs until hatching, and in the purple salamander *(Gyrinophilus porphyriticus)* and the dusky salamander *(Desmognathus fuscus)* the young may remain with the mother several days after hatching.[1010,1011]

In the plethodontid salamanders there is a broad spectrum of egg-laying sites, and this is the only family in which many species of many genera have achieved independence of standing water. The function of the adult with the eggs has been matter for some speculation. The adult female coils closely about the eggs and in terrestrial sites must elevate the relative humidity next to the eggs. Some students have suggested that the female's dermal secretions contain a substance that prevents fungal growth, but fungus attacks only dead eggs.[434] When with their eggs, female plethodontid salamanders do not eat, but their presence serves to ward off such active predators as carabid beetles. Although it is not clear why the primitive aquatic species of plethodontid salamanders assumed the habit of remaining with the eggs, in the terrestrial species this habit seems to prevent desiccation or even contribute moisture to the eggs, and is probably a major asset in the extensive terrestrial radiation of the Plethodontidae.

Anurans have developed many means of remaining with their eggs, and it is likely that they offer more than protection from possible predators. Although there are few tropical salamanders, there is a great variety and abundance of tropical frogs and toads, and the greatest variety of parental care is seen in species of the humid tropics.

In the terrestrial leptodactylid frogs, the attendant parent (the male in this case) probably functions much as does the guarding female in plethodontid salamanders. The terrestrial Darwin's frog (*Rhinoderma darwini;* Leptodactylidae) of Chile is a terrestrial breeder, and the male guards the eggs; as the tadpoles begin to struggle within the egg, the male places them in his gular pouch, where they pass their larval life.[142] *Hylambates brevirostris,* a rhacophorid, also carries the eggs in the buccal cavity; in this case it is the female that assumes the protective role. The Australian leptodactylid *Rheobatrachus silus* is aquatic, and the female carries the large, yolk-rich eggs and young in her stomach, at least until the four-limbed stage is reached.[183] In *Crinia darlingtoni,* a leptodactylid of forests of southern beech *(Nothofagus),* males house the developing eggs in lateral inguinal pouches.

In some anurans the eggs are carried in pockets in the back, as in *Pipa pipa,* in which the male moves them onto the dorsum of the female during fertilization. In a number of species of *Dendrobates* and *Phyllobates* tadpoles travel on the backs of the male or female parent, and they remain attached to the female dorsum in species of *Cryptobatrachus;* the responsible parent eventually carries the young to water.[202] Similarly species of *Sooglossus* of the Seychelles carry their tadpoles on the back of the male.

In the semiterrestrial midwife toad (*Alytes obstetricans;* Discoglossidae) of southwestern Europe, the male entwines the strings of eggs about his trunk region and carries them until they are ready to hatch, at which point he enters the water and the tadpoles escape from the egg membranes.

Parental care is seen in many Neotropical hylid frogs. In species of *Hemiphractus* the eggs are kept on the back of the female present, and the larval stage is passed in the egg. In the species of *Cryptobatrachus* of the montane forests of Colombia, the eggs are placed in small depressions on the back of the female and in the species of *Fritziana* the eggs are contained in a single dorsal concavity. In species of the genus *Gastrotheca* there is a dorsal marsupium (not ventral, as the generic name suggests); the eggs remain in this pouch (Fig. 11.4) until young

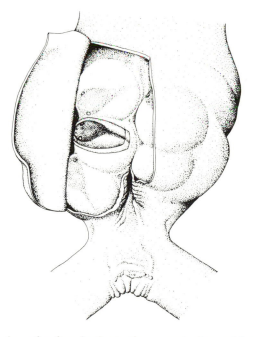

Fig. 11. 4. Dorsal view of a female *Gastrotheca ceratophrys* with the left side of the brood pouch exposed (Duellman, 1970).

frogs emerge or, in *G. marsupiatum,* until the tadpoles hatch, in which case meta-morphosis occurs in water. These hylid species occur generally in humid montane forests, which lack abundant standing water.[259]

Several species of Neotropical hylid frogs have disk-shaped or bell-shaped gills by which they attach to the dorsum of the female (Fig. 11.5). In early develop-mental stages these gills envelop the embryo like an allantois and later serve to retain the small frog until the four-legged stage. These gills are vascularized and apparently absorb oxygen from the mother.[764] One must assume that some nutrients are also obtained from the parent.

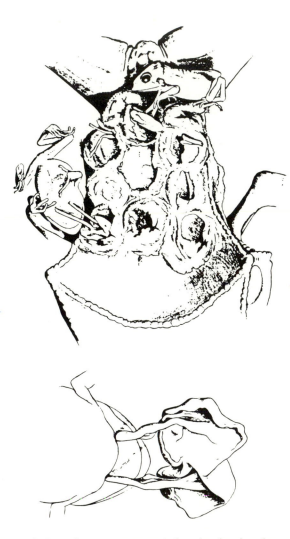

Fig. 11.5. Dorsum of *Hemiphractus panamensis* female, showing the young froglets held to the dorsal pits by the strands of tissue leading to the gills (Duellman, 1970).

Parental Care in Reptiles

Very few reptiles exhibit an awareness of their own eggs and young. In most cases gravid females dig a depression in the ground, deposit their eggs and cover them. The site is not uncovered, nor does the female usually remain nearby. The eggs are warmed by solar radiation and convection of heat through the soil, and they usually hatch the same season. Although some lizards lay their eggs under bark, most reptiles deposit them in the sand or soil, or at least cover them by leaf mold. The substrate is frequently moist, but eggs are never laid in water. Many turtles moisten the ground with urine and dig a flask-shaped hole with the hind legs when the soil is soft; such urine may also provide some moisture for the eggs. Most species build individual nests, but a few may lay eggs in communal nests.

It is unusual for the parents to remain at the nest after eggs are laid. Among skinks (Scincidae), females of the oviparous species of *Eumeces* coil about their eggs during incubation, and the Floridian *Neoseps reynoldsi* also broods its clutch of two eggs. Females of species of *Eumeces* not only remain with the eggs, defending them from potential enemies, but occasionally move and rotate them and release fluid from the bladder, which moistens the soil near the eggs. At hatching time the mother skink may also assist the young in their struggle to emerge from the shell, and she remains with them for 10 or more days after hatching.[281]

The females of species of "glass snakes," *Ophisaurus ventralis* (an anguid lizard), coil about the eggs until they hatch. This habit may possibly be found in a number of lizards; when eggs are discovered, it is usually by accident, and a brooding female, should there be one, might very likely be disturbed.

Among snakes, only boids and elapids exhibit any amount of parental care of the eggs. Of the Boidae, the Old World pythons coil about their clutch of eggs, covering them completely. *Python molurus* of southeast Asia and India (and possibly several other species) are thermogenic, so that their bodies raise the temperature of the eggs several degrees above the ambient. The king cobra *(Ophiophagus hannah)* is well known for its habit of building a nest of vegetative debris and "guarding" the site after egg-deposition. Individual females seem to vary in their degree of aggressiveness at the nest, and their role may be more than defense of the eggs. At times they rearrange the covering over the nest, perhaps to preserve the cover and maintain moisture within the nest.

Crocodilians lay their eggs in a nest of damp decaying vegetation, sometimes 100 or 200 m from water. The adults remain nearby throughout the incubation, and the female is especially aggressive toward intruders. Mammalian predators, such as bears, viverrids and raccoons, prey on eggs of crocodilians, and monitors (*Varanus* spp.) seek and consume eggs of these huge reptiles. The young call from within the eggs and can be heard at 10 m or more from the nest. As hatching nears, one or both parents may assist their young in escaping from the shells; they are either picked up in the mouth and released in the nearby water, or they may proceed to water under their own power. They stay in the vicinity of their parents for about another two months. There are predators of young crocodilians, and biparental protection must greatly increase their chances of survival. Parental care in crocodilians is known in species in Africa, Asia, Australia, and the New World and persists in captivity. It is surely an important aspect of their lives and may well account for their long and successful history.[470,756,828]

The conspicuous lack of concern that reptiles show for the next generation may be a consequence of the special features of the reptilian egg and the site in which it is laid. The shell allows the egg to be laid in soil or under the bark of a tree, so that both the shell and the immediate environment not only retard water loss but even augment the water content of the egg. Also, an egg buried or otherwise concealed is less open to predation than one visible. The amphibian egg, on the other hand, although frequently covered and concealed, is not in complete contact with the substrate and is therefore much more exposed to the risks of desiccation and predation.

Parental care, as it is most commonly recognized, includes postfertilization activities that enhance survival of the young, and here it specifically excludes ovoviviparity and viviparity. Parental care is widespread among ectotherms, occuring in a variety of situations. Parental care is shown by both sexes, but males commonly assume most or all of this activity. This care in fishes extends from simple nest-guarding (which may include fanning of eggs and removal of detritus) to oral brooding, specialized brood pouches for the eggs and guarding of small young.

Theoretical advantages to parental care always include an ultimate increase in survival of the young (despite a possible destruction of all the offspring should the nurturing parent be killed). Possible advantages include increased egg-production (when the male nurtures). Also, when the male parent provides the care, selection becomes especially rigorous; he must succeed not only in obtaining a mate but in defending the eggs as well.

Delayed Births

In all classes of vertebrates there are some species that experience an extended period between mating and birth. The storage of sperm with *delayed fertilization* is known to follow mating in some kinds of bony fish, amphibians, reptiles, birds and mammals; *delayed implantation* (or embryonic diapause) is a regular feature in the gestation of several orders of mammals. These modifications clearly result in birth at the most appropriate time in some vertebrates, but the function is obscure in others.

Sperm Storage

Sperm formation and mating may occur long before ovogenesis, and mating is then not immediately followed by fertilization. The sperm is stored in the female's reproductive tract and may survive to fertilize ova months or even years later. In some species secretions of glycogen or lipids seem to nourish the male gametes, which remain immobile in modified pits, folds or crypts in the female reproductive tract.[455]

Several kinds of elasmobranchs house sperm in the oviducal gland; in these fish, sperm apparently enter the egg just prior to shell deposition. Among bony fish, sperm storage is best known in the guppies, mollies, swordtails and their allies (Poeciliidae). Sperm become embedded in pits in the surface of the single median ovary and periodically enter the ovary, where they fertilize ripe ova in the lumen. They may remain stored for up to eight months.[1019]

Among anurans, fertilization is external in virtually all species. Ova are fertilized internally in the species of *Nectophrynoides* (Bufonidae) of Africa and in the bell toad, *Ascaphus truei* (Ascaphidae), of western North America. In the latter species sperm remain viable for months in the oviduct. In all families of salamanders, except the Cryptobranchidae and the Hynobiidae, sperm are contained in a small sperm packet, which the female picks up with her cloacal lips; thus internal fertilization is widespread among the Caudata. Sperm have been found throughout the year in the sperm reservoir in the female common newt, *Notophthalmus viridescens* (Salamandridae), and the salamander *Desmognathus fuscus* (Plethodontidae), both of eastern North America. There is no obvious adaptive significance to delayed fertilization in amphibians. Perhaps it provides flexibility of mating time in animals that require a humid environment for sexual encounters, thus ensuring reproduction in periods of irregular precipitation.

In contrast to amphibians, delayed fertilization in many reptiles has a fairly apparent adaptive significance. Delayed fertilization is best known among reptiles and is apparently a regular feature in some species in all living orders. The tuatara *(Sphenodon punctatus)*, the sole modern rhynchocephalian, mates about nine months before egg-laying, and sperm must be housed in the female genital tract for that period. Marine turtles mate in the water shortly before the females crawl onto beaches to lay eggs; because the formed eggs are shelled at the time of mating, sperm from one mating must fertilize the next clutch, which matures two years later. Sperm storage is reported for several species of pond turtles (Emydidae). Among Squamata numerous species of both lizards and snakes mate some time before ovulation, and viable sperm may subsequently remain in the oviduct for as long as six years in some snakes. In the green anole *(Anolis carolinensis)* sperm can be found in folds in the vaginal tube throughout the year, and in an Australian skink *(Hemiergis peronii)* sperm survive in the oviduct from the time of mating in autumn (February/March) to ovulation in the spring (September/October), six or seven months later.[311] The Australian gekkonid lizard, *Phyllodactylus marmoratus,* mates in late autumn; ovulation and fertilization occur the following spring. In these species sperm are stored in lamellae in the upper oviduct, and presumably fertilization occurs as unshelled eggs enter the oviduct.[542] Sperm storage occurs in the reproductive tract of female Iguanidae, Agamidae, Chamaeleonidae, Gekkonidae and Eublepharidae. The prolonged retention of viable sperm in the oviduct is a regular feature in the breeding cycle of many species of snakes, especially in the widespread family Colubridae and vipers (Viperidae).

After mating, some species of birds retain sperm for short periods; sperm are stored in the oviduct and near the uterovaginal junction. In the bobwhite quail

(Colinus virginianus) and the ring dove *(Streptopelia risoria),* sperm survive for about one week, but in the ring-necked pheasant *(Phasianus cochicus)* and turkey *(Meleagris gallopava)* storage may persist for three or six weeks, respectively. The quail and the dove both form pairs, providing many opportunities for mating, but in the turkey and the pheasant polygyny prevails, and heterosexual meetings are much less common.

Sperm storage and delayed fertilization in mammals are characteristic only of the insectivorous bats (Microchiroptera). Autumnal mating is a regular event in many species of the Vespertilionidae, and sperm survive in the uterus (in contact with, or with their heads embedded in the uterine epithelium) and sometimes also in the oviduct. Ovulation and fertilization follow in the spring, about five months later, more-or-less simultaneously within a female colony. Because many bats segregate into male and female groups during hibernation, males and females do not necessarily associate when ovulation occurs; in such species autumnal mating ensures a high incidence of pregnancy. In other species, however, both sexes remain in the same caves, and they mate in both autumn and spring and sporadically in the winter. In the noctule *(Nyctalus noctula)* of Europe, the single young are born in June, and mating follows from midsummer to October. In this bat, sperm is stored not only in the copulatory plug in the female but also in the cauda epididymides of the male.

Sperm storage may last for more than one week, but less than two weeks in the dog, but more than 30 days in the hare *(Lepus europaeus).* Sperm storage permits mating weeks or months before the next ovulation, and this sequence raises two questions: First, in the absence of the surge of estrogens and LH associated with ovulation and the drive for copulation, what conditions the female for copulation? Second, when the sperm are quiescent in the genital tract, what stimulates their arousal just at the time of ovulation?

The advantages of delayed fertilization may include the very brief period during which an ovum can be fertilized. In the domestic hen, this may be 15 minutes, and in mammals for which such information is available ova must be fertilized within 12 hours of ovulation. Storage of sperm for weeks and months in bats that hibernate or migrate and hibernate might constitute an accommodation to sexual segregation at the season of ovulation.

Delayed Implantation

In many mammals the blastocyst remains in a condition of suspended development in the uterus, and later implants on the uterine wall. The blastocysts, together with the uterus, form a placenta and development proceeds. This phenomenon prolongs gestation, frequently as long as one year, so that parturition occurs shortly before the next mating. Delayed implantation is common in some migratory or amphibious mammals, species in which the sexes are more-or-less isolated most of the year. Implantation may be delayed in species having a postpartum estrus; if mating and fertilization occur during lactation, implantation is delayed for a longer interval than occurs in nonlactating females. The underlying causes of this difference may not be the same as in species that normally experience a prolonged delayed implantation. Nursing, the physical stimulation of the nipples, increases pituitary output of prolactin and lowers FSH secretion. Prolactin delays implantation, whereas FSH increases estrogen output, which in turn promotes implantation.

Such facultatively delayed implantation occurs in a number of genera of macropodid marsupials, the kangaroos. In many polyestrous species, gestation is shorter than a complete estrous cycle; in these species pregnancy and lactation do not alter the schedule of the postpartum estrus, including ovulation. During lactation a fresh corpus luteum attests to postpartum ovulation. At this time, if there was a postpartum mating, a dormant blastocyst lies in the uterus. The end of nursing, whether by weaning or death of the young, seems to cause the corpus luteum of pregnancy to enlarge substantially, and its activity in turn either stimulates the blastocyst to resume growth or increases the receptivity of the endometrium. In nature, mortality of nursing young occurs with some frequency; with a blastocyst already in the uterus, gestation resumes promptly. If there were no blastocyst, a new estrus cycle followed by a normal gestation would take twice the time to produce another young.[138] There is a postpartum estrus and mating in the tammar *(Protemnodon eugenii),* a wallaby of Australia, and implantation is suppressed by the stimulus of nursing. By the time the young (or joey) is weaned, the corpus luteum has become quiescent, and remains so until the following breeding season. Thus in the tammar implantation may be delayed for 11 months.[909]

Delayed implantation is a feature in the reproduction of many species of temperate and cool climate mustelid carnivores. Mating takes place in the summer, and parturition the following spring, so that there is a prolonged gestation, from 200 to 350 days. In the fisher *(Martes pennanti)* parturition is in spring, followed by a postpartum estrus, and blastocysts implant the following winter. Similarly the stoat or ermine *(Mustela erminea),* a Holarctic weasel, mates in late spring; blastocysts are dormant until the following winter or early spring. In the long-tailed weasel *(Mustela frenata)* young are born in spring, and estrus follows the end of lactation in July. The blastocyst lies dormant until the following spring and implants about 3½ weeks before parturition. The diminutive *Mustela rixosa* has a prompt development with no delayed implantation, even having two or three broods a year in some areas. In the western spotted skunk *(Spilogale putorius latifrons),* implantation (or nidation) is delayed about 200 days, during which period progesterone production is minimal. The corpus luteum increases progesterone secretion several days before implantation; increased photoperiod apparently stimulates the anterior pituitary to increase LH secretion, which is suspected as being the pituitary hormone initiating luteal activity shortly before implantation. Neither hysterectomy nor removal of the unimplanted blastocysts affects luteal activity. As in other mustelid carnivores, the placenta does not control luteal function, and produces little if any progesterone.[683,684]

There is a tendency for dormant blastocysts to implant in times of short days in winter; in the marten *(Martes americanus)* implantation occurs experimentally under short but increasing photoperiods in late February in the wild. In the feral state, implantation occurs in January in the fisher *(Martes pennanti)* and the wolverine *(Gulo gulo),* shortly after the shortest days of the year, but in *Mustela frenata* blastocysts implant in March. Most commonly implantation in mustelid carnivores occurs as days become longer, but some species respond to decreasing daylength.

Delayed implantation prevails in seals (Phocidae) and sea lions (Otariidae) but not in the walrus (Odobenidae). The migratory species aggregate in the high latitudes during the late winter or early spring, and adult males and females do not associate at other times. The harbor seal *(Phoca vitulina)* gives birth to young in late spring, and estrus and mating are delayed until after the end of lactation, in late summer. In this species implantation seems to be confined to a brief period at the end of November and early December. The bearded seal *(Erignathus barbatus)* mates after the young are weaned, and the blastocysts implant about 10 weeks later. Lactation lasts roughly three weeks, so that mating occurs not long after parturition. The northern fur seal *(Callorhinus ursinus)* is probably the most thoroughly studied pinniped. Migration is precisely adjusted for the subsequent social and reproductive events. Parturition occurs within a day and a half after of females reaching their summer home, and mating and ovulation occur within the same week, in late June or early July. Implantation is delayed until after lactation ceases but there is no causal relationship, for primiparous females also exhibit delayed implantation. The arrested development of the blastocyst in pinnipeds is an accommodation to (1) a restricted period for mating followed by a rather solitary period and (2) a precisely timed postmigratory parturition. In most species, mating and ovulation occur within several days after parturition, usually during lactation. Mating occurs after lactation in some species that nurse the pup for a brief period (e.g., the hooded seal, *Cystophora cristata*). The corpus luteum is quiescent prior to implantation; it enlarges and presumably releases progesterone after lactation. Implantation is not directly related to the end of nursing in pinnipeds.

The brown bear *(Ursus arctos)* and the Nearctic black bear *(U. americanus)* mate in early or midsummer. Implantation occurs five months later, in November, in the black bear; very small young are born in Feburary, after a gestation of about eight months.

The nine-banded armadillo *(Dasypus novemcinctus)* has a prolonged mating season. Young females breed in late autumn, but older females mate in midsummer; in either case the blastocysts implant in late November or early December. Prior to implantation, the corpora lutea actively produce progesterone. In contrast to corpora lutea in marsupials, the postimplantation corpora lutea of armadillos are not greatly enlarged over their condition during diapause of the blastocyst.

The roe deer *(Capreolus capreolus)* mates in midsummer, whereas late-autumn mating is the rule in other temperate deer. The blastocyst remains in the lumen of the uterus until December, when it implants; fawns are born in the spring, as with other deer. During the blastocyst stage the corpus luteum is active and does not change markedly after implantation. Unlike delayed implantation in other orders, however, rapid growth of the early embryo in the roe deer is seen prior to implantation. Delayed implantation is unknown in other species of Artiodactyla.

There is a delay in implantation (which may include a slow rate of embryonic development) in several species of Microchiroptera. The Neotropical fruit bat *Artibeus jamaicensis* has a delay during which cell division is retarded but not arrested, thus differing slightly from the "reproductive diapause" of some carnivores and marsupials.[302] In *Natalus stramineus* the delay is from 8 to 10 months, and from 9 to 10 months in *Macrotus waterhousii*. The latter species mates in the autumn; although implantation occurs promptly, development of the embryo is greatly retarded until the following spring.[138] The bent-winged bat *(Miniopterus schreibersi),* a vespertilionid bat of Eurasia, and *Macrotus californicus,* a phyllostomatid bat of western North America, both mate and ovulate in autumn.[113,138] In *M. schreibersi,* sperm are stored for up to six days in the microvilli and cilia of the lining of the uterotubal junction (where the oviduct enters the uterus). Release of a single ovum (always from the left ovary) in October is followed by fertilization and delayed implantation during hibernation.[715]

In many species with delayed implantation there is a postpartum estrus, and implantation is concomitant with a sudden growth of corpora lutea, suggesting that luteal activity is required for implantation. Implantation does not follow the experimental administration of either progesterone or progesterone and estrogen, so the precise stimulus remains obscure.[278]

Breeding Seasons

A breeding season must be defined somewhat arbitrarily. For oviparous vertebrates, the breeding season may be regarded as the time of oviposition. In many oviparous groups, however, sperm may be stored for weeks or even years, and egg-laying may be separated from mating. In such animals there are (at least) two periods of reproduction: mating and oviposition. The two sexes may differ in times of fecundity: males may be capable of mating for an extended period, and females may be receptive for a restricted period. In species with sperm storage (as in some bats and reptiles) mating may be bimodal, spring and autumn, while births are restricted to a brief period each year. In species living more than one year but breeding annually, there may be two overlapping peaks of mating, older females becoming sexually active somewhat earlier than females breeding for the first time; this occurs in some midlatitude lizards as well as in some mammals, such as the armadillo. If breeding seasons are truly adaptive, most births should occur at about the same time in any given region. To some extent this is true, but different major taxa breed at various times: salamanders and warblers, for example, differ in their environmental demands, and their times of oviposition differ too. Even within a group of families, such as passerine birds, while most nest and rear young in the spring, a few, such as the cedar waxwing *(Bombycilla cedrorum)* and the goldfinch *(Spinus tristis),* tend to nest late in the summer.

Regularity of reproductive patterns tends to reflect the predictability of seasonal environmental cues. Changes in daylength are regular, and in many animals increasing daylength stimulates gonadal recrudescence. Some autumnal breeders may respond positively to decreasing daylength. The role of photoperiod seems to be important for males, and females of many vertebrates are aroused partly by the males' courtship behavior. In regions where rainfall varies from year to year and the quality of plant food is not so predictable as is daylength, reproductive seasons fluctuate with the rainfall. At lower latitudes, the amount of precipitation during the rainy season becomes increasingly important as a cue to annual reproduction.

Virtually all vertebrates live in fluctuating environments, and there is usually one time when conditions are optimal for nesting, development of the egg and feeding of the young. Because reproduction demands much energy from at least one parent, and growing young frequently have narrowly defined needs, their rearing is generally confined to these periods of optimal conditions. In regions where environmental conditions are irregularly optimal, such as in deserts, reproduction may be sporadic. Close to the equator, where environmental conditions tend toward only slight seasonal cycles, reproduction may continue throughout the year in a given population, whereas individuals within that population may have discrete times for breeding interrupted by periods of gonadal regression. Even in the tropics, however, seasonal rainfall or social interactions, or both, seem to produce recognizable reproduction seasons in many species.

Vertebrates in higher latitudes have increasingly restricted times for breeding. Even in human beings, who are always prepared to indulge in heterosexual play, there are seasonal peaks in pregnancies. Because growth of eggs and their incubation, gestation of young, production of milk and care of young all require energy far in excess of that needed for bare subsistence, reproduction generally takes place when food is abundant.

Among species of a family of even of a genus, there is a far greater variation in reproductive patterns than physical similarity would suggest. Even within closely related species of mammals, one may have delayed implantation and another direct development, one species may produce precocial young and another altricial young, whereas among congeneric species of reptiles one may be oviparous and another viviparous. Correlations of reproductive patterns and taxonomic position (and genetic similarity) are often difficult to establish because these modes are adaptive. The frequent lack of similarity among allied species has allowed these forms to diversify physiologically and ecologically, increasing the possibility for closely related species to dwell in overlapping areas. Such reproductive diversity may very well accelerate speciation and eventual morphologic diversity.

Because gonadal activity of males is frequently more extended than that of females, FSH levels probably determine mating times. Progesterone and estrogen reach maximal levels around ovulation, and most wild vertebrates mate at this time or close to it. In humans this synchrony of ovulation and mating has been questioned, mostly because of variations in reported strength of libido. A study of incidence of both intercourse and orgasm among married couples showed, however, a remarkably close correlation between mating and ovulation, with a peak around the 15th to the 19th day, and a low shortly after the onset of menstruation.[172,1034]

For many wild species the determining factors are not readily apparent and may not always include female hormonal levels: for species in which females store sperm, mating may occur before ovulation, and there is little information on the causes of receptivity in such cases.

Tropical Environments

Characterizing tropical environments and contrasting them to habitats in mid-latitudes are frustrating exercises, fraught with ambiguities and contradictions. Many temperate regions have summer temperatures exceeding those near the equator, and tropical variations in rainfall are matched by those far from the trop-ics. High tropical mountains are in some respects displaced temperate islands, and such mountains are subjected to low temperatures comparable to winter frigidity in midlatitudes. Although some tropical regions show little monthly variations in temperature, rainfall and insolation, most low-latitude environments are variable. Seasonal cycles in rainfall can equal those of temperate regions, and daily tem-perature extremes on equatorial mountains match seasonal variations in temper-ate regions.

The single consistent aspect of the tropics is the potentially regular insolation, *potentially* regular because it is altered by both seasonal and sporadic cloud cover. The regularity of sunrise and sunset, and the heating effect of the sun, prevent the development of the sharp seasonal extremes seen in temperate regions. Neverthe-less the differences between the equatorial tropics and midlatitude temperate regions are a matter of degree, and regions are not mutually exclusive and dis-crete: the tropics interdigitate with the temperate regions over broad latitudinal ranges on both sides of the equator. Not surprisingly, a wide variety of reproduc-tive patterns is found in the multifaceted tropics.

In subtropical regions of acyclic rainfall, reproduction for many vertebrates is closely associated with either rain itself or the flush of plant and insect growth produced by rain. Although a *correlation* sometimes exists between temperature fluctuations and variation in breeding times, the relationship varies among the classes. Fish, amphibians and reptiles are sensitive to and respond directly to changes in environmental temperatures. There is no demonstrably *direct* effect for birds and mammals; these vertebrates probably respond to temperature-induced changes in plants, insects and (for hibernators) soil temperature and texture.

Seasonal reproduction represents an adaptation to annual climatic changes or climatically related changes, such as plant growth after the onset of an annual rainy period. Presumably these physical or biotic features to which breeding is adjusted favor one or more aspects of the entire process. In species in which small young forage for themselves, their appearance usually coincides with production of substantial amounts of their food. Tadpoles hatch when algal growth resumes, and young lizards and insectivorous birds frequently hatch when small arthropods are abundant.

Close to the equator, breeding seasons show great diversity among resident species; optimal times for nesting and care of young vary among major taxa. Also, optimal times may prevail over longer periods, and breeding seasons *seem* to be ended by the refractory stage as much as by changes in environmental resources. Variations in temperature are of small magnitude in contrast to monthly fluctuations in rainfall. Here again, however, optimal conditions occur in dry periods for some vertebrates and rainy period for others. In regions with two annual wet–dry cycles, some vertebrates may breed twice a year, with all mature individuals participating both times. Other species breed two or three times a year, each population breeding only once annually, but the time depending on the geographic locality.

Among tropical ectotherms there seems to be no prevalent pattern of reproduction. Some species have relatively well-defined periods during which pregnancy or oviposition predominates or occurs exclusively, while spermatogenesis is continuous throughout the year. Even where tropical breeding seasons are readily apparent, the environmental significance, when there is such, is sometimes obscure.

In fishes inhabiting regions of seasonal rainfall, reproduction follows the onset of rains. An extreme case is the cyprinodont fish of South America and East Africa: their eggs lie in the bottom of dry temporary ponds, and development is arrested until rains flood the areas. Growth is rapid and spawning occurs again before the rainy season ends. These are annual fish with a complete population turnover each year.[71]

In low latitudes where there is always high humidity, anurans may breed throughout the year. This is especially true of those species that deposit eggs on land. In the New World tropics species of *Eleutherodactylus,* which breed on land, have extended, or more-or-less continuous, egg-laying, whereas species of *Hyla,* which breed in water, have their time of oviposition correlated with rainfall. *Bufo melanostictus* of Ceylon lays its eggs mostly in the early part of the monsoons but takes advantage of any heavy rain after an extremely dry spell.[549] As in species of *Scaphiopus,* incubation is brief, at times within 48 hours. In species of *Pseudophryne,* Australian anurans, egg-laying is protracted over almost two months; if early nests suffer an unseasonal dry spell and the eggs become desiccated, the generation is continued by later egg-laying.[1106]

In an aseasonal environment of Amazonian Ecuador, anurans have several schedules of reproduction. Some species breed more-or-less continuously. Dendrobatid frogs, for example, that lay eggs on land, breed throughout the year as do some other terrestrial breeders, while other species that oviposit in water tend to breed most actively after heavy rains. The latter are sporadic, so that the reproductive cycles of many of the anurans in this aseasonal climate, although irregular, follow extrinsic cues.[260]

Egg-laying in tropical salamanders, in contrast to that of lizards, is adjusted so that hatching occurs as the annual rains begin. One of the salamander parents coils about the clutch and remains throughout the dry season, protecting the eggs not only from predation but also from desiccation. This pattern characterizes species of the plethodontid genera *Pseudoeurycea, Parvimolge* and *Bolitoglossa. Bolitoglossa rostrata,* a tropical lungless salamander (Plethodontidae) inhabiting seasonally rainless montane meadows in Guatemala, lays eggs at the onset of the dry season (November or early December), and the female remains with the eggs for about six months or until the resumption of the rains. Males are fecund throughout the year.[1051]

In many tropical oviparous lizards and snakes the onset of egg-laying occurs with the first rainfall.[354] Among some species of the iguanid genus *Anolis,* eggs develop with the approach of a rainy season. Ovogenesis may not directly depend on rainfall, for there are annual changes in temperature and daylength coincident with the annual maxima in precipitation. Oviposition, however, may be a response to rain and the resultant soft, moist soil, for a delay in rain seems to cause a retention of eggs in the oviduct in some anoles. The anole *Anolis aeneus* of Grenada, in the Lesser Antilles, is typical: maximal occurrence of oviducal eggs closely parallels early heavy rainfall. This relationship between rainfall and egg-laying is not universal.[962] Another anole, *Anolis acutus* of St. Croix, also in the Lesser Antilles, breeds throughout the year; ovigerous females occur in every month, and an adult female lays an average of one egg each month.[869] *Tropidurus hispidus,* a Neotropical iguanid lizard, inhabits xeric habitats in Venezuela and has an egg-laying period confined to part of the rainy season.[834] In *Agama agama* of West Africa breeding occurs in all seasons, but both sexes show increased reproductive activity at the end of the dry season and the start of the rains.[211] Of three species of skinks on the Ivory Coast, *Mabuya buettneri* lays eggs at the end of one rainy season, and they hatch four months later. *M. maculilabris* generally lays eggs during the dry season, and in *Panaspis nimbaensis* both egg-laying and hatching occur in the same rainy season. Consequently, in the same habitat, hatching of these species is not exactly coincident but nevertheless does provide for the growing period of the young to fall during the rains, when insect food is most likely to be abundant (Fig. 11.6).[52] In the tropical scincoid lizard, *Emoia atrocostata,* reproduction has no discrete seasonal peak.[3] In contrast the Oriental gecko, *Hemidactylus flaviviridus,* has a reproductive season between October and May.[877] Reproduction in the Malaysian snake *(Homalopsis buccata)* continues throughout the year with no discrete peak in sexual activity.[89] The live-bearing skink, *Mabuya striata,* in Tanzania breeds all year, whereas the sympatric egg-laying agamid lizard, *Agama cyanogaster,* lays eggs only during the rainy season, and its eggs hatch at the onset of the *dry* season.[52]

Probably the correlation of rainfall and egg-laying has a different significance among the various tropical oviparous species. Although some students have suggested that young lizards that hatch at the beginning of the rainy season find insect food especially abundant at that time, others believe that moist soil provides water needed for the proper development of the embryo within the egg. Eggs of snakes are frequently laid in moist ground, where they not only avoid desiccation, but also absorb water during embryonic development. Eggs of the colubrid *Spalerosophis cliffordi,* when kept wet one day in four during incubation, increase almost one-quarter over the original weight in contrast to no significant increase for eggs kept in saturated air. There was no difference in size or weight of the young from these two groups of eggs.[244]

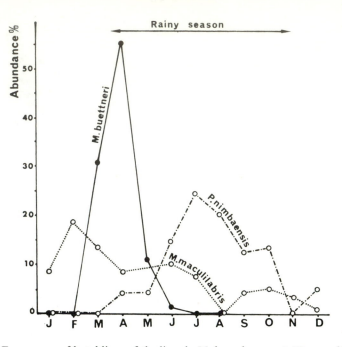

Fig. 11.6. Emergence of hatchlings of the lizards *Mabuya buettneri, M. maculilabris* and *Panaspis nimbaensis* on the Ivory Coast in relation to the rainy season (Barbault, 1976).

Egg-laying in turtles of arid and semiarid regions is stimulated by rain. The rain seems to moisten soil sufficiently for digging; all turtles nevertheless release urine in the soil before excavating the egg-chamber, and *Chelodina expansa* of New Guinea and Australia muddies its newly laid eggs so that a shell of dirt adheres to them.

Breeding seasons of birds may extend throughout the year in certain low-latitude habitats, but, even within a given species, not all individuals breed without interruption. Many nonmigratory birds of Central America (10°–15°N) breed in the spring, as do the bulk of migratory species that nest far to the north. For example, hummingbirds, which feed on nectar, deviate from this pattern: they live in the high mountains and nest mostly in autumn and spring, at the time when most flowering plants bloom. In midlatitudes many passerine birds have an annual period of reproduction, with two or more broods being reared within a single reproductive season of the year. It is sometimes difficult to distinguish between the one extreme of a prolonged or continuous reproductive "season" in the tropics and the other of a discrete season of two or more successive broods, as sometimes occurs in birds of temperate regions. Thus there is no clear-cut distinction in reproductive seasons of tropical and temperate birds.[936]

The irregular reproduction of birds in Australian deserts is dependent upon rainfall.[479,480,907] In certain arid areas of central Australia ducks tend to move nomadically and breed where rivers, flowing from the water-rich mountains, flood the flatland. Such flooding is irregular and depends on occasional heavy rains. Thus, breeding of some of the pond ducks (Anatidae) in this region occurs whenever and wherever flooding occurs; breeding here is predicated on water, but the rainfall has occurred elsewhere. In the gray teal *(Anas gibberifrons)*, for example, courtship commences a day or two after flooding, and eggs may be laid within two weeks of first flooding.[906] In East Africa, where precipitation is sporadic and unpredictable, weaver finches (Ploceidae) and other small passerine birds begin to nest shortly after a soaking rain. Some observers suggest that even the sight and sound of falling rain stimulate gametogenesis in some desert-dwelling birds.

The woolly opossum *(Caluremys derbianus)* of Middle America breeds from the dry season to the early part of the period of heavy rains. The degu *(Octodon degus)* of Chile breeds in September, at the end of the annual rainy season, when plant growth is maximal. In Madagascar the tenrec *(Hemicentetes semispinosus)* breeds mostly during the rainy season, from November to May.[358]

Increase in ambient temperature and decrease in precipitation are often concurrent, and frequently accompany a decline in vertebrate reproduction in arid regions. In much of western North America where a dry rainless summer is the rule, many small mammals breed in the spring. On the irrigated lands, such as alfalfa fields and wet pastures, jack-rabbits *(Lepus californicus)*, pocket gophers *(Thomomys bottae)* and voles *(Microtus californicus)* may extend reproduction well into the summer. Prolonged breeding is seen in captive *Microtus californicus* when plant growth is maintained by artificial watering.

In the Rift Valley of Kenya wild ungulates live under a climate of substantial annual fluctuations in rainfall. Browse species for Grant's and Thompson's gazelles, Coke's hartebeest and giraffe flourish in years of greater-than-normal rainfall; pregnancy and birth rates in a given year correlate positively with rainfall of the previous year.[292] Similarly in the African buffalo *(Syncerus caffer)* there is a seasonal cycle of female fecundity associated with a similar cycle of precipitation and food. The Queen Elizabeth National Park in Western Uganda lies on the equator and photoperiod is constant but, with a gestation of about 340 days, the buffalo there have bimodal peaks of conception and birth that follow the rainfall (Fig. 11.7).

In some tropical regions reproduction in small mammals is apparently correlated with seasonal rains, but the relationship between rain and breeding is not always the same. In some species the young are produced during the rains, when there is a maximal resurgence of plant growth, and in others the arid season is the time of greatest reproductive activity.

Eonycteris spelaea, a cave-dwelling fruit bat of Malaysia, breeds through the year; adult males are fecund continuously, and more than one-half of the females are gravid or lactating at any given time. This species is polyestrous and experiences a postpartum estrus that frequently results in pregnancy.[73] In contrast *Pteropus geddiei,* a fruit bat, and *Miniopterus australis,* an insectivorous bat, in the New Hebrides both have well-marked breeding seasons.[689] In the insectivorous bat, *Hipposideros caffer,* at 0° 27′N in Africa, males exhibit no gonadal cycle of

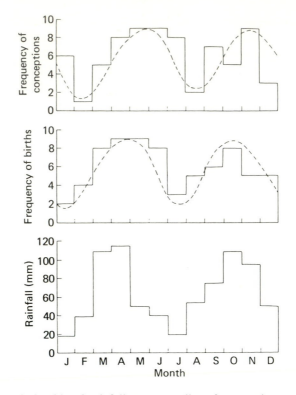

Fig. 11.7. The relationship of rainfall to seasonality of conceptions and births in the African buffalo *(Syncerus caffer)* in Queen Elizabeth National Park, Uganda (Grimsdell, 1973; courtesy of The Journal of Reproduction and Fertility).

regression and recrudescence, but females become gravid in November and give birth to a single young in February or March. Young of *H. caffer* are born in the season of greatest rainfall, when the insect food is most abundant.[737]

Seasonality at Midlatitudes

Spring at midlatitudes is a season of birth for both plants and animals. It is a time not only of thawing ice and melting snow but also of rising diurnal and nocturnal temperatures and increasing daylength. Plants, dependent on warm air, unfrozen water and solar radiation to break their winter dormancy, provide a sudden and varied fare for insects and vertebrates. Virtually every physical and biologic feature of the environment undergoes some change.

The increasing daylength and more direct insolation distinguish the midlatitude spring from the perpetually warm climate of the tropics. Shorter days and less direct sunlight retard most plant growth and arthropod activity in the autumn. Thus the annual pulsations of autumnal dormancy and vernal rebirth are reverse sides of the same coin, and together they produce the features that distinguish the temperate regions from the tropics. The distinctions are clouded but not obliterated by variations in elevation, precipitation and seasonal winds and storms. The depressing elements of winter, together with the stimulating factors of spring, both account for the concentration of much reproductive activity in the spring. This seasonality is even more pronounced in the very high latitudes.

In temperate latitudes most amphibians lay eggs in the spring. Anurans, with external fertilization, mate during oviposition, and the two acts are simultaneous and coordinated. In the spring the oviducts hypertrophy and secrete the gelatinous coat of the eggs. Frogs and toads usually spawn once a year in temperate regions, and breeding seems to be controlled by both rains and rising temperatures in the spring. Many kinds of anurans congregate in ponds. Males arrive first and, by their calls, attract females to the water. In addition to rising temperatures and rain, algal growth produces an odor to which some frogs respond: the Eurasian *Rana temporaria* and the Nearctic *R. clamitans* both seem attracted to the olfactory cue of algae. Spadefoot toads (*Scaphiopus* spp.), dwellers of more arid temperate areas, emerge and breed in temporary ponds after an extended dry spell is broken by a heavy rain. Their gonads ripen during the dormant period.

The breeding cycle of the dusky salamander *(Desmognathus fuscus)* is typical of many temperate salamanders: vitellogenesis begins in late summer and is completed by the following spring.[1011] Many plethodontid salamanders mate in the autumn; sperm is stored in the oviduct, and ovulation and egg-laying follow in the spring. In *Plethodon welleri* courtship takes place in the autumn, and in the following spring eggs are laid in a moist rotting log.[780]

Some amphibians breed more than once a year. The plethodontid salamander *Desmognathus ochrophaeus* apparently breeds in both spring and autumn, but a given female does not necessarily breed twice a year. Timing of egg-laying in the red-spotted newt *(Notophthalmus viridescens)* depends on the hibernation site of the female, those individuals that winter in ponds breeding prior to those that migrate to ponds from terrestrial hibernacula. The smooth newt *(Triturus vulgaris)* of Europe breeds in both spring and autumn; schedules of oviposition vary according to whether a given indiviudal hibernates on land or in water, the latter group breeding early in the year. In higher latitudes the red-backed salamander *(Plethodon cinereus)* has two peaks of oviposition but only a single annual peak to the south. The Nearctic green frog *Rana clamitans* is known to spawn twice, once in spring and once in summer. The spring peeper *(Hyla crucifer)* moves to ponds in the spring rains, and may make a breeding migration, call and deposit eggs under the same stimulus in the autumn. Circumstantial evidence suggests several small clutches in some tropical species: some tropical hylid, leptodactylid and dendrobatid frogs have developing ova of two or more sizes.

In temperate latitudes most reptiles are spring-breeders, but variations among different species are legion. In some species, mating may occur either in autumn or spring, with egg-laying in late spring (and birth of young in spring or summer). Clearly gonadal activity of males and females is not always parallel. Generally testes and ovaries begin to grow after summer refractoriness, and in many species gamete development takes place during hibernation. In some viviparous reptiles spermatogenesis in late summer is followed by mating and ovulation in fall. In the anole *(Anolis carolinensis)* in the United States, spermatogenic activity begins in the winter and continues throughout the summer, and sperm storage does not occur. Thus spermatogenesis begins when daylength is decreasing and extends through the period of increasing daylength.[311]

The live-bearing alligator lizard *(Gerrhonotus coeruleus)* mates and ovulates from late April to mid-June in coastal Washington; the single annual brood is born in late August or early September.[1057] Similarly the viviparous skink *(Leiolopisma zelandica)* of New Zealand has an extended period of ovogenesis in winter (April to September) and mates in early October; the young are born in January.[67] In at least some viviparous iguanid lizards, however, gametes ripen in the summer and mating occurs in the fall: both *Sceloporus jarrovii* and *S. cyanogenys* mate in the fall and give birth the following spring.[198,348] These two species of *Sceloporus* do not hibernate. In the Nearctic natracine snakes, all of which are viviparous, mating occurs in the spring, and the young are born the same year in the summer.

Birth (hatching) may be delayed in some oviparous reptiles; the young may remain in the egg over the winter and hatch the following spring. This seems to be a common event in the painted turtle *(Chrysemys picta)* and some other turtles of North America. *Chelodina expansa,* a chelid turtle of New Guinea and Australia, lays eggs in autumn, from March to May, which remain in the nest over the winter to hatch about one year later. Likewise the tuatara lays its eggs from October to December, the New Zealand summer; the incubation occupies about 14 months, the eggs hatching at the end of the second summer.

Two herbivorous reptiles of western North America, the chuckwalla *(Sauromalus obesus)* and the desert tortoises (*Gopherus* spp.), adjust their reproduction to the amount of precipitation: in years of light rainfall, when there is a poor growth of annual forbs, reproduction may be suspended.[746] Even in insectivorous species, sparse rainfall may retard breeding: the zebra-tailed lizard *(Callisaurus draconoides),* an iguanid lizard of the Great Basin, may not lay eggs when rainfall is light and plant growth and insect populations subnormal.[530]

The gonadal cycle in many species of mid- or high-latitude birds is strongly influenced by photoperiod. Frequently, in migratory species, territoriality is displayed prior to the arrival of females on the breeding grounds; neighboring males conceivably stimulate each other to song and display. When females appear on the scene, there has already been considerable ovarian development, and the courtship by males further stimulates the females to yolk deposition, ovulation and egg-laying. In several species of birds, stimulation by the male is essential to complete ovarian development and oviposition. The white-crowned sparrow *(Zonotrichia leucophrys gambelli),* a migratory fringillid of North America, has an annual cycle of relatively fixed and frequently mutually exclusive series of events, which are regulated by annual changes in photoperiod (Fig. 11.8).[286] Probably the same general pattern exists in many migratory passerine birds. In the house sparrow *(Passer montanus saturatus)* in Hong Kong there are two distinct periods of testicular recrudescence, in April and July, between which times the testes regress.[161] The underlying cause for this bimodal pattern is not apparent.

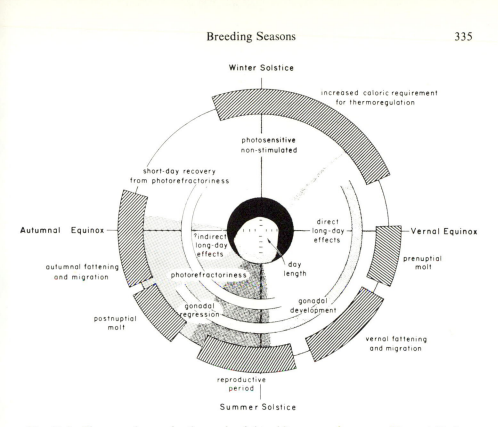

Fig. 11.8. The annual reproductive cycle of the white-crowned sparrow *(Zonotrichia leucophrys gambelli)* and its relationship to molt, migration and exogenous cues (Farner, 1964; courtesy of The American Scientist).

Pigeons and doves (Columbidae) tend to have a prolonged nesting period, even in temperate regions. Because the young subsist on "milk" that the parents produce, these birds are not dependent on the environment's providing a food appropriate for young birds.

The refractory period, a time when the gonads do not respond to stimuli, is characteristic of many avian reproductive cycles. It serves to suppress reproduction at a time when daylength would be stimulatory but in which the environmental conditions might not be suitable for the growth or development of the young.

Although most birds at temperate latitudes are clearly responsive to photoperiod and breed in the spring, breeding may be extended in years of summer rains when summers are usually dry. The California quail *(Lophortyx californica)* produces a single clutch in most years, but extended rains that result in heavy herbaceous growth and high soil moisture often result in a second (summer) brood. Intensity of reproductive activity seems to be inversely correlated with the abundance of phytoestrogens which accumulate in the quail's plant food in a dry year.[589] In Arizona, in the Sonoran Desert, there is annually a rain in March or

April, and in some years a second rain in July or August. Abert's towhee *(Pipilo aberti)* nests regularly in the spring, about 10–15 days after the first rain, and may nest again if there is a midsummer rain; rain seems to be an important Zeitgeber for reproduction in this as in some other desert birds.[645]

Some shearwaters (Procellariidae) have predictable and narrow breeding periods. The slender-billed shearwater *(Puffinus tenuirostris)*, on Fisher Island of the Furneaux Group, lays 85 percent of the eggs for the entire season within a span of three days, with no significant difference in egg-laying from one year to another or from one colony to another.[904]

Vernal reproduction typifies many small mammals at middle and high latitudes. Shrews, bats and many rodents characteristically produce young in the spring, and the same pattern is found in the Southern Hemisphere. Vespertilionid bats typically mate in autumn, and parturition occurs in the spring. Frequently there are bimodal patterns of breeding in which sexual activity is resumed in the fall. Such bimodal cycles characterize some species of *Peromyscus* in North America, and the vole *Microtus montebelli* in Japan has regular peaks of reproduction in May and September.[526] A midsummer lull in breeding may reflect dietary changes, for in years of food abundance voles (*Microtus* spp. and *Clethrionomys* spp.) and mice (*Peromyscus* spp. and *Apodemus* spp.) may continue reproduction from spring until the following winter.[305] A bimodal reproductive pattern in the Nearctic white-footed mouse *(Peromyscus leucopus)* results from reproduction in older females in April and May, and later in younger females, breeding for the first time in August (Fig. 11.9).[184] Although experimentally heat may cause temporary sterility in males, this is not common in the wild. In the red kangaroo *(Megaleia rufa)*, however, hot droughts apparently cause not only a decline in pregnancy but a reduction of spermatogenesis and libido in males.[761]

There is an increase in both onset and duration of breeding in some wild mice when food is especially abundant.[177] Estrogenic properties are known for a large number of common plants; some of these materials are identical with estrogens in animals, and other plant materials that have estrogenic effects are chemically unlike steroidal estrogens.[569] Although ingestion of some presumed "phytoestrogens" stimulates gonadal growth in small or moderate amounts, in large amounts they may prevent the release of gonadotropins from the adenohypophysis.[87] Voles (*Microtus* spp.) are stimulated by sprouting green food in the diet, and apparently cease breeding as green forbs become mature and dry.[751,752] By continuous cropping of forbs and grasses, voles maintain the production of sprouting shoots; under such conditions they may continue to breed for greatly extended periods. Reproduction in many temperate regions is dependent partly on irregular precipitation. When rains cease, accumulation of certain phenolic compounds in grasses inhibits gonadal activity in both sexes. The Levant vole *(Microtus guentheri)* of the Near East produces abnormally large numbers of ova when fed a diet of fresh legumes, and *Microtus montanus* of western North America comes into estrus a day after a diet of sprouting wheat.[104,752] Extracts of sprouting wheat cause an increase in delta cells, a source of gonadotropins from the anterior pituitary.[436] Experimentally, sprouting wheat fed to voles was followed by more frequent postpartum matings and greater viability of young.[750]

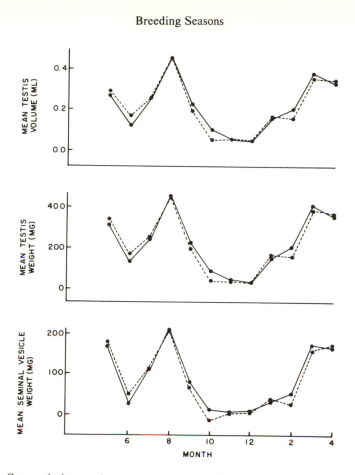

Fig. 11.9. Seasonal changes in mass and volume of testes and mass of seminal vesicles of white-footed mice *(Peromyscus leucopus)* in West Virginia (Cornish and Bradshaw, 1978).

Historic changes in reproductive activity point to other, still undetected, ultimate factors in timing reproductive seasons. Prior to the introduction of myxomatosis in England, the rabbit *(Oryctolagus cuniculus)* bred mostly in early spring and ceased in July; subsequent to the establishment of myxomatosis, however, breeding extended into August. The relationship of this mosquito-borne virus to the breeding season of rabbits is not apparent.

Typical patterns of vernal birth are not seen in all vertebrates. Some insectivores and mice regularly breed in autumn and winter, but the significance of such timing is obscure. In the eastern United States (as far north as Pennsylvania) the short-tailed shrew *(Blarina brevicauda)* may breed and produce young as early as January.[169] The shrew-mole *(Neurotrichus gibbsi)* of the North American Pacific Coast also breeds in January and February, even at 1500 m and above in the Sierra Nevada. The lowland shrew-mole *(Urotrichus talpodes)* of Japan similarly breeds mostly in January and February.[1116] In Kyushu, southern Japan, both

Apodemus speciosus and *A. argenteus* breed from September or October to March or April.[1117] The golden mouse *(Ochrotomys nuttalli),* a crecitid rodent of the southeastern United States, typically breeds in the winter. Autumnal reproduction of the piñon mouse *(Peromyscus truei)* is positively correlated with a heavy crop of seed from the piñon pine *(Pinus monophylla).*[608]

The bobcat *(Lynx rufus)* is a seasonal breeder, but parturition may occur from June until late August. This cat is polyestrous and may experience several ovulations in a season before fertile mating, which accounts for an extended season of birth (Fig. 11.10).[200]

Among primates there tends to be an increased seasonality of both mating and birth with an increase in latitude. A number of tropical monkeys and apes experience fertile matings throughout the year, but the Japanese macaque *(Macaca fuscata),* at more than 30°N, mates mostly in January and February and young are born from May to September. The rhesus monkey *(Macaca mulatta)* introduced on Cayo Santiago (in the Lesser Antilles), at 18°N, has a well-marked testicular cycle in which maximal size is correlated with the period of copulation and minimal size with the period of births. Most of the well-documented reproductive patterns of equatorial primates reveal almost continuous breeding and birth, but some seasonality occurs in regions where rainfall is cyclical.

Seasonality in conception rates in human beings is well established; peaks vary with latitude and social factors, but annual cycles are constant within a given

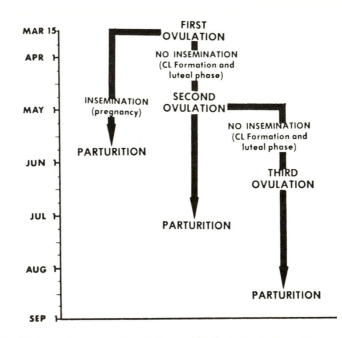

Fig. 11.10. Estimated pattern of ovulation and births in the bobcat *(Lynx rufus)* (Crowe, 1975).

society. Time of conception is determined by subtracting nine months from the birth date without correcting for "natural" prenatal deaths or induced abortions. In Sweden, England and Wales the peak of conception is in the summer, but in Japan, on the other hand, the peak occurs in April. In some central and western Europe countries a smaller peak occurs in December; these peaks are not related to temperature, and one worker suggests that this is in celebration of Christmas with its Teutonic fertility-rite aspects (Fig. 11.11). Interestingly dates of marriage and conception of first-born are not related.[632] There is a clear-cut and highly suggestive inverse relationship between temperature and incidence of conception in Hong Kong (Fig. 11.12), but the underlying cause is not definitely known.[160] Among yam growers in Western Abelam, New Guinea, where the temperature and humidity are relatively constant, births are concentrated between October and March. Because male status increases with the number and size of ceremonial yams that a man produces, and because sexual abstinence is believed to enhance the growth of his yams, conception is almost restricted to the period during which yams are not cultivated.[884]

Photoperiod influences the seasonality of reproduction in virtually all species of endothermic vertebrates studied, but in the tropics the endogenous nature of the reproductive cycle seems less sensitive to light than to some other exogenous factors, especially rainfall. Rainfall may have either a direct or indirect effect on reproduction in tropical vertebrates. Indirect effects are most commonly seen

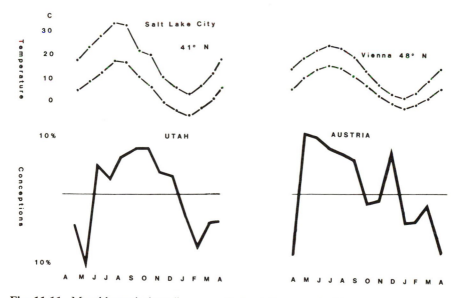

Fig. 11.11. Monthly variations (in percent) about the mean incidence of conception for Vienna and Salt Lake City (Utah). The annual pattern for Austria is similar to the curves for Sweden, England and Wales (including both legitimate and illegitimate births) (MacFarlane, 1970).

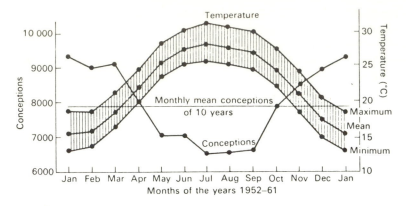

Fig. 11.12. The inverse relationship between human conception rates and air temperature in Hong Kong (Chang et al., 1963).

through food, but some oviparous vertebrates depend directly on water in ponds or in soil for an appropriate site for oviposition. In nature the variable factors that influence timing of reproduction do not necessarily fluctuate in unison, and it is very difficult to identify the critical seasonal changes in the environment that determine the onset and end of breeding activity. Some tropical birds breed two or even three times a year but nevertheless tend to follow a rhythm characteristic of the species. In some desert areas of Africa and Australia, breeding is irregular, and occurs in response to changes produced by sporadic rains. These areas are also at rather low latitudes with correspondingly moderate seasonal changes in photoperiod. The effect of photoperiod, as a Zeitgeber, is less important than it is in the north temperate and Arctic regions. The changes in daylength, which initiate gonadal growth, vernal molt and hyperphagia in late winter at high latitudes are also associated with the seasonal changes in plant and insect growth in the spring.

Temperature-Related Reproduction

Ectotherms are much more sensitive to environmental temperatures than are birds and mammals and much less responsive to changes in photoperiod. When in water, the body temperature of fishes and amphibians is with very few exceptions the same as that of the water.

Seasonal reproduction in fishes of midlatitudes is frequently predicated on temperature changes. Spring-breeding species respond to rising temperatures, and those that spawn in the autumn mate as temperatures fall. In the pumpkinseed *(Lepomis gibbosus)*, a centrarchid fish, gonadal recrudescence occurs with rising temperatures in the spring: testicular growth in the laboratory occurs between 11.5° and 14.0°C and ovarian growth between 14.0° and 16.5°C. The effect of photoperiod seems to be secondary.[139] More than 90 percent of the spawning of the striped bass *(Morone saxatilis)* in the California Sacramento River occurs between 17.2° and 20.0°C, although the exact timing varies with the rate of melt-ing snow in the watershed above the river. The shield darter *(Percina peltata),* a Nearctic percid, spawns in riffles of small streams in the spring, when the water temperature is 10°C or warmer.[760] Whitefish (Coregonidae) spawn when the water drops to a critical minimum temperature: in the Nearctic coregonid *Leu-cichthys artedi,* spawning occurs in late autumn at water temperatures of 4°C or lower, and annual variation in spawning dates reflect yearly variations at the time 4°C is reached.[509]

The zebrafish *(Brachydanio rerio)* has a series of ovulatory cycles at 26°C, but egg-laying ceases when the temperature drops to 22.5°C. Species of rockfish *(Sebastes)* of the Pacific are ovoviviparous, mating in October and producing lar-val young from January to July, depending on the species. Parturition occurs ear-lier in the most southerly occurring species, and two species off the California coast *(S. goodei* and *S. paucispinis)* produce young twice a year. Mating is appar-ently controlled by temperature in these species of *Sebastes.*[1087]

Numerous amphibians seem to respond to temperature changes in temperate climates, and like fishes they are both spring and autumn breeders. Complicating the situation is multiple spawning in many species.

Temperature may control the reproductive period in amphibians that are essentially aquatic. In the plethodontid salamander, *Eurycea multiplicata,* gonadal development (and mating and egg-laying) differs apparently in response to normally different annual temperature cycles: in a population inhabiting a stream that was frequently frozen in the winter, reproduction extended from Jan-uary to March but in a pond population with water temperature seasonally con-stant between 14° and 18°C, reproduction extended from September to January (Fig. 11.13). A Neotropical toad, *Bufo variegatus,* breeds in the Chilean Andes at 49° south latitude; amplexus and oviposition occur at air temperature of 7.0°C and water temperature of 11.5°C, in the second week of September, apparently shortly after emergence from hibernation. In the Chilean Andes at 1750 m *Bufo arunco* breeds in small creeks which are warmed by hot springs, and in these waters breeding takes place in very early spring.[355]

In temperate regions reproduction of reptiles is closely tied to temperatures, although the controlling mechanism is not well understood. In a number of Squa-mata of temperate regions, testes are of greatest size immediately after emergence of species from hibernation, and mating occurs some weeks later, when testes have regressed considerably. In colubrid snakes of temperate regions males emerge from hibernation somewhat ahead of the females, and courtship occurs after females appear. In some species both sexes actively participate in courtship move-ment, and mating occurs following several days of courting and continuous asso-ciation of a pair. The garter snakes (*Thamnophis* spp.) of the Nearctic region hibernate throughout most of their ranges. In Manitoba (at 50°–55°N latitude) *Thamnophis sirtalis* mates soon after emergence from hibernation in late April and early May.[6] Experimentally this species mated when held in darkness for several months at 5°C and then moved to 25°C with a 12L:12D regiment[412] In warmer regions to the south, however, the same species may mate in both autumn

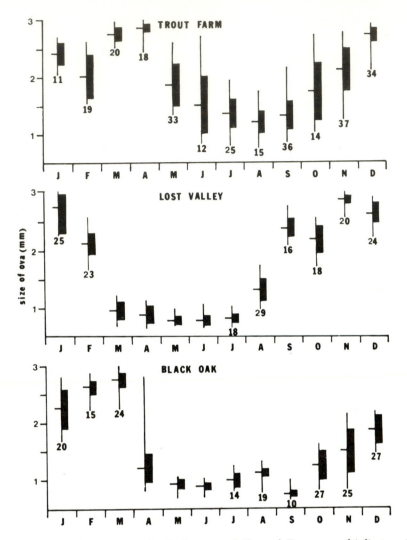

Fig. 11.13. Monthly mean ova size in three populations of *Eurycea multiplicata griseogaster:* the Black Oak population is from a stream frequently frozen in winter; the Trout Farm had little temperature variation; and the Lost Valley population is from ponds with temperatures from approximately 14° to 18°C (Ireland, 1976).

and spring. Upon emergence from hibernation, the red-sided garter snake *(Thamnophis sirtalis parietalis)* is very sensitive to changes in temperature, and both courtship and mating become intense at from 25°C to 30°C (Fig. 11.14). At this same time testosterone levels are low relative to levels found later in the year, and courtship occurs independently of testosterone buildup, but these reptiles seem to be refractory to the high levels of hormone that accumulate in the autumn.[7] In

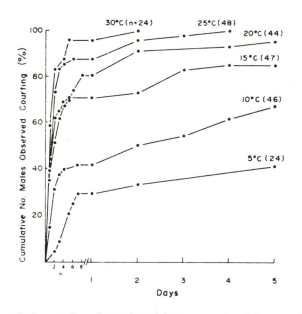

Fig. 11.14. Cumulative number of courting male garter snakes *(Thamnophis sirtalis par-ietalis)*, expressed as a percentage of the total, when transferred from 1.5°C to several high temperatures (Hawley and Aleksiuk, 1975; courtesy of The Canadian Journal of Zoology).

captive brown snakes *(Storeria dekayi)*, a natricine of eastern North America, courtship follows when they are removed from a temperature of 7°C in spring and kept in light at 25°C. In the wild they mate in early spring, when they emerge from their hibernacula.

A midsummer gonadal refractory period in some temperate climate lizards gradually weakens in winter, and gametogenesis is produced by increases in temperature. Under experimental conditions, *Anolis carolinensis* males are variably sensitive to changes in temperature and photoperiod: low temperatures promote testicular growth in autumn and winter, and short days (less than 13.5 h) effect testicular regression (Fig. 11.15). Apparently the rise in temperature in the spring acts on neural centers rather than the gonads or associated reproductive structures.[197]

Reproductive times of birds are generally independent of temperature, but they are sensitive to changes in food supply. The great tit *(Parus major)* readily nests in artificial nestboxes, and details of its breeding cycle have been carefully followed for many years at Oxford, England. A typical clutch of 10 eggs is equal in weight to the female and is laid in 10 days. The females begin to lay after there is an adequate supply of caterpillars, the staple diet of the young in the spring. Hatching of caterpillars is temperature-dependent, and indirectly so is oviposition by the great tit: in years of warm spring (1 March–20 April) first eggs may be laid as early as 10 April but as late as 10 May in a cold spring.[810]

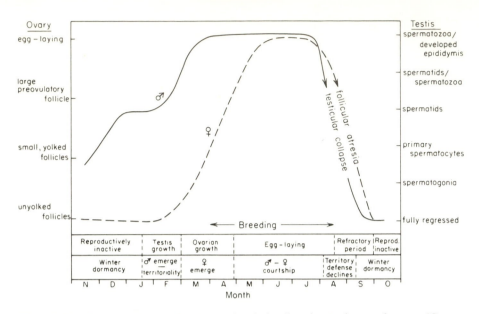

Fig. 11.15. The annual reproductive cycle of the lizard, *Anolis carolinensis* (Crews, 1975, copyright American Association for the Advancement of Science).

Mammals that hibernate usually undergo gametogenesis during the winter and are prepared to mate soon after emergence. Some ground squirrels (*Citellus* spp.) do not become fecund when maintained at high temperatures, which prevent their dormancy at the normal time. In some ectotherms, including some anurans, spermatozoa may mature before the start of hibernation, and there is then no testicular activity or growth during dormancy. Breeding seasons are usually brief in species that hibernate and are punctuated by a midsummer refractory period. That these are genuinely refractory phases can be shown when the gonads fail to respond to the injection of gonadotropins.

There is not always a close correlation between testes' size and mating. Among many hibernating sciurids, mating occurs very shortly after emergence and when the testes have attained their maximal size, as is the case with the mantled ground squirrel *(Citellus lateralis)* and the marmot *(Marmota flaviventris).*[14,51,677] In the prairie dog *(Cynomys leucurus),* however, testes are large when hibernation is ended in early March but, when mating begins some four weeks later, the testes are much smaller. When the prarie dog mates, the seminal vesicles reach maximal size and the cauda epididymides are filled with sperm.[44]

Temperate species of bats of the families Vespertilionidae and Rhinolophidae hibernate. Spermatogenesis occurs in summer and by the time males enter dormancy the testes have regressed and the cauda epididymides are swollen with sperm. Plasma testosterone levels are high in midsummer in *Myotis lucifugus;* accessory reproductive glands (e.g., seminal vesicles) enlarge in autumn as the testes regress and as testosterone levels decline. Mating sometimes occurs in autumn and more regularly in spring; in either season testes are small and testosterone levels low.[382]

The chipmunk *(Tamias striatus)* of eastern North America seems the only hibernator that produces two broods a year. After emergence in the spring some old females have fertile matings, and in the summer both young and old females participate in a second period of breeding (Fig. 11.16).[949]

Biennial Reproduction

In some vertebrate species a given female gives birth every other year. Although there is an annual breeding season in which all males participate, only about one-half of the mature females may mate in any one year. The phenomenon of biennial breeding is poorly documented, and may be more frequent than literature suggests.

Alpine populations of *Salamandra atra* carry developing eggs and the young are born from one to three years after ovulation and fertilization. Two tropical lungless salamanders, *Bolitoglossa rostrata* and *B. subpalmata,* oviposit biennally.[671] Females of a number of species of reptiles produce young every other year, the nonproductive year for any given female apparently being a period of respite during which fat stores are replenished.[1051] Biennial reproduction is found in many species of vipers (Viperidae) in both the Old and the New Worlds. Certain species of boreal vipers (*Crotalus* spp. and *Viperus berus)* produce young on a biennial schedule;[1012] the cottonmouth *(Agkistrodon piscivorous),* a Nearctic viper, produces sperm throughout the year, and females with ovaries with matur-

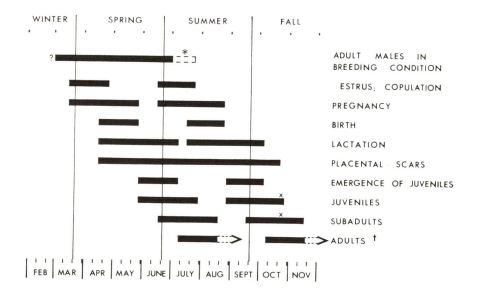

Fig. 11.16. Major features of the breeding seasons of the chipmunk, *Tamias striatus,* the only hibernator known to produce two broods of young annually (Smith and Smith, 1972; courtesy of The Canadian Journal of Zoology).

ing eggs are heavily inseminated. Mating apparently occurs at any season. Females, on the other hand, have a biennial reproductive cycle. Similarly the Gila monster *(Heloderma suspectum),* the poisonous lizard of southwestern United States and Mexico, lays eggs every second year. The loggerhead turtle *(Caretta caretta)* nests every second year but may lay three or four times in a single season.

Biennial reproduction is uncommon among birds but, in species characterized by very slow growth, prolonged parental care extends beyond the onset of the next breeding season. The huge condors breed every second year, since the young is still dependent on its parents until well past age one year. Some large marine birds (e.g., albatross and frigate birds) are biennial breeders. The king penguin *(Aptenodytes patagonicus)* breeds within a protracted nesting period, and about 15 months elapse between one egg-laying and the next. Some females are early, and others late, layers, and a given individual nests two years out of three. The royal albatross *(Diomedea epomophora)* on Campbell Island nests annually, but any given individual breeds biennially: reproduction begins in October and the young remain in the nest until the following October, by which time the other reproductive segment has arrived on the nesting grounds.[1086]

Extended parental care in several large mammals precludes annual pregnancies, and these species breed in alternate years. In mammals ovulation may be suppressed during lactation, but there seems to be much interspecific variation in this matter. In most species of finback whales (*Balaenoptera* spp.) gestation is nearly 12 months and mating is biennial. Few females experience a fertile postpartum estrus, but occasionally whalers capture one that is both gravid and lactating. In some if not most species of bears (Ursidae), breeding is biennial: normally mating occurs in July, but in alternate years an adult female is caring for small cubs, and does not mate. Nursing continues until autumn. At least some of the large cats (e.g., *Felis concolor*) mate in alternate years, but this is not a clear-cut pattern in tropical species that have no discrete season for mating.

Extended gestation in other large mammals prevents annual breeding. Gestation in the Indian elephant *(Elephas maximus)* is more than 630 days, and some rhinoceroses, tapirs and camels have gestations of more than one year, precluding annual mating. Gestation in the walrus occupies 12 months and the nursing 18–24 months, so that mating is biennial.

Females of the yellow-bellied marmot *(Marmota flaviventris)* at 3400 m in Colorado produce young every second year; this is apparently limited by the time available to accumulate fat, whereas females at lower elevations produce litters annually. Similarly the Olympic marmot *(Marmota olympus)* at timberline in the Olympic National Park (Washington) breeds bienially; in a barren year a female is anestrous, presumably also a reflection of fat deficiency.[14]

Socially Synchronized Reproduction

It is apparent that many stimuli for sexual activity are psychological. Although seasonal gonadal recrudescence is a response to changes in photoperiod and weather in many species, there are species in which females require the additional stimulus of a courting male.

The female anole *(Anolis carolinensis)*, an iguanid lizard, responds to courtship by increased gonadal recrudescence. The courtship activity seems to be an essential prerequisite to egg-laying, and the presence of a castrated male does not effect ovulation.[311]

Certain tropical bird species have rather discrete seasons for nesting, but these seem sometimes to lack any relationship with time of year. In some species there is a possibility of reproductive synchronization through social factors.

The sooty tern *(Sterna fuscata)* has a regular breeding cycle in any given part of its geographic range but differs markedly from one area to another. One population in the Hawaiian Islands at 21° N breeds in the spring (on Manana Island), and another begins its reproductive cycle in autumn. There is no obvious cause for this dichotomy; the social requirement that the group breed as a unit seems to be of greater importance than a regional climatic factor. On Ascension Island at 8° S the same tern breeds at a different time every year, each cycle being separated by 9.7 months from the next. In low latitudes, where external cues differ only slightly with the seasons, intervals between breeding cycles could be determined by the length of the refractory period and the stimulation of the population within itself.

In the seasonally uniform tropical rain forest of the New Hebrides, the golden whistler *(Pachycephala pectoralis)*, a nonmigratory passerine bird, has a discrete, restricted breeding season. There is no synchrony with populations of the same species in Australia or elsewhere at the same latitude.[41] The silver gull *(Larus novaehollandiae)* may have either autumn or spring breeding and may even breed in both seasons; this same pattern is seen in the noddy *(Anous stolidus)* and in some other gulls and terns of Australia. Generally these bimodal species breed in September to November in the south but in February or March to May in the north. The red-billed tropic bird *(Phaethon aethereus)* breeds continuously on Daphne Island of the Galápagos, with eggs and young occurring each month of the year. On nearby South Plaza Island, breeding occurs between August and January. Apparently a high concentration of this bird and its nests on Daphne Island lead to intense competition for nest sites, and this may contribute to a delay in nesting for some pairs. The possible effects of food on the density and reproductive cycle of this bird is not known.[954] The swiftlet *(Collocalia escuelenta)* in a given region has a discrete breeding season, two or three times a year, but the seasons lack synchrony from one region to another.[687] Very possibly regional variations in breeding seasons of tropical birds reflect annual cycles in their food organisms.

Spawning Synchronized with Tides

Several fish spawn in synchrony with cycles of marine tides. Perhaps the best known is the grunion *(Leuresthes tenuis)*, a species of silverside of southern California. Spawning grunion ride a wave up to a high point in wet sand and, before the wave recedes, a female buries herself tail down in the sand where she releases her eggs. Males, clustering about her, release milt which runs down her body to the ova. The spawning fish return to the sea on the next high wave. Until the next high tide, some two weeks later, the developing eggs are moist but undisturbed by wave action. When the subsequent high tide exposes the grunion fry and carries them to sea, they are sufficiently developed to continue life as free-living fish.

In New Zealand the galaxiid *Galaxias maculatus* spawns in synchrony with the tide. In autumn adults move from rivers to brackish grassy flats and spawn at high tide. As the tide recedes, the fertilized eggs settle about the bases of the grasses, and hatching occurs two weeks later at the next high tide.[672] The mummichog *(Fundulus heteroclitus),* a common cyprinodont fish of Atlantic coastal marshes, spawns at night during high tides at either a full- or a new moon. Eggs are placed among leaves of sa salt marsh grass *(Spartina alterniflora),* and fry emerge about two weeks later at the next high tide.[997] These fish may spawn several times each season, and distinct size groups of fry represent each spawning period (Fig. 11.17).

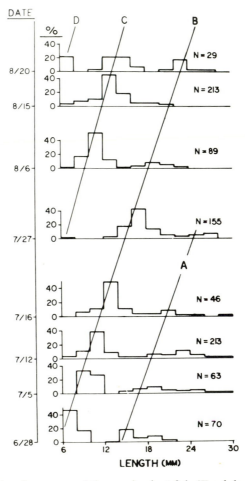

Fig. 11.17. Successive size groups of the cyprinodont fish *(Fundulus heteroclitus);* diagonal lines (A–D) apparently indicate growth of spawning groups (Taylor et al., 1979).

Summary

In addition to adaptation to factors critical to survival, reproduction may be responsive to certain environmental cues preparing the organism for events to come. Best known are the many examples of sensitivity to changes in daylength, or photoperiod. Changes in gonadal growth frequently occur with progressive increases or decreases in day length, and gonadal recrudescence is the most common beginning of a season of reproduction. This effect may differ between the sexes in mammals, so that the seasonal development of secondary sexual features and the maturation of sperm may occur as both sexes experience a peak in sex drive, and ovulation or implantation may occur weeks or months later. In a very general way those species that mate in the autumn respond with an increase in gonadal activity as the day is shortened and vernal mating is activated by longer daylengths. Hibernating mammals constitute an interesting exception. As the daylight shortens in the autumn, they enter winter dormancy but, when males emerge in the spring, their testes are large and the epididymides are packed with sperm. They had been exposed only to autumnal photoperiodic changes; no one seems to have demonstrated if they are sensitive to shortened days of the autumn.

In the overwhelming majority of birds, the increase in daylength is a stimulus to gonadal activity. Many birds can respond to changes of several minutes in daylength. The response is not always to increase in daylength, for some species show gonadal activity under *decreasing* daylight. Many birds, like other vertebrates, adjust their breeding season to rainfall, especially in lower latitudes, but this could in fact be a response to food produced by seasonal rains.

In birds males are more sensitive than females to photostimulations and, although females are affected by increasing daylight of late winter, they seem to require the stimulus of courtship to bring them to the point of nest-building and ovulation. When the female responds to the reproductive state of the male, the pair becomes synchronized in the reproductive process. This is especially important in species in which both individuals care for the young. In colonial nesting birds, breeding of a large group is frequently synchronized, apparently the presence of courting males affecting not only their own mates but any females that may observe them. Experimentally males may stimulate their mates through voice, visual signals and touch.

Although many birds and mammals of the midlatitudes undergo gonadal recrudescence in late winter and mating in spring, there are a number of exceptions. These include insectivorous bats, many carnivores and some rodents and insectivores. Such deviations from the prevalent pattern involve delayed fertilization, delayed implantation, sperm storage in the cauda epididymides and mating and parturition in the autumn and winter.

One of the conspicuous features of the major reproductive parameters is their variability: superimposed on the genetically fixed reproductive features of a species is a phenotypic flexibility permitting a broad spectrum of responses. We have seen the powerful role of food in determining not only growth and sexual maturity, but also litter size and duration of breeding season. There are now well-established roles of social interaction (e.g., the Bruce effect, the Whitten effect and others) on sexual maturity and ovulation. Litter size may sometimes vary with age, so that mean litter size in a population may reflect its age structure. Daylength not only is an important Zeitgeber in nature but experimentally can alter ovulation rates and sexual maturity in females and growth of testes and seminal vesicles in males. This information greatly increases our understanding of reproductive patterns, but ironically increases the known complexity of controlling factors and raises new questions. To what extent is the general pattern of greater litter and brood size at high latitudes a genetically determined feature, and to what extent are they affected by daylength, food and other exogenous factors? Such questions indicate the direction of some future research.

Suggested Readings

Crump ML (1974) Reproductive strategies in a tropical anuran community. University of Kansas Museum of Natural History, Lawrence

Fitch HS (1970) Reproductive cycles in lizards and snakes. University of Kansas Museum of Natural History, Lawrence

Matthews LH (1955) The evolution of viviparity in vertebrates. Mem Soc Endocrin 4: 129–144

Nalbandov AV (1976) Reproductive physiology of mammals and birds. Freeman, San Francisco

Rowlands IW (1966) Comparative biology of reproduction in mammals. Zool Soc London Symposia, No. 15. Academic, London

Sadleir RMFS (1969) The ecology of reproduction in wild and domestic mammals. Methuen, London

van Tienhoven A (1968) Reproductive physiology of vertebrates. Saunders, Philadelphia

Wimsatt WA (1975) Some comparative aspects of implantation. Biol Reproduction 12: 1–40

12. Growth

Growth begins with fertilization of the ovum, although initial growth is rather slow in some vertebrates. Embryonic growth and postnatal development are discussed in detail in numerous textbooks, and these topics generally lie beyond the realm of vertebrate natural history. There are included in this chapter, however, distinctive adaptive patterns that characterize reproductive specializations of vertebrates. In egg-laying ectotherms, growth reflects the light, oxygen and temperature to which the eggs are exposed, and growth rates are very responsive to these factors. Viviparous ectotherms, of which there are many, have developed a variety of behavioral and physiologic procedures that tend to remove growth and growth rates from total dependency on the environment. Viviparity in fishes, amphibians and reptiles constitutes an adaptation for several situations.

Specializations in Early Growth

Egg-laying can be distinguished by the place and time of fertilization: *ovuliparity* is the process by which fertilization occurs in an aqueous medium after the egg(s) leaves the body (as in many fish and amphibians); *oviparity* refers to eggs laid after fertilization which is internal (as in some fish and in reptiles and birds). Sometimes both these processes are called oviparity. When the egg is retained within the uterus but receives no nutrition from the mother other than that in the egg itself, the process is *ovoviviparity*. When some sort of placenta develops to transport nutrients from the mother to the embryos, the condition is *viviparity*. In nature these distinctions are sometimes difficult to make.

In a sense, monotreme mammals are ovoviviparous, for the egg is retained in the uterus for most of the embryo's growth, and some nutrients from the uterine wall nourish the embryo. Subsequently the egg is laid and incubated by the female, and it hatches in its nest. In some oviparous lizards, eggs are retained a few days after fertilization, during which time development commences.

Marsupial development can be regarded as modified ovoviviparity: the embryo develops for more than one-half of its intrauterine life solely from the yolk. After "hatching" in utero, it receives additional nourishment from a brief placental attachment. In eutherian mammals, yolk is greatly reduced and, during virtually all of its intrauterine life, the embryo is nourished by a placenta.

The amniotic egg allows deposition of eggs on rather dry land. It contains nutrients and provides for removal of metabolic wastes from the embryo (Fig. 12.1). It is initially provided with water, and, as fat of the yolk is consumed, additional water is produced. As incubation proceeds, water passes from the egg, which gradually loses weight. Eggs of reptiles and birds have a shell that retards the passage of water; these types of eggs are cleidoic (= locked) eggs, although water *does* pass through the shell in some species. Eggs of many lizards, snakes and turtles acquire water from the surrounding soil and rapidly increase in weight after oviposition. Eggs of the desert-dwelling small-spotted lizard, *Eremias guttulata,* absorb 270–350 percent of their initial weight in water, such increase occurring shortly after they are laid.[782] In many reptiles the newly hatched young weigh more than do the freshly laid eggs, indicating that water (and perhaps oxygen) taken in during development had become part of the embryonic tissue. As the embryo develops, metabolic heat not only raises the temperature slightly above that of the surrounding soil but increases the vapor pressure above that of

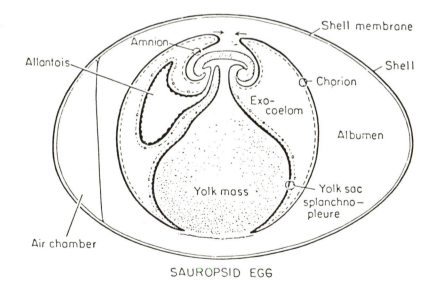

SAUROPSID EGG

Fig. 12.1. Cleidoic egg of a reptile, showing the fetal membranes in an early embryonic stage (Luckett, 1977; courtesy of Plenum Press).

the air next to the egg. Presumably this differential in vapor pressure accounts for water loss from the egg during development. The eggshells of turtles, crocodilians and birds contribute some calcium to the demand for bone-building of the embryo.

Before or during egg-laying, the bird that is to incubate the eggs develops a broodpatch. The broodpatch develops in response to an increase in circulating prolactin: feathers molt and the skin becomes highly vascularized so that the brooding bird can place the breast directly against the eggs.[516] In male phalaropes (Phalaropodidae) the broodpatch develops in response to secretion of testosterone and prolactin. Males of other species may assume care of the eggs without the development of a broodpatch. Within hours of laying her egg the female king penguin *(Apternodytes patagonicus)* returns to sea, and the male alone broods the egg about 25–30 days, more than one-half the total incubation period, or until the female returns and broods the egg until hatching. The male emperor penguin *(A. forsteri)* alone incubates the egg until hatching, about 65 days.[586]

The incubating bird not only maintains the appropriate temperature and humidity for the eggs, but it shifts their position frequently. At the periphery of the nest, the eggs are usually at a temperature lower than that in the center. Apparently rotation early in incubation reduces adhesion of the membranes; later, as the egg develops asymmetry in weight, each egg settles after parental rotation, the heavier part being near the bottom of the nest. With embryonic growth the developing chick comes to occupy that part of the egg oriented dorsad, namely, the blunt part. As hatching approaches, the tarsal joints lie ventrally on the pointed end, while the bill is directly dorsally, next to the air chamber at the top of the egg. Before hatching, the embryo begins to breathe, and at roughly this point the bill ruptures the shell membranes, one or two days before the egg is pipped. The embryo can call from its shell at this stage and gives a distress cry if the egg is moved to an improper orientation. The distress call prompts the incubating parent to restore the egg so that the heavy end is ventrad. A taped embryonic distress cry causes the adult to rearrange its egg.[254]

Hatching is a systematic process during which the embryo alternately pecks at the top of the shell and pushes with its legs against the bottom. As its tarsi push the embryo dorsad, it also rotates slightly within its shell and, as it turns, the chick pecks a circular fracture in the upper end of the shell. When the lid thus formed ruptures, the chick can thrust itself through the hole.[1052]

Although a given clutch may be laid and incubated over several days, the eggs are usually synchronous in hatching. In the rhea *(Rhea americana),* a Neotropical ratite bird, the male incubates; the female continues to add eggs for 10 or more days from the onset of incubation, but hatching is compressed within a two- to three-hour period.[289] In some species of precocial birds, pulmonary movements of the embryo produce sounds audible to eggs in contact, and such "clicking" seems to stimulate the hatching process of those eggs that are slightly slower in their hatching schedule. Eggs lack synchrony in hatching when kept out of contact in an electric incubator.[1052] As hatching approaches, the infant bird pierces the membrane at the air chamber; the shell is more pervious to water than is the membrane, giving the bird access to a greater supply of oxygen. Thus the bird uses its lungs and air sacs before hatching and while oxygen is still entering through the chorioallantoic membranes.

Temperature of the egg during incubation varies with species; in some there is a thermal circadian rhythm with a lower temperature occurring at night. Generally as the day of hatching approaches, the mean egg temperature increases. The incubating parent can protect the eggs from excessive heat by standing over them and shading them from solar heat or, or by panting and gular flutter, lower the temperature of the blood passing through the broodpatch.[842]

Young birds have been described as *altricial* or *precocial*. Altricial young are born naked and blind and unable to walk or fly; precocial young are covered with down, with eyes open, and able to walk and feed themselves within minutes or hours of hatching. There are several intermediate conditions involving various degrees of natal feathering and helplessness. Birds with precocial young lay eggs with a relatively large amount of fat, which permits a longer prehatching period than for altricial young. Those species in which hatchlings are altricial produce eggs that are 20 percent yolk, whereas eggs of precocial species have from two to three times this much yolk. The incubation time of an egg is a function of its weight, regardless of whether the egg produces an altricial or precocial young.[842]

Early growth of birds is extremely rapid, especially in altricial species. Among passerine birds, for example, virtually full size is attained before they are capable of flight, and they must fly before they leave their nest, sometimes at less than two weeks of age.

Gestation

The duration of gestation is one of the least variable segments of the reproductive cycle; for most mammals it is genetically fixed (constant for a given population) and determined by the rate of embryonic development. In species with delayed fertilization, gestation can be variable. In at least some heterothermic bats, low ambient temperatures may arrest or retard embryonic development.[837]

Variations in embryonic development of mammals are greatest in the very early stages, but there is a steady continuous growth rate after the embryo has attained about one-third its natal size. Different groups of mammals each have their own rates of fetal development, which presumably are adaptive. The growth rate, or the *specific fetal growth velocity,* can be calculated for the latter two-thirds of pregnancy by $W^{1/3} = a (t - t_0)$, in which velocity is indicated by a, the fetal weight by W, the time from conception by t, and the intersection of the linear part of the slope (representing the latter two-thirds of pregnancy) with the time axis by t_0. The specific fetal growth velocity is determined by plotting the weight to the one-third power of embryos of known conception date, adjusting for species with delayed implantation.[462] Although the growth velocity is unrelated to adult size, it is slowest among some primates (Fig. 12.2).

There are endless variations in length of gestation among various mammalian groups. Very frequently variations are adaptive in nature and do not necessarily

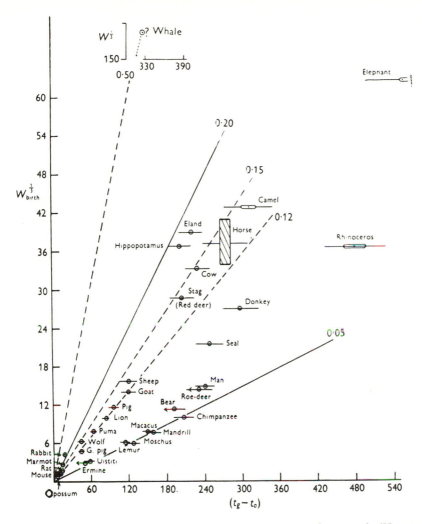

Fig. 12.2. The specific fetal growth velocity of some groups of mammals (Huggett and Widdas, 1951; courtesy of The Journal of Physiology).

follow phylogenetic lines. Rabbits (e.g., *Sylvilagus* or *Oryctolagus*) have relatively short gestation and the young are born nude and blind, whereas hares *(Lepus)* have a longer gestation, and their young are born fully furred with the eyes open. Hystricomorph rodents are for the most part born in a precocious condition, following a long gestation, more than 200 days for several species.[1082]

Many bats are characterized by relatively slow fetal development and long gestations. In the pipistrelle *(Pipistrellus pipistrellus)*, pregnancy is followed by a relatively constant core temperature, but at ambient temperatures of from 5° to

14°C and without food, pregnant females become torpid, and then gestation is extended by the duration of torpor. This unusual response to low ambient temperatures is known for other heterothermic bats and must have great survival value in nature.

Tropical bats of the families Pteropidae and Phyllostomatidae are advanced at birth—their eyes are open and the pelage is well developed. Post-natal growth is slower, however, than in the relatively altricial young of most Microchiroptera, which have a shorter gestation. Young big brown bats *(Eptesicus fuscus),* a temperate species, double their natal weight in three weeks, and growth ceases at seven weeks.[220] Insectivorous bats reared in cool caves develop more slowly than do those in warmer caves. This possibly reflects the poor thermoregulatory ability of these altricial bats.[837]

Gestation in whalebone whales (Mysticeti) is uniform and uncorrelated with size. Most species migrate to polar regions in summer and feed on the abundant krill; at summer's end, when they are fattened, they move to warmer latitudes. Both mating and parturition occur in winter in the warm seas, and gestation (10–11 months) is tied to the annual cycle of feeding and migration. In the spring they once again move to regions of cool waters and long days, and the young grow rapidly in the presence of an abundant food supply. Lactation is also adjusted to these seasonal cycles and lasts less than one year, ending before the whales return to lower latitudes and winter fasting. The sperm whale *(Physeter catodon),* a toothed whale (Odontoceti), has a gestation of 14–15 months. Pregnancy may occur once every four years, but some females ovulate and mate midway in lactations and thereby shorten the four-year cycle. It is not a plankton feeder and does not make a regular annual migration characteristic of the whalebone whales.

Early growth of young mammals is related to the amount and quality of milk. Growth is very rapid in marine mammals. The blue whale produces an estimated 130 gallons of milk daily. Milk of cetaceans contains 40–50 percent fat in contrast to about 2 percent in human milk. Lactation is prolonged in many larger mammals, lasting from as long as two years in many species of cattle to three years in proboscideans. In the huge whalebone whales (Mysticeti) the young nurse for only 10 or 11 months, but the sperm whale calf nurses for two or more years.[937]

Blue whales double their birth weight in 7 days. A colt from birth requires 60 days to double its weight, and a pig 14 days. A human baby will not double its birth weight much before six months, smaller infants having more rapid postnatal growth than larger ones.

Many small mice have life spans of about one year, during which the major events of maturation follow one another rapidly, with most reproduction occurring before maximal size is reached (Fig. 12.3). The prairie vole *(Microtus ochrogaster),* a typical annual rodent, attains 90 percent of its growth within the first six weeks in the laboratory. In the wild, early growth rates vary with the season of birth.[445] In many annual species of small mammals (both insectivores and rodents), growth is very slow in winter when most individuals are nonparous subadults, but rapid the following spring. Typically the Nearctic red-backed vole *(Clethrionomys gapperi)* in the northern part of its range shows almost no growth

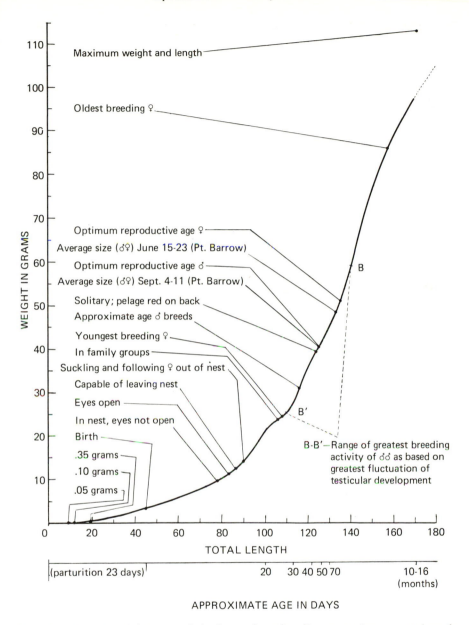

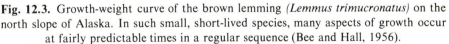

Fig. 12.3. Growth-weight curve of the brown lemming *(Lemmus trimucronatus)* on the north slope of Alaska. In such small, short-lived species, many aspects of growth occur at fairly predictable times in a regular sequence (Bee and Hall, 1956).

in the winter but rapidly increases in size and weight in the spring (Fig. 12.4), when snow melts and herbaceous growth resumes.[965]

Some rodents reflect marked local variations in growth. Pocket gophers *(Thomomys talpoides)* in the alpine life zone in Wyoming are significantly larger than individuals at lower elevations, in Canadian and transition life zones. This greater alpine growth is correlated with relatively greater protein content of plant foods in that area, and size differences under such circumstances may be a phenotypic reflection of nutrition.[1033] This may also be an illustration of Bergmann's Rule.

Growth rates vary in the ground squirrel genus *Citellus:* development is rapid in three hibernating species, in contrast to the relatively slow growth rate and late weaning date for *Citellus leucurus,* a nonhibernator that does not have a seasonal fat cycle (Fig. 12.5). *C. lateralis* becomes independent of the mother at an early date and accumulates fat before an extended hibernation. *C. mohavensis* and *C. tereticaudus* are hibernators that occupy a less severe climate than does *C. lateralis* and have growth rates and weaning dates intermediate to *C. lateralis* and *C. leucurus.*[802]

Efficiency of use of food is maximal during youth and generally decreases with age. A population of half-grown individuals increases more rapidly in biomass than a population of full-grown adults. The phenomenon is considered by fisher-

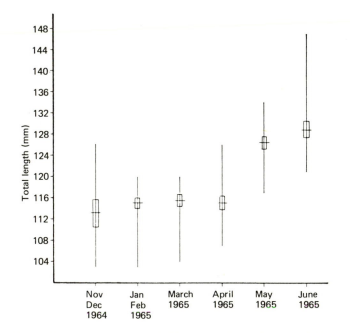

Fig. 12.4. Changes in growth (total length) in the red-backed vole *(Clethrionomys gapperi)* in Alberta, Canada. Vernal acceleration in growth is typical of many small mammals in temperate climates (Stebbins, 1976).

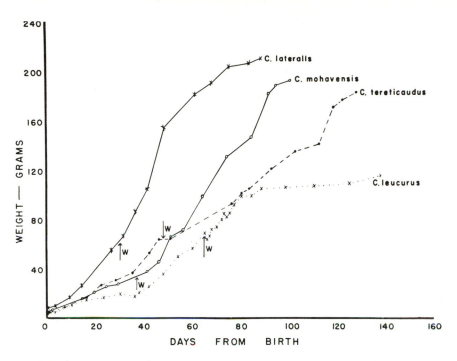

Fig. 12.5. Growth rates and weaning times of four species of ground squirrels *(Citellus);* *C. lateralis* is a high montane hibernator, *C. mohavensis* and *C. tereticaudus* are desert-dwelling hibernators and *C. leucurus* is a nonhibernator of the deserts (Pengelley, 1966).

ies' biologists in deciding on size limits of fishes that can be legally taken: by capturing fish at a sexually mature but rapidly growing size, most efficient use is made of the conversion of food materials to protein (i.e., annual increase in weight).

With increased size (or age), animals become less efficient in converting protein to growth and more efficient in accumulating fat. Therefore, from birth, there is a gradual decline in growth rate. In ectotherms and mammals there may be substantial growth after sexual maturity. This relationship, [(size at sexual maturity)/(maximal size)], approaches unity in such species as the Pacific salmon (*Oncorhynchus* spp.) and long-tailed shrews (*Sorex* spp.) that die shortly after reproduction, but it may approximate one-third in especially precocious species, such as voles (*Microtus* spp.).

Growth rates in fishes are variable and depend on environmental factors, especially food and temperature. Usually fishes grow more rapidly in warmer waters, but longevity is greater in colder waters. An abundance of food generally results in more rapid growth of young fishes but not in greater maximal size. When food is a limiting factor, there is an inverse relationship between growth rate and biomass. A large biomass makes a greater demand on the food supply, assuming a constant age structure of the population; on the other hand, a reduction in biomass should leave a greater food supply for the remaining individuals.[91,335] In an unfished population of perch *(Perca fluviatilis)*, the adults increased a mean of about 5 g annually, but their growth accelerated to double that rate, or about 10 g annually, after fishing began. As the perch were reduced, however, trout *(Salmo trutta)* increased about five times over their prior abundance and, because they ate much of the same foods taken by the perch, they might have prevented the perch from realizing maximal growth rates. In both these examples the observed increase in growth rates may have resulted from: (1) greater food availability with the decrease in biomass or (2) the increase of younger individuals, or both factors acting in concert.[583]

The relationship between age and growth rate is illustrated by the pike *(Esox lucius)* in a lake (Windermere) in the British Isles. Between 1930 and 1938 in a mature (unfished) population, the young grew more slowly than they did from 1951 to 1955, when most of the fish were younger (Fig. 12.6). In the perch *(Perca fluviatilis)* growth rates vary with several factors, and higher temperatures are correlated with faster growth. In a field experiment a markedly reduced density of perch was followed by a noticeable increase in their growth rate, presumably because of the resulting increase in food per individual fish.

In ectotherms there is a widespread tendency for species that live in colder habitats to develop more rapidly than those living in warmer regions, when kept at the same temperatures. In the anuran genus *Rana,* for example, when reared under controlled laboratory conditions, boreal species have more rapid rates of development than have taxa from lower latitudes. This physiologic feature tends to compensate for a reduction in metabolism with a decline in temperature.

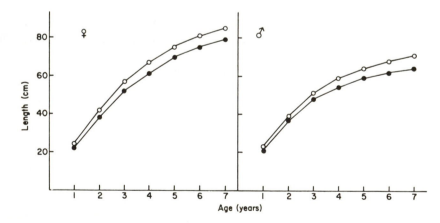

Fig. 12.6. Growth of pike *(Esox lucius)* in Great Britain by age classes before (closed circles) and after (open circles) removal of old, mature individuals (Frost and Kipling, 1967; courtesy of The Journal of Animal Ecology).

Ambystoma tigrinum breeds in ponds at a broad range of elevations and matures more slowly in montane regions. In western Colorado at 2880 m larvae transform in their first summer when about 50 mm long, but at 3097 m they spend the first winter in their natal ponds and transform at 70–80 mm in their second summer. At higher elevations some individuals are neotenic. *Pseudotriton montanus,* a permanently aquatic plethodontid salamander, metamorphoses at one and a half years of age at 200 m elevation but not until two and a half years at 600 m. The size at metamorphosis is the same in both populations.[129]

Larval salamanders may remain in their natal ponds and become sexually mature while retaining their gills. In some species, metamorphosis (loss of gills) may follow reproduction, as in some kinds of Ambystomidae, and larvae reproduction is then called *neoteny.* In certain salamanders (Proteidae) the gills are never lost, either in nature or experimentally, and the phenomenon is then called *paedogenesis.* Some workers make no distinction between facultative neoteny and permanent paedogenesis. Anurans, which undergo major morphologic changes in metamorphosis, never breed in the tadpole state. The ambystomid salamander *Dicamptodon copei* is not known to metamorphose, whereas the sympatric *D. ensatus* has normal growth and metamorphosis. *D. copei* is relatively insensitive to thyroxin and matures in the gilled form. With these two species, failure to transform must be determined by factors other than or in addition to age and environment.[771]

Generally in fishes, growth is more rapid in short-lived species: some species of Pacific salmon (*Oncorhynchus* spp.) grow rapidly and die in five or six years, whereas the long-lived sturgeon (*Ascipenser* spp.) grows at slow but relatively constant rate.

In reptiles that hibernate, growth is punctuated by interruptions during winter rest. In the western fence lizard *(Sceloporus occidentalis)* growth begins after hatching in mid- and late summer and stops with winter dormancy; the greatest rate of growth is between the first and second periods of hibernation, after which it is negligible.[296] In Georgia the eastern fence lizard *(Sceloporus undulatus)* may grow during its first winter; smaller individuals enter hibernation later, emerge more frequently in the winter and are active earlier in the spring than are older animals.[194] Rapid growth in the Nearctic tortoise, *Gopherus agassizi,* may occur in an abundance of food.[487] This tortoise grows rapidly between mid-April and early July with virtually no growth at other periods, and the amount of growth is positively correlated with winter rainfall and production of annuals on which it feeds. (Fig. 12.7).[685]

Generally ectothermous vertebrates grow throughout their lives, when food and living conditions are adequate, and very old individuals sometimes reach great size. As with birds and mammals, however, growth of ectotherms is most rapid in early life. Growth (i.e., annual increase in weight) in the colubrid snake *Spalerosophis cliffordi,* for example, decreases with age from 63 percent in the first year to 35 percent in the fourth and 16 percent in the eighth for males; these data are slightly greater for females until the ninth or tenth year.[244]

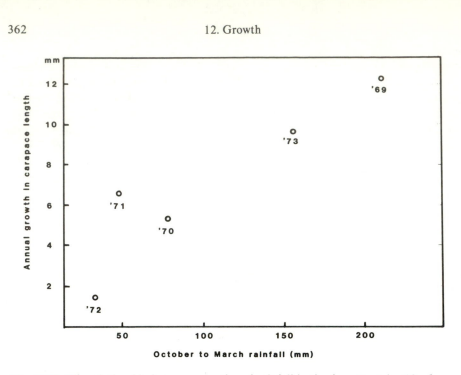

Fig. 12.7. The relationship between growth and rainfall in the desert tortoise *(Gopherus agassizi)* (Medica et al., 1975).

Age of Sexual Maturity

Among the several classes of vertebrates there is great variation in the age of first reproduction. Some workers have suggested that sexual maturity is attained at about the first 10 percent of maximal longevity, or that maximal longevity is about 10 times the age at sexual maturity. Perhaps this is a useful rule of thumb, but there are not reliable data for statistics demonstrating both these events for very many species. In many species females first breed at an earlier age than do males, and in some species females have greater longevity.

The data for fishes are more detailed than for most vertebrates, for many species have long been commercially exploited, and successful management depends on sound biological knowledge. Although fishes seem not to follow any general pattern, small species, which mature at a small size, tend to have a short reproductive life and brief life span. Among short-lived species, those living four to six years at most, sexual maturity is frequently reached in the second year. Anchovies (*Engraulis* spp.), horsemackerel (*Trachurus* spp.) and cod and whiting (*Gadus* spp.) all fit this common pattern. The South American cyprinodont *Nothobranchius guentheri,* however, breeds when one month old; this is an annual species, with a complete turnover in population every year.[71]

Among three common viviparous perch (Embiotocidae), coastal fishes of the North Pacific, sexual development is highly variable. *Cymatogaster aggregata,* a

California species, is sexually mature at birth. In contrast the tule perch *(Hysterocarpus traski)* born in May, mates the following August; sperm are stored in crevices in the ovary until fertilization the following January.[133] The barred surfperch *(Amphistichus argenteus)*, a larger species, attains sexual maturity at age two years. In that species, males live a maximum of six years, but females may survive until nine years of age.[152]

Species of rockfish (*Sebastes* spp.; Scorpaenidae) are also live-bearing fishes of coastal waters of the North Pacific. In these fishes sexual maturity in more northern waters is attained at smaller mean sizes: in the western Gulf of Alaska they mature at about age 10 years but farther south sexual maturity is reached at 11–15 years.[1087] Some live-bearing fishes seem to live past their reproductive age, at least in captivity. In species of mollies, guppies, swordtails and their relatives (Poeciliidae), females produce fewer young as they approach the end of their life span.[955,1019]

The pike *(Esox lucius)*, a voracious predator of the Northern Hemisphere, becomes sexually mature at a certain length rather than at a given age. Samples from several localities in Scotland, Ireland, Russia, Canada and the United States began to breed between 38 and 46 cm in length but attained the size at one to five years of age.[319]

Among amphibians many small salamanders mature at 2 or 3 years, but their longevity is seldom known. The red-backed salamander *(Plethodon cinereus)*, a lungless amphibian of eastern North America, is sexually mature in its third year of life as is the red salamander *(Pseudotriton ruber)*, another lungless species.[129,743] On the other hand, some tropical plethodontid salamanders mature more slowly. *Bolitoglossa rostrata* matures at about 2.5–3 years (♂ ♂) and 3.5–4 years (♀ ♀), and *B. subpalmata* matures at from 4 to 9 years (♂ ♂) and from 9 to 14 years (♀ ♀).[1051] *Triturus helveticus,* a European newt, may metamorphose in its first or second year or remain in its natal ponds and become neotenic (Fig. 12.8);

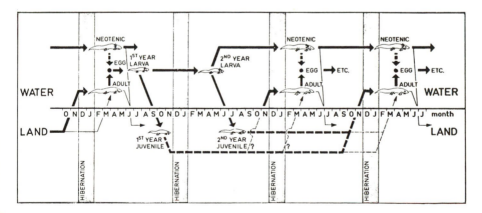

Fig. 12.8. Generalized life cycle of the newt, *Triturus helveticus.* Solid lines represent observed patterns, and the broken line, not observed (van Gelder, 1973).

in either case it first breeds in its third year.[1045] In the smooth newt *(Triturus vulgaris)* the majority of several populations in Britain reproduced for the first time between 6 and 7 years. In the bullfrog *(Rana catesbeiana),* a large species of North America, sexual maturity is achieved in the third or fourth year in the northeast (where winters are cold) but in the mild climate of California, where it has been introduced, it may breed at age 1 year. The tropical *Rana erythraea* of the Philippines in contrast breeds at age six months (males) or nine months (females).

The diversity of reptiles is reflected in their different patterns of reproductive cycles. Although there is a general positive correlation between sexual maturity, size and apparent longevity, there is a remarkable divergence among the major groups.

Lizards are the most thoroughly studied reptiles as they are frequently abundant and easily captured. Some small species are precocious. The tiny *Takydromus tachydromoides,* a lacertid lizard of eastern Asia, breeds soon after emergence from its first hibernation. Another small lacertid, *Eremias guttulata,* of arid regions of Israel, becomes sexually mature at eight months and experiences a complete population turnover annually.[782] The Nearctic iguanid lizard *Urosaurus ornatus* becomes sexually mature in its first year, and the chameleon, *Chamaeleo lateralis,* of Madagascar lays eggs when three months old.[95] Many species in the temperate regions breed first near the end of their second year, at which time they are about two-thirds grown. This is true for most of the investigated species of *Sceloporus,* small Nearctic iguanids, but some larger iguanids mature at 4 or 5 years. The eyed gecko *(Oedura ocellata)* of Australia reaches sexual maturity in its third year.[143] Similarly the black shore skink *(Leiolopisma suteri),* a nocturnal lizard of New Zealand, matures at 33 months, or near the end of its third year.[1017]

Developmental rates of snakes seem to resemble those of lizards. Many colubrid snakes mature in their second or third years. Maturation of ova in the Nearctic diamond-backed water snake *(Nerodia rhombifera)* requires about 2½ years; males are sexually mature in their third year.

At the other end of the developmental spectrum, large crocodilians grow and mature slowly. The American alligator *(Alligator mississippiensis)* matures at a length of about 4–5 feet, at which time it is entering its fifth year. Maximal longevity seems to lie between 30 and 40 years, at which age they rarely, if ever, exceed 18 feet in length. The Nile crocodile becomes sexually mature at age 12–15 years.[756]

Fresh-water turtles attain sexual maturity as the growth rate declines; some workers believe that reproduction begins at a fixed age, whereas others believe it to be predicated on a certain size. Blanding's turtle *(Emydoidea blandingii)* matures at about 12 years and has an estimated maximal life span of 40 years.[361] The alligator snapping turtle, *Macrochelys temmincki,* reaches sexual maturity between its 11th and 13th year and has a known maximal life span of 36 years. Although most turtles are rather slow to develop, some species breed as early as their second or third year. The chicken turtle *(Deirochelys reticularia)* of North America becomes fecund by the third year.[338] The cooter *(Chrysemys florida),* another Nearctic fresh-water turtle, becomes sexually active in its third year.[337]

Exogenous factors may affect rates of sexual development. In the wild the diamond-backed terrapin *(Malaclemys terrapin)* begins to lay eggs in its seventh year. But when kept in captivity at 80° C and fed throughout the year, egg-laying commences at age 4 years.

Unquestionably the most unhurried species is the tuatara, the sole rhynchocephalian, which persists on several small islands off the coast of New Zealand. Growth of the tuatara *(Sphenodon punctatus)* is very slow: sexual maturity is reached at about 20 years of age, and growth seems to continue until about 50 years of age. Maximal life span is about 100 years.

Many species of large birds are sexually mature at age 2–10 years but do not breed for the first time until several years later. Some raptors breed first in their third year, and many gulls, shearwaters, gannets and other seabirds delay initial reproduction until 4–10 years of age. The emperor penguin *(Aptenodytes forsteri)* begins to breed when 4–6 years of age.[586] The royal albatross *(Diomedia epomophora)* first breeds at age 9–11.[1086] The mechanism controlling this delay is not known. It is assumed that limited food for both adults and young creates nutritional problems for the entire population when large numbers of young need to be fed. Since one or both parents must devote much time and energy to the care and nutrition of the young, perhaps only older adults, being generally more efficient hunters and fatter, are fully capable of foraging for both themselves and their offspring. This pattern results in an extremely low reproductive rate, and often occurs when adults are relatively long-lived. When the annual mortality of adults is less than 5 percent, a substantial percentage will survive 50 or more years, and a low reproductive rate is adequate to sustain a population.[1086] In birds in which sexual maturity is delayed beyond the first year, females breed at an earlier age than do males, following the pattern found in most vertebrates. Parrots, cranes and penguins are among those groups that mature slowly, but definite data for wild individuals are meager.

Among migratory passerine birds, at the time an individual attains *potential* sexual maturity, it is probably experiencing a period of decreasing daylength (late summer) and may not be exposed to a stimulating photoperiod for another 6 or 7 months. Thus many small passerine species first breed when nearly one year of age, and many larger passerines—although possibly sexually mature at 10 or 11 months of age—may not breed until the end of their second year, nesting being based on scarcity of nesting sites, preponderance of older and more vigorous males and possibly other such density-dependent factors.

Sexual maturity in most mammals occurs near the age at which growth slows, so that the two events, growth and sexual activity, *tend* to be mutually exclusive. Elephants begin reproduction at about age 10 years, near when growth is completed, and the tiny soricid shrews breed at nearly their maximal size, when about 11 months old, only to die a few weeks later. Some small mice are exceptions, for they become fertile and mate before they are 2 months old. The initial onset of reproduction in the voles (*Microtus* spp.) is partly a function of the season of birth; individuals born in the spring breed at about 1 month of age; those born in late summer grow more slowly and reproduce first in the following spring, at 6 months of age.

In the laboratory mouse sexual maturity depends partly on social factors. Female mice maintained in a bisexual atmosphere and exposed to odors from males mature more rapidly than when kept in isolation or in a unisexual environment.[1048] In the cuis *(Galea musteloides)*, a Neotropical rodent, the presence of a male induces estrus in the female. It seems likely that, in a population of wild rodents, a paucity of males during periods of low populations delays sexual maturity in females.[1082]

Among the pinnipeds there is considerable variation in age of sexual maturity. The walrus *(Odobenus rosmarus)* breeds first at 5 or 6 years, or perhaps longer for bulls. In seals (Phocidae) and sea lions (Otariidae) sexual maturity in bulls may occur before they participate extensively in breeding, for in polygynous species older bulls dominate the harems. Bachelor bulls seldom mate, but harem bulls may mate 6–10 times daily for several weeks. In the crabeater seal *(Lobodon carcinophagus)* of the Antarctic, ovulation and mating may occur before 1 year of age, but such precocity is not found in other phocids. The bearded seal *(Erignathus barbatus)* becomes sexually mature at about 7 years. Some species in the family breed at 3–5 years of age.

Females of the harp seal *(Pagophilus groenlandicus)* first breed when 6 years old, and sexual maturity is reached at 7 or 8 years in the ringed seal *(Pusa hispida)*. The harp seal and the ringed seal sexually mature at a very slow rate and have an estimated life span of more than 40 years, and these two statistics are probably correlated among the Pinnipedia.[902] It is interesting to note that the harp seal, which has been killed for its fur for many years, today experiences a faster growth rate and earlier sexual maturity than in prior decades (Fig. 12.9).[903]

Similarly recent years have seen faster growth and earlier sexual maturity in the southern elephant seal *(Mirounga leonina)*. In an undisturbed population of elephant seals females became gravid for the first time when from 4 to 7 years of age, but in a heavily hunted population all females conceived first at age 3 years. In the northern fur seal *(Callorhinus ursinus)*, one of the species of sea lions (Otariidae), there has been an acceleration of sexual maturity. In both the northern fur seal and the southern elephant seal, hunting pressure is confined to males, suggesting that more rapid growth and earlier sexual maturity may have resulted from lowered pressure on available food and are thus density-dependent.[903]

The same phenomenon is seen in some cetaceans, in which both sexes are hunted. Sexual maturity in finback whales (*Balaenoptera* spp.) had been attained at about 10 years of age until 1930; between 1930 and 1960, however, the age of first breeding for females declined to 5 or 6 years. Inasmuch as the *size* at sexual maturity (about 20–22 m) has not changed, the difference results from faster growth in recent years. Presumably this change reflects greater supplies of food per whale with the decline in total population size.[324]

The approach of sexual maturity in human beings is measured in terms of increasing urinary levels of FSH and LH. These hormones increase gradually from 2 to 8 years: FSH more than doubles and LH increases more than 10-fold between the ages of 2 and 10.[853] The onset of the menstrual cycle does not necessarily indicate fertility. Among some aboriginal groups in Samoa and elsewhere in the South Pacific, frequent intercouse in the early adolescent years rarely

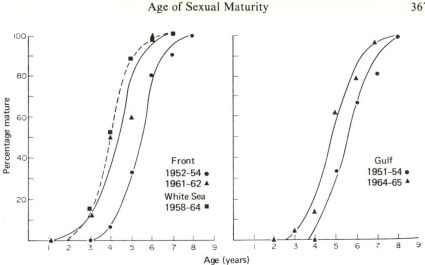

Fig. 12.9. Reduction in the age of sexual maturity in female harp seal *(Pagophilus groen-landicus)* in the northwest Atlantic (Gulf, Front) and in the White Sea (Sergeant, 1973; courtesy of The Journal of Reproduction and Fertility).

results in pregnancy; several workers have suggested that ovulations may not always occur in the initial menstrual cycles.[797] In the human female, sexual maturity (as measured by the time at which menarche occurs) depends rather on body mass (about 48 kg), not height or age; ovulation usually commences several years later.[317] As in other mammals, the age at which the critical body mass is attained seems to depend on nutrition. In developed countries the age of sexual maturity (as suggested by menarche) has declined steadily over the past century (Fig. 12.10), presumably as a result of increased protein intake.[922]

Both sexes of *Myotis nigricans,* a Neotropical vespertilionid bat, become sexually mature as early as 4 months of age in Paraguay.[742] In temperate regions many vespertilionid bats may mate in the autumn of the year of their birth, at about age 4–6 months. Female pallid bats *(Antrozous pallidus)* breed in their first year, one year earlier than initial breeding of males. *Hipposideros caffer,* a tropical insectivorous bat with a discrete breeding season, attains adult size at 3 months but does not breed until the next reproductive season, at 10 months.[737] A tropical fruit bat, *Eonycteris spelaea,* of West Malaysia, a nonseasonal breeder, matures at between one and two years of age.[73] The Japanese pipistrelle *(Pipistrellus abramus)* in Kyushu (southern Japan) is sexually mature and mates in October of the year of its birth, when about 4 months old.

Sexual maturity in the larger bears seems to occur between age three and four years. In contrast, the female stoat or ermine *(Mustela erminea),* one of the smallest carnivores, may experience estrus and mate at less than two months of age. The Nearctic pronghorn *(Antilocapra americana)* usually breeds in its second year, but some females may mate at five or six months of age and produce young in their first year.

Among vertebrates in general there seems to be no universally applicable relationship between sexual maturity and longevity. The age of first reproduction is

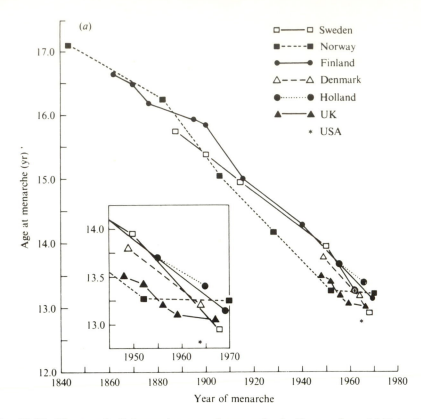

Fig. 12.10. The trend of decreasing age of menarche in Europe from 1840 to 1970 (Eveleth and Tanner, 1976; courtesy of Cambridge University Press).

affected by growth rates, hibernation, migration and density-dependent social factors. Although it is difficult to demonstrate, initial breeding in some birds is apparently determined by food available to the entire population. Among many vertebrates there tends to be a correlation between sexual maturity and size. Size is in turn a function of growth rate. This is highly variable and both density-dependent (reflecting food supplies) and sometimes temperature-dependent.

Viviparity in Ectotherms

Among ectotherms variations in live-bearing range from clear-cut cases of ovoviviparity to genuine viviparity (or euviviparity). To separate these phenomena from several examples of extrauterine parental protection of ova, live-bearing is considered to include those cases in which ova are fertilized within the reproductive tract (either within the ovary itself or in the oviduct or a modified part of it) and retained there until released as either larva or metamorphosed young. Retention of a fertilized ovum within the female's reproductive tract is a step toward viviparity; it involves physical care of the egg and embryo at the very least and, in some cases, elaborate specializations of both mother and fetus for the transfer of nutrients and metabolic wastes.

Internal fertilization is necessary for internal development of a fetus and has arisen independently many times among unrelated familes of fishes, some amphibians and all reptiles. Usually there is an intromittent organ for the transfer of sperm or spermatophore. Live-bearing vertebrates may date from the Devonian; the placoderm *Rhamphodopsis* possessed claspers and may have brought forth well-formed young. In some fishes internal fertilization occurs with oviparity. Members of the blenny genus *Starksia* (family Clinidae) transmit sperm by an intromittent organ formed from an anal papilla and the first anal spine but are egg-layers. Similarly the Pantodontidae of African lakes has an intromittent organ, and the female may store sperm, but it is oviparous. Some species of both fresh-water and marine sculpins (family Cottidae) have a well-developed anal papilla, and in such species fertilization is internal. In *Clinocottus analis*, a tide-pool sculpin of the eastern Pacific, fertilization is internal, but cleavage begins after the eggs are laid. In the cottid genus *Myoxocephalus* fertilization is internal, but some development occurs before oviposition, representing a stage toward ovoviviparity.

Viviparity in Fishes

In elasmobranchs the pelvic fins are modified to form claspers and function as intromittent organs. They may contain cartilaginous rods and highly vascularized erectile tissue. During mating, the claspers become turgid and enlarged, the sides overlapping or rolled in a scroll-like fashion, forming a groove, and the tips may have spines that effect a long intromission. During copulation of *Squalus canicula*, a dogfish, the male coils about the cloacal region of the female, the claspers being fixed so as to form a 90° angle with the long axis of the body and with one clasper inserted into the female's cloaca (Fig. 12.11). Sperm pass from the urogenital

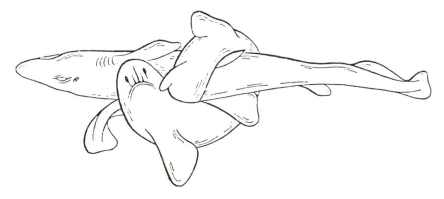

Fig. 12.11. Coilus in the Shark, *Squalus canicula* (Gilbert and Heath, 1972, after Bolau, 1881; courtesy of Pergamon Press).

papillae into the groove of the clasper and are washed down by seawater, and a sticky fluid is secreted by goblet cells in the epithelial lining of the siphon sac. This secretion probably lubricates the clasper prior to intromission and may also stimulate uterine contractions.[342]

Although corpora lutea occur in oviparous elasmobranchs, they have special functions in maintaining embryos in live-bearing fishes and attain their ultimate development in placental mammals. It has been suggested that luteal hormones stimulate both the formation of the egg-case (in oviparous forms) and the thickening of the uterine lining and the uterine element of the yolksac placenta in viviparous elasmobranchs. In live-bearing species the fertilized ovum is contained in a temporary shell, from which the embryo emerges while still in the uterus. The uterus is highly vascularized and provided with villi or furrows during pregnancy. In live-bearing species of *Mustelus* (a shark) the eggshell is membranous, and the yolksac becomes interdigitated with the uterine lining and shrinks throughout pregnancy. In some apparently ovoviviparous species, the eggs may be surrounded by fluids with some nutrient content. In *Trygon* (a sting ray), for example, fats occur in the uterine fluid. In some viviparous elasmobranchs the liver loses weight during gestation, such loss being a decrease in liver fats, which are presumably used for embryonic nutrition.[12]

In viviparous rays the lower part of the oviduct does not form any placenta-like parts but is specialized to secrete nutrients, which are taken in through the mouth or gills. In both the ray *Pteroplatea* and the whale shark *Cetorhinus,* extremely long villi or trophonemata extend from the uterus and appear to provide secretions to nourish the young. In *Pteroplatea* trophonemata actually enter the spiracle and reach the embryonic gut. Viviparous elasmobranchs include the hammerhead sharks (*Sphyrna* spp.), the smooth dogfish (*Mustelus* spp.), the gray sharks (*Carcharinus* spp.) and the blue shark *(Prionace glauca).*

In teleosts the ovaries may be hollow or solid, and in viviparous groups the ovaries join so that there is a single median hollow ovary. In viviparous teleosts the oviduct is frequently replaced by a gonoduct. The gonoduct receives the intromittent organ, but it does not serve to house the developing embryos as it does in the elasmobranchs. In most live-bearing teleosts fertilization occurs in the ovary, within which the development of the embryos may proceed. In such cases ovulation does not occur, suggesting either a change in production of trophic hormones from the pituitary or an alteration in follicular sensitivity to trophic hormones.[12,1022]

Among the Cyprinodontiformes (Cyprinodontes) there is an extensive variety of reproductive patterns, and within several families there is a transition from oviparity with internal fertilization through apparent ovoviviparity to viviparity. In the Neotropical Tomeuridae and the oriental Horaichthyidae, males have well-developed gonopodia formed from the anal fin and fertilization is internal, but both families are oviparous. In *Horaichthys setnai* the males use a long gonopodium to implant spermatophores near the female's genital orifice. Each spermatophore bears a circlet of tiny hooks at one end, and at the hooked end the spermatophore ruptures; sperm escapes, and enters the oviduct, where it may either be stored or proceed to the ovary. A series of ovipositions may follow a single mating.

There are four families of Neotropical cyprinodontiform fishes (Goodeidae, Jenynsiidae, Poeciliidae and Anablepidae) that exhibit some kind of live-bearing. In these four families there is a single hollow ovary, the oviduct is replaced by a gonoduct, and a gonopodium is formed from the anal fin to serve as an intromittent organ. Sperm is stored in the epithelial covering of the ovary, and fertilization occurs within the follicle. In the Jenynsiidae and Goodeidae, development takes place within the lumen of the hollow ovary following a brief development in the follicle. In the Goodeidae, nutrients enter the embryo through finger-like extensions (trophotaeniae) from the hind gut (Fig. 12.12). In the Goodeidae and the Jenynsiidae the ovarial epithelium becomes both glandular and highly vascularized. The yolksac is small, but in the young embryo the pericardial cavity is greatly enlarged, and is presumably a transitory respiratory organ. As the embryo develops, both yolksac and pericardial sac shrink, and the trophotaeniae enlarge and assume a nutritive function. In the jenynsiid embryos early development parallels that in the Goodeidae, but the former develop rather spacious opercular openings, and the spongy lining of the ovary closely approaches the gills of the embryo; the ovarian tissues invade not only the adjacent opercular opening but also the mouth, but the opercle on the side away from the ovarian lining develops normally. In these two families the embryos float in a single ovarian cavity, and dead embryos serve as nutrients for their surviving siblings.[1022]

In the Poeciliidae and Anablepidae the embryos develop within the follicle, so that each individual is separated from the others of the brood. In small embryos the pericardial sac is large and vascularized and lies against the vascularized lining of the follicle. In anablepid embryos the villi of the gut are greatly enlarged so as to increase their absorptive surface. In the molly (a poeciliid), sperm are stored on the surface of the ovary, which is pitted with funnel-like depressions, each leading to a ripening follicle. When ova are mature, sperm are released and

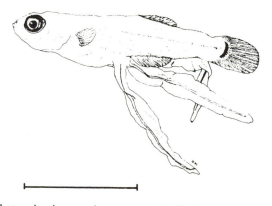

Fig. 12.12. Trophotaeniae in an unborn young *Skiffia francesae* (bar = 5 mm) (Kingston, 1978).

fertilization occurs within the follicle. Development within the follicle requires about one month.[1019]

Various species differ in the amount of nutrient received from maternal fluids. Both the follicular wall and the yolksac are highly vascularized and facilitate an exchange of oxygen, nutrients and metabolic wastes. In oviparous fishes that develop entirely from nutrients in the yolk, the young fish weighs roughly two-thirds the weight of the fertilized ovum, indicating that about one-third of the material in the yolk is utilized for metabolic energy during development. Many species of poeciliid fishes have a remarkably constant weight during their embryonic life, showing that the energy needed for metabolism is provided by follicular fluids. In two species *(Heterandria formosa* and *Aulophallus elongatus)* the full-term embryo is much heavier than the freshly fertilized ovum.[901]

The blennioid eel-pout (*Zoarces viviparus;* Zoarcidae) of the eastern North Atlantic releases eggs, which are fertilized in the uterus, where the embryos develop. Following ovulation, corpora lutea develop and remain until after the birth of the young four months later. In this species embryos take some nourishment orally, in the form of uterine secretions and the remains of dead embryos. Surfperches (Embiotocidae) are mostly marine teleosts that range from southern California around the North Pacific to Japan. All are viviparous and remarkable in the very advanced state of development of the young at birth. Embryonic viviparous perch (Embiotocidae) have highly vascularized fins, which are expanded between the rays or spines (Fig. 12.13); these serve for prenatal gas exchange and regress at birth.[77] Gestation in the barred surfperch *(Amphistichus argenteus)* is about five months. Larger females give birth earlier than younger ones, suggesting that smaller (and younger) females mate later than older ones.[152] At the other end

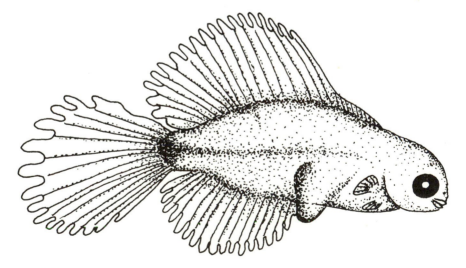

Fig. 12.13. Prenatal young of *Hypsurus caryi,* a viviparous perch (Embiotocidae), showing the vascularized extensions of the membranes of the median fins (Behrens, 1977).

of the spectrum, biologically, lie species of Scorpaenidae (scorpean fish and rock-fish), in which there is fertilization within the ovary. Rockfish (*Sebastes* spp.) produce thousands of tiny young and are truly ovoviviparous.[1087]

True viviparity is found in the little-known Baikal cod (Comephoridae) and many Brotulidae. The viviparous species of brotulids possess a conspicuous intromittent organ, and fertilization occurs within the ovary. The brotulid *Dinematichthys ilucoeteoides* of the Indian Ocean produces 250–450 young per brood. Early development takes place within the follicle but, after the tailbud stage, the embryos are discharged into the lumina of the paired ovaries. Intrauterine embryos of the viviparous brotulid *Parabrotula dentiens* have elongated trophotaeniae from the cloacal area (as in the embryos of the Goodeidae), suggesting a prolonged embryonic life for this species. Fish of the little-known deep-sea family Aphyonidae are viviparous, as are species of the ophidioid genus *Oligopus*.[1109] The sole-surviving coelacanth, *Latimeria chalumnae,* is ovoviviparous. Quite probably other obscure marine fishes will prove to be viviparous.[942]

One looks in vain for underlying patterns of viviparity among fishes. Not only is this mode of reproduction common to many fresh-water and marine teleosts quite unrelated to one another, but viviparity characterizes a number of elasmobranchs. True viviparity is the ultimate development of prenatal parental care. It is efficient in insuring against loss of ova (except in the death of a gravid female) but is expensive in demanding prolonged nourishment of the embryos. Viviparity, together with nest-guarding, is common among shoreline and coastal fishes, which occupy an environment rich in both nutrients and predators, and seems to be the exception among epipelagic species.[682] Viviparity in the fishes in which it occurs has had a long history, and its origins are obscured in the relatively unknown conditions of the past.

Viviparity in Amphibians

Although most amphibians are oviparous, live-bearing occurs in the Apoda, some Caudata and a few Anura. Viviparity has developed many times independently, and frequently live-bearers and egg-layers are closely related.

All caecilians (Apoda) have an intromittent organ and presumably internal fertilization. The semiaquatic *Ichthyophis* of the Oriental region lays eggs in burrows in stream banks. Gills develop early in life but are almost completely absorbed before hatching; young larvae are aquatic and retain gill slits and small gill remnants. Some terrestrial caecilians are viviparous, as are all members of the fully aquatic family Typhlonectidae. In caecilians fully developed young are much larger than are the fertilized eggs. No placentation develops, but embryos have large plumose or saccular gills and a very thin, vascularized integument for intrauterine gaseous exchange. At birth the young sheds and may eat its thin skin, and subsequently develops the rather cornified covering characteristic of terrestrial amphibians. Yolk of terrestrial caecilians is insufficient to nourish embryos to birth; mature eggs are about 4 mm in diameter, and the embryo becomes many times larger. The young lie stretched out within the uterus, and development continues until the full-grown embryo is 75–120 mm long. This mode of development is found in the African genera *Geotrypetes* and *Schistometopum* and the Neo-tropical *Chthonerpeton* and *Gymnopis* and several other genera.[791,1066] After the yolk is exhausted and the embryos escape from the eggs, the young ingest uterine secretions, and the gut may be filled with a thick creamy material with fat droplets and cellular fragments from the uterine lining, collectively called "uterine milk." Large oil droplets are in the uterine endometrium in late stages of pregnancy, in which case the uterus is highly vascularized. In caecilians the embryonic gall blad-der is very large.[791]

Several families of salamanders (Caudata) exhibit transitions from oviposition in water to egg-laying on land and finally to ovoviviparity. In the Holarctic Sal-amandridae there is a transition from aquatic egg-laying, in ponds or slow streams, to true viviparity. In the Nearctic genera *Taricha* and *Notophthalmus* and in their Old World relatives, *Triturus* and *Cynops,* gilled larvae hatch from eggs laid in water. A departure from this basic pattern is seen in the western Palaearctic *Salamandra: S. salamandra* and *S. atra* are both viviparous. *S. sal-amandra* produces 15–25 or more gilled larvae after a gestation of about three months. The montane *S. atra* occurs in the European Alps from about 1000 m to above timberline, and the young metamorphose in utero. Although up to 50 eggs may be produced in a given season, only two to four young are born, the nonviable ova nourishing the surviving embryos. As in the viviparous caecilians, the gills of larval *S. atra* become very large and lie in contact with the uterus, only to dis-appear before birth. Gestation may extend from 12 months at lower elevations to more than three years above timberline.

A similar sort of viviparity occurs in the olm *(Proteus anguineus)* of southern Austria and eastern Yugoslavia. The olm ovulates numerous eggs of which only one in each oviduct is destined to develop, or it may lay eggs. As in *Salamandra atra* the nonviable eggs nourish the growing embryos. In *Hydromantes genei* young are produced from eggs retained in the uterus, but nothing is known about their prenatal nutrition.

Three species of the African bufonid, *Nectophrynoides,* are viviparous.[876] Young are nourished by "uterine milk" as in viviparous caecilians, but in the toads this substance contains mucoproteins. Adaptations include a long, vascularized tail and a well-developed gut. Intrauterine growth results in an increase in size of 400–500 times over that of the fertilized egg. The uterus is highly vascularized, and corpora lutea persist throughout gestation.[1112] The viviparous Nimba toad *(Nectophrynoides occidentalis)* mates in October, near the end of the rainy season. The eggs number 2–10 and contain little yolk. After ovulation they remain in the uterus, and the young ingest secretions from the uterine epithelium. During the arid season (November–March), the corpora lutea are most active and apparently retard embryonic growth. When the female becomes active in April, at the onset of the seasonal rains, the corpora lutea become smaller, progesterone production declines and embryonic growth accelerates. Progesterone in this amphibian is believed to prepare the uterus for pregnancy, and, experimentally, ovariectomy does not terminate pregnancy except in the first year of the female's reproductive life.[1113]

Eleutherodactylus jasperi, an arboreal leptodactylid that lives in and among bromeliads of Puerto Rico, is a live-bearer. Gravid females have three to six developing ova from April to August, and the young are born fully transformed. In other species of *Eleutherodactylus,* eggs are laid in terrestrial sites, and metamorphosis occurs prior to emergence. Apparently in *E. jasperi* the eggs are simply retained within the reproductive tract until hatching to protect them from desiccation in the xeric habitat in which they live, and the condition is probably ovoviviparity. The reproductive tract of gravid females is highly vascularized, providing at least a source of oxygen and a means of waste removal for the embryos.[255,1067]

Although there is no single adaptive significance to live-bearing in amphibians, viviparity seems to provide some advantage in species in which it occurs. The alpine *Salamandra atra* can carry the intrauterine young into places of proper temperatures throughout the year; this situation is an advantage but clearly not essential, for some high-elevation amphibians are oviparous. Species of the caecilian *Gymnophis* and the viviparous toads *(Nectophrynoides)* give birth to young at the onset of the rainy season.

Viviparity in Squamata

Two aspects of reptiles facilitate live-bearing: (1) body temperatures (at normal activity), which are generally much higher than those in amphibians and (2) the circulatory system, which provides a high level of oxygen. These two features enable relatively high rates of metabolism. Combined high levels of oxygen and metabolism enable growth and nutrition of intrauterine young, and not surprisingly some amphisbaenians, snakes and lizards have independently exploited this improved internal environment to protect and prolong the prenatal life of their young.

There are some obvious cases of ovoviviparity in the Squamata. These include those that are facultatively oviparous but may retain the eggs for part of the embryonic development. In the Nearctic green snake *(Opheodrys vernalis),* there is a tendency for boreal populations to retain the eggs for longer embryonic growth than in southern populations. Similarly the exceptional retention of eggs in a normally oviparous species (a python, for example) is clearly ovoviviparity.

Some live-bearing reptiles have special structures or procedures that furnish maternal contribution to the prenatal development of the young, and such species must be considered viviparous. Generally viviparous reptiles have a thin covering about the eggs, the reptilian shell being absent, and in a few cases there is evidence of substantial maternal contribution to embryonic growth. There may be a simple chorioallantoic placenta in some reptiles and a yolksac placenta in others, but studies are mostly anatomic and descriptive, and the degrees of maternoembryonic exchange and hemotrophic nutrition are poorly known.[298]

Viviparity and Temperature. Some examples of reptilian viviparity seem to be adaptations to seasonal or nocturnal low temperatures. Like the adult, eggs are ectothermic: eggs laid in shallow soil are warm during the day but, at high latitudes or high elevations, become cool at night. If eggs are retained within the body of the female, she can retreat at night with her clutch under a rock or within a log, where temperatures do not reach the lows found in surface soils. In addition the female can bask in the day, thus keeping the eggs at an optimal temperature for development. In temperate climates viviparous forms frequently mate and ovulate in autumn, and embryonic development can either proceed on warm days in the winter or be suspended until the onset of higher daily means in spring. On the other hand, some of the most boreal reptiles are oviparous, and viviparity is seen in many tropical groups. There is clearly more than temperature involved in reptilian viviparity.

In the geckos (Gekkonidae) viviparity occurs only in three diplodactyline genera in New Zealand; geckos in Australia (including species of Diplodactylinae) are all oviparous, as are geckos everywhere else in the world. New Zealand is perhaps the coolest area of the world inhabited by the generally tropical or subtropical geckos, so viviparity in these lizards seems to be an adaptation to a cool environment.

Lizards of the family Anguidae are found in many temperate and tropical regions of the world, and both oviparous and viviparous species occur—sometimes within a single genus. The mesoamerican montane alligator lizards, *Gerrhonotus,* are viviparous, as are some montane species of the more aboreal *Abronia* spp., suggesting that live-bearing in these species is a response to reduced temperatures at high elevations. In the United States *G. coeruleus* is viviparous and is frequently sympatric with the oviparous *G. multicarinatus.*[1057] The glass "snakes" (*Ophisaurus* spp.) are all oviparous, but all other anguid lizards, including many tropical species, are live-bearers. Among the many species of Anguidae, live-bearing seems associated with low temperatures in only a few cases.

The Lacertidae is a large family with many species in temperate and tropical regions of the Old World. One species, *Lacerta (Zootoca) vivipara,* is live-bearing in northern Europe and oviparous around the Mediterranean Sea. As will be pointed out below, this species is truly ovoviviparous. Most other species in this very large family are egg-layers.

Among the many species of chameleons (Chamaeleontidae) there are both egg-layers and live-bearers. Within the genus *Chamaeleo* there are live-bearing species in the cooler, more austral or montane environments. For example, *C. hohnelli, C. bitaeniatus* and *C. jacksoni* of upland areas produce small young, whereas *C. chamaeleon* of Mediterranean lowlands and the species in the tropics of the Oriental region are all oviparous. The most southerly species, in the genus *Microsaura,* are all viviparous. In this family live-bearing is perhaps correlated with low temperatures.

Among skinks (Scincidae) there is the greatest assemblage of live-bearing species. Viviparity here occurs in many genera and usually follows phylogenetic lines: within a given genus, species are generally either all viviparous or all oviparous. In the very large genus *Eumeces* there are many tropical and temperate oviparous species. There are two viviparous species (*dicei* and *lynxae*) in the mountains of Mexico, but montane species in the United States are all oviparous. In the large and widespread genus *Mabuya,* both sorts of breeding are found, and many of the live-bearing species occur in the tropics of mesoamerica, Asia and Africa. In southwest Tanzania the skink *Mabuya striata* is viviparous and breeds throughout the year, whereas the oviparous agamid, *Agama cyanogaster,* lays eggs only during the rainy season. Because the eggs of *Agama* hatch at the onset of the dry period, viviparity in this case may enable the skink to protect its eggs from desiccation and breed in both rainless and rainy seasons.[857] Among the 32 species of the skink genus *Lygosoma,* the four live-bearing species occur in the extremely arid regions of Somalia and eastern Kenya.[367] As in *Eumeces,* most species of *Scincella* are oviparous, but the montane *S. himalaynum* is a live-bearer. Some entirely viviparous genera (e.g., *Lipinia, Isopachys* and *Tropidophorus*) are confined to the tropics. In Australia skinks are numerous, and viviparity is typical of those species in the southern mountains and in Tasmania. The giant skink *(Leiolopisma grande)* of New Zealand nourishes the embryos with a well-developed placenta. Immediately after birth of the young, the mother eats the placenta and licks up the amniotic fluid. Unlike young of oviparous skinks, newly born young of viviparous species lack an egg tooth. Thus, as in the Anguidae, there are many viviparous skinks in cool regions, but not all species in cool climates are live-bearers, and many live-bearing species occur in tropical lowlands.

In the large family Iguanidae of the New World, oviparity prevails, but some species are live-bearers. In some cases viviparity is associated with cool environments, but there are many exceptions. In many mesoamerican species of *Sceloporus,* including the tropical *S. malachiticus,* viviparity is the rule, and viviparity also occurs in *S. torquatus,* which is found from the southwestern United States to Guatemala. The more boreal species of *Sceloporus,* including species that live up to 2700 m in the mountains in the United States, are all oviparous. Viviparity undoubtedly permits *Liolaemus multiformis* to exist above 4500 m in the Andes of Peru.[798] It is perhaps significant that the most boreal species of the horned lizard, *Phrynosoma douglassi,* is viviparous. Within the mesoamerican genus *Corythophanes,* the highland *C. percarinatus* is viviparous while the lowland *C. cristatus* and *C. hernandesi* are both egg-layers. Although in this family live-bearing typifies some tropical montane species, oviparity is the rule in the boreal and montane species.

The Colubridae is by far the largest and most diverse group of snakes. Viviparity occurs in a number of genera, and in at least two *(Elaphe* and *Nerodia)* both modes of reproduction occur. In the mountains of Mexico *Conopsis nasus* and *Toluca lineata* are viviparous, whereas species of the closely related genera *Gyalopion* and *Ficimia* of the lowlands are all egg-layers. Other viviparous colubrids are discussed below.[298]

The Elapidae includes a large number of tropical and subtropical snakes noted for their potent toxin. In the New World it is represented by the brilliantly colored coral snakes, and in the Old World the elapids are cobras, kraits and a small

Oriental coral snake, and mambas (*Dendroaspis* spp.) in Africa. Among all these species, only the spitting cobra *(Haemachatus haemachates)* is viviparous. Only in the Australian elapids is viviparity common, and many of these are known to be either montane or to occupy the cool southern regions.

Viviparity in Aquatic Reptiles. For reptiles that spend most of their time in water, sometimes far from land, and move on land clumsily and with effort, egg-laying can be a major problem. It is not surprising to find viviparity the rule among more aquatic snakes.

The sea snakes (Hydrophiidae) comprise a number of species in several genera of highly spcialized marine serpents. They occur in warm seas from the Orient to Australia, on the east coast of Africa and with an isolated representative on the west coast of mesoamerica. Piscivorous and thoroughly at home in water, they are awkward on land. Except for species of *Laticauda* of the Oriental region, all hydrophiids are viviparous.

Aquatic colubrids of the subfamily Homalopsinae are all viviparous; they live in rivers and coastal waters of southeast Asia to Australia and include species of *Cerberus, Enhydris* and *Homalopsis.* The Neotropical colubrid *Helicops* is also aquatic and viviparous. Also the highly aquatic Acrochordidae of the tropical Orient consist of species that spend little time on land, and all are viviparous.

An aquatic habit does not necessarily result in viviparity. The rainbow snake, *Farancia ergtrogramma,* a colubrid of the southeastern United States, is aquatic but an egg-layer, as are the aquatic *Grayia smithi* and *Lycondonomorphus rufulus* of Africa, *Trotanorhinus variabilis* of Cuba as well as almost all of the Old World natricines, some of which are aquatic.

Viviparity and Geographic Distribution. The Nearctic Natricinae (Colubridae) are all viviparous and deserve special consideration. Among the Palaearctic natricines only *Natrix annularis* of Taiwan and the warm regions of China is viviparous. In North America this subfamily is widespread and diverse, and includes the large Nearctic genus *Thamnophis* (garter snakes and ribbon snakes), *Regina* (including *Liodytes*) of the southeastern United States, *Seminatrix, Nerodia, Storeria, Virginia, Clonophis, Tropidoclonion* and *Adelophis.* These snakes occur from the tropics to some cool-temperate regions of Canada, from lowland swamps in the southeastern United States to deserts in the west, and in high mountains, and all species are live-bearers. Neither cool climate nor aquatic habits can easily account for viviparity in these snakes. It seems far more plausible that a boreal Palaearctic natricine, perhaps a species of *Nerodia,* entered the New World across a Bering connection and subsequently radiated to produce the present array of Nearctic natricines. Quite probably viviparity enabled the first immigrants to enter the New World, but today their descendents are sympatric with many oviparous colubrids.[754] Although some species of *Thamnophis* are the most boreal reptiles in North America, this is not a result of viviparity, for the most northerly species in the Holarctic Region, *Natrix natrix,* is oviparous.

In the Boidae, New World members (the various boids, the anaconda and their allies) are live-bearers, but the role of the yolk and any maternal contribution to development are unknown. The Old World boids are all oviparous.

Viviparity Without Apparent Ecologic Significance. There are many live-bearing lizards and snakes for which there is no apparent adaptive meaning. The Xenosauridae of China and Mexico, for example, are viviparous, but little is known of their breeding habits or biology. The Cordylidae, a group of the tropical and subtropical Ethiopian region, consists of viviparous species. Likewise all species of the New World family Xantusiidae are live-bearers. The Uropeltidae, fossorial snakes of India and Ceylon, are all viviparous. Among live-bearing Colubridae, there are a number in tropical Africa, the Oriental region and South America; live-bearing is not correlated with any known biologic feature of these species.[298]

The common name "viper" literally means live-bearer, and most species of Viperidae are viviparous, but there seems to be no ecologic significance to viviparity in these snakes. The Japanese *Trimeresurus okinavensis* retains the eggs for a protracted period so that hatching follows oviposition within a few days or even hours; in this species the eggshell is very thin and membranous. *Trimeresurus stejnegeri* of China and Taiwan is a live-bearer. The Japanese and Korean species of *Agkistrodon (blomhoffi, caliginosus* and *saxatilis)* are all live-bearers, as are those of the New World, *A. contortrix* and *A. piscivorus;* in China, however, *A. acutus* is an egg-layer. In Central America the bushmaster *(Lachesis muta)* is the only egg-layer in the New World pit vipers. The majority of Old World vipers (*Vipera* spp. and *Bitis* spp.) are viviparous.

The legless lizard, *Anniella pulchra,* of the southwest coast of California, mates in the spring and gives birth to one to four large young in the fall. The large size of the newborn young (65–80 mm total length) indicates viviparity.[699]

Reproductive patterns among amphisbaenians are generally poorly known. Several species are egg-layers, and at least two *(Trogonophis wiegmanni* and *Monopeltis capensis)* are viviparous.[1058] *Bipes* spp. are probably also live-bearers. Viviparity is unknown among crocodilians, chelonians and the sole-surviving rhyncocephalian. Although crocodilians are fully as aquatic as the Homalopsinae and Acrochordidae, they are still compelled to return to land to lay eggs. Many turtles are aquatic and even pelagic, yet all must lay eggs on land.

Maternal Contribution in Reptilian Viviparity. The simplest step toward live-bearing is the brief retention of a fertilized ovum and, when kept until hatching, the condition is ovoviviparity. The Old World *Lacerta (Zootoca) vivipara* is shown experimentally to be truly ovoviviparous. Transfer I^{131}, Na^{24} and P^{32} from mother to embryo is slight. Fresh ova excised from the uterus and incubated in physiologic serum hatch successfully.[790] The eggshell of *Lacerta vivipara* is thin, and the shell membrane lies against a vascularized uterine lining. The yolk is large, and apparently only oxygen and carbon dioxide pass across the membrane. In oviparous populations the eggs hatch about a day after oviposition.

In the reptilian chorioallantoic placenta there is commonly some erosion of both uterine and chorionic tissues, permitting close association of the fetomaternal circulations.[1081] In more specialized types there are ridges and folds in the maternal tissue, with capillaries coming to the surface. Finally there is the development of discrete elliptical areas of both maternal and fetal tissues, the thickened chorionic ectoderm lying against the highly folded glandular and vascularized maternal tissue. In the Australian skinks yolk content in the viviparous species is about two-thirds of that in the oviparous forms; early in fetal life the yolksac develops its placentation and, as the yolk diminishes, a chorioallantoic placenta develops. In some live-bearing skinks the egg is much smaller than in oviparous species of comparable size and with relatively little yolk, but it grows during intrauterine development. Presumably amino acids and proteins as well as oxygen are received from the mother through the chorioallantoic placenta (Fig. 12.14).[623,700]

Since a corpus luteum and the secretion of progesterone are essential for the maintenance of mammalian pregnancy, there has been much speculation about

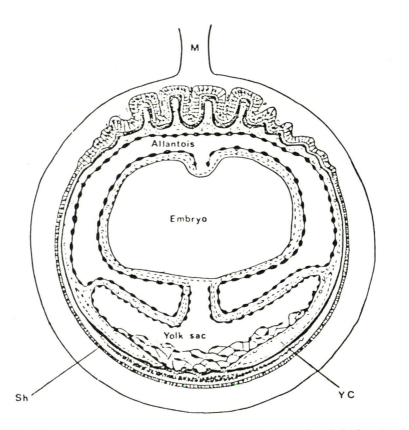

Fig. 12.14. Cross-section of the uterus of a gravid skink, *Chalcides chalcides,* showing the chorioallantoic placenta (Luckett, 1977; courtesy of Plenum Press).

hormonal control of pregnancy in reptiles. All viviparous amphibians and reptiles form a long-lived corpus luteum, which is apparently capable of secreting progesterone in all species in which it has been studied.

Placentation in Nearctic natricine snakes is similar to the chorioallantoic placenta described for Australian skinks. There is a rapid transmission of Na^{22} and I^{131} from the maternal to the fetal tissue in *Nerodia cyclopion* and *N. sipedon,* and sodium transport across the placenta of these species approaches the efficiency of the mammalian placenta.[179] The garter snake *(Thamnophis sirtalis),* another Nearctic natricine, also nourishes the embryos through a chorioallantoic placenta: yolk provides the embryo with fat, and apparently water and amino acids enter through the placenta. In most of the live-bearing species of vipers, embryos are nourished by a chorioallantoic placenta.

In the sea snake *Enhydrina schistosa,* embryos are held in "uterine chambers" of the upper region of the oviduct. The endometrium of this area lies closely appressed to the chorioallantoic epithelium, and both tissues are highly vascularized; hemotropic nutrition seems almost certain under such circumstances.[527]

Xantusiid lizards are among the most specialized reptiles for viviparity: broods are small, gestation is prolonged and contact between maternal and fetal circulation is intimate. Females of *Xantusia vigilis* reach sexual maturity in their third year. A single ovum matures in each ovary, and two young are born three months after mating. A fresh ovum weighs about 0.115 g and a newborn young 0.230 g, indicating a substantial amount of nutrition coming from the mother. Early in the life of the developing embryo the thin eggshell dissolves, and the embryo receives nourishment through a chorioallantoic placenta, which is appressed to the oviduct. Corpora lutea remain for the first two months of gestation. Shortly before parturition, embryonic membranes and their contents protrude from the cloaca. The mother pulls the membranes, stimulating and assisting the embryo in its escape. Holding the membranes in her jaws, the female pulls them from the cloaca and consumes them, lapping up droplets of fluid that fall on the ground.[186,700]

In the viviparous *Sceloporus jarrovi,* mating occurs in autumn but little embryonic development occurs until spring. A placenta forms early in April in this montane species, and the gravid females bask frequently until young are born in late June.[348] The cool-adapted viviparous geckos of New Zealand develop a chorioallantoic placenta in which there is a close association of fetomaternal circulation.

Summary

Viviparity in fishes occurs in an apparently random manner among various sharks and numerous families of teleosts. Live-bearing does not seem to be associated with any environmental factor. It occurs in many fresh-water forms and in both coastal and benthic marine groups but is scarce in epipelagic species. Among a diversity of amphibians and reptiles various degrees of viviparity occur. These range from purely facultative ovoviviparity, as seen in the Palaearctic lizard *Lacerta vivipara,* to viviparity in the xantusiid lizard *Xantusia vigilis,* in which the mother even eats the placental membranes. Inasmuch as the newborn young of an oviparous species weighs about two-thirds as much as a fertilized ovum (dry weights), approximately one-third of the energy of the ovum is spent in embryonic metabolic activity. Thus when the young exceeds two-thirds the weight of the new egg, one may assume that there has been some maternal contribution after fertilization. In a few groups of live-bearers (e.g., jenynsiid fishes, the olm and *Salamandra atra*), dead eggs or siblings nourish the developing young. In such cases young are surely heavier than two-thirds the weight of the fertilized ovum, but maternal contribution may cease with ovulation.

Among the carefully investigated reptilian live-bearers, viviparity is the rule and ovoviviparity the exception. For most species the degree of maternal contribution is not known. It is easier to recognize the various forms of viviparity than to explain conditions under which they developed, or to assign to them a definite environmental origin. Viviparity is frequently considered to be a response to cool or cold temperatures, such as are found at high elevations or high latitudes, but the occurrence of live-bearing among tropical reptiles suggests that such an explanation is too simple. The viviparous species of the iguanid *Sceloporus,* although montane, occur closer to the equator than do many of the oviparous species of that genus. A similar pattern is seen in the Anguidae. Among the natricine colubrid snakes, Nearctic forms are genuinely viviparous, whereas almost all the Old World species are egg-layers; indeed the most boreal snake in Europe, *Natrix natrix,* is an egg-layer.

It is realistic to consider that viviparity in amphibians and reptiles may offer a competitive advantage over a similar egg-laying species in a cool climate. But it is simplistic to assume that viviparity arose under the stress of low temperatures, or that it is prerequisite to life in a cool region. The frequency of viviparous reptiles at high elevations is more conspicuous at lower latitudes. If this is actually so, live-bearing could result not only from lower temperatures at high elevations but also from chilling effects of dense cloud cover so characteristic of many tropical mountain ranges. In the tropics, viviparity may be a means of protecting eggs from desiccation in the dry season.

Viviparity is clearly an adaptation to aquatic habits in snakes. It is characteristic of the extremely specialized aquatic groups (Homalopsinae, Acrochordidae and Hydrophiidae), but most of the casually aquatic species are oviparous.

The transition from oviparity to viviparity places additional costs on the female parent, but advantages lie in reduction of early mortality of the offspring. Live-bearing, either with or without postovulatory nutrition from the mother, delays birth. In an ovoviviparous species the mere act of supporting and carrying the embryo is expensive. If the species is viviparous, the cost can become substantial, not only in terms of nutrients provided by the mother but also as measured by the gradual impairment of her freedom of movement.

Viviparity in mammals seems not profoundly unlike live-bearing in fish, amphibians and reptiles, insofar as the hormonal relationship is concerned: variations are as great within a class as they are between classes. The mammalian placenta is far more complex, and the devices for ensuring the success of internal fertilization are more highly developed than in the lower classes.

Suggested Readings

Gallien L (1959) Endocrine basis for reproductive adaptations in amphibia. In: Gorbman A (ed) Comparative endocrinology. Wiley, New York, pp. 479–487

Greer AE (1977) The systematics and evolutionary relationships of the scincid lizard genus Lygosoma. J Nat Hist 11:515–540

Kehl R, Combescot C (1955) Reproduction in the reptilia. In The comparative physiology of reproduction and the effects of sex hormones in vertebrates. Cambridge University Press, Cambridge, England

Matthews, LH (1955) The evolution of viviparity in vertebrates. In: Jones C, Eckstein P (eds) The comparative physiology of reproduction and the effects of sex hormones in vertebrates. Cambridge University Press, Cambridge, England

Miller, MR (1959) The endocrine basis for reproductive adaptations in reptiles. In: Gorbman A (ed) Comparative endocrinology. Wiley, New York, pp. 499–516

13. Community and Population Density

In Chapter 11 population structure was discussed in relation to reproduction. Reproduction is a major factor in changes in density, which in turn influence reproductive rates; this is one of the best known density-dependent phenomena. The density of a population reflects the interaction of communal physical and biotic elements. This chapter describes some examples of the relationships between community structure, exogenous factors and fluctuations in density.

Community Structure

Many habitats which are rather clearly distinct from adjacent habitats have characteristic groups of species. Throughout the world, in terrestrial environments, a combination of a given soil type and climate will produce a certain combination of plants and with it a characteristic animal community.

Trophic Levels

Some students restrict the community to members of a certain major taxon, such as the avian or the butterfly community. However, considering the totality of animal life in a given habitat, one can separate a community into several trophic or feeding levels; this has advantages over a taxonomically defined community. Herbivores draw on food produced by plants, and there are many sorts of herbivores among annelid worms, arthropods and all classes of vertebrates. In almost any habitat, herbivores constitute most of the biomass (their total weight), except that their mass is greatly exceeded by the plants on which they feed. Herbivores have many predators: a crow or a kestrel may feed on grasshoppers, or a large cat may take numbers of impala.

This two-level trophic system is usually too simple to represent reality: many insects, such as dragonflies and tiger beetles, prey on herbivorous arthropods, and such predatory insects are in turn captured by many kinds of fish, amphibians, reptiles, birds and mammals. Predators are less frequent and as a group have less biomass than herbivores. That is to say, higher trophic levels have fewer individuals and less biomass per unit area because when a vole eats grass or when a hawk eats a vole: (1) some of the food ingested contributes to growth, (2) some is used for metabolic activity and reproduction and (3) some is undigested. Thus the productivity of each trophic level must be considerably less than the level on which it feeds, lest the balance of biomass be disturbed (which sometimes occurs). Also the efficiency of converting energy (or "ecologic efficiency") is about 10 percent. The trophic web becomes a tangle indeed when one considers the effect of internal and external parasites, which not only feed on both herbivores and predators but may transmit pathogens, which sometimes greatly reduce densities of the host. Thus there are sound reasons for including all major taxa in the community concept and avoiding the taxonomic community concept.

Interaction Among Species

The frequent assumption is that most species are members of a community; i.e., that they engage in a regimen of apparent interactions with other species, and changes in populations of one kind affect the other species. The community is therefore presumed to function as a unit. Interactions clearly exist in some well-known pairs (or groups) of species. However, among species at the same trophic level specific interactions, usually in the form of incomplete competition, are more often presumed than known; there are not many examples of well-documented unequivocal competition. More commonly, apparent instances of exclusion are only assumed to represent potential competition.[148,363]

Although there is no known way to make a numerical comparison of interspecific relationships in natural (wild) populations, it is clear that all species in a community do not relate equally to each other. A group of species of granivorous birds, for example, would be more likely to affect one another than they would insectivorous species. Insectivorous mammals (e.g., shrews) may seek some of the same food as ground-foraging insectivorous birds, such as some thrushes, but they would probably not compete with treetop insectivorous birds, such as vireos or warblers. Clearly there are different degrees of interaction among species in a community, and there is no reason to suppose that all species therein interact with all other species. Most commonly, any one species in a community affects the welfare of only a small number of other species, and the best-established examples of community interactions are between pairs of competing species or between pairs of predator–prey species.[344,364] This situation raises doubt about the validity of the community as a *totally* interacting web of interdependent species.

Community Changes

The community concept is restricted if one also considers the long-term constancy of a community's species composition. Biogeographic literature is replete with observations of species that change their geographic ranges, probably following natural geographic shifts in climate and vegetation. Because the factors that seem to limit geographic distribution do not act uniformly on all species in a community, the species composition is in continuous flux. The roe deer *(Capreolus capreolus),* for example, was known in Norway from a few scattered records in the latter part of the last century, but by 1940 there was a broad area over southeastern Norway in which breeding populations existed. In northeastern North America the cardinal *(Richmondena cardinalis)* has extended its range to the north, and the coyote *(Canis latrans)* has moved from its former prairie range to the Atlantic coast. In each of these examples there have not been major geographic shifts of entire communities but rather movements of individual species.

Some species are ubiquitous and genuine members of several communities within their geographic ranges. The short-tailed shrew *(Blarina brevicauda)* of eastern North America occurs in various forest types as well as in grasslands and is usually abundant. Most regions have some successful vertebrates that are members of several communities.

In addition to the movement of species into communities where they had not hitherto occurred, entire communities have changed their character and geographic range. Faunas moved hundreds of miles during the Pleistocene, the last such major shift having occurred only some 10,000 years ago. These changes characterized not only glaciated regions but the mid-latitudes and tropics as well. Although it is impossible to trace the movements of all species as communities moved during the Pleistocene, it is reasonable to assume that these movements were similar to those in historic times. That is, individual species moved in response to subtle (or drastic) climatic changes, and all species did not necessarily move at the same rate or even in the same direction. Thus the communities that we recognize today have not existed unchanged since ancient times but are rather recent, and experience slow but continuous change.

In addition to seasonal changes in temperature, precipitation, light and the climatic alterations that occurred throughout the Pleistocene, there are irregular and unpredictable environmental fluctuations. Well-recognized changes result from events that alter the flora and the soil. Hurricanes and forest fires disturb sequential floristic changes, and recovery may take many decades.[108] Not only does fire destroy food and cover for vertebrate populations but, when followed by heavy rains, the resulting soil erosion may cause permanent alteration of community composition. In addition to catastrophic winds and fire, droughts affect forest growth directly and lower the resistance of trees to insect attacks, which in turn may destroy extensive stands of mature trees. Along beaches, a tidal wave not only obliterates shoreline vegetation but destroys virtually all the plant growth on a small island.

Far more drastic changes in community structure are caused annually by migration and hibernation, but communities reform the following year, and such disturbances are temporary.

To summarize: (1) Communities evolve with time, and their composition becomes altered by additions and disappearances of member species. Most communities as we know them today are probably rather young. (2) Interactions among the member species are not equal but tend to be more powerful among species that either compete or have a predator–prey or a host–parasite relationship. Most species in a community do not fall into either of these relationships and seem to have a loose rather than a rigid relationship with one another. This does not deny the existence of the community but rather points out that it is more of a concept that an entity; that it is not a kind of superorganism that acts as a unit but is rather a loose assemblage of species, which may relate to each other, but all of which ultimately respond to the soil and the prevailing climatic factors. The community is a useful concept, for it helps us to evaluate the obvious inter-specific relationships among its interacting members.

Species Diversity

The relative abundance of different species is species diversity. A consistent feature of well-studied communities is that a small number of species make up most of the biomass; therefore most species are more-or-less uncommon. This relationship, seen in both terrestrial and aquatic communities, forms the basis for the earlier statement that interactions among member species are not equal. The disparate abundance of species within a community has been well known for a long time; it was perhaps first demonstrated for birds, for which a careful observer can obtain accurate data (Fig. 13.1). The high frequency of occurrence of a small

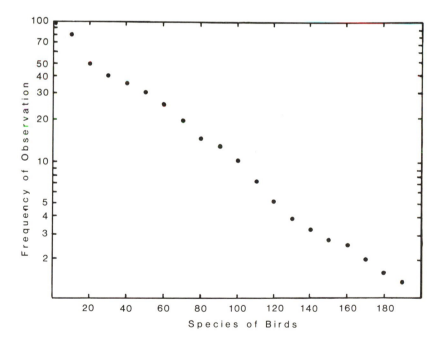

Fig. 13.1. Species diversity, an illustration of Raunkiaer's law of frequency, based on frequency of occurrence (vertical axis) of 194 species of birds in Doniphan County, Kansas, plotted on semilog scale. The bird species are arranged on the horizontal axis, in decreasing order of abundance (from left to right). Every tenth species is indicated by a point (Linsdale, 1928).

number of species and the relative scarcity of most species follow Raunkiaer's Law of frequency established for plants.[605]

Competition

Competition occurs when two individuals use a resource in limited supply; interspecific competition is the utilization of a limited resource by two (or more) species. There are many examples of animals using the same resource. Except when there are marked sexual differences in resource utilization (as in anglers), members of the same species and age tend to use the same (or similar) resources, and many sympatric species make similar demands on the environment. The difficulty in demonstrating competition lies in the establishment of a limitation of supply. For example, all vertebrates need oxygen, but the supply in the atmosphere has been adequate to date, and we do not compete for it.

The most intense competition is intraspecific. Within a given species, different age groups and both sexes may seek the same food. Terrestrial herbivores, such as voles and deer, often feed on identical materials. Among browsing ungulates, adults can reach higher than can the young and, if browse is limited, the young may starve. In the black-tailed deer *(Odocoileus hemionus),* moreover, the metabolic requirements of male is greater than that of female fawns; when food shortages occur, growing males suffer greater mortality than do females.[617]

Niches. The extent of competition can be studied by identifying possible realms or "niche dimensions" common to the activities of two (or more) species. There are not many published dimensions to compare, and most life history studies are not sufficiently precise about environmental utilization to be useful in defining the niche. All vertebrates require food and space, and the use of these dimensions is frequently not described with sufficient clarity to determine if sympatric species are using the same food or space. For heliothermous reptiles a basking site may be an important niche dimension; most birds require specific nest sites, and soil type is critical for most fossorial vertebrates. The extent of interspecific competition can be studied by identifying possible realms or niche dimensions common to the activities of two (or more) species.

A niche is living space that contains the requirements for a given kind of animal: the food, nesting and resting sites, proper temperature and humidity ranges as well as the absence of harmful elements. Niches are not fixed or rigid, nor are niches of different species necessarily mutually exclusive. Niches overlap when closely allied species compete; in such examples one of the two withdraws from part or all of the area of overlap.

This is illustrated by the occurrence and niche of each of several mice in the Japanese archipelago. Species of *Clethrionomys* are normally forest-dwellers and *Microtus* live in grassy meadows. This is their pattern of separation on the large island of Honshu. On Hokkaido, however, *Microtus* is absent, and there *Clethrionomys rufocanus* is a common grassland species. Thus when the two species live in the same region, *Microtus* causes *Clethrionomys* to remain in forested cover. This phenomenon has been experimentally reproduced with Nearctic species of voles.[148,363]

Populations of lizards are extremely high on Poor Knights Islands (New Zealand): several species attain densities of 1000 or more per acre, but ecologic and temporal segregations are clear-cut (Fig. 13.2), and competition is probably slight.[1088]

Ecologists point out that many vertebrates are separated by their activity periods, but such temporal isolation does not preclude competition when they utilize the same limited resource. Time is not a resource of the same character as food or space, and partitioning of time does not in itself reduce competition or amount to resource partitioning. If owls and hawks both forage for voles (*Microtus* sp.), which happen to be of a low density, the raptors may compete and their temporal separation will not eliminate this competition. In contrast, a nocturnal gecko may subsist on nocturnal insects and thus not compete with diurnal lizards. Temporal separation in oviposition of amphibians also may well result in avoidance of competition among their larvae. *Ichnotropis squamulosa* and *I. capensis*, two African lizards, are annual sympatric species of comparable size: they avoid the possibility of competition by a staggering of their life cycles so that adults do not occur together (Fig. 13.3).[120]

Competitive Exclusion. That two species filling the same niche do not occupy that niche together (simultaneously) gives rise to the concept of competitive exclusion. Should two species occur together over a protracted period, it is then assumed that the resources of the niche are sufficient to prevent competition. When one resource (a dimension of the niche) becomes restricted in amount or degree, competition is assumed to occur; it is expected that one species will survive (or increase) at the expense of the other. In nature, proof of competition is almost always lacking, and the relationship between competition and competitive exclusion becomes somewhat circular. In the absence of clear evidence of competition, however, competitive exclusion is accepted by many students as presumptive evidence of potential competition.

A complex habitat can be assumed to have a relatively large number of niches. Generally more complex habitats hava a larger number of species in their fauna. This simple relationship of habitat diversity (abundance of niches) to faunal diversity (or species diversity) can be modified by the degree to which the vertebrate species overlap in their niche requirements: (1) a preponderance of generalists in the fauna tends to reduce the total number of species, whereas (2) a preponderance of specialists tends to increase the total number of species, or faunal or species diversity. An increase in the number of generalists increases niche overlap (interspecific competition) or the degree to which different species exploit the same resource. But as in the exploitation of any resource, competition decreases as the resource increases; or, if food is the major aspect of niche overlap, an increase in food reduces niche overlap and permits an increase in either (1) numbers of individuals or (2) the number of species. In either case there is an increase in biomass.

The increase in numbers of species with decreasing latitude reflects not only increased complexity of habitats (i.e., more niches per unit area) but a higher incidence of specialized species. The increased specialization *reduces* interspecific

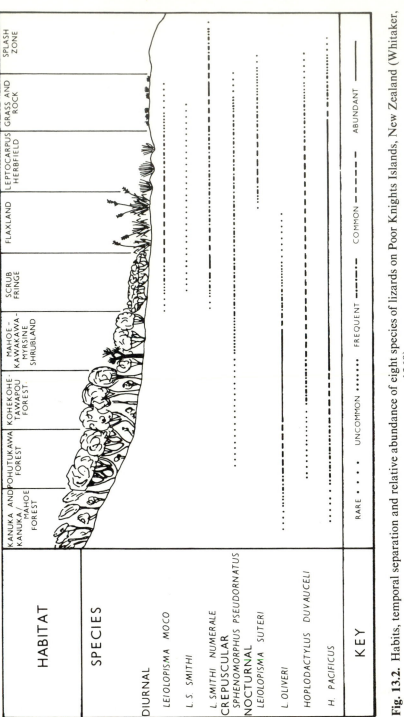

Fig. 13.2. Habits, temporal separation and relative abundance of eight species of lizards on Poor Knights Islands, New Zealand (Whitaker, 1968).

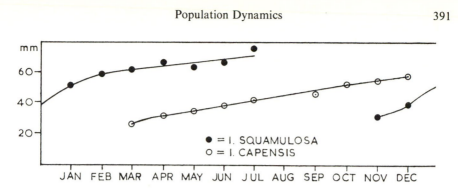

Fig. 13.3. Growth curves for two sympatric species of the lacertid lizard genus *Ichnotropis* of southern Africa. Each species is annual, and egg-laying is staggered so that adults of one species occur only with young of the other species (Broadley, 1967).

competition. Presumably the moderate annual climatic changes of the tropics, in contrast to the extreme seasons of higher latitudes, have allowed the development of a large number of specialists, each with a niche distinct from those of its relatives. In addition to the relatively equable annual weather cycle, the Pleistocene mesic and xeric weather cycles alternately fractured and united tropical rain forests, greatly accelerating speciation. Some workers believe tropical species' diversity to have resulted from "tropical productivity," but the mechanics of this proposal are unclear. Resource partitioning (i.e., lack of utilization of the same dimension of a niche) eliminates competition for the resource in question.

Professor G. Evelyn Hutchinson suggests that a ratio of 1.28:1 or more (in a linear dimension) between two species is enough to let them coexist.[474] This useful rule of thumb is applied by many students in field investigations of possible competition between two species, and seems generally applicable. Frequently the mouth width is taken as a measurement of food ingested; it indicates the maximal size of food item eaten and may be meaningful for many fish, amphibians and reptiles, which generally swallow their food whole. It seems not to apply to raptorial birds, which normally tear their prey to small pieces as they eat. That is to say, the 1.28:1 ratio in a linear measurement may permit coexistence but does not preclude competition; presumably the two species could compete for items which both were able to ingest and which were in limited supply. The 1.28:1 ratio is significant only for adults of both species; adults of a smaller species might compete with immatures of a larger species, when their sizes were close.

Population Dynamics

Having considered community structure, which relates to the number and kinds of species in an environment, we now take up the question of population structure, the number of individuals of a given species. Students of population changes, or population dynamics, measure characteristic numerical data for a species and attempt to observe and predict population trends.

Density

An important population characteristic is its *density*. A population of 1000 mice can be distributed over 1 hectare (about 2.5 acres) or 10 hectares (or any other area). On 1 hectare the same population has a density 10 times that of the population on 10 hectares. In neither example is the population necessarily distributed evenly over the area: the distribution may be somewhat clustered or contagious, or it may be random (in which event it is nevertheless not uniformly dispersed). Perhaps other statistics or parameters remain constant as this imaginary population is subjected to varying degrees of compression, or density. Density may be reflected eventually in other commonly counted or measured features, such as birth and growth rates; these are then *density-dependent* effects.

Intraspecific competition is an important function of population density. If a mouse species is a strict herbivore, a higher density will subject the vegetation to greater use. If the vegetation is abundant and if, on one hectare, it is adequate to provide food for 1000 mice, such a high density will not put pressure on the food source. On the other hand, if the vegetation on one acre is insufficient to support 1000 mice, the mice will depress the amount of fresh growth (but perhaps not the root systems). As a consequence, individual mice compete for the available supply of food. In competition, some mice will obtain enough food and others will not, an effect of, among other things, their innate variability. The effect of the mouse on the vegetation will depend on the parts of the plant eaten. Frequently voles (*Microtus* spp.) consume vegetative parts and, in feeding, destroy part of the plant growth, whereas wood mice (*Apodemus* spp.) and deer mice (*Peromyscus* spp.) tend to favor seeds of plants, and do not affect the growth or health of the plant itself.

Natality

A second prominent aspect of populations is *birth rate,* or *natality,* and there are several ways of indicating this feature. The simplest and least satisfactory is by the number of young born to the total population per unit time. Should the species produce a single brood per year, one year is a suitable time unit. But if the mouse is polyestrous, as are most small mice, a shorter interval is preferred, for it will not only produce a more sensitive measure of, but also indicate seasonal changes in, birth rate. The birth rate, however, is a function of females, and a more meaningful measure is to indicate the number of young per unit time per females in the population (or per 1000 females). This refinement is a real improvement, because sex ratios are rarely exactly 50:50. (In a parthenogenetic species, in which there is only one sex, there is no sex ratio.) Many females in a given population, however, are not sexually mature, and so birth rate can most logically be described as the number of young per unit time per unit population (e.g., 1000) of sexually mature females.

Mortality

The span of each individual's life is punctuated by death, and *death rate,* or *mortality,* is a variable aspect in many populations. All vertebrates eventually contribute to this statistic, but there may be sexual differences in longevity. An

extreme example of this is the brown antechinus *(Antechinus stuartii)*, an Australian marsupial insectivore, in which all males die within three weeks of mating.[114]
One can measure the number of individuals that die per population unit (1000,
say) per unit time; in a short-lived (annual) species one might decide to measure
death rate by months but annually for animals with greater longevity. Within any
given species, different age groups may have distinctive mortality rates: typically
species that produce larger numbers of young or ova (such as codfish, Gadidae)
suffer conspicuously high mortality in the first few weeks of life. The plaice *(Pleu-
ronectes platessa)*, a flatfish of the North Sea, lays large numbers of eggs, but
mortality in the first several months eliminates all but 1 individual in 10,000.

Death rates are sometimes expressed by the converse, *survivorship,* and survivorship curves have been calculated for some vertebrates. The painted turtle
(Chrysemys picta) of eastern North America loses a large part of its annual production early in life, in the egg stage. However, by the time the young reach a
pond (usually a few feet distant), chances for survival increase dramatically.[337]
Like many chelonians, the painted turtle is long-lived, and some individuals attain
an estimated 40 years. The tiger salamander *(Ambystoma tigrinum)* experiences
a drastic reduction in the first months of life (Fig. 13.4), but adults have a greatly
reduced mortality rate per unit time.[17] The survivorship curve should include some
indication of the age of sexual maturity (Fig. 13.5), for high mortality *prior* to
reproduction not only depresses the numbers of the generation in question but also
limits its reproductive potential, or the actual increase of the next generation.

A limitless number of physical and biologic fluctuations affect survival of all
wild vertebrates, and a natural situation is so incredibly complex that a contrived
model is but a crude approximation of nature and of limited value in predicting
changes in natural populations. Although laboratory studies serve to isolate some
distinct behavioral cause-and-effect relationships, they are almost always designed
to elicit responses to exaggerations or distortions of situations in nature. Concentrations of captive mice, for example, produce behavioral changes that result in

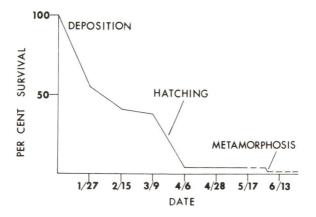

Fig. 13.4. Survival of eggs and larvae of the salamander *Ambystoma tigrinum* (Anderson
et al., 1971; courtesy of The Ecological Society of America).

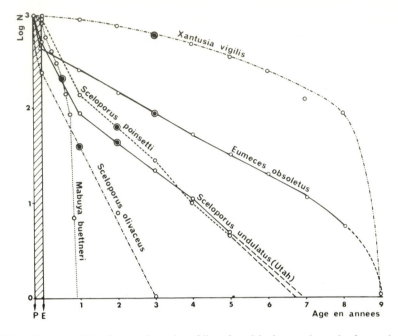

Fig. 13.5. Survivorship of several species of lizards, with the age (years) of sexual maturity (⊙) (Barbault, 1975).

altered mortality rates, but such concentrations in laboratory experiments usually greatly exceed those in natural situations and cannot be applied to most changes in mortality rates in nature. Many laboratory experiments show that various types of "stressful situations" (stressful in anthropocentric terms) result in a hypertrophy of adrenocortical tissue in mice. Voles (*Microtus* spp.) when subjected to crowding or periodic fighting show greater growth of the adrenal cortex than do unstressed individuals. However, the experimentally "stressed" voles may have adrenal cortices comparable to those of wild voles.

Birth and mortality rates act together to produce a change in density, or a *rate of increase* or *rate of decrease*. Regular seasonal reproductive cycles may be modified by environmental factors as well as by density-dependent factors. Most animals tend to vary in birth rate from year to year and from place to place.

Biomass and Carrying Capacity

Another major feature of population is *biomass,* the total mass of the population. The environmental requirements of a species vary not only with its density but also with its total weight. The concept of *carrying capacity* is closely associated with biomass. A given area of a certain habitat may provide resources adequate for a species at a certain population level, but not normally above that level. Possibly, however, a larger population could be sustained in the same area if smaller individuals—but ones constituting the same biomass—were substituted for larger ones. In other words, a certain habitat can support a certain biomass. Actually a natural situation cannot be so simply described, for voles, rabbits, deer and elephants—all herbivores—affect different parts of the vegetation.

In territorial species population densities tend to be inversely proportional to body size, because territory and home range tend to increase with body size. Carrying capacity, being a feature of the environment, changes with the productivity of the environment: the high latitudes have greater carrying capacities for most species in the summer but decrease in the winter. Carrying capacities of deserts increase with an increase in rainfall (or irrigation). Sometimes carrying capacity is expressed as the ability of a species to exploit (or fill) its environment; such a measure, however, bypasses (1) environmental variability and (2) the demands made by other species at the same trophic level.

Obviously, carrying capacity is affected not only by food but by tree-holes for many birds, squirrels and mice; by loose soil for fossorial animals; by loose rocks for the nonfossorial crevice-seekers; and by many other special aspects, not the least important of which are competitors for these factors.

The carrying capacity determines the level at which a mature (adult) population may stabilize; in a temperate latitude, the vernal prereproductive population may indicate this level. Among annually reproducing vertebrates, recruitment is restricted to the season of natality, whereas mortality tends to be more continuous. Consequently annual peaks in population density occur at the end of the season of natality. Some species of grouse (Tetraonidae) and partridge (Phasianidae) have been carefully monitored, and fluctuations are conspicuous. The partridge *(Perdix perdix)* of Scotland exhibits marked autumnal variations in density (Fig. 13.6), densities that do not reflect variations in numbers of breeding pairs in the

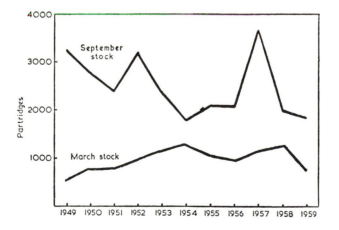

Fig. 13.6. Annual changes in estimated populations of the partridge *(Perdix perdix)* on a 1454-hectare area in West Hampshire (United Kingdom). The late summer (September) densities reflect summer weather and growing conditions and not the spring population (Blank and Ash, 1962; courtesy of Blackwell Scientific Publications).

spring but rather survival of young in June and July. Sunny, dry summers favor survival and are generally followed by high densities in autumn. It follows logically that autumnal populations fluctuate rather independently of the breeding populations.[96]

The overwintering "capital" population, that density which survives a period of restricted carrying capacity, is the population that produces the annual increment. Among species that can be followed visually, territoriality plays a role in determining vernal densities. This is apparent in some tree squirrels, beavers and many birds, which regulate their densities at the onset of reproduction.

Of all vertebrates, population densities are best known for birds. Because most avian species are diurnal and vocal, it is possible to determine the numbers of various species per unit area and to compare (1) densities of a given species in different areas, (2) densities of various species in the same habitat and (3) changes in density of a given species over time. Perhaps because bird communities are more complex (with more species) than are those of other terrestrial vertebrates, more emphasis is placed on the first two questions. The question of density fluctuations of birds over time has not received the attention that has been given to mammalian populations' cycles.

For hole-nesting birds, flying squirrels or some other cavity dwellers, the abundance of holes is an essential determinant of carrying capacity. In northern Europe the pied flycatcher *(Ficedula hypoleuca)* and the great tit *(Parus major)* nest in holes; when provided with nest boxes, their breeding populations increase (Fig. 13.7). Similar increases have been noted for other hole-nesting species, from tree-nesting ducks to small falcons. The increase in nesting pairs, however, does not exactly follow an increase in artificial nest boxes, for, if the sites are too close, territoriality may affect spacing of nesting pairs (Fig. 13.8).

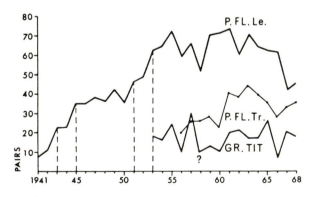

Fig. 13.7. Changes in density of nesting pairs of the pied flycatcher *(Ficedula hypoleuca)* and the great tit *(Parus major)* at Lemsjöholm (Le., heavy lines) and Träskvik (Tr., light line), Finland. Vertical broken lines indicate dates when nest boxes were increased at Lemsjöholm (von Haartman, 1971; courtesy of Academic Press).

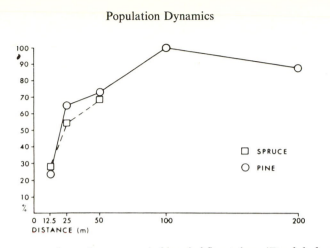

Fig. 13.8. Percentage of nest boxes occupied by pied flycatchers *(Ficedula hypoleuca)* at different spacings of nest boxes in Hyytiälä, south Finland (von Haartman, 1971; courtesy of Academic Press).

Age Structure

The relationship of biomass to density introduces another essential feature of population, that of *age structure,* or the relative abundance of different age groups within a population. The proportion of individuals of prereproductive and reproductive ages will affect the birth rate of the population. Also, because young individuals grow more rapidly than do older animals, a population of the former exhibits a more rapid increase in biomass.

A population of 1000 mature, old halibut *(Hippoglossus stenolepis)* will have a much greater biomass than will the same number of young halibut. They will differ in their demands on the environment as well as in their population characteristics. In a human population of 5-year-olds, the demand for food will be much less than in an equal number of people of 20 years of age. This difference reflects biomass, among other factors (the nutritional requirements not being related *solely* to mass).

A dramatic illustration of the significance of age structure to density is seen in the early history of the Pacific halibut fishery. When commercial fishing started in 1910, the population was mature, consisting mostly of large fish. The annual harvest had been measured by catch-per-unit-effort, and this statistic dropped from 266 pounds in 1915 to 95 pounds in 1925; during this period the percentage of small fish (12 pounds or less) increased. Thus the age structure changed from one of a small number of large fish to a large number of small (but sexually mature) fish. More rapid growth rates characterize smaller fish, and the annual increment in total population (expressed as a percentage of the average population) changed from 2.2 (1918–1923) to 14.8 (1931–1937).[137] The sea forms an immense habitat and the Pacific halibut population must consist of many separate populations which might be affected by unknown factors in addition to age structure.

Biomass reflects the relationships of growth rate and population density. When populations are maximal, maximal growth may not be achieved; the mean size of the adults may not reach that characteristic for the species, although the biomass may be at the carrying capacity for that environment.

These features are peculiar to populations but not to individuals. A population is described in terms of its birth rate, or natality; death rate, or mortality; density; biomass; and age structure. The increase or decrease in density is a function of all these elements.

Density Fluctuations

A thorough grasp of the structure of communities and interactions (inter- and intraspecific) should enhance our understanding of changes in population density. In an evaluation of the relationships among vertebrate prey species and between them and their predators, the information should suggest the relative roles of (1) community structure, (2) density-dependent (intraspecific) factors and (3) non-biologic extrinsic factors in density changes. These kinds of data have been incorporated in mathematical models, but a computer model must be tested to verify its similarity to reality. One difficulty lies in the dilemma of creating a model sufficiently general to fit theory yet specific enough to fit reality.

A major feature of populations is their change in density with time: density changes in vertebrates can be annual, cyclic but not annual, or sporadic. Various regulatory mechanisms control the upper limit of densities, but there is no completely effective mechanism for the lower limit, for populations and all members of a species sometimes reach a density of zero, or extinction.

Population fluctuations have been observed since the beginning of recorded history, and in this century many serious studies have homed in on the possible causes of major changes in population density. Some of the more conspicuous fluctuations have been observed in fishes and in small mammals, especially microtine rodents (Microtinae; voles and lemmings). These species dwell in high latitudes, frequently in relatively simple boreal or Arctic communities. The background of fluctuations of fishes is often ill-understood and presumably complex.

Density increase is a natural expression of a species' reproductive potential. Any factor that removes even a single animal is regulatory to some degree, i.e., it slows density increase. If a parasite or a predator selectively removes immature individuals, such regulation becomes especially effective, because not only are immature individuals about to enter their reproductive epoch but they are also the major element in biomass increase.

Reproductive Potential. Reproductive capacity (or potential) indicates a species' ability to increase its numbers. Factors influencing the reproductive capacity include survivorship, the age of sexual maturity, the number of young per brood (or eggs per clutch) and the number of broods per year.

Generally species with a relatively short life span have a rather high reproductive capacity and are demographically volatile. Among small rodents, microtines are most likely to experience extremes of density, whereas the numbers of cricetine mice and heteromyid mice are increasingly stable (Fig. 13.9).[315] This relationship between reproductive capacity and survival (an adaptive feature) is modified by food and its effect on growth and ovulation. Voles tend to be herbivores, which specialize in vegetative tissues and, as we have seen, as grasses and forbs mature and dry, voles cease to ovulate. Cricetine and heteromyid rodents are more generalized species, which frequently feed on insects and which tend to have more regular reproductive rates. The negative correlation between survival rate and reproductive capacity, however, does not obtain in some insectivores: various species of shrews (*Sorex* spp.) are annual species—often with single broods not greater in size than found in sympatric species of microtine rodents. Shrews, in

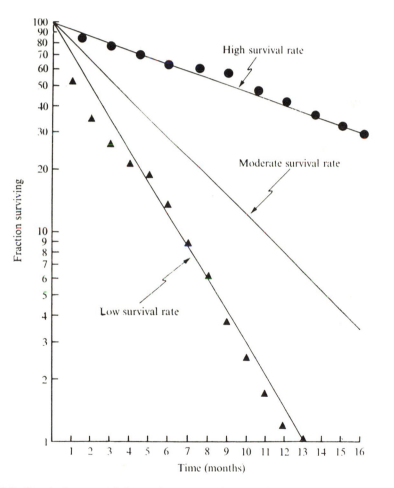

Fig. 13.9. Survival curves of the pocket mouse *(Perognathus formosus)* (●———●), the white-footed mouse *(Peromyscus leucopus)* (———), and the bank vole *(Clethrionomys glareolus)* (◄———►), plotted on semilog scale. Generally those species with greater survivorship exhibit lower reproductive capacity (French et al., 1975; courtesy of Cambridge University Press).

contrast to voles, have relatively few predators, and shrew populations disappear rather suddenly after reproducing once, whereas voles are highly palatable and subject to continuous predation.

Among annually reproducing vertebrates, recruitment (of new individuals) is restricted to the season of natality, whereas mortality tends to be more continuous. Consequently annual peaks in population density occur following the season of natality. A field population of the prairie vole *(Microtus ochrogaster),* when its diet was supplemented with commercial rabbit food, had faster growth, larger litters and higher densities than a control population, but autumnal populations ultimately declined to the same level (Fig. 13.10).[177]

In Finnish Lapland the vole, *Clethrionomys rufocanus,* breeds from May to September. Females born early in the year grow rapidly to sexual maturity in the same summer, and may themselves produce two or three broods by the end of their first summer. Such early fecundity, however, seems to be density-dependent, and decreases with increases in population density.[521] Isolated populations of mice may be distinctive in times of breeding. The Eurasian bank vole *(Clethrionomys glareolus)* on the British mainland has a prolonged season of sexual activity, extending from February to November or even through the entire winter, whereas

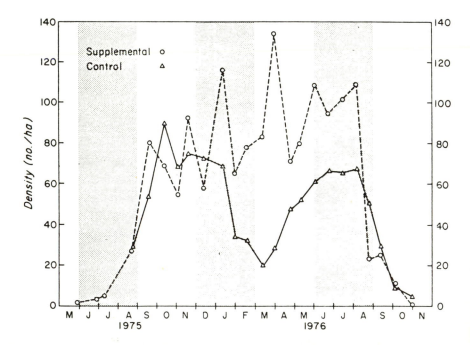

Fig. 13.10. Minimal monthly densities of the prairie vole *(Microtus ochrogaster)* with and without supplemental food (commercial rabbit food). Summer and winter quarters are shaded (Cole and Batzli, 1978).

on the offshore island of Skomer the same species has a clearly shorter season (Fig. 13.11). The insular forms are distinctive also in attaining a larger adult size and in having no sexually mature young in the year of their birth.[501] The apparently delayed sexual maturity of the island voles may reflect their seasonally late birth; juveniles appear a month later than in mainland populations and when photoperiod is not stimulating. Because the mean litter sizes on the mainland and Skomer populations are essentially the same, recruitment on the insular populations must be low, pointing to a low mortality rate. In voles, in which early sexual maturity and postpartum pregnancies are consonant with rapid increase in numbers, removal of the young places a major obstacle to increase in density.

The marked three- to four-year cycles of microtine rodents are not characterized by an increase in litter size but are frequently associated with the length of the breeding seasons.[393] When reproductive activity begins early in the year (mid-

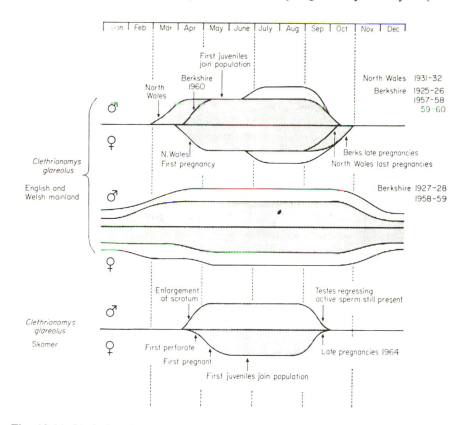

Fig. 13.11. Variations in reproductive seasons and age of sexual maturity in the bank vole *(Clethrionomys glareolus)*. Shaded areas indicate the breeding season, and unshaded areas the times that recently born young enter the breeding population. On the island of Skomer young become sexually mature only after overwintering (Jewell, 1966; courtesy of The Zoological Society of London).

or late winter), young of the year become sexually active at a few weeks of age, and overall recruitment is very high. At the risk of oversimplifying a complex situation, it appears that the increase in densities results from an extended breeding season with high recruitment and high survival of young. The relationship between extended reproductive seasons and increase in density is also well-documented for wild rabbits.[88] The rapid decline from population highs is marked not so much by an increase in mortality as by a reduction in recruitment.

Populations of vertebrates on small islands are unlike comparable mainland species in several ways. Insular populations of both small lizards and mice tend to be larger, have smaller clutch or brood sizes and are sometimes much more abundant than related taxa on the adjacent mainland. In addition the insular populations tend to be stable in density, and frequently small islands lack mammalian predators; the lack of population cycles on islands may be due to the absence of such predators as weasels and small cats.

Density-Dependent Fluctuations. Stated most simply, population density depends on the innate biologic features of the species, the regular and irregular fluctuations of the environment and the interactions between the species and other faunal elements. The density at any given time is a result of mortality, natality, immigration and emigration.

Some aspects of population behavior are positively or negatively correlated with density levels. This generality is important to students who attempt to determine the cause of density changes, for some causes include factors triggered by crowding. In nature one cannot easily separate fluctuating physical factors of the environment from strictly density-dependent biologic factors. For this reason most of the investigations have been either laboratory trials or observations under controlled field conditions.

Among a number of species of amphibians and fishes, crowding also inhibits reproduction; substances released in water retard sexual activity in others of the same species. Water from a crowded environment introduced into an aquarium inhibits reproduction in individuals even when they are not crowded. This effect extends across class boundaries: reproduction in the guppy *(Poecilia reticulata)* is inhibited by water from ponds where tadpoles of bullfrogs *(Rana catesbeiana)* were concentrated.[112]

There are many manifestations of "stress" in laboratory mammals, but stress in field populations is difficult to identify. The mechanism by which stress may regulate natural populations is nevertheless obscure. Adrenal hypertrophy, which is frequently associated with stress, is also associated with lowered resistance to disease. Also, when cages are moved or animals are handled, such stress in the laboratory rat is followed in several minutes by a three- to eight-fold increase in prolactin levels.[279] This phenomenon in wild rodents could strengthen the placental–uterine bond and thus enhance reproductive performance. In female laboratory rats, disturbance (= "stress") in proestrus is rapidly followed by greatly increased levels of luteinizing hormone, a relationship likely to disrupt the normal estrous cycle.[753] Thus, it is most difficult to assign a clear-cut role to stress in the regulation of cyclic densities.

Voles are relatively intolerant of one another, and the tendency toward fighting increases with density or crowding. This has led some workers to suspect that losses (mortality or emigration) are selective (nonrandom) during the decline in vole populations, and also that the sudden increase in mortality reflects behavioral responses to crowding. The surviving individuals, according to this "polymorphic behavior hypothesis" will be those that are most aggressive and most prone to be widely spaced from one another. The less aggressive will emigrate. The thesis is plausible but the evidence equivocal. Although emigrations are a feature of some irruptions in lemming populations, there is no obvious emigration in other species; nor is it apparent that the lemming emigrants are less or more aggressive than the individuals that do not emigrate.

Some students have suggested that dispersal affects the nature of population trends in cyclic rodents, and that dispersal increases as the population approaches the carrying capacity of the environment. This type of dispersal is *saturation dispersal,* and may differ from dispersal at lower densities, or *presaturation dispersal,* which is a characteristic of immature or young adults.[597] It is theorized that saturation dispersal reduces pressure on the species niche as some individuals move into less desirable habitats. Possibly these two sorts of dispersal grade into each other. Evidence points to an increase in mortality among dispersing over nondispersing animals.[599] It is logical that individuals that leave familiar cover are subjected to greater predation, and this is well established for muskrats *(Ondatra zibethica)* and some small mice.

Experimental evidence suggests that dispersal may retard population growth. Populations confined to small islands or artificial enclosures frequently have rapid initial increases in density but later become stable. The absence of dispersal (or emigration) may not be the only factor favoring rapid population growth, for confinement of the population also excludes mammalian predators. Inasmuch as there is a preponderance of immature and subadult individuals (those about to become sexually mature or those reproducing for the first time) among emigrants, dispersal can reduce rapid increase of biomass, density and reproductive capacity in the parent population.[501] The movement of dispersing individuals into an established population may not occur in nature, but experimentally such invading animals experience difficulty in invading the existing social hierarchy.

The red squirrel *(Tamiasciurus hudsonicus)* ranges across the coniferous forests of North America, subsisting largely on the seeds of spruce (*Picea* spp.) and pines (*Pinus* spp.). The mast from these trees varies annually, but populations of this squirrel remain at relatively stable densities. Territorial defense seems to preserve numbers sufficiently sparse so that in most crop years food scarcity does not become a limiting factor. No doubt mortality is high in dispersing young.[871] Densities are greatest in spruce forests, which produce the most food (Fig. 13.12). In the Nearctic gray squirrel *(Sciurus carolinensis),* births are adequate to cause an increase in density, but a dispersal of young individuals preserves a stability in population.[1007]

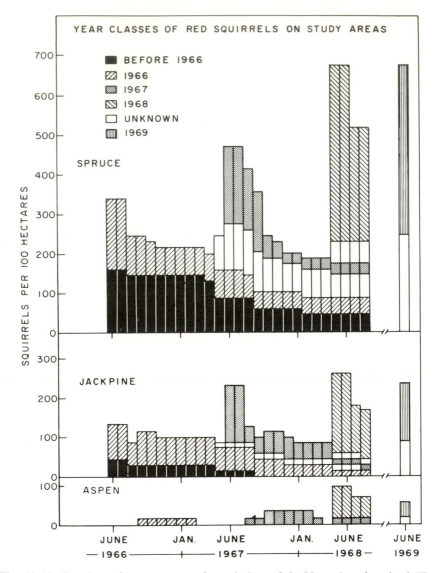

Fig. 13.12. Density and age structure of populations of the Nearctic red squirrel *(Tamiasciurus hudsonicus)* in three forest types. Density seems to depend on the amount of food produced by the dominant tree (Rusch and Reeder, 1978, copyright Ecological Society of America).

Predation. Causes for decreasing density are usually obscure. In the first place, the time at which density ceases to increase and begins to decrease is seldom recognized until it is past. In a weekly or monthly monitored population, small variations in size of censused samples tend to obscure initial changes or trends. Second, because there is frequently more than one possible, plausible or operative factor involved, a single cause is not immediately apparent.

Some of the causes that retard the population growth may also cause population decline, and predation is an example. Predation is sometimes severe during the declining phase of a cycle, for predators may reach maximal abundance just as the prey species is declining and has reached especially vulnerable densities.

As populations of herbivores increase, so do numbers of predators. In some instances there are larger broods of predators in years of prey abundance, and raptorial birds seldom fail to appear when small rodents reach peak populations. Characteristically predator populations peak after the herbivores have reached maximal densities, and this situation casts doubt on the effectiveness of predators in limiting prey populations.

The pressure of a carnivore on numbers of its prey can be described as "limiting," "controlling" or "regulating," without a clear definition being given of these words. When predators are removed artificially by man, prey populations usually increase, suggesting that previously predators had a role in depressing numbers of their prey.

The role of the predator is to capture its prey and, while predators may not directly *limit* maximal numbers of their prey, every individual prey removed has an effect on the prey population. The age of the prey removed is a critical factor: an immature female taken from the population removes, in theory, its potential offspring as well, whereas the death of a male or an aged or diseased individual has less of an impact on the total population.

In the Kaibab Plateau of southwestern North America, removal of large carnivores, especially the mountain lion *(Felis concolor)*, was followed by a great increase of deer *(Odocoileus hemionus)*. Prior to predator control, the role of carnivores was apparently subtle. In cyclic species, because predator populations reach maximal numbers *after* prey populations have begun to decline, predators can be especially effective in depressing the declining prey populations. In Australia, when droughts drastically reduce numbers of the European rabbit *(Oryctolagus cuniculus)*, the surviving individuals concentrate where they can find pockets of green vegetation; there predators take substantial numbers of the remaining rabbits.[536] In contrast, in the Timbavati Private Nature Preserve near Kruger National Park in southern Africa, mammalian carnivores preyed most heavily on young and half-grown ungulates, and more mature individuals were more capable of avoiding predators. Despite the numbers of young ungulates taken by lions, leopards and other large predators, starvation also accounted for the deaths of many mature ungulates.[437] Therefore predation did not limit population density in this instance but presumably did alter the age distribution.

On the north coastal plain of Alaska, the brown lemming *(Lemmus sibiricus)* was at a cyclical low in the spring of 1951, and its main avian predators were scarce or absent. Snowy owls *(Nyctea scandiaca)*, for example, were scarce and did not breed that year. In 1952 and 1953, as brown lemmings became abundant, jeagers (*Stercorarius* spp.) and owls appeared, nested abundantly and preyed heavily on the lemmings. Also foxes (*Vulpes fulva* and *Alopex lagopus*), virtually absent in 1951, arrived in small numbers in 1952 and 1953 and were common in 1954. Apparently the prey limited the predator population rather than the reverse.[823]

The number of voles *(Microtus californicus)* taken by mammalian carnivores in the San Francisco Bay Area (determined by recovery of skull fragments in scats) fluctuated with but lagged behind a peak in vole density (Fig. 13.13). Mammalian predation amounted to 88 percent of the mortalities in a peak population of voles. When voles declined, the carnivores fed on other rodents, and the carnivore density remained unchanged.[799] The role of avian predators in this cycle was not assessed, but presumably hawks and owls occurred. Raptorial birds are often present when vole densities are high; being active both day and night, voles are vulnerable to both hawks and owls in addition to mammalian carnivores.

There is a tendency for predators to suppress densities of prey species so that intraspecific competition is delayed. Because prey species (e.g., microtine rodents) frequently are more fecund than are most predators, the increase in predators lags behind that of its prey, and predation is maximal after the prey species has passed its peak abundance.

Diseases and Parasites. The possible role of diseases and parasites does not receive the attention it probably deserves. Pathogenic transmission increases with the density of the host, and many diseases are known to decimate populations of wild vertebrates. Theoretically a well-adjusted parasite does not destroy its host species lest it destroy its own livelihood. Naturally this is a difficult concept to apply to the individual host, say, someone dying of malaria. Another real diffi-

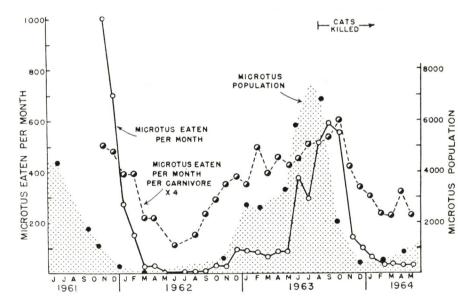

Fig. 13.13. Predation by small mammalian carnivores on voles *(Microtus californicus)* during a three-year cycle in the San Francisco Bay area, California, on a 14-hectare plot. To obtain the number of voles eaten per carnivore per month, divide the left-hand scale by 4 (Pearson, 1966; courtesy of Blackwell Scientific Publications, Ltd.).

culty in studying the possible role of diseases and parasites in population cycles is that few students who can recognize diseases and parasites in wild animals also study vertebrate populations. Although diseases of wild vertebrates are best studied when they appear in human populations, some are known to reduce numbers of wild vertebrates. Paramount among these diseases are plague, tularemia, various species of malaria (including those in birds and reptiles) and several rickettsial diseases.

Changes in densities of populations are clearly the effect of several interactions. Among the more important are (1) density-dependent reproductive aspects, (2) environmentally induced changes in reproduction, (3) age structure and (4) marked seasonal or irregular alterations in carrying capacity. Predation always has some depressing effect; this undoubtedly retards, at least slightly, the rate of increase, and may at times accelerate the rate of decrease of prey populations. Also the increase in prey populations seems invariably to attract predators, and prey abundance may cause an increase in reproduction of predators.

The past 40 years have seen an increase in the number of students following population phenomena, and the analysis of population dynamics has been enhanced by the vigor of observations on both natural populations and carefully manipulated or experimental situations. The study of population changes has not suffered from a lack of innovative thinking on the subject. It is nevertheless apparent that there is as yet no all-inclusive pattern describing density changes, even with a single species. Perhaps patterns of population dynamics are as diverse as the species that manifest conspicuous changes in density.

Suggested Readings

Berry RJ, Southern HN (1970) Variation in mammalian populations. Zoological Society of London No. 26

Elton CS (1942) Voles, mice and lemmings: problems in population dynamics. Oxford University Press, London

Golley FG, Petrusewicz K, Ryskowski L (eds) (1975) Small mammals: their productivity and population dynamics. International Biological Programme 5. Cambridge University Press, London

Grant PR (1972) Interspecific competition among rodents. Ann Rev Ecol Systemat 3:79–106

Lack D (1954) The natural regulation of animal numbers. Clarendon Press, Oxford, England

Lack D (1966) Population studies of birds. Clarendon Press, Oxford, England

Le Cren ED, Holdgate ML (eds) (1962) The exploitation of natural animal populations. British Ecological Society Symposium No. 2. Wiley, New York

Pianka ER (1977) Reptilian species diversity. In: Gans C (ed) Biology of the reptilia. Academic, New York

Snyder DP (1978) Populations of small mammals under natural conditions. University of Pennsylvania, Pymatuning Laboratory of Ecology, Special Publication Series, 5

Vandermeer JH (1972) Niche theory. Ann Rev Ecol Systemat 3:107–132

Weatherley AH (1972) Growth and ecology of fish populations. Academic, New York

References

1. Ackman RB, Burgher RC (1963) Component fatty acids of the milk of the Grey (Atlantic) seal. Can J Biochem Physiol 41: 2501–2505
2. Airapetyants ES, Konstantinov AI (1970) Echolocation in animals. Israel Program of Scientific Translations, Jerusalem
3. Alcala A, Brown WC (1967) Population ecology of the tropical scincoid lizard, *Emoia atrocostata,* in the Philippines. Copeia 596–604
4. Aleksiuk M (1968) Scent-mound communication, territoriality, and population regulation in beaver (*Castor canadensis* Kuhl). J Mamm 49: 759–762
5. Aleksiuk M (1977) Cold-induced aggregative behavior in the red-sided garter snake. Herpetol 33: 98–101
6. Aleksiuk M, Gregory P (1974) Regulation of seasonal mating behavior in *Thamnophis sirtalis parietalis.* Copeia: 681–689
7. Aleksiuk M, Stewart KW (1971) Seasonal changes in the body composition of the garter snake *(Thamnophis sirtalis parietalis)* at northern latitudes. Ecology 52: 485–490
8. Alexander NJ, and Fahrenbach WH (1969) The dermal chromatophores of *Anolis carolinensis* (Reptilia, Iguanidae). Am J Anat 126: 41–56
9. Ali MA (1975) Vision in fishes; new approaches in research. Plenum, New York
10. Amadon, D. (1950) The Hawaiian honeycreepers (Aves, Drepaniidae). Bull. Amer. Mus. Nat. Hist., 95: 151–262
11. Amoroso EC (1959) The biology of the placenta. Fifth conference on gestation. In: Villee CA (ed) Gestation. Josiah Macy, Jr. Foundation, New York
12. Amoroso EC (1960) Viviparity in fishes. Symposium. Zool Soc London 1: 153–181
13. Amoroso EC, Matthews LH (1952) Reproduction and lactation in the seal. Rept. Second International Congress of the Physiology and Pathology of Animal Reproduction and of Artificial Insemination. Copenhagen
14. Andersen DC, Armitage KB, Hoffmann RS (1976) Socioecology of marmots: female reproductive strategies. Ecology 57: 552–560
15. Anderson FS (1948) Contributions to the biology of the ruff (*Philomachus pugnax* L.). Dansk Ornithol For Tids 42: 125–148
16. Anderson HT (1966) Physiological adaptations in diving vertebrates. Physiol Rev 46: 212–243
17. Anderson JD, Hassinger DD, Dalrymple GH (1971) Natural mortality of eggs and larvae of *Ambystoma t. tigrinum.* Ecology 52: 1107–1112

18. Andraesson, S. (1973) Seasonal changes in diel activity of Cottus poecilopus and C. gobio (Pisces) at the Arctic Circle. Oikos 24: 16–23
19. Aoki I, Kuroki T (1975) Alarm reaction of three Japanese cyprinid fishes, *Tribolodon hakonensis, Gnathopogon elongatus elongatus* and *Rhodeus ocellatus ocellatus*. Bull Jpn Soc Fish 41: 507–513
20. Arata AW, Vaughn JB, Thomas ME (1967) Food habits of certain Colombian bats. J Mamm 48: 653–655
21. Archer AL, Glen RM (1969) Observation on the behavior of two species of honey-guides *Indicator variegatus* (Lesson) and *Indicator exilis* (Cassin). Los Angeles County Museum Contributions in Science No. 160
22. Arnott HJ, Best ACG, Ito S, et al (1974) Studies on the eyes of catfishes with special reference to the tapetum lucidum. Proc R Soc London B 186: 13–36
23. Aschoff J (1960) Exogenous and endogenous in circadian rhythms. Cold Spring Harbor Symposium Quantitative Biology 25: 11
24. Aschoff J, von Holst D (1960) Schlafplatzflüge der Dohle, *Corvus monedula* L. Proceedings of the 12th International Ornithological Congress. Helsinki 55
25. Aschoff J, Wever R (1962) Beginn und Ende der Täglichen Aktivität freilebender Vögel. J Ornithol 103: 2–27
26. Ashworth US, Ramaiah GD, Keyes MC (1966) Species differences in the composition of milk with special reference to the northern fur seal. J Dairy Sci 49: 1206–1211
27. Asplund KK, Love CH (1964) Reproduction cycles as controlled by temperature and light in the iguanid lizards *Urosaurus ornatus* and *Uta stansburiana* in southeastern Arizona. J Morphol 115: 27–34
28. Assenmacher I, Tixier-Vidal A, Astier H (1965) Effets de la sous-alimentation et du jeûne sur la gonadostimulation du canard. Ann Endocrinol (Paris) 26: 1–26
29. Atz JW (1964) Intersexuality in fishes. In: Armstrong CN, Marshall AJ (eds) Intersexuality in vertebrates including man. Academic, New York, pp 145–232
30. Augee ML, Ealey EHM (1968) Torpor in the echidna, *Tachyglossus aculeatus*. J Mamm 49: 446–454
31. Axelrod DL (1960) The evolution of flowering plants. In: The evolution of Life I. University of Chicago, pp 227–305
32. Axelrod DI (1970) Mesozoic paleogeography and early angiosperm history. Bot Rev 36: 277–319
33. Axelrod DI (1972) Ocean-floor spreading in relation to ecosystematic problems In: Allen RT, Jemis FC (eds) A symposium on ecosystematics. University of Arkansas Museum Occasional Paper No. 4: 15–76
34. Axelrod DI (1976) History of the coniferous forests, California and Nevada. University of California Publications in Botany. Vol. 70
35. Axelrod DJ (1979) Age and origin of Sonoran desert vegetation. Occasional Papers of the California Academy of Sciences, No. 132
36. Axelrod DJ, Raven P (1972) Evolutionary biogeography viewed from plate tectonic theory. In: Behnke JA (ed) Challenging biological problems: directions toward their solutions. Oxford University Press, Oxford, England, pp 218–236
37. Backus RH (1958) Sound production by marine animals. J Underwater Acoust 8: 191–202
38. Bailey JA (1967) Rate of food passage of caged cottontails. J Mamm 49: 340–342
39. Baker JR (1938) The relation between latitude and breeding season in birds. Proc Zool Soc London 108: 557
40. Baker JR (1947) The seasons in a tropical rain forest. Part B. Lizards (Emoia). J Linnean Soc London (Zool) 41: 243–247
41. Baker JR, Marshall AJ, Harrison TH (1940) The seasons in a tropical rain forest (New Hebrides). Part 5. Birds (Pachycephalus). J Linnean Soc London (Zool) 41: 50–70
42. Baker WW, Marshall SG, Baker VB (1968) Autumn fat deposition in the evening bat *(Nycticeius humeralis)*. J Mamm 49: 314–317

43. Baker-Dittus AM (1978) Foraging patterns of three sympatric killifish. Copeia: 383–389
44. Bakko EB, Brown LN (1967) Breeding biology of the white-tailed prairie dog, *Cynomys leucurus*, in Wyoming. J Mamm 48: 100–112
45. Balinsky JB, Choritz EL, Coe CGL, et al (1967) Amino acid metabolism and urea synthesis in naturally aestivating *Xenopus laevis*. Comp Biochem Physiol 22: 59–68
46. Ballinger RE, Marion KR, Sexton OJ (1970) Thermal ecology of the lizard, *Anolis limifrons*, with comparative notes on three additional Panamanian anoles. Ecology 51: 246–254
47. Balsano JS, Darnell RM, Abramoff P (1972) Electrophoretic evidence of triploidy associated with populations of the gynogenetic teleost *Poecilia formosa*. Copeia: 292–297
48. Bancroft TL (1918) Some further notes on the life-history of *Ceratodus (Neoceratodus) forsteri*. Proc R Soc Queensland 30: 91–94
49. Bang BG (1968) Olfaction in Rallidae (Gruiformes), a morphological study of thirteen species. J Zool Soc London 156: 97–107
50. Banner AH, Helfrich P, Piyakarnchana T So (1966) Retention of ciguatera toxin by the red snapper, *Lutjanus bohar*. Copeia: 297–301
51. Barash DP (1974) The evolution of marmot societies: a general theory. Science 185: 415–420
52. Barbault R (1975) Dynamique des populations de lézards. Bull d'Ecol 6: 1–22
53. Barbault R (1976) Population dynamics and reproductive patterns of three African skinks. Copeia: 483–490
54. Berdach JE, Winn HE, Menzel DW (1959) The role of the senses in the feeding of the nocturnal reef predators, *Gymnothorax moringa* and *G. vicinus*. Copeia: 133–139
55. Bardach JE, Case J (1965) Sensory capabilities of the modified fins of squirrel hake *(Urophycis chuss)* and searobins (*Prionotus carolinus* and *P. evolans*). Copeia: 194–206
56. Barnes AT, Case JF (1974) The luminescence of lanternfish (Myctophidae); spontaneous activity and responses to mechanical, electrical, and chemical stimulation. J Exp Marine Biol Ecol 15: 203–221
57. Barnett SA, Mount LE (1967) Resistance to cold in mammals. In: Rose AH (ed) Thermobiology. Academic, New York
58. Barrett R, Maderson PFA, Meszler RM (1970) The pit organs of snakes. In: Gans C, Persons TS (eds) Biology of the Reptilia. Morphology B 2: 277–300
59. Barry TW (1962) Effect of late seasons on Atlantic brant reproduction. J Wildlife Management 26: 19–26
60. Barthelemy H (1929) Recherches biométriques et expérimentales sur l'hibernation, la maturation et la surmaturation de la grenouille rousse ♀ *(Rana fusca)*. CR Acad Sci Paris. T. 182: 1653–1654
61. Bartholomew GA (1966) A field study of temperature relations in the Galapagos marine iguana. Copeia: 241–250
62. Bartholomew GA, Tucker VA (1963) Control of changes in body temperature, metabolism, and circulation by the agamid lizard, *Amphibolurus barbatus*. Physiol Zool 36: 199–218
63. Bartholomew GA, Leitner P, Nelson JE (1964) Body temperature, oxygen consumption, and heart rate in three species of Australian flying foxes. Physiol Zool 37: 179–198
64. Bartholomew GA, Howell RT, Cade TJ (1957) Torpidity in the white-throated swift, Anna hummingbird, and the poorwill. Condor 59: 145–155
65. Bartholomew GA, Hudson JW (1960) Aestivation in the Mohave ground squirrel, *Citellus mohavensis*. In Mammalian Hibernation. Bull Museum of Comparative Zoology at Harvard College 124: 193–208
66. Barthalmus GT, Bellis ED (1969) Homing in the northern dusky salamander, *Desmognathus fuscus fuscus* (Rafinesque). Copeia: 148–153
67. Barwick RE (1959) Life history of the common New Zealand skink, *Leiolopisma zelandica*. Trans R Soc NZ 86: 331–380
68. Batts BS (1961) Intertidal fishes as food of the common garter snake. Copeia: 350–351
69. Batzli GO (1975) The role of small mammals in arctic ecosystems. In: Small mammals: their productivity and population dynamics. International Biological Programme 5. Cambridge University Press, Cambridge, England, pp 243–268

70. Batzli GO, Pitelka F (1975) Vole cycles: test of another hypothesis. Am Natur 109: 482–487
71. Bay EC (1966) Adaption studies with the Argentine pearl fish, *Cynolebias bellottii,* for its introduction into California. Copeia: 839–846
72. Beauchamp GK, Doty RL, Moulton DG, et al (1976). The pheromone concept in mammalian chemical communication: a critique. In: Doty RL (ed) Mammalian olfaction, reproductive processes, and behavior. Academic, New York, pp 143–160
73. Beck AJ, Lim BL (1973) Reproductive biology of *Eonycteris spelaea* Dobson (Megachiroptera) in West Malaysia. Acta Tropica 30: 251–260
74. Becker HE, Cone RA (1966) Light-stimulated electric responses from skin. Science 154: 1051–1053
75. Bedárd J (1967) Coexistence, coevolution and convergent evolution in seabird communities: a comment. Ecology 57: 177–184
76. Bee JW, Hall RE (1956) Mammals of northern Alaska on the Arctic Slope. University of Kansas Museum of Natural History, Miscellaneous Publication No. 8
77. Behrens DW (1977) Fecundity and reproduction of the viviparous perches *Hypsurus caryi* (Agassiz) and *Embiotoca jacksoni* Agassiz. Calif Fish and Game 63: 234–252
78. Belkin DA (1964) Variations in heart rate during voluntary diving in the turtle, *Pseudemys concinna.* Copeia: 321–330
79. Benedict FG, Lee RC (1938) Hibernation and marmot physiology. Carnegie Institute of Washington Publication No. 499, Waverly Press, Baltimore
80. Benes ES (1969) Behavioral evidence for color discrimination by the whiptail lizard, *Cnemidophorus tigris.* Copeia: 707–722
81. Bennett MVL (1968) Neural control of electric organs. In: Ingle D (ed) The central nervous system and fish behavior. University of Chicago, Chicago, pp 147–169
82. Bennett MVL (1971) Electric organs. In: Hoar WS, Randall BS (eds) Fish physiology. Academic, New York, Vol V, pp 347–491
83. Bennett MVL (1971) Electroreception. In: Hoar WS, Randall BS (eds) Fish physiology. Academic, New York, Vol V, pp 493–574
84. Ben-Shaul, DM (1962) The composition of the milk of wild animals. Int Zool Yearbook 4: 333–342
85. Bentley PJ, Lee AK, Main AR (1958) Comparison of dehydration and hydration of two genera of frogs (*Heleioporus* and *Neobatrachus*) that live in areas of varying humidity. J Exp Biol 35: 677–684
86. Bentley PJ, Knut Schmidt-Nielsen (1966) Cutaneous water loss in reptiles. Science 151: 1547–1549
87. Berger BJ, Sanders EH, Gardner PD, Negus NC (1977) Phenolic plant compounds as reproductive inhibitors in *Microtus montanus.* Science 195: 575–577
88. Bergerud AT (1967) The distribution and abundance of arctic hares in Newfoundland. Can Field Nat 81: 242–248
89. Berry PY, Lim GS (1967) The breeding pattern of the puff-faced water snake, *Homalopsis buccata* Boulenger. Copeia 307–313
90. Betz TW (1963) The gross ovarian morphology of the diamond-backed water snake, *Natrix rhombifera,* during the reproductive cycle. Copeia: 692–697
91. Beverton RJH, Holt SJ (1959) A review of the lifespans and mortality rates of fish in nature, and their relation to growth and other physiological characteristics. In: Wolstenholme GEW, O'Connor M (eds) Ciba foundation colloquia on aging. The lifespan of animals. Churchill, London, Vol 5, pp 142–180
92. Bishop SC (1941) The salamanders of New York. New York State Museum Bull 324
93. Blair WF (1960) The rusty lizard. University of Texas Press, Austin
94. Blake, BH (1972) The annual cycle and fat storage in two populations of golden-mantled squirrels. J Mamm 53: 157–167
95. Blanc F (1970) Le cycle reproducteur chez la femelle de *Chamaeleo lateralis.* Grey, 1831. Tananarive Malagasy Republic Universite de Madagascar Annales, Sev. Sci. de la Nature et mathematiques 7: 345–358

96. Blank TH, Ash JS (1962) Fluctuations in a partridge population, In: LeCren ED, Holdgate ML (eds) The exploitation of natural animal populations. British Ecological Society Symposium No. 2. Wiley, New York, pp 118–133

97. Blaylock, LA, Ruibal R, Platt-Aloia K (1976) Skin structure and wiping behavior of phyllomedusine frogs. Copeia 283–295

98. Blazka P (1958) The anaerobic metabolism of fish. Physiol Zool 31: 117–128

99. Blem, CR (1976) Patterns of lipid storage and utilization in birds. Am Zool 16: 671–684

100. Bligh J (1973) Temperature regulation in mammals and other vertebrates. North Holland Research Monographs, Frontiers of Biology, 30. North Holland, Amsterdam

101. Bligh J, Moore RE (ed) (1972) Essays on temperature regulation. North Holland, Amsterdam

102. Blumenstock DI (1966) Pleistocene and post-Pleistocene climatic variations in the Pacific area. A symposium. Bishop Museum Press, Honolulu

103. Bock CE, Lepthien LW (1976) Synchronous eruptions of boreal seed-eating birds. Am Nat 110: 559–571

104. Bodenheimer FS, Sulman F (1946) The estrous cycle of *Microtus guentheri* D. and A. and its ecological implications. Ecology 27: 255–256

105. Bogert C (1937) Birds collected during the Whitney South Sea Expedition. XXXIV. The distribution and migration of the long-tailed cuckoo (*Urodynamis taitensis* Sparrman). Am Mus Novitates 933: 1–12

106. Boley, RB, Kennerly Jr TE (1969) Cellulolytic bacteria and reingestion in the plains pocket gopher, *Geomys bursarius*. J Mamm 50: 348–349

107. Bolt JR (1969) Lissamphibian origins: possible protolissamphibian from the Lower Permian of Oklahoma. Science 166: 888–891

108. Bormann FH, Likens GE (1979) Catastrophic disturbance and the steady state in northern hardwood forests. Am Sci 67: 660–669

109. Bowmaker JK, Knowles A (1977) The visual pigments and oil droplets of the chicken retina. Vis Res 17: 755–764

110. Bowman RI (1961) Morphological differentiation and adaptation in the Galapagos finches. Univ Calif Publ Zool 58: 1–302

111. Bowman RI, Billeb SL (1965) Blood-eating in a Galápagos finch. Living Bird 4: 29–44

112. Boyd SH (1975) Inhibition of fish reproduction by Rana catesbeiana larvae. Physiol Zool 48: 225–234

113. Bradshaw GVR (1962) Reproductive cycle of the California leaf-nosed bat, *Macrotus californicus*. Science 136: 645–646

114. Braithwaite RW (1974) Behavioral changes associated with the population cycle in *Antechinus stuartii* (Marsupialia): Aust J Zool 22: 45–62

115. Branson BA (1966) Histological observations on the sturgeon chub, *Hybopsis gelida* (Cyprinidae). Copeia: 872–876

116. Brattstrom BH (1962) Homing in the giant toad, *Bufo marinus*. Herpetol 18: 176–180

117. Brattstrom BH (1962) Thermal control of aggregation behavior in tadpoles. Herpetol 18: 38–46

118. Brattstrom BH (1963) A preliminary review of the thermal requirements of amphibians. Ecology 44: 238–255

119. Briggs JC (1966) Zoogeography and evolution. Evolution 20: 282–289

120. Broadley DG (1967) The life cycles of two sympatric species of *Ichnotropis* (Sauria: Lacertidae). Zool Africana 3: 1–2

121. Brockway BF (1962) The effects of nest-entrance positions and male vocalizations on reproduction in budgerigars. Living Bird 1: 93–101

122. Brockway BF (1965) Stimulation of ovarian development and egg-laying by male courtship vocalization in budgerigars *(Melopsittacus undulatus)*. Anim Behav 13: 575–578

123. Brodie ED Jr (1968) Investigations on the skin toxin of the adult rough-skinned newt *Taricha granulosa*. Copeia: 307–313

124. Bromley PT (1969) Territoriality in pronghorn bucks on the National Bison Range, Moiese, Montana. J Mamm 50: 81–89

125. Bronson FH (1976) Urine marking in mice: causes and effects. In: Doty RL (ed) Mammalian olfaction, reproductive processes, and behavior; Academic, New York, pp 119–141

126. Brown CW (1968) Additional observations on the function of the nasiolabial grooves of plethodontid salamanders. Copeia: 728–731

127. Brown FA Jr (1972) The "clocks" timing biological rhythms. Am Sci 60: 756–766

128. Brown HA (1967) High temperature tolerance of the eggs of a desert anuran, *Scaphiopus hammondii*. Copeia: 365–370

129. Bruce RC (1978) A comparison of the larval periods of Blue Ridge and Piedmont mud salamanders *(Pseudotriton montanus)*. Herpetol 34: 325–332

130. Brundin L (1965) On the real nature of transantarctic relationships. Evolution 19: 496–505

131. Brundin L (1966) Transantarctic relationships and their significance as evidenced by chironomid midges. Kong Sven Vetens Haud 11: 1–72

132. Brundin L (1972) Phylogenetics and biogeography. System Zool 21: 69–79

133. Bryant GL (1977) Fecundity and growth of the tule perch, *Hysterocarpus traski,* in the lower Sacramento-San Joaquin Delta. Calif Fish and Game 63: 140–156

134. Budker P (1971) The life of sharks. Columbia University, New York

135. Bull JJ, Vogt RC (1979) Temperature-dependent sex determination in turtles. Science 206: 1186–1188

136. Bullock TH (1973) Seeing the world through a new sense: electroreception in fish. Am Sci 61: 316–325

137. Burkenwood MD (1948) Fluctuation in abundance of Pacific halibut. In: A symposium on fish populations. B Bingham Oceanographic Collection, Peabody Museum of Natural History, Yale University 11, Art 4

138. Burns JM, Baker RJ, Bleier WJ (1972) Hormonal control of "delayed" development: in *Macrotus waterhousii*. 1. Changes in plasma thyroxine during pregnancy and lactation. Gen Comp Endocrinol 18: 54–58

139. Bruns JR (1976) The reproductive cycle and its environmental control in the pumpkinseed, *Lepomis gibbosus* (Pisces: Centrarchidae), Copeia 449–455

140. Bury RB, Ernest CH (1977) *Clemmys*. Catalogue of American amphibians and reptiles 203

141. Busnel R-G (1963) Acoustic behavior of animals. Elsevier, Amsterdam

142. Busse K (1970) Care of the young by male *Rhinoderma darwini*. Copeia 395

143. Bustard HR (1971) A population study of the eyed gecko, *Oedura ocellata Boulenger,* in the Northern New South Wales. Australia. Copeia 658–669

144. Cade TJ (1964) The evolution of torpidity in rodents. Ann Acad Sci Fenn Biologica 71: 79–112

145. Calder WA (1972) Heat loss from small birds, analogy with Ohm's law and a re-examination of the "Newtonian Model." Comp Biochem Physiol 43A: 13–20

146. Calder WA, Schmidt-Nielsen K (1967) Temperature regulation and evaporation in the pigeon and the roadrunner. Am J Physiol 213: 883–889

147. Calder WA, Booser J (1973) Hypothermia of broadtailed hummingbirds during incubation in nature with ecolotical correlations. Science 180: 751–753

148. Cameron AW (1964) Competitive exclusion between rodent genera *Microtus* and *Clethrionomys*. Evolution 18: 630–634

149. Capranica RR, Frishkoph LS, Nevo E (1973) Encoding of geographic dialects in the auditory system of the cricket frog. Science 182: 1272–1275

150. Carey FG, Teal JM (1966) Heat conservation in tuna fish muscle. Proc Nat Acad Sci 56: 1464–1469

151. Carey FG, Teal JM, Kanwisher JW, Lawson KV, Beckett JS (1971) Warm-bodied fish. Am Zool 11: 135–143

152. Carlisle JG Jr, Schott JW, Abramson NJ (1960) The barred surfperch (*Amphistichus argenteus* Agassiz) in southern California. Calif Dept Fish Game Fish Bull 109

153. Carpenter CC (1957) Hibernation, hibernacula, and associated behavior of the three-toed box turtle *(Terrapene carolina triunguis)*. Copeia: 278–282

154. Carpenter FL (1974) Torpor in an Andean hummingbird: its ecological significance. Science 183: 545–547

155. Carr A (1967) So excellent a fishe. Natural History Press, Garden City, New York
156. Carroll RL, Currie PJ (1975) Microsaurs as possible apodan ancestors. Zool J Linnean Soc 57: 229–249
157. Carter GS, Beadle LC (1930) Notes on the habits and development of *Lepidosiren paradoxa*. J Linnean Soc (Zool) 37: 197–203
158. Carter GS, Beadle LC (1931) The fauna of the swamps of the Paraguayan Chaco in relation to its environment. II. Respiratory adaptations in the fishes. J Linnean Soc (Zool) 37: 327–368
159. Case JF, Strause LG (1978) Neurally controlled luminescent systems. In: Herring PJ (ed) Bioluminescence in action. Academic, New York, pp 331–366
160. Chang KSF, Chan ST, Low WD, et al (1963) Climate and conception rates in Hong Kong. Hum Biol 35: 366–376
161. Chan KMB, Lofts B (1974) The testicular cycle and androgen biosynthesis in the tree sparrow *Passer montanus saturatus*. J Zool 172: 47–66
162. Chapman DW (1966) Food and space as regulators of salmonid populations in streams. Am Nat 100: 345–357
163. Charles-Dominique P (1971) Eco-ethologie des prosimiens du Gabon. Biol Gabon 7: 121–228
164. Chave EH, Randell HA (1971) Feeding behavior of the moray eel, *Gymnothorax pictus*. Copeia: 570–574
165. Chew RM (1961) Water metabolism of desert-inhabiting vertebrates. Biol Rev 36: 1–31
166. Chew RM (1965) Water metabolism of mammals. In: Mayer WV, Van Gelder RG (eds) Physiological mammalogy. Mammalian reactions to stressful environments. Academic, New York, Vol 2, pp 43–178
167. Chew RM, Dammann AE (1960) Evaporative water loss of small vertebrates, as measured with an infrared analyzer. Science 133: 384–385
168. Chidester FE (1920) The behavior of *Fundulus heteroclitus* on the salt marshes of New Jersey. Am Nat 54: 551–557
169. Christian JJ (1969) Maturation and breeding of *Blarina brevicauda* in winter. J Mamm 50: 272–276
170. Chung SH, Pettigrew A, Anson M (1978) Dynamics of the amphibian middle ear. Nature 272: 142–147
171. Clark, DB, Gibbons JW (1969) Dietary shift in the turtle *Pseudemys scripta* (Schoepff) from youth to maturity. Copeia: 704–706
172. Clark JH, Zarrow MX (1971) Influence of copulation on time of ovulation in women. Am J Obstet Gynecol 109: 1083–1085
173. Cloudsley-Thompson JL (1961) Rhythmic activity in animal physiology and behaviour. Academic, New York
174. Cloudsley-Thompson JL (1971) The temperature and water relations of reptiles. Morrow, Watford, England
175. Cockrum E (1969) Migration in the guano bat, *Tadarida brasiliensis*. In: Contributions in Mammalogy, a volume honoring Professor E. Raymond Hall. Museum of Natural History, The University of Kansas, Miscellaneous Publication 51: 303–336
176. Colbert EH, Breed WJ, Elliot DH, et al (1970) Triassic tetrapods from Antarctica: evidence for continental drift. Science 199: 1197–1201
177. Cole, FR, Batzli GO (1978) Influence of supplemental feeding on a vole populations. J Mamm 59: 809–819
178. Collette BB (1961) Correlations between ecology and morphology in anoline lizards from Havana, Cuba and southern Florida. Bull Mus Comp Zool at Harvard College 125: 135–162
179. Conaway CH, Fleming WR (1960) Placental transmission of Na^{22} and I^{131} in *Natrix*. Copeia: 53–55
180. Conte FP (1969) Structure and functional pathways of osmoregulation. Excretion and secretion by epithelial tissue. In: Fish physiology. Academic, New York, Vol 1
181. Conway-Morris S (1977) Aspects of the Burgess shale fauna, with particular reference to the non-arthropod component. J Paleont 51, Suppl. to no. 2, Pt. III: 7–8

182. Cook HW, Peterson AM, Simmons MM, et al (1970) Dall sheep *(Ovis dalli dalli)* milk. I. Effects of stage of lactation on the composition of the milk. Can J Zool 48: 629–633

183. Corben CJ, Ingram GJ, Tyler MJ (1974) Gastric brooding: unique form of parental care in an Australian frog. Science 186: 946–947

184. Cornish LM, Bradshaw WN (1978) Patterns in twelve reproductive parameters for the white-footed mouse *(Peromyscus leucopus)*. J Mamm 59: 731–739

185. Courtenay WR Jr (1971) Sexual dimorphism of the sound-producing mechanism of the striped cusk-eel, *Rissola marginata* (Pisces: Ophidiidae). Copeia: 259–268

186. Cowles RB (1944) Parturition in the yucca lizard. Copeia: 98–100

187. Cowles RB (1965) Hyperthermia, aspermia, mutation rates and evolution. Q Rev Biol 40: 341–367

188. Cowles RB, Bogert CM (1944) A preliminary study of thermal requirements of desert reptiles. Bull Amer Mus Nat Hist 83 261–296

189. Cracraft J (1972) The relationships of the higher taxa of birds: problems in phylogenetic reasoning. Condor 74: 379–392

190. Cracraft J (1973) Continental drift, paleoclimatology, and the evolution and biogeography of birds. J Zool 169: 455–545

191. Cracraft J (1973) Vertebrate evolution and biogeography in the Old World tropics: implications of continental drift and paleoclimatology. In: Tarling DH, Runcorn SK (eds) Implications of continental drift to the earth sciences. Academic, London, pp 373–393

192. Cracraft J (1974) Continental drift and vertebrate distribution. Ann Rev Ecol Systemat 5: 215–261

193. Crane JM Jr (1965) Bioluminescent courtship display in the teleost *Porichthys notatus*. Copeia: 239–241

194. Crenshaw JW Jr (1955) The life history of the southern spiny lizard, *Sceloporus undulatus* Latreille. Am Midl Nat 54: 257–298

195. Crescitelli F (1972) The visual cells and visual pigments of the vertebrate eye. In: Dartnall HJA (ed) Handbook of sensory physiology, VII/1, photochemistry of vision. Springer, Berlin, pp 245–363

196. Cressey RF, Lachner EA (1970) The parasitic copepod diet and life history of diskfishes (Echeneidae). Copeia 310–318

197. Crews D (1975) Psychobiology of reptilian reproduction. Science 189: 1059–1065

198. Crisp TM (1964) Studies of reproduction in the female ovoviviparous lizard *Sceloporus cyanogenys* Cope. Tex J Sci 16: 481–489

199. Crompton AW, Taylor CR, Jagger JA (1978) Evolution of homeothermy in mammals. Nature 272: 333–336

200. Crowe DM (1975) Aspects of ageing, growth and reproduction of bobcats from Wyoming. J Mamm 56: 177–198

201. Crump ML (1972) Territoriality and mating behavior in *Dendrobates granuliferous* (Anura: Dendrobatidae). Herpetol 28: 195–198

202. Crump ML (1974) Reproductive strategies in a tropical anuran community. University of Kansas Museum of Natural History. Misc. Publ. No. 61

203. Cuellar O (1966) Oviducal anatomy and sperm storage in lizards. J Morphol 119: 7–20

204. Cuellar HS, Cuellar O (1977) Refractoriness in female lizard reproduction: a probable circannual clock. Science 197: 495–497

205. Cyrus RV, Mahmoud IY, Klicka J (1978) Fine structure of the corpus luteum of the snapping turtle. *Chelydra serpentina*. Copeia: 622–627

206. Czopek J (1959) Skin and capillaries in European common newts. Copeia: 91–96

207. Daan S (1973) Activity during natural hibernation in three species of vespertilionid bats. Neth J Zool 23: 1–71

208. Daiber FC (1956) A comparative analysis of the winter feeding habits of two benthic stream fishes. Copeia: 141–151

209. Dalquest WW (1947) Notes on the natural history of the bat *Corynorhinus rafinesquii* in California. J Mamm 28: 17–30

210. Daniel JC Jr (1970) Dormant embryos of mammals. Biosci 20: 411–415
211. Daniel PM (1960) Growth and cyclic behavior of the West African lizard, *Agama agama africana*. Copeia: 94–97
212. Dantzler WH, Schmidt-Nielsen B (1966) Excretion in fresh-water turtle *(Pseudemys scripta)* and desert tortoise *(Gopherus agassizii)*. Am J Physiol 211: 198–210
213. Darevsky IS (1966) Natural parthenogenesis in a polymorphic group of Caucasian rock lizards related to *Lacerta saxicola* Eversmann. J Ohio Herpetol Soc 5: 115–160
214. Darlington PJ Jr (1965) Biogeography of the southern end of the world. Harvard University Press, Cambridge. 236 pp
215. Das BK (1934) The habits and structure of *Pseudapocryptes lanceolatus*, a fish in the first stages of structural adaptation to aerial respiration. Proc R Soc London 115: 422–430
216. Das BK (1936) On ecology and bionomics of an air-breathing loach, Lepidocephalus guntea (Ham. Buch) with a review on air breathing fishes. XII Congrès International de Zoologìa, Lisbonne 1935: 866–920
217. Davis DE (1955) Breeding biology of birds. In: Wolfson A (ed) Recent studies in avian biology. University of Illinois Press, Urbana, pp 264–308
218. Davis DE (1967) The annual rhythm of fat deposition in woodchucks *(Marmota monax)*. Physiol Zool 40: 391–402
219. Davis WH (1959) Disproportionate sex ratios in hibernating bats. J Mamm 40: 16–19
220. Davis WH (1970) Hibernation: ecology and physiological ecology. In: Wimsatt WA (ed) Biology of bats. Academic, New York, Vol 1, pp 265–300
221. Dawe AR, Spurrier WA (1969) Hibernation induced in ground squirrels by blood transfusion. Science 163: 298–299
222. Dawe AR, Spurrier WA, Armour JA (1970) Summer hibernation induced by cryogenically preserved blood "trigger." Science 168: 497–498
223. Dawe AR, Spurrier WA (1972) The blood-borne "trigger" for natural mammalian hibernation in the 13-lined ground squirrel and the woodchuck. Cryobiol 9: 163–172
224. Dawkins R (1976) The selfish gene. Oxford University Press, New York
225. Dawson WR, Shoemaker VH, Licht P (1966) Evaporative water losses of some small Australian lizards. Ecology 47: 589–594
226. Deevey ES (1961) Recent advances in Pleistocene stratigraphy and biogeography. In: Blair WF (ed) Vertebrate speciation. University of Texas, Austin, pp 594–623
227. Deevey ES (1947) Life tables for natural populations of animals. Qu Rev Biol 22: 283–314
228. de Kock LL, Robinson AE (1966) Observations on a lemming movement in Jämtland, Sweden, in autumn 1963. J Mamm 47: 490–499
229. Demian JJ, Taylor DH (1977) Photoreception and locomotion rhythm entrainment by the pineal body of the newt, *Notophthalmus viridescens* (Amphibia, Urodela, Salamandridae). J Herpetol 11: 131–139
230. Dengo G (1973) Estructure geologica, historia tectonica y morphologia de America Central. 2nd Ed. Institute Centroamericano de Investigacion y Tecnologia Industrial. Guatemala
231. Denisen DM, Kooyman GL (1973) The structure and function of the small airways in pinniped and sea otter lungs. Resp Physiol 17: 1–10
232. Dent JN (1956) Observations on the life history and development of *Leptodactylus albilabris*. Copeia: 207–210
233. DeRosa CT, Taylor DH (1978) Sun-compass orientation in the painted turtle, *Chrysemys picta* (Reptilia, Testudines, Testudinidae). J Herpetol 12: 25–28
234. De Ruiter L (1963) The physiology of vertebrate feeding behaviour: towards a synthesis of the ethological and physiological approaches to problems of behavior. Z Tierpsychol 20: 498–516
235. Dessauer HC (1953) Hibernation in the lizard, *Anolis carolinensis*. Proc Soc Exp Biol Med 82: 351–353
236. DeVooys CGN (1971) The stimulation of the ornithine cycle of Xenopus laevis Daudin by ammonia in the environment. Neth J Zool 21: 403–411
237. Dial BE (1978) The thermal ecology of two sympatric, nocturnal *Coleonyx* (Lacertilia: Gekkonidae). Herpetol 34: 194–201

238. Dice LR, Clark PJ (1953) The statistical concept of home range as applied to the recapture radius of the deer mouse *(Peromyscus)*. Contributions from the Laboratory of Vertebrate Biology, University of Michigan, No. 62

239. Dietz M (1972) Erdkröten können UV-Licht sehen. Naturwissenschaften 59: 316

240. Dietz RS, Holden JC (1970) Reconstruction of Pangaea: breakup and dispersion of continents, Permian to present. J Geophys Res 75: 4939–4956

241. Dijkgraaf S (1967) Biological significance of the lateral line organs. In: Cahn PH (ed) Lateral line detectors. Indiana University, Bloomington, pp 83–95

242. Disler NN (1971) Lateral line sense organs and their importance in fish behavior. (Izdatel'stvo Akademii Nauk SSSR, Moskva, 1960). Israel Program for Scientific Translations, Jerusalem

243. Dixon JR (1957) Geographic variation and distribution of the genus Tomodactylus in Mexico. Tex J Sci 9: 379–409

244. Dmi'el R (1967) Studies on reproduction, growth, and feeding in the snake *Spalerosophis cliffordi* (Colubridae). Copeia: 332–346

245. Dole JW (1972) The role of olfaction and audition in the orientation of leopard frogs, *Rana pipiens*. Herpetol 28: 258–260

246. Dole JW (1972) Evidence of celestial orientation in newly metamorphosed *Rana pipiens*. Herpetol 28: 273–276

247. Dolk HE, Postma N (1927) Uber die Haut- und die Lungenatmung von *Rana temporaria*, Z Vergleich Physiol 5: 444–517

248. Donn WL, Shaw DM (1977) Model of climate evolution based on continental drift and polar wandering. Geolog Soc Am Bull 88: 390–396

249. Dorf E (1959) Climatic changes of the past and present. Contributions of the University of Michigan Museum of Paleontology 13: 181–210

250. Doty RL (1976) Reproductive endocrine influences upon human nasal chemoreception: a review. In: Doty RL (ed) Mammalian olfaction, reproductive processes and behavior. Academic, New York, pp 295–321

251. Doty RL, Ford M, Preti G, et al (1975) Changes in the intensity and pleasantness of human vaginal odors during the menstrual cycle. Science 190: 1316–1317

252. Douglass EL, Friedl WA, Pickwell GV (1976) Fishes in oxygen-minimum zones: blood oxygenation characteristics. Science 191: 957–959

253. Dowling JE, Ripps H (1970) Visual adaptation in the retina of the shark. J Gen Physiol 56: 441–451

254. Drent R (1973) The natural history of incubation. In: Farner DS (ed) Breeding biology of birds. Natl Acad Sci, Washington, D.C., pp 262–311

255. Drewry GE, Jones KL (1976) A new ovoviviparous frog, *Eleutherodactylus jasperi* (Amphibia, Anura, Leptodactylidae), from Puerto Rico. J Herpetol 10: 161–165

256. Dryden GL (1968) Growth and development of *Suncus murinus* in captivity on Guam. J Mamm 49: 51–62

257. Dryden GL, Conaway CH (1967) The origin and hormonal control of scent production in *Suncus murinus*. J Mamm 48: 420–428

258. Dubost G (1968) Apercu sur le rythme annuel de reproduction des muridés du nord-est du Gabon. Biol Gabonica 4: 227–239

259. Duellman WE (1970) The hylid frogs of Middle America. Monograph of the Museum of Natural History, The University of Kansas, Number 1

260. Duellman WE (1978) The biology of an equatorial herpetofauna in Amazonian Ecuador. University of Kansas. Museum of Natural History, Miscellaneous Publication 65

261. Duellman WE, Savitzky AH (1976) Aggressive behavior in a centrolenid frog, with comments on territoriality in anurans. Herpetol 32: 401–404

262. Dunbar MJ (1949) The Pinnipedia of the arctic and subarctic. Bull Fisheries Research Board Can 82.

263. Duncker HR (1972) Structure of avian lungs. Resp Physiol 14: 44–63

264. Duncker HR (1974) Structure of the avian respiratory tract. Resp Physiol 22: 1–19

265. Eakin RM (1970) A third eye. Am Sci 58: 73–79

266. Dunn ER (1926) The salamanders of the family Plethodontidae. Smith College, Northampton, Massachusetts

267. Eadie WR (1948) Shrew-mouse predation during low mouse abundance. J Mamm 29: 35–37

268. Eakin RM (1973) The third eye. University of California, Berkeley

269. Eckardt MJ, Whimster IW (1971) Skin homografts in the all female gekkonid lizard, *Hemidactylus garnotii*. Copeia: 152–154

270. Eisenberg JF, Kleiman DG (1972) Olfaction communication in mammals. Ann Rev Ecol Systemat 3: 1–32

271. Elliot DH, Colbert EH, Breed WJ, et al (1970) Triassic tetrapods from Antarctica: evidence for continental drift. Science 169: 1197–1201

272. Elsner R (1965) Heart rate response in forced verses trained experimental dives in pinnipeds. Hvalradets Skrifter 48: 24–29

273. Emlen ST (1967) Migratory orientation in the indigo bunting, *Passerina cyanea*. Auk 84: 309–342; 463–489

274. Emlen ST (1970) Celestial rotation: its importance in the development of migratory orientation. Science 170: 1198–1201

275. Emlen ST (1975). Migration: orientation and navigation. In: Farner DS, King JR (eds) Avian biology. Academic, New York, Vol 5

276. Emlen ST, Wiltschko W, Demong NJ, Wiltschko R, Bergman S, (1976) Magnetic direction finding: evidence for its use in migratory indigo buntings. Science 193: 505–508

277. Emlen ST, Oring LW (1977) Ecology, sexual selection, and the evolution of mating systems. Science 197: 215–223

278. Enders AC (ed) (1963) Delayed implantation. University of Chicago

279. Euker JS, Meites J, Kiegle GD (1975) Effects of acute stress on serum LH and prolactin in intact, castrate and dexamethasone-treated male rats. Endocrinol 96: 85–92

280. Evans KJ (1966) Responses of the locomotor rhythms of lizards to simultaneous light and temperature cycles. Comp Biochem Physiol 19: 91–103

281. Evans LT (1959) A motion picture study of maternal behavior of the lizard. *Eumeces obsoletus* Baird and Girard. Copeia: 103–110

282. Evans PR (1966) Migration and orientation of passerine night migrants in northeast England. J Zool 150: 319–369

283. Eveleth PB, Tanner JM (1976) Worldwide variation in human growth. Cambridge University Press, Cambridge, England

284. Farner DS (1959) Photoperiodic control of annual gonadal cycles in birds. Am Assoc Adv Sci 55: 717–750

285. Farner DS (1961) Comparative physiology: photoperiodicity. Ann Rev Physiol 23: 71–96

286. Farner DS (1964) The photoperiodic control of reproductive cycles in birds. Am Sci 52: 137–156

287. Farner DS (ed) (1972) Breeding biology of birds. Proceedings of a symposium on breeding behavior and reproductive physiology of birds. Natl Acad Sci, Washington, D.C.

288. Farner DS, Follett BK (1979) Reproductive periodicity in birds. In: Barrington EJW (ed) Hormones and evolution. Academic, London pp 829–872

289. Faust R (1960) Brutbiologie des Nandus (Rhea americana) in Lefangenschaft. Verhandl Dtsch Zool Gschaft 42: 398–401

290. Ferguson DE (1963) Orientation in three species of anuran amphibians. Ergebnisse der Biologie 26: 128–134

291. Ferris VR, Goseco CG, Ferris JM (1976) Biogeography of freeliving soil nematodes from the perspective of plate tectonics. Science 193: 508–510

292. Field CR, Blankenship LH (1973) Nutrition and reproduction of Grant's and Thompson's gazelles, Coke's hartebeest and giraffe in Kenya. J Reprod Fertil Suppl 19: 287–301

293. Finerty JP (1979) Cycles in Canadian lynx. Am Nat 144: 453–455

294. Finley RB Jr (1969) Cone caches and middens of *Tamiasciurus* in the Rocky Mountain Region. Museum of Natural History, University of Kansas, Misc Pub 51: 233–273

295. Fisher KC (1964) On the mechanism of periodic arousal in the hibernating ground squirrel. In: Suomalainen P (ed) Mammalian hibernation II. Ann Acad Sci Fenn IV Biol 71: 143–156

296. Fitch HS (1940) A field study of the growth and behavior of the fence lizard. University of California Publications in Zoology 44 (2): 151–172

297. Fitch HS (1958) Home ranges, territories, and seasonal movements of vertebrates of the Natural History Reservation. University of Kansas Publications, Museum of Natural History 11: 63–326

298. Fitch HS (1970) Reproductive cycles in lizards and snakes. University of Kansas Museum of Natural History, Misc Pub

299. Fitch HS (1972) Ecology of *Anolis tropidolepis* in Costa Rican cloud forest. Herpetol 28: 10–21

300. Fitch HS (1973) Observations on the population ecology of the Central American iguanid lizard *Anolis cupreus*. Caribbean J Sci 13: 215–228

301. Fleharty ED, Krause ME, Stennett DP (1973) Body composition, energy content and lipid cycles of four species of rodents. J Mamm 54: 426–438

302. Fleming TH (1971) *Artibeus jamaicensis:* delayed embryonic development in a neotropical bat. Science 171: 402–404

303. Flock Å (1967) Ultrastructure and function in the lateral line organs. In: Cahn PH (ed) Lateral line detectors. Indiana University, Bloomington, pp 163–197

304. Flock Å (1971) The lateral line organ mechanoreceptors. In: Hoar WS, Randall DS (eds) Fish physiology V. Academic, New York, pp 241–263

305. Flowerdew JR (1973) The effect of natural and artificial changes in food supply on breeding in woodland mice and voles. J Reprod Fertil Suppl 19: 259–269

306. Folk GE Jr (1957) Twenty-four hour rhythms of mammals in a cold environment. Am Midl Nat 91: 153–166

307. Fooden J (1972) Breakup of Pangaea and isolation of relict mammals in Australia, South America, and Madagascar. Science 175: 894–898

308. Forman GL (1973) Studies of gastric morphology in North American Chiroptera (Emballonuridae, Noctilionidae and Phyllostomatidae). J Mamm 54: 909–923

309. Fox HM, Vevers G (1960) The nature of animal colours. Sidgwick and Jackson, London

310. Fox W (1952) Seasonal variations in the male reproductve system of Pacific coast garter snakes. J Morphol 90: 481–553

311. Fox W (1958) Sexual cycle of the male lizard, *Anolis carolinensis*. Copeia: 22–29

312. Foxon GEH (1964) Blood and respiration. In: Moore JA (ed) Physiology of the Amphibia. Academic, New York, pp 151–209

313. Frair W, Ackman RG, Mrosovsky N (1972) Body temperature of *Dermochelys coriacea:* warm turtle from cold water. Science 177: 791–793

314. Freeland WJ, Janzen DH (1974) Strategies in herbivory by mammals: the role of plant secondary compounds. Am Nat 108: 269–289

315. French NR, Stoddart DM, Bobek B (1975) Patterns of demography in small mammal populations. In Small mammals: their productivity and population dynamics. International Biological programme 5. Cambridge University Press, Cambridge, England, pp 73–102

316. Fricke HW (1970) Die ökologische Spezialisierung der Eidechse *Cryptoblepharus boutoni cognatus* (Boettger) auf das Leben in der Gezeitenzone (Reptilia, Skinkidae). Oecologia 5: 380–391

317. Frisch RE, McArthur JW (1974) Menstrual cycles: fatness as a determinant of minimum weight for height necessary for their maintenance or onset. Science 185: 949–951

318. Fromme HF (1961) Untersuchungen über das Orientierungs-vermögen nächtlich ziehender Kleinvögel *(Erithacus rubecula, Sylvia communis)*. Z Tierpsychol 18: 204–220

319. Frost WE, Kipling C (1967) A study of reproduction, early life, weight-length relationship and growth of pike, *Esox lucius* L., in Windermere. J Anim Ecol 36: 651–693

320. Funakoshi K, Uchida TA (1975) Studies on the physiological and ecological adaptation of temperate insectivorous bats. I. Feeding activities in the Japanese long-fingered bats, *Miniopterus schreibersi fuliginosus*. Jpn J Ecol 25: 219–234

321. Funakoshi K (1978) Studies on the physiological and ecological adaptation of temperate insectivorous bats. II. Hibernation and winter activity in some cave-dwelling bats. Jpn J Ecol 28: 237–261

322. Galambos R (1942) Cochlear potentials from bats by supersonic sounds. J Acoust Soc Am 14: 41–49

323. Gallien L (1959) Endocrine basis for reproductive adaptations in Amphibia. In: Gorbman A (ed) Comparative endocrinology. John Wiley, New York, pp 479–487

324. Gambell R (1973) Some effects of exploitation on reproduction in whales. J Reprod Fertil Suppl 19: 533–553

325. Gameson ALH, Griffith SD (1959) Six months' oxygen records for a polluted stream. Water Waste Treatment J Jan/Feb 1959

326. Gans C (1970a) Respiration in early tetrapods—the frog is a red herring. Evolution 24: 723–734

327. Gans C (1970b) Strategy and sequence in the evolution of the external gas exchangers of ectothermal vertebrates. Forma et functio 3: 61–104

328. Gans C, de Jongh HJ, Farber J (1969) Bullfrog *(Rana catesbeiana)* ventilation: how does the frog breathe? Science 163: 1223–1225

329. Garg RK, Pandha SK, Thapliyal JP (1967) Further studies on the development of eggs of the garden lizard, *Calotes versicolor.* Copeia: 865–867

330. Garrick LD (1979) Lizard thermoregulation: operant responses for heat at different thermal intensities. Copeia: 258–266

331. Gaskin DE (1970) The origins of the New Zealand fauna and flora: a review. Geogr Rev 60: 414–434

332. Gaunt AS, Gans C (1969) Diving bradycardia and withdrawal bradycardia in Caiman crocodilus. Nature 223: 207–208

333. Gentry JB, Smith MH (1968) Food habits and burrow associates of *Peromyscus polionotus.* J Mamm 49: 562–565

334. Gerald JW (1966) Food habits of the long-nosed dace, *Rhinichthys cataractae.* Copeia: 478–485

335. Gerking SD (1959) Physiological changes accompanying aging in fishes. In: Wolstenholme GEW, O'Connor M (eds) Ciba foundation colloquia on aging. The lifespan of animals. Churchill, London, Vol 5, pp 142–180

336. Giaja J (1938) L'homeothermie actualités scientifique et industrielles. No. 576. Herman, Paris

337. Gibbons JW (1968) Population structure and survivorship in the painted turtle, *Chrysemys picta.* Copeia: 260–267

338. Gibbons JW (1969) Ecology and population dynamics of the chicken turtle, *Deirochelys reticularia.* Copeia: 669–676

339. Gibson RN (1965) Rhythmic activity in a littoral fish. Nature 207: 544–545

340. Gibson RN (1971) Factors affecting the rhythmic activity of *Blennius pholis* L. (Teleostei). Anim Behav 19: 335–343

341. Gibson RN (1978) Lunar and tidal rhythms in fish. In: Thorpe J (ed) Rhythmic activity of fishes. Academic, New York

342. Gilbert PW, Heath GW (1972) Clasper-siphon mechanism in *Squalis acanthias* and *Mustelus canis.* Comp Biochem Physiol 42A: 97–119

343. Girgis S (1961) Aquatic respiration in the common Nile turtle, *Trionyx triunguis* (Forskal). Comp Biochem Physiol 3: 206–217

344. Glasser JW (1979) The role of predation in shaping and maintaining the structure of communities. Am Nat 113: 631–641

345. Godet R (1961) Étude expérimentale de la formation de mucus tégumentaire et de la réalisation du cocon chez le Proptère. CR Acad Sci, Paris 252: 2451–2452

346. Goel HC (1966) Sound production in *Clarias batrachus* (Linnaeus). Copeia: 622–624

347. Goin OB, Goin CJ (1962) Amphibian eggs and the montane environment. Evolution 16: 364–371

348. Goldberg SR (1971) Reproductive cycle of the ovoviviparous iguanid lizard *Sceloporus jarrovi* Cope. Herpetol 27: 123–131
349. Good R (1964) The geography of flowering plants. Longmans, London
350. Gordon MS, Boëtius J, Evans DH, et al (1968) Additional observations on the natural history of the mudskipper, *Periophthalmus sobrinus*. Copeia: 853–857
351. Gordon MS, Schmidt-Nielsen K, Kelly HM (1961) Osmotic regulation in the crab-eating frog *(Rana cancrivora)*. J Exp Biol 38: 437–445
352. Goris RC (1963) Observations on the egg-crushing habits of the Japanese four-lined rat snake, *Elaphe quadrivirgata*. Copeia: 573–574
353. Goris RC, Terashima S (1976) The structure and function of the infrared receptors of snakes. In: Iggo A, Ilyinsky OB (eds) Progress in brain research. Somatosensory and visceral receptor mechanisms. Elsevier, Amsterdam, Vol 43, pp 159–170
354. Gorman GC, Licht P (1974) Seasonality in ovarian cycles among tropical lizards. Ecology 55: 360–369
355. Gorman J (1968) Breeding of an Andean toad in hot springs. Copeia: 167–170
356. Gorman RR, Ferguson JH (1970) Sun-compass orientation in the western toad *Bufo boreas*. Herpetol 26: 34–45
357. Görner P (1973) The importance of the lateral line system for the perception of surface waves in the clawed toad, *Xenopus laevis*, Daudin. Experientia (Basel)29: 295–296
358. Gould E, Eisenberg JF (1966) Notes on the biology of the Tenrecidae. J Mamm 47: 660–686
359. Govardovskii VI, Zueva LV (1977) Visual pigments of chicken and pigeon. Vis Res 17: 537–543
360. Graham JB (1974) Aquatic respiration in the sea snake *Pelamis platurus*. Resp Physiol 21: 1–7
361. Grahan TE, Doyle TS (1977) Growth and population characteristics of Blanding's turtle, *Emydoidea blandingii*, in Massachusetts. Herpetol 33: 410–414
362. Grant D, Anderson O, Twitty V (1968) Homing orientation by olfaction in newts *(Taricha rivularis)*. Science 160: 1354–1356
363. Grant PR (1972) Interspecific competition among rodents. Ann Rev Ecol Systemat 3: 79–106
364. Grant PR (1978) Competition between species of small mammals. In: Snyder DP (ed) Populations of small mammals under natural conditions. University of Pennsylvania, Pymatuning Laboratory of Ecology, Special Publication Series, Vol 5, pp 38–51
365. Graul WD, Derrickson SR, Mock DW (1977) The evolution of avian polyandry. Am Nat III: 812–816
366. Grayson DK (1977) Pleistocene avifaunas and the overkill hypothesis. Science 195: 691–693
367. Greer AE (1977) The systematics and evolutionary relationships of the scincid lizard genus *Lygosoma*. J Nat Hist 11: 515–540
368. Gressitt JL (1956) Some distributional patterns of Pacific island faunas. Syst Zool 5: 11–32
369. Griffin DR (1958) Listening in the dark. Yale University Press, New Haven
370. Griffin DR (1970) Migrations and homing of bats. In: Winsatt WA (ed) Biology of bats. Academic, New York, Vol 1, pp 233–264
371. Griffin DR, Suthers RA (1970) Sensitivity of echolocation in cave swiftlets. Biol Bull 139: 495–501
372. Griffin DR, Webster FA, Michael C (1960) The echolocation of flying insects by bats. Anim Behav 8: 141–154
373. Griffin DR, Buchler ER (1978) Echolocation of extended surfaces. In: Schmidt-Koenig K, Keeton WT (eds) Animal migration, navigation and homing. Springer, New York
374. Griffiths TA (1978) Muscular and vascular adaptations for nectarfeeding in the glossophagine bats *Monophyllus* and *Glossophaga*. J Mamm 59: 414–418
375. Grimsdell JJR (1973) Reproduction in the African buffalo, *Syncerus caffer*, in western Uganda. J Reprod Fertil Suppl 19: 303–318
376. Grove AT, Pullan RA (1963) Some aspects of the Pleistocene paleogeographie of the Chad Basin. African Ecology and Human Evolution. Viking Fund Publ Anthropol 36: 230–245

377. Grubb JC (1973) Orientation in newly metamorphosed Mexican toads, *Bufo valliceps*. Herpetol 29(2): 95–100
378. Grubb TC Jr (1972) Smell and foraging in shearwaters and petrels. Nature 237: 404–405
379. Gruber SH (1977) The visual system of sharks: adaptations and capability. Am Zool 17: 453–469
380. Gruber SH, Hamasaki DI, Bridges CDB (1963) Cones in the retina of the lemon shark *(Negaprion brevirostris)*. Vis Res 3: 397–399
381. Guimond RW, Hutchison VH (1976) Gas exchange of the giant salamanders of North America. In: Hughes GM (ed) Respiration in amphibious vertebrates. Academic, New York, pp 313–338
382. Gustafson AW, Shemesh M (1976) Changes in plasma testosterone levels during the annual reproductive cycle of the hibernating bat, *Myotis lucifugus,* with a survey of plasma testosterone levels in adult male vertebrates. Biol Reprod 15: 9–24
383. Gwinner, E (1977) Cirannual rhythms in bird migration. Ann Rev Ecol Systemat 8: 381–405
384. Hafner MS, Hafner DJ (1979) Vocalization of grasshopper mice (genus *Onychomys*). J Mamm 60: 85–94
385. Hahn WE (1967) Estradiol-induced vitellogenesis and concomitant fat mobilization in the lizard *Uta stansburiana*. Comp Biochem Physiol 23: 83–93
386. Hailman JP (1967) Oil droplets in the eyes of adult anuran amphibians: a comparative survey. J Morphol 148: 453–468
387. Hailman JP (1977) Optical signals. Indiana University Press, Bloomington
388. Halpern EA, Lowe CH (1968) Metabolism of the iguanid lizard *Uta stansburiana* in the supercooled state. Physiol Zool 41: 113–124
389. Halstead LB (1968) The pattern of vertebrate evolution. Freeman, San Francisco
390. Hamilton A (1976) The significance of patterns of distribution shown by forest plants and animals in tropical Africa for the reconstruction of Upper Pleistocene palaeoenvironments: a review. In: van Zinderen Bakker EM (ed) Palaeoecology of Africa and of the surrounding island and Antarctica. Balkema, Cape Town, Vol 9
391. Hamilton TH, Barth RH Jr (1962) The biological significance of seasonal change in male plumage appearance in some New World migration bird species. Am Nat 94: 129–144
392. Hamilton WJ Jr (1940) The feeding habits of larval newts with reference to availability and predilection of food items. Ecology 21: 351–356
393. Hamilton WJ Jr (1937) The biology of microtine cycles. J Agric Res 54: 779–790
394. Hamilton WJ III (1973) Life's color code. McGraw-Hill, New York
395. Hampton IFG, Whittow GC (1976) Body temperature and heat exchange in the Hawaiian spinner dolphin, *Stenella longirostris* Comp Biochem Physiol 55A: 195–197
396. Haneda Y, Tsuji FI (1971) Light production in the luminous fishes *Photoblepharon* and *Anomalops* from the Banda Islands. Science 173: 143–145
397. Hansen RM, Cavender BR (1973) Food intake and digestion by blacktailed prairie dogs under laboratory conditions. Acta Theriologica 18: 191–200
398. Hansen RM, Dearden BL (1975) Winter foods of mule deer in Piceance Basin, Colorado. J Range Management 28: 298–300
399. Hansen RM, Clark RC (1977) Food of elk and other ungulates at low elevations in northwestern Colorado. J Wildlife Management 41: 76–80
400. Hansen RM, Clark RC, Lawhorn W (1977) Foods of wild horses, deer and cattle in the Douglas Mountain Area, Colorado. J Range Management 30: 116–118
401. Hara, T (1971) Chemoreception. In: Hoar WS, Randal DJ (eds) Fish physiology. Academic, New York,Vol 5, pp 79–120
402. Hara T (1975) Olfaction in fish. Progr Neurobiol 5: 271–335
403. Harden-Jones FR (1968) Fish migration. Arnold, London,
404. Harrington RW Jr (1959) Photoperiodism in fishes in relation to annual sexual cycle. In: Withrow RB (ed) Photoperiodism and related phenomena in plants and animals. Am Assoc Adv Sci 55 Washington D.C., pp 651–667

405. Harrington RW Jr (1971) How ecological and genetic factors interact to determine when self-fertilizing hermaphrodites of *Rivulus marmoratus* change into functional secondary males, with a reappraisal of the modes of inter-sexuality among fishes. Copeia: 389–432

406. Hasler AD (1960) Guideposts of migrating fishes. Science 132: 785–792

407. Hasler AD (1966) Underwater guideposts; homing of salmon. University of Wisconsin Press, Madison

408. Hasler AD, Wisby WJ (1951) Discrimination of stream odors by fishes and relation to parent stream behavior. Am Nat 85: 223–238

409. Hasler AD, Scholz AT, Horrall RM (1978) Olfaction imprinting and homing in salmon. Am Sci 66: 347–355

410. Hassinger DD, Anderson JD (1970) The effect of lunar eclipse on nocturnal stratification of larval *Ambystoma opacum*. Copeia: 178–179

411. Hawes ML (1977) Home range, territoriality, and ecological separation in sympatric shrews. *Sorex vagrans* and *Sorex obscurus*. J Mamm 58: 354–367

412. Hawley AWL, Aleksiuk M (1975) Thermal regulation of spring mating behavior in the red-sided garter snake *(Thamnophis siratalis parietalis)*. Can J Zool 53: 768–776

413. Hayes SR (1976) Daily activity and body temperature of the southern woodchuck, *Marmota monax monax*, in northwestern Arkansas. J Mamm 57 (2): 291–299

414. Heath JE (1962) Temperature-independent morning emergence in lizards of the genus Phrynosoma. Science 138: 891–892

415. Heath JE (1966) Venous shunts in the cephalic sinuses of horned lizards. Physiol Zool 39: 30–35

416. Heatwole H (1978) Adaptations of marine snakes. Am Sci 66: 594–604

417. Heatwole H, Seymour R (1975) Pulmonary and cutaneous oxygen uptake in sea snakes and a file snake. Comp Biochem Physiol 51A: 399–405

418. Heatwole H, Seymour R (1976) Respiration of marine snakes. In: Hughes, GM (ed) Respiration of amphibious vertebrates. Academic, New York

419. Heatwole H, Veron JEN (1977) Vital limit and evaporative water loss in lizards (Reptilia, Lacertilia): a critique and new data. J Herpetol 11: 341–348

420. Heiligenberg W (1975) Theoretical and experimental approaches to spatial aspects of electrolocation. J Comp Physiol 103: 247–272

421. Heiligenberg W (1977) Principles of electrolocation and jamming avoidance in electricfish. Studies in Brain Function. Springer, Berlin, Vol 1

422. Hennessy DF, Owings DH (1977) Snake species discrimination and the role of olfactory cues in the snake-directed behavior of the California ground squirrel. Behaviour 65: 115–124

423. Henshaw RE, Underwood LS, Casey TM (1972) Peripheral thermoregulation: foot temperature in two arctic canines. Science 175: 988–990

424. Heppner FH (1970) The metabolic significance of differential absorption of radiant energy by black and white birds. Condor 72: 50–59

425. Héroux O (1961) Climatic and temperature induced changes in mammals. Rev Can Biol 20: 55–68

426. Herreid CF II, Kessel B (1967) Thermal conductance in birds and mammals. Comp Biochem Physiol 21: 405–414

427. Herring PJ, Moran JG (1978) Bioluminescence in fishes. In: Herring PJ (ed) Bioluminescence in action. Academic, New York, pp 273–329

428. Hershkovitz P (1958) A geographic classification of Neotropical mammals. Fieldiana: Zoology 36: 583–620

429. Hertz PE (1974) Thermal passivity of a tropical forest lizard, *Anolis polylepis*. J Herpetol 8: 323–327

430. Heusner AA, Roberts JC, Smith RM (1971) Circadian patterns of oxygen consumption in *Peromyscus*. J Appl Physiol 30: 50–55

431. Heyer WR (1969) The adaptive ecology of the species groups of the genus *Leptodactylus* (Amphibia, Leptodactylidae). Evolution 23: 421–428

432. Highfill DR, Mead RA (1975a) Sources and levels of progesterone during pregnancy in the garter snake, *Thamnophis elegans*. Gen Comp Endocrinol 27: 389–400
433. Highfill DR, Mead RA (1975b) Function of corpora lutea of pregnancy in the viviparous garter snake, *Thamnophis elegans*. Gen Comp Endocrinol 27: 401–407
434. Highten R, Savage T (1961) Functions of the brooding behavior in the female red-backed salamander, *Plethodon cinereus*. Copeia: 95–98
435. Hildemann WH, Walford RL (1963) Annual fishes—promising species as biological control agents. J Trop Med Hyg 66: 163–166
436. Hinkley R Jr (1966) Effects of plant extracts in the diet of male *Microtus montanus* on cell types of the anterior pituitary. J Mamm 47: 396–400
437. Hirst SM (1969) Predation as a regulatory factor of wild ungulate populations in a Transvaal lowveld nature reserve. Zool Africana 4: 199–230
438. Hirth HF (1966) Weight changes and mortality of three species of snakes during hibernation. Herpetol 22: 8–12
439. Hissa R, Palokangas R (1970) Thermoregulation in the titmouse (*Parus major* L.). Comp Biochem Physiol 33: 941–953
440. Hobson ES (1965) Diurnal-nocturnal activity of some inshore fishes in the Gulf of California. Copeia: 941–302
441. Hock RJ (1951) The metabolic rates and body temperatures of bats. Biol Bull 101: 289–299
442. Hock RJ (1955) Photoperiod as stimulus for onset of hibernation. Fed Proc Soc Exp Biol 14: 73–74
443. Hock RJ (1960) Seasonal variations in physiologic functions of Arctic ground squirrels and black bears. In: Mammalian Hibernation. Bull Museum of Comp Zool Harvard College 124: 155–171
444. Hock RJ (1964) Terrestrial animals in the cold: reptiles. In Handbook of physiology, Section 4. Adaptation to the Environment. American Physiological Society, Washington, D.C., pp 357–359
445. Hoffmeister DF, Getz LL (1968) Growth and age classes in the prairie vole, *Microtus ochrogaster*. Growth 32: 57–69
446. Hoffstetter R (1972) Relationships, origins, and history of the ceboid monkeys and caviomorph rodents: a modern reinterpreation. Evol Biol 6: 323–347
447. Holzapfel RA (1937) The cyclic character of hibernation in frogs. Q Rev Biol 12: 65–84
448. Homan WP (1968) Osmotic regulation in *Rana pipiens*. Copeia: 876–878
449. Homma K, Wilson WO, Siopes TD (1972) Eyes have a role in photoperiodic control of sexual activity of Coturnix. Science 178: 421–423
450. Hopkins CD (1972) Sex differences in electric signaling in an electric fish. Science 176: 1035–1037
451. Hopkins CD (1974) Electrical communication in fish. Am Sci 62: 426–437
452. Hora S (1935) Ecology and bionomics of the gobioid fishes of the Gangetic Delta. Comptes Rendus XIIᵉ Congreś International de Zoologie: 841–865
453. Houk JB, Geibel JJ (1974) Observation of underwater tool use by the sea otter, *Enhydra lutris* Linnaeus. Calif Fish and Game 60: 207–208
454. Howard RD (1978) The evolution of mating strategies in bullfrogs, *Rana catesbeiana*. Evolution 32: 850–871
455. Howarth B Jr (1974) Sperm storage: as a function of the female reproductive tract. In: Johnson AD, Foley CW (eds) The oviduct and its functions. Academic, New York, pp 237–270
456. Howell DJ (1974) Acoustic behavior and feeding in glossophagine bats. J Mamm 55: 293–308
457. Howell DJ (1976) Weight loss and temperature regulation in clustered versus individual *Glossophaga soricina*. Comp Biochem Physiol 53A: 197–199
458. Hsü CY, Liang HM (1970) Sex races of *Rana catesbeiana* in Taiwan. Herpetol 26: 214–221
459. Hubbs, CL, Hubbs LC (1932) Apparent parthenogenesis in nature in a form of fish of hybrid origin. Science 76(1983): 628–630

460. Hubbs CL, Bailey RM (1947) Blind catfishes from artesian waters of Texas. Occasional Papers of the Museum of Zoology, University of Michigan, no. 499

461. Hudson JW (1967) Variations in the patterns of torpidity of small homeotherms. In: Mammalian hibernation III. American Elsevier, New York, pp 30–46

462. Huggett AS, Widdas WF (1951). The relationship between mammalian fetal weight and conception age. J Physiol (London) 114: 306–317

463. Hughes GM (1965) Comparative physiology of vertebrate respiration. Harvard University Press, Cambridge

464. Hughes GM, Shelton G (1958) The mechanics of gill ventilation in three freshwater teleosts. J Exp Biol 35: 807–823

465. Hughes GM, Singh BN (1970) Respiration in an air breathing fish, the climbing perch, *Anabas testudineus.* J Exp Biol 53: 281–298

466. Hulet WH, Musil G (1968) Intracellular bacteria in the light organ of the deep sea anglerfish, *Melanocetus murrayi.* Copeia: 506–512

467. Hultén E (1937) Outline of the history of arctic and boreal biota during the Quarternary period. Bokförlags Aktiebolaget Thule, Stockholm

468. Humphrey SR, Kunz TH (1976) Ecology of a Pleistocene relict, the western big-eared bat *(Plecotus townsendii),* in the southern Great Plains. J Mamm 57: 470–494

469. Hungate RE (1975) The rumen microbial ecosystem. Ann Rev Ecol Systemat 6: 39–66

470. Hunt WH (1976) Maternal behavior in the Morelet's crocodile, *Crocodylus moreleti.* Copeia: 763–764

471. Hunter JR, Hasler AD (1965) Spawning association of the redfin shiner, *Notropis umbratilis,* and the green sunfish, *Lepomis cyanellus.* Copeia: 265–281

472. Hurlbert SH (1970) The post-larval migration of the red-spotted newt, *Notophthalmus viridescens* (Rafinesque). Copeia: 515–528

473. Hutchinson GE (1959) Homage to Santa Rosalia, or, why are there so many kinds of animals. Am Nat 93: 145–159

474. Hutchinson GE (1978) An introduction to population ecology. Yale University Press, New Haven.

475. Hutchison VR (1963) Critical thermal maxima in salamanders. Physiol Zool 36: 92–125

476. Hutchison VH, Dowling HG, Vinegar A (1966) Thermoregulation in a brooding female Indian python, *Python molurus bivattus.* Science 151: 694–696

477. Iles TD, Holden MJ (1969) Bi-parental mouth brooding in *Tilapia galilaea.* J Zool (London) 158: 327–333

478. Imber MJ (1973) The food of the grey-faced petrels (*Pterodroma macroptera gouldi* (Hutton), with special reference to diurnal migration of their prey. J Anim Ecol 42: 645–662

479. Immelmann K (1963) Drought adaptation in Australian desert birds. Proceedings of the XIII International Ornithological Congress (1962) 2: 649–657

480. Immelman K (1973) Role of the environment in reproduction as source of "predictive" information. In: Farner DS (ed) Breeding biology of birds. Natl Acad Sci, Washington, D.C., pp 121–257

481. Inger R, Marx H (1961) The food of amphibians. Exploration du Parc national de l'Upemba 64. Mission G.F. de Witte et al., Imprimerie Hayes, Brussels

482. Ireland PH (1976) Reproduction and larval development of the graybellied salamanders *Eurycea multiplicata griseogaster.* Herpetol 32: 233–238

483. Irving L (1969) Principles and further problems in the study of dormancy and survival. In: Dormancy and survival. 23rd Symposium for the Society of Experimental Biology 551–564

484. Irving L, Hart JS (1955) The metabolism and insulation of seals as bare-skinned mammals in cold water. Can J Zool 35: 497

485. Isaakian LA, Felberbaum RA (1949) Physiological investigation of the yellow ground squirrel *(Citellus fulvus)* at the time of entering estivation. In Bykov KM (ed) Experience in the study of the periodic changes of physiological functions in the organism. Acad Med Sci Press, Moscow: 194–206

486. Iverson JB (1978) Reproductive cycle of female loggerhead musk turtles *(Sternotherus minor minor)* in Florida. Herpetol 34: 33–39

487. Jackson CG Jr, Trotter JA, Trotter TH, Trotter MW (1970) Accelerated growth rate and early maturity in *Gopherus agassizi* (Reptilia: Testudines). Herpetol 32: 139–145

488. Jaeger EC (1949) Further observations on the hibernation of the poor-will. Condor 51: 105–109

489. Jackson DC, Allen J, Strupp PK (1976) The contribution of non-pulmonary surfaces to CO_2 loss in 6 species of turtles at 20°C. Comp Biochem Physiol 55A: 243–246

490. Jameson EW Jr (1952) Food of deer mice, *Peromyscus maniculatus* and *P. boylei,* in the northern Sierra Nevada, California. J Mamm 33: 50–60

491. Jameson, E.W., Jr (1954) Insects in the diet of pocket mice, *Perognathus parvus.* J Mamm 592–593

492. Jameson EW Jr (1964) Patterns of hibernation of captive *Citellus lateralis* and *Eutamias speciosus.* J Mamm 45: 455–460

493. Jameson EW Jr (1965) Food consumption of hibernating and nonhibernating *Citellus lateralis.* J Mamm 46: 634–640

494. Jameson EW Jr, Mead RA (1964) Seasonal changes in body fat, water and basic weight in *Citellus lateralis, Eutamias speciosus* and *E. amoenus.* J Mamm 45: 359–365

495. Jameson EW Jr, Allison A (1976) Fat and breeding cycles in two montane populations of *Sceloporus occidentalis* (Reptilia, Lacertilia, Iguanidae). J Herpetol 10: 211–220

496. Jameson EW Jr, Heusner AA, Arbogast R (1977) Oxygen consumption of *Sceloporus occidentalis* from three different elevations. Comp Biochem Physiol 56A: 73–79

497. Janssens PA (1964) The metabolism of the aestivating African lungfish. Comp Biochem Physiol 11: 105–117

498. Jarvik E (1955) The oldest tetrapods and their forerunners. Sci Monthly 80: 141–154

499. Jenni DA (1974) The evolution of polyandry in birds. Am Zool 14: 129–144

500. Jenssen TA (1967) Food habits of the green frog, *Rana clamitans,* before and during metamorphosis. Copeia: 214–218

501. Jewell PA (1966) Breeding season and recruitment in some British mammals confined on small islands. Zool Soc London Symp 15: 89–116

502. Johansen K (1963) Temperature regulation in the nine-banded armadillo *(Dasypus novemcinctus mexicanus).* Physiol Zool 36: 126–144

503. Johansen K (1966) Air breathing in the teleost Symbranchus marmoratus. Comp Biochem Physiol 18: 383–395

504. Johansen K (1970) Air breathing in fishes (361–411) In: Hoar WS, Randall DJ (eds) Fish physiology. Academic, New York

505. Johansen K, Lenfant C, Hanson D (1968) Cardiovascular dynamics in the lungfishes. Z Vergleich Physiol 59: 157–186

506. Johansen K, Lomholt JP (1974) Control of breathing in *Amphipnous cuchia,* an amphibious fish. Resp Physiol 21: 325–340

507. Johansen K, Krog J (1959) Diurnal body temperature variations and hibernation in the birch-mouse, *Sicista betulina.* Am J Physiol 196: 1200–1204

508. Johansson B, Senturia JB (1972) Seasonal variations in the physiology and biochemistry of the European hedgehog *(Erinaceus europaeus)* including comparisons with non-hibernators, guinea-pig and man. Acta Physiol Scand Suppl 380

509. John KR (1956) Onset of spawning activities of the shallow water cisco, *Leucichthys artedi* (Le Sueur) in Lake Mendota, Wisconsin, relative to water temperature. Copeia: 116–118

510. Johns MA, Feder HH, Komisaruk BR, et al (1978) Urine induced reflex ovulation in anovulatory rats may be a vomeronasal effect. Nature 272: 446–448

511. Johnels AG, Svensson GSO (1954) On the biology of *Protopterus annectens* (Owen). Ark Zool 7: 131–164

512. Johnson GE (1931b) Hibernation of the thirteen-lined ground squirrel *Citellus tridecemlineatus* (Mitchell). Biol Bull 59: 114–127

513. Johnson JL, McBee RH (1967) The porcupine caecal fermentation. J Nutrition 91: 540–546

514. Johnson MK, Hansen RM (1977) Foods of coyotes in the lower Grand Canyon, Arizona. J Arizona Acad Sci 12: 81–83

515. Johnson ML (1952) Herpetological notes from northeastern Brazil. Copeia: 283–284

516. Jones RE (1969) Hormonal control of incubation patch development in the California quail *Lophortyx californicus*. Gen Comp Endocrinol 13: 1–12

517. Judd FW, Herrera J, Wagner M (1978) The relationship between lipid and reproductive cycles of a subtropical population of *Peromyscus leucopus*. J Mamm 59: 669–676

518. Kaczymarski F (1966) Bioenergetics of pregnancy and lactation in the bank vole. Acta Theriologica 11: 409–417

519. Kahl MP Jr (1963) Thermoregulation in the wood stork, with species reference to the role of the legs. Physiol Zool 36: 141–151

520. Kalabukov NE (1960) Comparative ecology of hibernating mammals. In: Mammalian hibernation. Bull Mus Comp Zool, Harvard College 124: 45–74

521. Kalela O (1957) Regulation of reproductive rate in subarctic populations of the vole *Clethrionomys rufocanus* (Sund.). Suomalaisen Tiedeakatemian (Annales Academiae Scientiarum Fennicae), ser. A, IV. Biologica 34

522. Kallman KD (1962) Gynogenesis in the teleost *Mollienesia formosa* (Girard) with a discussion of the detection of parthenogenesis by tissue transplant. J Genet 58: 7–24

523. Kalmijn AJ (1971) The electrical sense of sharks and rays. J Exp Biol 55: 371–383

524. Kamiya T, Pirlot P (1975) Comparative gastric morphology of Old World bats. J Mamm Soc Jpn 6: 145–154

525. Kammermeyer KE, Marchinton RL (1976) Notes on dispersal of male white-tailed deer. J Mamm 57: 776–778

526. Kaneko Yukibumi (1976) Reproduction of Japanese field voles, *Microtus montebelli* Milne-Edwards, at Iwakura, Kyoto, Japan. Jpn J Ecol 26: 107–114

527. Katsurirangan LR (1951) Placentation in the sea-snake, *Enhydrina schistosa* (Daudin). Proc Indian Acad Sci 34: 1–32

528. Kawaji N, Shiraishi S (1979) Birds on the North coast of the Sea of Ariake. II. The relation between food habits of sandpipers and invertebrates in the substrate. J Faculty Agriculture, Kyushu University 23: 163–175

529. Kawamichi Takeo (1976) Hay territory and dominance rank of pikas *(Ochotona princeps)*. J Mammal 57: 133–148

530. Kay FR, Miller BW, Miller CL (1970) Food habits and reproduction of *Callosaurus draconoides* in Death Valley, California. Herpetol 26: 431–436

531. Kayser Ch, Vincendon G, Frank R, et al (1964) Some external (climatic) and internal (endocrine) factors in relation to production of hibernation. Ann Acad Sci Fenn Ser. A. IV Biologica 71/19: 271–282

532. Keast JA, Marhall AJ (1954) The influence of drought and rainfall on reproduction in Australian desert birds. Proc Zool Soc London 124: 494–499

533. Keast JA, Erk FC, Glass B (eds) (1972) Evolution, mammals, and southern continents. State University of New York, Albany

534. Keeton WT (1971) Magnets interfere with pigeon homing. Proc Nat Acad Sci 68: 102–106

535. Kehl R, Combescot C (1955) Reproduction in the reptilia. In: The comparative physiology of reproduction and the effects of sex hormones in vertebrates. Cambridge University Press, Cambridge, England, pp 57–74

536. Keith LB (1974) Some features of population dynamics in mammals. Proc Int Congr Game Biol 11: 17–58

537. Kelley WE, Atz JW (1964) A pygidiid catfish that can suck blood from a goldfish. Copeia: 702–704

538. Kellogg WN (1961) Porpoises and sonar. University of Chicago Press, Chicago

539. Kendall WA, Sherwood RT (1975) Palatability of leaves of tall fescue and reed canarygrass and some of their alkaloids to meadow voles. Agron J 67: 667–671

540. Kennicott R (1857) The quadrupeds of Illinois. Part I. Patent Office Reports for 1856: 52–110

541. Khalil F, Abdel-Messeih G (1962) Tissue constitutents of reptiles in relation to their mode of life-II. Lipid content. Comp Biochem Physiol 6: 171–174

542. King M (1977) Reproduction in the Australian gekko *Phyllodactylus marmoratus* (Gray). Herpetol 33: 7–13

543. King JR (1963) Autumnal migratory fat deposition in the white-crowned sparrow. Proceeding of the XIII International Ornithological Congress 2: 940–949

544. King JR (1967) Adipose tissue composition in experimentally induced fat deposition in the white-crowned sparrow. Comp Biochem Physiol 21: 393–403

545. King JR, Farner DS (1965) Studies in fat deposition in migratory birds. Ann NY Acad Sci 131, 1: 422–440

546. King JS (1970) Photoregulation of food intake and fat metabolism in relation to avian sexual cycles, pp. 365–385 In: Benoit J, Assenmacher I (eds) La photorégulation de la reproduction chez les oiseaux et les mammiféres. Paris

547. King JS, Farner DS (1965) Studies of fat deposition in migrating birds. Ann NY Acad Sci 131: 422–440

548. Kingston DI (1978) *Skiffia francesae,* a new species of goodeid fish from western Mexico. Copeia: 503–508

549. Kirtisinghe P (1957) The amphibia of Ceylon. Privately published, Ceylon

550. Kleerekoper H (1967) Some aspects of olfaction in fishes, with special reference to orientation. Am Zool 7: 385–395

551. Kluge AG, Eckardt MJ (1969) *Hemidactylus garnotii* Duméril and Bibron, a triploid all-female species of gekkonid lizard. Copeia: 651–664

552. Kluijver HN (1951) The population ecology of the great tit, *Parus m. major L.* Ardea 39: 1–135

553. Kluyver HN (1963) The determination of reproductive rates in Paridae. Proceedings of the XIII International Ornithological Congress (1962) 2: 706–716

554. Knepton JC Jr (1954) A note on the burrowing habits of the salamander *Amphiuma means means.* Copeia: 68

555. Knudsen EI (1975) Spatial aspects of the electric fields generated by weakly electric fish. J Comp Physiol 99: 103–118

556. Kolata, GB (1974) !Kung hunter-gatherers: feminism, diet and birth control. Science 185: 932–934

557. Kooyman GL, Drabek CM (1968) Observations on milk, blood, and urine constituents of the Weddell seal. Physiol Zool 41: 187–194

558. Kramer DL, Graham JB (1977) Synchronous air breathing, a social component of respiration in fishes. Copeia: 689–697

559. Kramer G (1959) Experiments on bird orientation and their interpretation. Ibis 99: 106–227

560. Krebs CJ (1966) Demographic changes in fluctuating populations of *Microtus californicus.* Ecol Monogr 36: 239–273

561. Krebs CJ, Gaines M, Keller B, et al (1973) Population cycles in small mammals. Science 179: 35–41

562. Kreithen ML, Eisner T (1978) Detection of ultraviolet light by the homing pigeon. Nature 272: 347–348

563. Krogh A (1941) The comparative physiology of respiratory mechanisms. University of Pennsylvania Press, Philadelphia

564. Krazanowski A (1964) Three long flights by bats. J Mamm 45: 152

565. Kühme W. (1963) Chemisch ausgelöste Brutpflege und Schwarmverhalten bei *Hemichromis bimaculatus* (Pisces). Z Tierpsychol 20: 688–704

566. Kurtén B (1960) Rates of evolution in fossil mammals. Cold Spring Harbor Symposia on Quantitative Biology 24: 205–215

567. Kurtén B (1968) Pleistocene mammals of Europe. Aldine, Chicago

568. Kurtén B (1969) Continental drift and evolution. Sci Am 220: 54–64
569. Labov JB (1977) Phytoestrogens and mammalian reproduction. Comp Biochem Physiol 57A: 3–9
570. Lack D (1948) Natural selection and family size in starlings. Evolution 2: 95–110
571. Lack D (1959) Migration across the sea. Ibis 101: 374–399
572. Lack D (1968) Ecological adaptations for breeding in birds. Methuen, London
573. Lackey JA (1978) Reproduction growth, and development in high-latitude and low-latitude populations of *Peromyscus leucopus* (Rodentia). J Mamm 59: 69–83
574. Landry SO Jr (1970) The Rodentia as omnivores. Q Rev Biol 45: 351–372
575. Lanyon WE, Fish WR (1958) Geographic variations in the vocalizations of the western meadowlark. Condor 60: 339–341
576. La Pointe JL, Rodriguez EM (1974) Fat mobilization and ultrastructural changes in the peritoneal fat body of the lizard, *Klauberina riversiana,* in response to long photoperiod and exogenous estrone or progesterone. Cell Tiss Res 155: 181–193
577. Larsson TB, Hansson L, Nyholm E (1973) Winter reproduction in small rodents in Sweden. Oikos 24: 475–476
578. Lasiewski RC (1963) Oxygen consumption of torpid, resting, active and flying hummingbirds. Physiol Zool 36: 122–140
579. Lasiewski RC, Weathers W, Bernstein MH (1967) Physiological responses of the giant hummingbird, *Patagona gigas.* Comp Biochem Physiol 23: 797–813
580. Laws RM (1953) The elephant seal (*Mirounga leonina* Linn.). I: Growth and age. Falkland Islands Dependencies Survey Sci Rep 8: 1–62
581. Laws RM (1958) Growth rates and ages of crabeater seals, *Lobodon carcinophagus,* Jacquinot and Pucheran. Proc Zool Soc London 130: 275–288
582. Laws RM (1959) Accelerated growth in seals, with special reference to the Phocidae. Norsk Hvalfangst Tidende 9: 425–438
583. Le Cren ED (1958) Observations on the growth of perch *(Perca fluviatilus)* over twenty-two years with special reference to the effects of temperature and changes in population density. J Anim Ecol 27: 287–334
584. Lee AK, Mercer EH (1967) Cocoon surrounding desert-dwelling frogs. Science 157: 87–88
585. Leggett WC (1977) The ecology of fish migrations. Ann Rev Ecol Systemat 8: 285–308
586. Le Maho Y (1977) The emperor penguin: a strategy to live and breed in the cold. Am Sci 65: 680–693
587. Lenfant C, Johansen K, Gregg GC (1964) Respiratory properties of blood and pattern of gas exchange in the lungfish. Resp Physiol 2: 1–22
588. Lenfant C, Jonansen K, Hanson D (1970) Bimodal gas exchange and ventilation-perfusion relationship in lower vertebrates. Fed Proc 29: 1124–1129
589. Leopold A, Erwin M, Oh J, et al (1976) Phytoestrogens: adverse effects on reproduction in California quail. Science 191: 98–100
590. Li Hui-Lin. (1952) Floristic relationships between eastern Asia and eastern North America. Trans Am Phil Soc 42: 371–429
591. Li Shichun, Liu, Bingqian. (1978) Cold resistance and thermoregulation of nestlings of the eastern great reed warbler. Acta Zool Sin 24: 251–254
592. Licht LE (1967) Death following possible ingestion of toad eggs. Toxicon 5: 141–142
593. Licht LE, Low B (1968) Cardiac response of snakes after ingestion of toad parotoid venom. Copeia: 547–551
594. Licht P (1971) Regulation of the annual testis cycle by photoperiod and temperature in the lizard *Anolis carolinensis.* Ecology 52: 240–252
595. Licht P (1972) Problems in experimentation on timing mechanisms for annual physiological cycles in reptiles. In: South FE et al (ed) Hibernation—hypothermia, prospectives and challenges. Elsevier, Amsterdam, pp 681–711
596. Lidicker W (1973) Regulation of numbers in an island population of the California vole, a problem in community dynamics. Ecol Monogr 43: 271–302

597. Lidicker W (1975) The role of dispersal in the demography of small mammals. In: Petruse-wicz K et al (eds) Small mammals: their productivity and population dynamics. Cambridge University Press, London, pp 103–128

598. Lidicker W (1976) Experimental manipulation of the timing of reproduction in the California vole. Res Popul Ecol 18: 14–27

599. Lidicker W (1978) Regulation of numbers in small mammal populations—historical reflections and a synthesis. In: Snyder DP Populations of small mammals under natural conditions. University of Pennsylvania, Pymatuning Laboratory of Ecology, Special Publication Series, Vol 5, pp 122–141

600. Liebman PA, Granada AM (1971) Microspectrophotometric measurements of visual pigments of turtle, *Pseudemys scripta* and *Chelonia mydas*. Vis Res 11: 105–114

601. Liem KF (1967) Functional morphology of the integumetary, respiratory, and digestive systems of the symbranchoid fish *Monopterus albus*. Copeia: 375–388

602. Ligon JD (1971) Late summer-autumnal breeding of the piñon jay in New Mexico. Condor 73: 147–153

603. Ligon JD (1974) Green cones of the piñon pine stimulate late summer breeding in the piñon jays. Nature 250: 80–83

604. Lim Boo-Liat (1970) Distribution, relative abundance, food habits, and parasite patterns of giat rats *(Rattus)* in West Malaysia. J Mamm 51: 730–740

605. Linsdale, JM (1928) A method showing relative frequency of occurrence of birds. Condor 30: 180–184

606. Linsley EG, Usinger RL (1959) Linnaeus and the International Code of Zoological Nomenclature. Syst Zool 8: 39–47

607. Lissmann HW, Machin KE (1958) The mechanisms of object location in *Gymnarchus niloticus* and similar fish. J Exp Biol 35: 415–486

608. Llewellyn JB (1978) Reproductive patterns in *Peromyscus truei truei* in a pinyon-juniper woodland of Western Nevada. J Mamm 59: 449–451

609. Lloyd HG (1970) Variation and adaptation in reproductive performance. In: Berry RJ, Southern HN (eds) Variation in mammalian populations. Zool Soc London Symp 26: 165–188

610. Lloyd JE (1977) Bioluminescence and communication. In: Sebeok TA (ed) How animals communicate. Indiana University Press, Bloomington

611. Lockington WN (1877) The long-jawed goby. Am Nat 11: 474–478

612. Lofts B, Marshall AJ (1960) The experimental regulation of *Zugunruhe* and the sexual cycle in the brambling *Fringilla montrifringella*. Ibis 102: 209–214

613. Lofts B (1974) Reproduction. In: Lofts B (ed) Physiology of the Amphibia. Academic, New York, pp 107–218

614. Lofts B, Murton RK (1968) Photoperiodic and physiological adaptations regulating avian breeding cycles and their ecological significance. J Zool Soc London 155: 327–394

615. Löhrl H (1963) The use of calls to clarify taxonomic relationships. Proceedings of the XIII International Ornithological Congress (1962) 1: 544–552

616. Longhurst WM, Oh HK, Jones MB, et al (1968) A basis for the palatability of deer forage plants. Trans North Am Wildlife Nat Res Conf, pp 181–192

617. Longhurst WM, Connolly GE, Browning BM, et al (1979) Food interrelationships of deer and sheep in parts of Mendocino and Lake Counties, California. Hilgardia 47, no. 6: 190–247

618. Lord RD Jr (1960) Litter size and latitude in North America mammals. Am Midl Nat 64: 488–499

619. Love RM (1970) The chemical biology of fishes. Academic, London

620. Lowe CH, Wright JW, Cole CJ, et al (1970) Natural hybridization between the teiid lizards *Cnemidophorus sonorae* (Parthenogenic) and *Cnemidophorus tigris* (Bisexual). System Zool 19: 114–127

621. Lucey EC, House EW (1977) Effect of temperature on the pattern of lung ventilation and on the ventilation-circulation pattern relationship of the turtle, *Pseudemys scripita.* Comp Biochem Physiol 57A: 239–243

622. Lutz, B (1972) Geographical and ecological notes on Cisandine and Platine frogs. J Herpetol 6: 83–100

623. Luckett WP (1977) Ontogeny of amniote fetal membranes and their application to phylogeny. In: Hecht MK et al (eds) Major patterns of vertebrate evolution. Plenum, New York, pp 439–516

624. Lyman CP (1954) Activity, food consumption and hoarding in hibernators. J Mamm 35: 545–552

625. Lyman CP (1963). Hibernation in mammals and birds. Am Sci summer: 127–138

626. Lyman CP (1970) Thermoregulation and metabolism in bats. In: Winsatt WA (ed) Biology of bats. Academic, New York, Vol 1, pp 301–330

627. Lyman CP, Chatfield PO (1950) Mechanism of arousal in the hibernating hamster. J Exp Zool 114: 491–516

628. Lynch GR (1973) Seasonal changes in the thermogenesis, organ weights, and body composition in the white-footed mouse, *Peromyscus leucopus.* Oecologia 13: 363–376

629. Macan TT (1963) Freshwater ecology. Wiley, New York

630. MacArthur R (1971) Patterns of terrestrial bird communities. In: Farner DS, King JR Avian biology. Academic, New York, Vol 1

631. MacArthur R, Wilson EO (1967) The theory of island biogeography. Monographs in Population Biology 1. Princeton University Press, Princeton

632. Macfarlane WV (1970) Seasonality of conception in human populations. Biometeorology 4: 167–182

633. MacMillen RE (1965) Aestivation in the cactus mouse, *Peromyscus eremicus.* Comp Biochem Physiol 16: 227–248

634. MacPhee C (1960) Postlarval development and diet of the largescale sucker, *Catostomus macrocheilus,* in Idaho. Copeia: 119–125

635. Maeda K (1974) Eco-éthologie de la grande noctule, *Nyctalus lasiopterus,* á Sapporo, Japon. Mammalia 38: 461–487

636. Magnussen H, Willmer H, Scheid P (1976) Gas exchange in air sacs: contribution to respiratory gas exchange in ducks. Resp Physiol 26: 129–146

637. Major PF (1973) Scale feeding behavior of the leatherjacket, *Scomberoides lysan* and two species of the genus *Oligophetes* (Pisces: Carangidae). Copeia: 151–154

638. Malfait BT, Dinkelman MG (1972) Circum-Caribbean tectonic and igneus activity and the evolution of the Caribbean plate. Geol Soc Am Bull 83: 251–272

639. Marcellini DL (1976) Some aspects of the thermal ecology of the gecko *Hemidactylus frenatus.* Herpetol 32: 341–345

640. Markham CD, McLain DR (1977) Sea surface temperature related to rain in Ceara, northeastern Brazil. Nature 265: 320–323

641. Marler P (1957) Specific distinctiveness in the communication signals of birds. Behaviour 11: 13–39

642. Marler P (1965) Communication in monkeys and apes. In: De Vore I (ed) Primate Behavior. Holt, Rinehart, New York, pp 544–584

643. Marshall AJ (1948) The breeding cycle of an equatorial bat *(Pteropus giganteus)* of Ceylon. Proc Linnean Soc London 159: 103–111

644. Marshall FHA (1936) Sexual periodicity and the causes which determine it. Phil Trans B 220: 423–456

645. Marshall JT Jr (1963) Rainy season nesting in Arizona. Proc of the XIII International Ornithological Congress (1962) 2: 620–622

646. Marshall NB (1967) Sound-producing mechanisms and the biology of deep-sea fishes. In: Tavolga WN (ed) Marine bio-acoustics. Pergamon, New York, Vol 2, pp 123–133

647. Martin AA (1970) Parallel evolution in the adaptive ecology of leptodactylid frogs of South America and Australia. Evolution 24: 643–644

648. Martin AA, Cooper AK (1972) The ecology of terrestrial anuran eggs, genus *Crinia* (Leptodactylidae). Copeia: 163–168

649. Martin GR (1974) Colour vision in the tawny owl *(Strix aluco)*. J Comp Physiol Psychol 86: 133–141

650. Martin PS, Harrell BE (1957) The Pleistocene history of temperate biotas in Mexico and eastern United States. Ecology 38: 468–480

651. Martin PS, Wright HE Jr (1967) Pleistocene extinctions; the search for a cause. Yale University Press, New Haven

652. Martinet L, Raynaud F (1973) Prolonged spermatozoa survival in the female hare uterus: explanation of superfetation. In: Hafez ESE, Thibault CG (eds) The biology of spermatozoa. INSERM Int. Symposium, Nouzilly 1973. Karger, Basel, pp 134–144

653. Martinsen DL (1968) Temporal patterns in the home ranges of chipmunks *(Eutamias)*. J Mamm 49: 83–91

654. Martof BS (1962) Some observations on the role of olfaction, among salientian Amphibia. Physiol Zool 35: 270–272

655. Maruyama N, Yukitoshi T, Nobuyuki K (1975) Seasonal changes of food habits of the sika deer in Omote-Nikko J Mamm Soc Jpn 6: 163–173

656. Maslin TP (1971) Conclusive evidence of parthenogenesis in three species of *Cnemidorphorus* (Teiidae). Copeia: 156–158

657. Maslin TP (1971) Parthenogenesis in reptiles. Am Zool 11: 361–380

658. Matthews GVT (1968) Bird navigation, 2nd ed. Cambridge University Press, Cambridge

659. Mautz WJ, Lopez-Forment W (1978) Observations on the activity and diet of the cavernicolous lizard *Lepidophyma smithii* (Sauria: Xantusiidae). Herpetol 34: 311–313

660. May RM (1977) Mathematical models and ecology. In: Goulden CE (ed) Changing scenes in natural sciences, 1776–1976. Philadelphia Academy of Natural Sciences Special Publication 12, Philadelphia

661. Mayhew WW (1964) Photoperiodic responses in three species of the lizard genus *Uma*. Herpetol 20: 95–113

662. Mayhew WW (1965) Adaptations of the amphibian, *Scaphiopus couchi,* to desert conditions. Am Midl Nat 74: 97–109

663. Mayhew WW, Weintraub JD (1971) Possible acclimatization in the lizard *Sceloporus orcutti*. J Physiol 63 (3): 336–340

664. Maynard Smith J (1977) Parental investment: a prospective analysis. Anim Behav 25: 1–9

665. Mayr E (1944) Wallace's line in the light of recent zoological studies. Q Rev Biol 19: 1–14

666. McAuliffe JR (1978) Seasonal migrational movements of a population of the western painted turtle *Chrysemys picta bellii* (Reptilia, Testudines, Testudinidae). J Herpetol 12: 143–149

667. McBee RH (1970) Metabolic contributions of caecal flora. Am J Clin Nutrition 23: 1514–1518

668. McCance RA (1971) The regulation of voluntary food intake. Proc Nutrition Soc 30: 103–149

669. McCarley H (1966) Annual cycle, population dynamics and adaptive behavior of *Citellus tridecemlineatus*. J Mamm 47: 294–316

670. McClintock MK (1971) Menstrual synchrony and suppression. Nature 229 (5282): 244–245

671. McDiarmid RW, Worthington RD (1970) Concerning the reproductive habits of tropical plethodontid salamanders. Herpetol 26: 57–70

672. McDowall RM, Whitaker AH (1975) Freshwater fishes. In: Kuschel G (ed) Biogeography and ecology in New Zealand. Junk, The Hague

673. McGavin M (1978) Recognition of conspecific odors by the salamander *Plethodon cinereus*. Copeia: 356–358

674. McGinnis SM, Moore RG (1969) Thermoregulation in the boa constrictor, *Boa constrictor*. Herpetol 25: 38–45

675. McGinnis SM, Whittow GC, Ohata CA, et al (1972) Body heat dissipation and conservation in two species of dolphins. Comp Biochem Physiol 43A: 417–423

676. McIntosh TK, Barfield RJ, Geyer LA (1978) Ultrasonic vocalizations facilitate sexual behavior of female rats. Nature 272: 163–164

677. McKeever S (1964) The biology of the golden-mantled ground squirrel, *Citellus lateralis*. Ecol Monogr 34: 383–401

678. McLean EK (1970) The toxic actions of pyrrolizidine *(Senecio)* alkaloids. Pharmacol Rev 22: 429–483

679. McMillan JP (1972) Pinealectomy abolishes the circadian rhythm of migratory restlessness. J Comp Physiol 79: 105–112

680. McNab BK (1978) The evolution of endothermy in the phylogeny of mammals. Am Nat 112: 1–21

681. McNab BK, Auffenberg W (1976) The effect of large body size on the temperate regulation of the Komodo dragon, *Varanus komodoensis*. Comp Biochem Physiol 55A: 345–350

682. Mead GW, Bertlesen E, Cohen DM (1964) Reproduction among deep sea fishes. Deep Sea Res 11: 569–596

683. Mead RA (1968) Reproduction in western forms of the spotted skunk (genus *Spilogale*). J Mammal 49: 373–390

684. Mead RA, Swannick A (1978) Effects of hysterectomy on luteal function in the western spotted skunk *(Spilogale putorius latifrons)*. Biol Reprod 18: 379–383

685. Medica PA, Bury RB, Turner FB (1975) Growth of the desert tortoise *(Gopherus agassizi)* in Nevada. Copeia: 639–643

686. Medway L (1959) Echolocation among *Collocalia*. Nature 184: 1352–1353

687. Medway L (1962) The swiftlets *(Collocalia)* of Niah Cave, Sarawak. Ibis 104: 45–66; 228–245

688. Medway L (1970) Breeding of the silvered leaf monkey, *Presbytis cristata*, in Malaya. J Mamm 51: 630–632

689. Medway L (1971) Observation of social and reproductive biology of bentwinged bat *Miniopterus australis* in northern Borneo. J Zool (London) 165: 261–273

690. Medway L (1972) Reproductive cycle of the flat-headed bats *Tylonycteris pachypus* and *T. robustula* (Chiroptera; Vespertilioninae) in a humid equatorial environment. Zool J Linnean Soc 51: 36–61

691. Meier AH, Burns JT (1976) Circadian hormone rhythms in lipid regulation. Am Zool 16: 649–659

692. Meive AH, Ferrell BR (1978) Avian endocrinology. In: Brush AH (ed) Chemical zoology, X. Aves. pp 213–271

693. Menaker M (1962) Hibernation-hypothermia: an annual cycle of response to low temperature in the bat *Myotis lucifugus*. J Cell Comp Physiol 59: 163–173

694. Menzies JI (1976) Handbook of common New Guinea frogs. Wau Ecology Institute Handbook No. 1. Wau, Papua New Guinea

695. Merrill ED (1926) The correlation of biological distribution with the geological history of Malaysia. Third Pan-Pacific Science Congress, Tokyo

696. Meszler RM, Webster DB (1968) Histochemistry of the rattlesnake facial pit. Copeia: 722–728

697. Migdalski EC (1957) Contribution to the life history of the South American fish, *Arapaima gigas*. Copeia: 54–56

698. Migula P (1969) Bioenergetics of pregnancy and lactation in European common vole. Acta Theriologica 14: 167–179

699. Miller CM (1944) Ecological relations and adaptations of the legless lizards of the genus *Anniella*. Ecol Monogr 14: 289–291

700. Miller MR (1948) The seasonal histological changes occurring in the ovary, corpus luteum and testis of the viviparous lizard *Xantusia vigilis*. University of California Publications Zoology 47: 197–223

701. Miller RR (1958) Origin and affinities of the freshwater fish fauna of western North America. In: Hubbs CL (ed) Zoogeography. Am Assoc Adv Sci Publ 51, Washington DC
702. Miller RS (1969) Competition and species diversity. Brookhaven Symposium Quant Biol 22: 63–70
703. Milne A (1961) Definition of competition among animals. Soc Exp Biol (London) Symposia 15: 40–61
704. Moehn LD (1974) The effects of quality of light on agonistic behavior of iguanid and agamid lizards. J Herpetol 8: 175–183
705. Mohr CO (1947) Table equalivalent of populations of North American small mammals. Am Midl Nat 37: 223–249
706. Möhres FP (1953) Über die Ultraschallorientierung der Hufeisennasen (Chiroptera-Rhinlophinae) Z Vergleich Physiol 34: 547–588
707. Möhres FP (1956) Über die Orientierung der Hughunde. Z Vergleich Physiol 38: 1–29
708. Moir RJ (1968) Ruminant digestion and evolution. In: Handbook of physiology, section 6: alimentary canal, vol. 5 Bile; digestion and ruminal physiology: 2673–2694
709. Moller P (1976) Electric signals and schooling behavior in a weakly electric fish, *Marcusenius cyprinoides* L. (Mormyriformes). Science 193: 697–699
710. Moore FR (1977) Geomagnetic disturbance and the orientation of nocturnally migrating birds. Science 196: 682–684
711. Moore GA (1950) The cutaneous sense organs of barbeled minnows adapted to life in the muddy waters of the Great Plains region. Trans Am Microsc Soc 69: 69–95
712. Moore WS, Miller RR, Schultz RJ (1980) Distribution, adaptation and probable origin of an all-female form of *Poeciliopsis* (Pisces: Poeciliidae) in northeastern Mexico. Evolution 24: 789–795
713. Moreau RE (1950) The breeding season in African birds. I. Land birds. Ibis 92: 223–267
714. Moreau RE (1963) Vicissitudes of the African biomass in the late Pleistocene. Proc Zool Soc London 141: 395–421
715. Môri T, Uchida TY (1980) Sperm storage in the reproductive tract of the female Japanese long-fingered bat, *Miniopterus schreibersi fuliginosus*. J Reprod Fertil 58: 429–433
716. Morin JG, Harrington A, Nealson K, Krieger N, Baldwin TD, Hastings JW (1975) Light for all reasons: versatility in the behavioral repertoire of the flashlight fish. Science 190: 74–76
717. Morris L, Morrison P (1964) Cyclic responses in dormice, *Giis glis,* and ground squirrels, *Spermophilus tridecemlineatus,* exposed to normal and reversed yearly light schedules. *Science in Alaska:* Proceedings of the 15th Alaska Science Conference, A A A S: 40–41
718. Morrison P (1966) Insulative flexibility in the guanaco. J Mamm 47: 18–23
719. Morrison P, Rosenmann M, Estes JA (1974) Metabolism and thermoregulation in the sea otter. Physiol Zool 47: 218–229
720. Morton ES (1977) On the occurrence and significance of motivation-structural roles in some bird and mammal sounds. Am Nat 111: 855–869
721. Morton ES, Tenaza R (1977) Signalling behavior of apes with special reference to vocalization. In: Sebeak TA (ed) How animals communicate. Indiana University Press, Bloomington
722. Moulton DG (1967) Olfaction in mammals. Am Zool 7: 421–429
723. Moy-Thomas JA (revised by) Miles RS (1971) Palaeozoic Fishes. Saunders, Philadelphia
724. Moyle V (1949) Nitrogenous excretion in chelonian reptiles. Biochem J 44: 581–584
725. Mrosovsky N (1971) Hibernation and the hypothalamus. Appleton-Century-Crofts, New York
726. Mrosovsky N, Fisher KC (1970) Sliding set points for body weight in ground squirrels during the hibernation system. Can J Zool 48: 241–247
727. Mrosovsky N, Pritchard PCH (1971) Body temperatures of *Dermochelys coriacea* and other sea turtles. Copeia: 624–631
728. Mrosovsky N, Barnes DS (1974) Anorexia, food deprivation and hibernation. Physiol Behav 12: 265–270

729. Mueller CC, Sadlier RMFS (1977) Changes in the nutrient composition of milk of black-tailed deer during lactation. J Mamm 58: 421–423
730. Mueller CF (1969) Temperature and energy characteristics of the sagebrush lizard *(Sceloporus graciosus)* in Yellowstone National Park. Copeia: 153–160
731. Muir BS, Buckley RM (1967) Gill ventilation in *Remora remora*. Copeia: 581–586
732. Muir BS, Kendall JI (1968) Structural modifications in the gills of tunas and some other oceanic fishes. Copeia: 388–398
733. Müller P (1973) The dispersal centres of terrestrial vertebrates in the Neotropical Realm. Junk, The Hague
734. Muntz WRA (1972) Inert absorbing and reflecting pigments. In: Dartnall HJA (ed) The handbook of sensory physiology, VII/I. Springer Verlag, Berlin
735. Murrish DE, Schmidt-Nielsen K (1970) Exhaled air temperature and water conservation in lizards. Resp Physiol 10: 151–158
736. Mushinsky HR, Hebrard JJ (1977) Food partitioning of five species of water snakes in Louisiana. Herpetol 33: 162–166
737. Mutere FA (1970) The breeding biology of equatorial vertebrates: reproduction in the insectivorous bat, *Hipposideros caffer*, living at 0°27'N. Proceedings of the Second International Bat Research Conference. Bijdragen Tot de Dierkunde 40: 56–58
738. Muth A (1977) Body temperatures and associated postures of the zebra-tailed lizard, *Callisaurus draconoides*. Copeia: 122–125
739. Muul I (1969) Photoperiod and reproduction in flying squirrels. Glaucomys volans. J Mamm 50: 542–549
740. Müller P (1973) The dispersal centres of terrestrial vertebrates in the Neotropical Realm. Biogeographica 2
741. Myers GS (1951) Fresh-water fishes and East Indian Zoogeography. Stanford Ichthyol Bull 4: 11–21
742. Myers P (1977) Patterns of reproduction of four species of vespertilionid bats in Paraguay. University of California Publications in Zoology 107: 1–41
743. Nagel JW (1977) Life history of the red-backed salamander, *Plethodon cinereus,* in Northeastern Tennessee. Herpetol 33: 13–18
744. Nagy JG, Steinhoff HW, Ward GM (1964) Effects of essential oils of sagebrush on deer rumen microbial function. J Wildlife Management 28: 785–790
745. Nagy JG, Tengerdy RP (1967) Antibacterial action of essentials oils of *Artemisia* as an ecological factor. I. Antibacterial action of the volatile oils of *Artemisia tridentata* and *Artemisia nova* on aerobic bacteria. App Microbiol 15: 819–821
746. Nagy K (1973) Behavior, diet and reproduction in a desert lizard, *Sauromalus obesus*. Copeia: 93–102
747. Nagy K (1977) Cellulose digestion and nutrient assimulation in *Sauromalus obesus,* a plant-eating lizard. Copeia: 355–362
748. Nakamura Tsukasa (1969) Physiological and ecological studies on bird migration with special reference to body lipids. Bull Faculty Education, Yamanaski University 4: 131–163
749. Nathan JM, James VG (1972) The role of protozoa in the nutrition of tadpoles. Copeia: 669–679
749a Neave DJ, Wright BS (1968) Seasonal migration of the harbor-porpoise *(Phocoena phocoena)* and other cetacea in the Bay of Fundy. J Mamm 49: 259–264
750. Negus NC, Berger PJ (1971) Pineal weight response to a dietary variable in *Microtus montanus*. Experientia 27: 215–216
751. Negus NC (1972) Environmental factors and reproductive processes in mammalian populations. In: Velardo NT, Kasprow BA (eds) Biology of reproduction, basic and clinical studies. Third Pan American Congress of Anatomy, New Orleans
752. Negus NC, Berger PJ, Forslund LG (1977) Reproductive strategy of *Microtus montanus*. J Mamm 58: 347–353

753. Neill JD (1970) Effect of "stress" on serum prolactin and luteinizing hormone levels during the estrous cycle of the rat. Endocrinol 87: 1192–1197

754. Neill WT (1964) Viviparity in snakes: some ecological and zoogeographic considerations. Am Nat 98: 35–55

755. Neill WT (1950) An estivating bowfin. Copeia: 240

756. Neill WT (1971) The last of the ruling reptiles: alligators, crocodiles and their kin. Columbia University Press, New York

757. Nelson CE, Cuellar HS (1968) Anatomical comparison of tadpoles of the genera *Hypopachus* and *Gastrophryne*. Copeia: 423–424

758. Nelson DR, Johnson RH (1972) Acoustic attraction of Pacific reef sharks: effect of pulse intermittency and variability. Comp Biochem Physiol 42A: 85–95

759. Nelson GJ (1969) The problem of historical zoogeography. System Zool 18: 243–246

760. New JG (1966) Reproductive behavior of the shield darter, *Percina peltata peltata* in New York. Copeia: 20–28

761. Newsome AE (1973) Cellular degeneration in the testis of red kangaroos during hot weather and drought in central Australia. J Reprod Fertil Suppl 19: 191–201

762. Nichol JAC, Arnott HJ, Best ACG (1973) Tapeta lucida in bony fishes (Actinopterygii): a survey. Can J Zool 51: 69–81

763. Nikolsky GV (1963) The ecology of fishes. (Translated by L. Birkett.) Academic, London

764. Noble GK (1917) The systematic status of some batrachians from South America. Bull Am Mus Nat Hist 37: 793–814

765. Noble GK (1937) The sense organs involved in the courtship of *Storeria, Thamnophis* and other snakes. Bull Am Mus Nat Hist 73: 673–725

766. Noble GK, Schmidt A (1937) The structure and function of the facial and labial pits of snakes. Proc Am Philos Soc 77: 263–288

767. Norberg Å (1967) Physical factors in directional hearing in *Aegolius funereus* (Linné) (Strigiformes), with special reference to the significance of the asymmetry of the external ears. Ark Zool 20: 181–204

768. Norberg Å (1977) Occurrence and independent evolution of bilateral ear asymmetry in owls and implications on owl taxonomy. Philos Trans R Soc London B 280: 375–408

769. Norris KS (1966) Whales, dolphins, and porpoises. University of California Press, Berkeley

770. Novick A (1957) Orientation in paleotropical bats: I. Microchiroptera. Fed Proc 16: 95–96

771. Nussbaum RA (1970) *Dicamptodon copei*, n. sp., from the Pacific Northwest, U. S. A. (Amphibia: Caudata: Ambystomidae). Copeia: 506–514

772. O'Day WT (1973) Luminescent silhouetting in stomiatoid fishes. Los Angeles County Museum Contributions in Science No. 246

773. O'Day WT (1974) Bacterial luminescence in the deep-sea anglerfish *Oneirodes acanthias* (Gilbert, 1915) Los Angeles County Museum Contributions in Science No. 255

774. O'Day WT, Fernandez HR (1974) *Aristostomias scintillans* (Malacosteidae): a deep-sea fish with visual pigments apparently adapted to its own bioluminescence. Vis Res 14: 545–550

775. O'Day WT, Young RW (1978) Rhythmic daily shedding of outer segment membranes by visual cells in the goldfish. J Cell Biol 76: 593–604

776. Odum EP, Kuenzler EJ (1955) Measurements of territory and home range in birds. Auk 72: 128–137

777. Oh, Hi Kon, Jones MB, Longhurst WM (1968) Comparison of rumen microbial inhibition resulting from various essential oils isolated from relatively unpalatable plant species. App Microbiol 16: 39–44

778. Ohmart RD (1972) Physiological and ecological observations concerning the salt-secreting glands of the roadrunner. Comp Biochem Physiol 43: 311–316

779. Oksche A (1978) Evolution, differentiation and organization of hypothalamic systems controlling reproduction. In: Scott DE et al (eds) Brain-endocrine interaction. III. Neural hormones and reproduction. Karger, Basil

780. Organ JA (1960) Studies on the life history of the salamander *Plethodon welleri*. Copeia: 287–297

781. Orians GH (1969) On the evolution of mating systems in birds and mammals. Am Nat 103: 589–603

782. Orr Y (1974) Adaptation of the small-spotted lizard, *Eremias guttulata*. Israel J Med Sci 10: 285–286

783. Osgood DW (1970) Thermoregulation in water snakes studied by telemetry. Copeia: 568–570

784. O'Shea TJ, Vaughn TA (1977) Nocturnal and seasonal activities of the pallid bat, *Antrozous pallidus*. J Mamm 58: 269–284

785. Oshima K (1969) Electroencephalic olfactory response in adult salmon to waters traversed in the homing migration. J Fisheries Res Board Can 26: 2123–2133

786. Packard GC, Tracy CR, Roth JJ (1977) The physiological ecology of reptilian eggs and embryos, and the evolution of viviparity within the class Reptilia. Biol Rev 52: 71–105

787. Packard WC (1963) Observations on the breeding migration of *Taricha rivularis*. Copeia: 378–382

788. Packard WC (1960) Bioclimatic influences on the breeding migration of *Taricha rivularis* Ecology 41: 509–517

789. Packard WC (1963) Observation on the breeding migration of *Taricha rivularis*. Copeia: 378–382

790. Panigel M (1956) Contribution á l'etude de l'ovoviviparité chez reptiles: gestation et parturition chez le lizard vivipare *Zootoca vivipara*. Ann Sci Nat Zool Biol Anim 18: 569–668

791. Parker HW (1956) Viviparous caecilians and amphibian phylogeny. Nature 178: 250–252

792. Parkes AS (1976) Patterns of sexuality and reproduction. Oxford University Press, London

793. Patterson B, Pascual R (1972) The fossil mammal fauna of South America. In: Evolution, mammals and southern continents. State University of New York, Albany, pp 247–309

794. Payne RS (1972) Acoustic location of prey by barn owls. J Exp Biol 54: 535–573

795. Payne RS, McVay S (1971) Songs of humpback whales. Science 173: 585–599

796. Peaker M, Linzell JL (1975) Salt glands in birds and reptiles. Cambridge University Press, London

797. Pearl R (1939) The natural history of population. Oxford University Press, New York

798. Pearson OP (1954) Habits of the lizard *Liolaemus multiformis multiformis* at high altitudes in southern Peru. Copeia: 111–116

799. Pearson OP 1966. The prey of carnivores during one cycle of mouse abundance. J Anim Ecol 35: 217–233

800. Pearson OP (1977) The effect of substrate and of skin color on thermoregulation of a lizard. Comp Biochem Physiol 58A: 353–358

801. Pearson OP, Bradford DF (1976) Thermoregulation of lizards and toads at high altitudes in Peru. Copeia: 155–170

802. Pengelley ET (1966) Differential developmental patterns and their adaptive value in various species of the genus *Citellus*. Growth 30: 137–142

803. Pengelley ET (1967) The relation of external conditions to the onset and termination of hibernation and estivation. In: Mammalian hibernation III. American Elsevier, New York, pp 1–29

804. Pengelley ET, Fisher KC (1961) Rhythmical arousal from hibernation in the golden-mantled ground squirrel *Citellus lateralis tescorum*. Can J Zool 39: 105–120

805. Pengelley ET, Kelley KH (1966) A "circannian" rhythm in hibernating species of the genus *Citellus* with observations on their physiological evolution. Comp Biochem Physiol 19: 603–617

806. Penn GH Jr, Pottharst KE (1940) The reproduction and dormancy of *Terrapene major* in New Orleans. Herpetol 2: 25–29

807. Perdeck AC (1974) An experiment on the orientation of juvenile starlings during spring migration. Ardea 62: 190–195

808. Perdeck AC (1958) Two types of orientation in migrating starlings *Sturnis vulgaris* L. and chaffinches *Fringilla coelebs* L., as revealed by displacement experiments. Ardea 46: 1–37
809. Perdeck AC (1967) Orientation of starlings after displacement to Spain. Ardea 55: 194–202
810. Perrins CM (1973) Some effects of temperature on breeding of the great tit and Manx shearwater. J Reprod Fertil Suppl 19: 163–173
811. Perry JS (1971) The ovarian cycle of mammals. University Reviews in Biology, no. 13, Oliver and Boyd, Edenburgh
812. Peters RC, van Wijland F (1974) Electro-orientation in the passive electric catfish, *Ictalurus nebulosus*. J Comp Physiol 92: 273–280
813. Peterson EA (1966) Hearing in the lizard: some comments on the auditory capacities of a nonmammalian ear. Herpetol 22: 161–171
814. Pettigrew JD (1978) A role for the avian pecten oculi in orientation to the sun. In: Schmidt-Koenig K, Keeton WT (eds) Animal migration, navigation and homing. Springer, Berlin, pp 42–54
815. Pfeiffer W (1974) Pheromones in fish and amphibians In: Birch MC (ed) Pheromones. Frontiers of biology. North-Holland, Amsterdam, Vol 32, pp 269–296
816. Philip A, Bury RB, Turner FB (1976) Growth of the desert tortoise *(Gopherus agassizi)* in Nevada. Copeia: 639–643
817. Phillips JA (1979) Indirect body fat manipulation and its effect on hibernation cycles in *Citellus lateralis*. Can J Zool 57: 976–978
818. Pianka ER (1977) Reptilian species diversity. In: Gans C (ed) Biology of the Reptilia. Academic, New York, pp 1–34
819. Pickens PE, MacFarland WN (1964) Electric discharge and associated behaviour in the stargazer. Anim Behav 12: 362–367
820. Pietsch TW (1977) Dimorphism, parasitism, and sex: reproductive strategies among deep sea ceratoid anglerfishes. Copeia: 781–793
821. Piggins DJ (1970) Refraction of the harp seal, *Pagophilus groenlandicus* (Erxleben 1977). Nature 227 (5253): 78–79
822. Piiper J, Schied P (1975) Gas transport efficacy of gills, lungs and skin: theory and experimental data. Resp Physiol 23: 209–221
823. Pitelka FA, Tomich PQ, Treichel GW (1955) Ecological relations of jeagers and owls as lemming predators near Barrow, Alaska. Ecol Monogr 25: 85–117
824. Pivorunas A (1979) The feeding mechanism of baleen whales. Am Sci 67: 432–440
825. Pohl H (1961) Temperature regulation und Tagesperiodik des Stoflwechsels bei Winterschlafern. Z Vergleich Physiol 45: 109–153
826. Polder JJW (1971) On gonads and reproductive behavior in the cichlid fish *Aequidens portalegrensis* (Hensel). Neth J Zool 21: 265–365
827. Pond CM (1978) Morphological aspects and the ecological and mechanical consequences of fat deposition in wild vertebrates. Ann Rev Ecol Systemat 9: 519–570
828. Pooley AC (1962) The Nile crocodile, *Crocodilus niloticus:* Notes on the incubation period and growth rate of juveniles. Lammergeyer 2: 1–55
829. Porter LS Wingo (1978) Pleistocene pluvial climates as indicated by present day climatic parameters of *Cryptotis parva* and *Microtus mexicanus*. J Mamm 59: 300–338
830. Porter WP (1967) Solar radiation through the living body walls of vertebrates with emphasis on desert reptiles. Ecol Monogr 37: 273–296
831. Prevost J, Vilter V (1963) Histologie de la sécrétion oesophagienne du Manchot Empereur. Proceedings of the 13th International Ornithological Congress, 1962: 1085–1094
832. Price MRS (1977) The estimation of food intake, and its seasonal variation, in the hartebeest. East African Wildlife J 15: 107–124
833. Price MRS (1978) The nutritional ecology of Coke's hartebeest *(Alcelaphus buselaphus cokei)* in Kenya. J Appl Ecol 15: 33–49
834. Prieto AS, Leon JR, Lara G (1976) Reproduction in the tropical lizard, *Tropidurus hispidus* (Sauria: Iguanidae). Herpetol 32: 318–323
835. Pucke J, Umminger BL (1979) Histophysiology of the gills and dendritic organ of the marine catfish, *Plotosus lineatus,* in relation to osmoregulation. Copeia: 357–360

836. Pyke, GH (1979) The economics of territory size and time budget in the golden-winged sun-bird. Am Nat 114: 131–145
837. Racey PA (1973) Environmental factors affecting the length of gestation in heterothermic bats. J Reprod Fertil Suppl 19: 175–189
838. Rahm U (1970) Note sur la reproduction des sciuridés et muridés dans la forêt equatoriale au Congo. Rev Suisse Zool 77: 635–646
839. Rahn H (1939) Structure and function of placenta and corpus luteum in viviparous snakes. Proc Soc Exp Biol Med 40: 381–382
840. Rahn H (1956) The relationship between hypoxia, temperature, adrenalin release and melanophore expansion in the lizard, *Anolis carolinensis*. Copeia: 214–217
841. Rahn H, Rahn KB, Howell BJ, et al (1971) Air breathing of the garfish *(Lepisosteus osseus)*. Resp Physiol 11: 285–307
842. Rahn H, Ar A (1974) The avian egg: incubation time and water loss. Condor 76: 147–152
843. Rahn H, Howell BJ (1976) Bimodal gas exchange. In: Hughes GM (ed) Respiration of amphibious vertebrates. Academic, New York, pp 271–285
844. Rand AL (1954) The ice age and mammal speciation in North America. Arctic 7: 31–35
845. Raven PH, Axelrod DI (1972) Plate tectonics and Australasian paleobiogeography. Science 176: 1379–1384
846. Raven PH (1974) Angiosperm biogeography and past continental movements. Ann Missouri Bot Garden 61: 539–673
847. Raven PH (1975) History of the flora and fauna of Latin America. Am Sci 63: 420–429
848. Regal PJ (1967) Voluntary hypothermia in reptiles. Science 155: 1551–1553
849. Reid B, Williams GR (1975) The kiwi. In: Kuschel G (ed) Biogeography and ecology in New Zealand. Junk, The Hague
850. Reinboth R (ed) (1974) Symposium on intersexuality in the animal kingdom. Springer, Berlin
851. Rickleffs RE (1973) Fecundity, mortality and avian demography. In: Farner DS (ed) Breeding biology of birds. Natl Acad Sci Washington, D.C., pp 366–435
852. Rice DW, Kenyon KW (1962) Breeding cycles and behavior of Layson and black-footed albatrosses. Auk 79: 517–567
853. Rifkind AB, Kulin HE, Ross GT (1967) Follicle stimulating hormone (FSH) and luteinizing hormone (LH) in the urine of pre-pubertal children. J Clin Invest 12: 1123–1128
854. Rigley L, Marshall JA (1973) Sound production by the elephant-nose fish, *Gnathonemus petersi* (Pisces, Mormyridae). Copeia: 134–135
855. Ripley SD (1976) Rails of the world. Am Sci 64: 628–635
856. Robertson DR (1972) Social control of sex reversal in a coralreef fish. Science 177: 1007–1009
857. Robertson LAD, Chapman BM, Chapman RF (1965) Notes on the biology of the lizards *Agama cyanogaster* and *Mabuya striata striata* collected in the Rukwa Valley, Southwest Tanganyika. Proc Zool Soc London 145: 305–320
858. Rodieck RW (1973) The vertebrate retina: Principles of structure and function. WH Freeman, San Francisco
859. Romer AS (1947) Review of the Labyrinthodontia. Bull Mus Comp Zool 99: 1–368
860. Romer AS, Olson EC (1954) Aestivation in a Permian lungfish. Brevoria, Mus Comp Zool 30: 1–8
861. Rosen DE (1975) A vicariance model of Caribbean biogeography. Syst Zool 24: 431–464
862. Rosen DE, Greenwood PH (1970) Origin of the Weberian apparatus and the relationships of ostariophysan and gonorynchiform fishes. Am Mus Novitates 2428: 1–25
863. Rowan MK (1967) A study of the colies of Southern Africa. Ostrich 38: 63–115
864. Rowan W (1926) On photoperiodism, reproductive periodicity and annual migrations of birds and certain fishes. Proc Boston Soc Nat Hist 38: 149–189
865. Ruibal R (1959) The ecology of a brackish water population of Rana pipiens. Copeia: 315–322
866. Ruibal R (1961) Thermal relations of five species of tropical lizards. Evolution 15: 98–111
867. Ruibal R (1962) The ecology and genetics of a desert population of *Rana pipiens*. Copeia: 189–195

868. Ruibal R, Philibosian R (1970) Eurythermy and niche expansion. Copeia: 645–653

869. Ruibal R, Philibosian R, Adkins JL (1972) Reproductive cycle and growth in the lizard *Anolis acutus*. Copeia: 509–518

870. Rüppell W (1931) Zug der jugen Störche (*Ciconia c. ciconia* L.) ohne Führung der Alten? Vogelzug 2: 119–122

871. Rusch DA, Reeder WG (1978) Population ecology of Alberta red squirrels. Ecology 59: 400–420

872. Russell IJ (1976) Amphibian lateral line receptors. In: Llinás R, Precht W. Frog neurobiology. Springer, New York, pp 513–550

873. Rutledge JT (1974) Circannual rhythms of reproduction in male European starlings *(Sturnus vulgaris)*. In: Pengelley ET (ed) Circannual clocks. Academic, New York, pp 297–345

874. Ryan MJ (1978) A thermal property of the *Rana catesbeiana* (Amphibia, Anura, Ranidae) egg mass. J Herpetol 12: 247–248

875. Salthe SN, Duellman WE (1973) Quantitative constraints associated with reproductive mode in anurans. In: Evolutionary biology of the anurans. University of Missouri Press, Columbia, pp 229–249

876. Salthe SN, Mecham JS (1973) Reproductive biology of the amphibia. In: Lofts B (ed) Physiology of the Amphibia. Academic Press, New York, Vol 2

877. Sanyal MK, Prasad MRN (1967) Reproductive cycle of the Indian house lizard, *Hemidactylus flaviviridus* Ruppell. Copeia: 627–633

878. Sarnthein M (1978) Sand deserts during glacial maximum and climatic optimum. Nature 272: 43–46

879. Sauer EGF, Sauer EM (1960) Star navigation of nocturnal migrating birds. The 1958 planetarium experiments. Cold Spring Harbor Symposia on Quantitative Biology 25: 463–473

880. Savage JM (1974) The Isthmian Link and the evolution of Neotropical mammals. Los Angeles County Natural History Museum, Contributions in Science, Number 260

881. Savage JM (1966) The origins and history of the Central American herpetofauna. Copeia: 714–766

882. Sawicka-Kapusta K (1968) Annual fat cycle of field mice, *Apodemus flavicollis* (Melchoir, 1834). Acta Theriologica 13: 329–339

883. Scalia F, Winans SS (1976) New perspectives on the morphology of the olfactory system: olfactory and vomeronasal pathways in mammals. In: Doty RL (ed) Mammalian olfaction, reproductive processes, and behavior. Academic, New York, pp 7–28

884. Scaglion R (1978) Seasonal births in a Western Abelam Village, Papua New Guinea. Hum Biol 50: 313–323

885. Scheid P, Piiper J (1972) Cross-current gas exchange in avian lungs: effects of parabronchial air flow in ducks. Resp Physiol 16: 304–312

886. Schindler O (1947) Einiges vom Winterschlaf der Fische. Allgem Fischerei-Z 72 (1): 4–5

887. Schmidt RS (1965) Larynx control and call production in frogs. Copeia: 143–147

888. Schmidt RS (1970) Auditory receptors of two mating call-less anurans. Copeia: 169–170

889. Schmidt-Koenig K, Schlichte HJ (1972) Homing in pigeons with impaired vision. Proc Natl Acad Sci (USA) 69: 2446–2447

890. Schmidt-Nielsen K (1975) Animal physiology; adaptation and environment. Cambridge University Press, London

891. Schmidt-Nielson K, Dawson TJ, Hammel HT, et al (1965) The jackrabbit—a study in desert survival. Hvalrådets Skrifter 48: 125–142

892. Schmidt-Nielsen K, Schmidt-Nielsen B, Jarnum SA, et al (1957) Body temperature of the camel and its relation to water economy. Am J Physiol 188: 103–112

893. Schoener TW (1968) Sizes of feeding territories among birds. Ecology 49: 123–141

894. Schoener TW (1969) Optimal size and specialization in constant and fluctuating environments: an energy-time approach. Brookhaven Symposium in Biology 22: 103–114

895. Schoener TW (1971) Theory of feeding strategies. Ann Rev Ecol Systemat 2: 369–404

896. Schoener TW (1977) Competition and the niche. In: Gans C, Tinkle DW (eds) Biology of the reptilia. Ecology and Behavior A. Academic, New York, Vol 7, pp 35–136

897. Schultz RJ (1971) Special adaptive problems associated with unisexual fishes. Am Zool 11: 351–360

898. Schüz E (1953) Die Zugscheide des Weissen Storches nach den Beringungs-Ergebnissen. Bonner Zool Beit 4: 31–72

899. Schwartzkopff J, Winter P (1960) Zur Anatomie der Vogel-Cochlea unter natürlichen Bedingungen. Biol Zent 79: 607–625

900. Sculthorpe MA (1967) The biology of aquatic vascular plants. St. Martin's Press, New York

901. Scrimshaw NS (1945) Embryonic development in poeciliid fishes. Biol Bull 88: 233–246

902. Sergeant DE (1966) Reproductive rates of harp seals, *Pagophilus groenlandicus* (Erxleben). J Fisheries Res Board Can 23: 757–766

903. Sergeant DE (1973) Environment and reproduction in seals. J Reprod Fertil Suppl 19: 555–561

904. Serventy DL (1963) Egg-laying timetable of the slender-billed shearwater, *Puffinus tenuerostris*. Proceedings of the XIII International Ornithological Congress (1962) 1: 338–343

905. Serventy DL (1970) Torpidity in the white-backed swallow. Emu 70: 27–28

906. Serventy DL (1971) Biology of desert birds. In: Farner DS, King JR (eds) Avian biology. Academic, New York, Vol 1.

907. Serventy DL, Marshall AJ (1957) Breeding periodicity in western Australian birds: with an account of unseasonable nestings in 1953 and 1955. Emu 57: 99–126

908. Sexton OJ, Brown KM (1977) The reproductive cycle of an iguanid lizard *Anolis sagrei*, from Belize. J Nat Hist 11: 241–250

909. Sharman GB (1959) Marsupial reproduction. In: Biogeography and ecology in Australia. Monogr Biol 8: 332–368

910. Sharp HF (1967) Food ecology of the rice rat, *Oryzomys palustris* (Harlan), in a Georgia salt marsh. J Mamm 48: 557–563

911. Sharp AJ (1966) Some aspects of Mexican phytogeography. Ciencia (Mexico) 24: 229–232

912. Shaw WT (1925a) Duration of the aestivation and hibernation of the Columbian ground squirrel *(Citellus columbianus)* and sex relation to the same. Ecology 6: 75–81

913. Shaw WT (1925b) The seasonal differences of north and south slopes in controlling the activities of the Columbian ground squirrel. Ecology 6: 157–162

914. Shaw WT (1925c) The hibernation of the Columbian ground squirrel. Can Field-Nat 39: 56–61; 79–82

915. Shilov IA (1973) Heat regulation in birds, an ecological-physiological outline. Amerind, New Delhi

916. Shimoizumi Jukichi (1939) Studies on the hibernation of the Japanese dormouse, *Glirulus japonicus* (Schinz). Science Reports of the Tokyo University of Literature and Science (Tokyo Bunrika Daigaku), Section B, vol 4: 51–61

917. Shimoizumi Jukichi (1940) Studies on the hibernation of the Japanese dormouse, *Glirulus japonicus* (Schinz). (2) On the body temperature during hibernation. Science Reports of the Tokyo University of Literature and Science (Tokyo Bunrika Daigaku), Section B, vol 5: 21–26

918. Shimoizumi Jukichi (1959) Studies of the hibernation of bats. Science Report of the Tokyo Kyoiku Daigaku, Sect. B, 9: nos. 131, 136 and 137.

919. Shinn EA, Dole JW (1979) Lipid components of prey odors elicit feeding responses in western toads *(Bufo boreas)*. Copeia: 275–278

920. Shkolnik A, Borut A. (1969) Temperature and water relations in two species of spiny mice *(Acomys)*. J Mamm 50: 245–255

921. Shoemaker VH, Balding D, McClahahan LL (1972) Uricotelism and low evaporative water loss in a South American frog. Science 175: 1018–1020

922. Short RV (1976) The evolution of human reproduction. Proc R Soc London, Ser. B, 195: 3–24

923. Shlaifer A, Breder CM Jr (1940) Social and respiratory behavior of small tarpon. Zoologica 25: 493–512

924. Shul'man GE (1974) Life cycles of fish. Keter, Jerusalem

925. Sillman AJ (1973) Avian vision. In: Farner DS, King JR (eds) Avian Biology. Academic, New York
926. Sill WD (1968) The zoogeography of the Crocodilia. Copeia: 76–88
927. Silverstone PA (1973) Observations on the behavior and ecology of a Colombian poison-arrow frog, the kôkoé-pá (*Dendrobates histrionicus* Berthold). Herpetol 29: 295–301
928. Simberloff D (1976) Species turnover and equilibrium island biogeography. Science 194: 572–578
929. Simpson GG (1940) Review of the mammal-bearing Tertiary of South America. Proc Am Philos Soc 83: 649–709
930. Simpson GG (1947) Holarctic mammalian faunas and continental relationships during the Cenozoic. Bull Geol Soc Am 58: 613–688
931. Simpson GG (1950) History of the fauna of Latin America. Am Sci 38: 361–389
932. Simpson GG (1956) Zoogeography of West Indian land mammals. Am Mus Novitates 1759: 1–28
933. Simpson GG (1969) South American mammals. In: Fihkan EJ, et al (eds) Biogeography and ecology in South America. Monogr Biol 19: 879–909
934. Simpson GG (1978) Early mammals in South America: fact, controversy, and mystery. Proc Am Philos Soc 122: 318–328
935. Sivertsen E (1941) On the biology of the harp seal *(Phoca groenlandica)* Hvalradets Skrifter 26: 1–166
936. Skutch AF (1950) The nesting seasons of Central American birds in relation to climate and food supply. Ibis 92: 185–222
937. Slijper EJ (1962) Whales. Hutchinson, London
938. Smalley IJ (1978) Pleistocene land-sea correlations. Nature 272: 754–755
939. Smith CC (1968) The adaptive nature of social organization in the genus of tree squirrels *Tamiasciurus*. Ecol Monogr 38: 31–63
940. Smith CL (1971) Secondary gonochorism in the serranid genus *Liopropoma*. Copeia: 316–319
941. Smith CL, Atz EH (1973) Hermaphroditism in the mesopelagic fishes *Omosudis lowei* and *Alepisaurus ferox*. Copeia: 41–44
942. Smith CL, Rand CS, Schaeffer B, Atz JW (1975) *Latimeria*, the living coelacanth, is ovoviviparous. Science 190: 1105–1106
943. Smith EN (1975) Thermoregulation of the American alligator, *Alligator mississippiensis*. Physiol Zool 48: 177–194
944. Smith HW (1930) Metabolism of the lungfish *Protopterus aethiopicus*. J Biol Chem 88: 97–130
945. Smith HW (1931) Observations on the African lung-fish, *Protopterus aethiopicus*, and on the evolution from water to land environments. Ecology 12: 164–181
946. Smith HW (1935) Metabolism of the lungfish—I. General considerations of the fasting metabolism in the active fish. J Cell Physiol 6: 43–67
947. Smith H (1953) From fish to philosopher. Little Brown, Boston
948. Smith JC (1975) Sound communication in rodents. In: Bench RJ et al (eds) Sound reception in mammals. Symposia of the Zoological Society of London, Number 37. Academic, London, pp 317–330
949. Smith LC, Smith DA (1972) Reproductive biology, breeding seasons, and growth of eastern chipmunks. *Tamias striatus* (Rodentia: Sciuridae) in Canada. Can J Zool 50: 1069–1085
950. Smith MH, McGinnis JT (1968) Relationships of latitude, altitude, and body size to litter size and mean annual production of offspring in *Peromyscus*. Res Popul Ecol 10: 115–126
951. Smith NS, Buss IO (1975). Formation, function and persistence of the corpora lutea of the African elephant *(Loxodonta africana)*. J Mamm 56: 30–43
952. Smith RC, Tyler JE (1967) Optical properties of clear natural water. J Optic Soc Am 57: 589–595
953. Smith RE, Horwitz BA (1969) Brown fat and thermogenesis. Physiol Rev 49: 330–426

954. Snow DW (1965) The breeding of the red-billed tropic bird in the Galápagos Isands. Condor 67: 210–214
955. Sohn JJ (1977) The consequences of predation and competition upon demography of *Gambusia manni* (Pisces: Poeciliidae). Copeia: 224–227
956. Songdahl JH, Hutchison VH (1972) The effect of photoperiod, parietalectomy and eye enucleation on oxygen consumption in the blue granite lizard, *Sceloporus cyanogenys*. Herpetol 28: 148–156
957. Sorenson MW, Conaway CH (1968) The social and reproductive behavior of *Tupaia montana* in captivity. J Mamm 49: 502–512
958. Soulé M (1966) Trends in insular radiation of a lizard. Am Nat 100: 47–64
959. Spieth HT (1966) Hawaiiań honeycreeper, *Vestiaria coccinea* (Forster), feeding on lobeliad flowers, *Clermontia arborescens* (Mann) Hillebr. Am Nat 100: 470–473
960. Stager KE (1967) Avian olfaction. Am Zool 7: 415–420
961. Stamp NE, Ohmart RD (1978) Resource utilization by desert rodents in the Lower Sonoran Desert. Ecology 59: 700–707
962. Stamps JA (1977) Egg retention, rainfall and egg laying in a tropical lizard *Anolis aeneus*. Copeia: 759–764
963. Stamps JA, Crews DP (1976) Seasonal changes in reproduction and social behavior in the lizard *Anolis aeneus*. Copeia: 467–476
964. Standaert T, Johansen K (1974) Cutaneous gas exchange in snakes. J Comp Physiol 89: 313–332
965. Stebbins LL (1976) Overwintering of the red-backed vole at Edmonton, Alberta, Canada. J Mamm 57: 554–561
966. Stebbins LL (1977) Energy requirements during reproduction of *Peromyscus maniculatus*. Can J Zool 55: 1701–1704
967. Stebbins LL, Orich R (1977) Some aspects of overwintering in the chipmunk, *Eutamias amoenus*. Can J Sci 55: 1139–1146
968. Stebbins RC (1958) An experimental study of the "third" eye of the tuatara. Copeia: 183–190
969. Stebbins RC 1970. The effect of parietalectomy on testicular activity and exposure to light in the desert night lizard *(Xantusia vigilis)*. Copeia: 261–270
970. Stebbins RC, Eakin RM (1958) The role of the "third eye" in reptilian behavior. Am Mus Novitates 1870: 1–40
971. Stebbins RC, Kalk M (1961) Observations on the natural history of the mud-skipper, *Periophthalmus sobrinus*. Copeia: 18–27
972. Stebbins RC, Barwick RE (1968) Radiotelemetric study of thermoregulation in a lace monitor. Copeia: 541–547
973. Steen E (1967) Some remarks on the migrations of wild reindeer in Fennoscandia. Ark Zool 20: 175–179
974. Stell WK, Lightfoot DO, Wheeler TG, Leeper HF (1975) Goldfish retina: functional polarization of cone horizontal cell dendrites and synapses. Science 190: 989–990
975. Steven DM (1963) The dermal light sense. Biol Rev 38: 204–240
976. Strasburg DW (1959) Notes on the diet and correlating structures of some central Pacific echeneid fishes. Copeia: 244–248
977. Strum JM (1969) Photophores of *Porichthys notatus:* ultrastructure of innervation. Anat Rec 164: 463–478
978. Strumwasser F (1960) Some physiological principles governing hibernation in *Citellus beecheyi*. In: Mammalian hibernation. Bull. Mus. Comp. Zool. Harvard 124
979. Suga Nobuo (1967) Electrosensitivity of specialized and ordinary lateral line organs of the electric fish, *Gymnotus carapo*. In: Cahn P (ed) Lateral line detectors. Indiana University Press, Bloomington, pp 395–409
980. Sullivan T (1977) Demography and dispersal in island and mainland populations of the deer mouse, *Peromyscus maniculatus*. Ecology 58: 964–978

981. Suthers RA (1965) Acoustic orientation by fish-catching bats. J Exp Zool 158: 319–347
982. Suthers RA (1967) Comparative echolocation by fishing bats. J Exp Zool 158: 319–347
983. Swan H, Jenkins D, Knox K (1968) Anti-metabolic extract from the brain of *Protopterus aethiopicus*. Nature 217: 671
984. Szarski H (1977) Sarcopterygii and the origin of tetrapods. In: Hecht MK et al (eds) Major patterns in vertebrate evolution. Plenum, New York, pp 517–540
985. Taber CA, Wilkinson RF Jr, Topping MS (1976) Age and growth of hellbenders in the Niangua River, Missouri. Copeia: 633–639
986. Talbot FH, Penrith JJ (1962) Spearing behavior in feeding in the black marlin, *Istiompax marlina*. Copeia 468
987. Tanner JM (1962) Growth at adolescence. Blackwell, Oxford, England
988. Tarkkonen H (1972) Physiological variations in the brown adipose tissue of mice and some other small mammals. Reports from the Department of Zoology, University of Turku Nr 1
989. Tarling DH (1971) Gondwanaland, palaeomagnetism and continental drift. Nature 229: 17–21; 71
990. Tarling DH, Runcorn SK (1973) Implications of continental drift to the earth sciences. Academic, London
991. Tavolga WN (1971) Sound production and detection. In: Hoar WS, Randall DJ (eds) Fish physiology. Academic, New York, Vol 5, pp 135–205
992. Tavolga WN (1977) Mechanisms for directional hearing in the sea catfish *(Arius felis)*. J Exp Biol 67: 97–115
993. Taylor CR (1970) Strategies of temperature regulation: effect on evaporation in East African ungulates. Am J Physiol 219: 1131–1135
994. Taylor DH, Ferguson DE (1970) Extraoptic celestial orientation in the southern cricket frog *Acris gryllus*. Science 168: 390–392
995. Taylor HL, Walker JM, Medica PA (1967) Males of three normally parthenogenetic species of teiid lizards (Genus *Cnemidophorus)*. Copeia: 739–743
996. Taylor JA, Hadley ME (1970) Chromatophores and color change in the lizard, *Anolis carolinensis*. Z Zellforsch Mikrosk Anat 104: 282–294
997. Taylor MH, Leach GJ, DiMichele L (1979) Lunar spawning cycle in the mummichog, *Fundulus heteroclitus* (Pisces: Cyprinodontidae). Copeia: 291–297
998. Teal JM, Carey FG (1967) Skin respiration and oxygen debt in the mudskipper *Periophthalmus sobrinus*. Copeia: 677–679
999. Telford SR Jr, Campbell HW (1970) Ecological observations on an all female population of the lizard *Lepidophyma flavimaculatum* (Xantusiidae) in Panama. Copeia: 379–381
1000. Tenaza RR (1966) Migration of hoary bats on South Farallon Island, California. J Mamm 47: 533–535
1001. Terashima SI, Goris RC (1974) Electrophysiology of snake infrared receptors. In: Kerkut GA, Phillus JW (eds) Progress in neurobiology. Pergamon, Oxford, England, Vol 4, pp 311–332
1002. Test FH, McCann RG (1976) Foraging behavior of *Bufo americanus* tadpoles in response to high densities of micro-organisms. Copeia: 576–578
1003. Tester JR, Breckenridge WJ (1964) Winter behavior patterns of the Manitoba toad, Bufo hemiophrys, in northwestern Minnesota. In: Soumalainen P (ed) Mammalian hibernation II. Ann Acad Sci Fenn ser. A, IV Biologica 71: 423–431
1004. Thielcke G (1969) Geographic variation in bird vocalizations. In: Hinde RA (ed) Bird vocalizations. Cambridge University Press, Cambridge, England, pp 311–339
1005. Thomas DG, Dartnall AJ (1970) Premigratory deposition of fat in the red-necked stint. Emu 70: 87
1006. Thomas RDK, Olson EC (ed) (1980) A cold look at the warm-blooded dinosaurs. Select Sympos 28. Am Assoc Adv Sci, Washington, DC
1007. Thompson DC (1978) Regulation of a northern grey squirrel *(Sciurus carolinensis)* population. Ecology 59: 708–715
1008. Thompson AL (1931) On "abmigration" among ducks: an anomaly shown by the results of

bird-marking. Proceedings of the 7th International Ornithological Congress (Amsterdam; 1930): 382–388

1009. Thorpe WH, Griffin DR (1962) Ultrasonic frequency in bird song. Ibis 104: 220–227

1010. Tilley S (1972) Aspects of parental care and embryonic development in *Desmognathus ochrophaeus*. Copeia: 532–540

1011. Tilley S, Tinkle DW (1968) A reinterpretation of the reproductive cycle and demography of the salamander *Desmognathus ochrophaeus*. Copeia: 299–303

1012. Tinkle DW (1962) Reproductive potential and cycles in female *Crotalus atrox* from northwestern Texas. Copeia: 306–313

1013. Tinkle DW (1967) Home range, density, dynamics and structure of a Texas population of the lizard *Uta stansburiana*. In: Milstead WH (ed) Lizard. Ecology symposium. University of Missouri Press, Columbia, pp 5–29

1014. Tinkle DW, Irwin LN (1965) Lizard reproduction: refractory period and response to warmth in *Uta stansburiana* females. Science 148: 1613–14

1015. Todd ES, Ebeling AW (1966) Aerial respiration in the long jaw mudsucker, *Gillichthys mirabilis* (Teleostei: Gobiidae). Biol Bull 130: 265–268

1016. Tomich PQ (1969) Mammals in Hawaii; a synopsis and notational bibliography. Bernice P. Bishop Museum Spec Publ 57, Honolulu

1017. Towns DR (1975) Reproduction and growth of the black shore skink, *Leiolopisma suteri* (Lacertilia: Scincidae), in northeastern New Zealand. NZ J Zool 2: 409–423

1018. Trautman MB (1957) The fishes of Ohio. Ohio State University Press, Columbus

1019. Tseng RWH (1972) Aspects of ovarian maturation in the molly, *Mollienesia latipinna*. Biol Bull Taiwan Normal Univ 7: 138–146

1020. Tsuji FJ, Haneda Y, Lynch RV III, et al (1971) Luminescence cross-reactions of Porichthys luciferin and theories on the origin of luciferin in some shallow-water fishes. Comp Biochem Physiol 40: 163–180

1021. Tucker VA (1966) Diurnal torpor and its relation to food consumption and weight changes in the California pocket mouse *Perognathus californicus*. Ecology 47: 245–252

1022. Turner CL (1940) Viviparity in teleost fishes. Sci Monthly (London) 65: 508–518

1023. Tuttle MD (1967) Predation by *Chrotopterus auritus* on geckos. J Mamm 48: 319

1024. Tuttle MD (1968) Feeding habits of *Artibeus jamaicensis*. J Mamm 49: 787

1025. Tuttle MD (1978) Population ecology of the gray bat *(Myotis grisescens)*: philopatry, timing and patterns of movement, weight loss during migration, and seasonal adaptive strategies. Occasional Papers of the Museum of Natural History, University of Kansas. No. 54

1026. Twente JW, Twente JA (1965) Regulation of hibernating periods by temperature. Proc Natl Acad Sci 54: 1058–1061

1027. Twente JW, Twente JA (1965) Effect of core temperature upon duration of hibernation of *Citellus lateralis*. J Appl Physiol 20: 411–416

1028. Twente JW, Twente JA (1967) Seasonal variation in the hibernation behavior of *Citellus lateralis*. In: Mammalian hibernation III. American Elsevier, New York, pp 47–63

1029. Twente JW, Twente JA (1970) Arousing effects of trophic hormones in hibernating *Citellus lateralis*. Comp Gen Pharmacol 1: 431–436

1030. Twente JW, Twente JA, Moy RM (1977) Regulation of arousal from hibernation by temperature in three species of *Citellus*. J Appl Physiol 42(2): 191–195

1031. Twitty VC 1955. Field experiments on the biology and genetic relationships of the California Species of *Triturus*. J Exp Zool 129: 129–148

1032. Twitty VC (1959) Migration and speciation in newts. Science 139: 1735–1743

1033. Tyron CA, Cunningham HN (1968) Characteristics of pocket gophers along an altitudinal transect. J Mamm 49: 699–705

1034. Udry JR, Morris NM (1968) Distribution of coitus in the menstrual cycle. Nature 220: 593–596

1035. Udvardy MDF (1963) Zoogeographical study of the Pacific Alcidae. In Pacific Basin Biogeography, (Tenth Pacific Science Congress). Bishop Museum Press: 85–111

1036. Udvardy MDF (1975) A classification of the biogeographical provinces of the world. Occasional Paper No. 18. International Union for Conservation of Nature and Natural Resources. Morges, Switzerland

1037. Ultsch GR (1976) Eco-physiological studies of some metabolic and respiratory adaptations of sirenid salamanders. In: Hughes GM (ed) Respiration in amphibious vertebrates. Academic, New York, pp 287–312

1038. Umminger BL (1969) Physiological studies on supercooled killifish *(Fundulus heteroclitus)* II. Serum organic constituents and the problem of supercooling. J Exp Zool 172: 409–424

1039. Underwood H (1977) Circadian organization in lizards: the role of the pineal organ. Science 195: 587–589

1040. Uzzell TM (1964) Relations of the diploid and triploid species of *Ambystoma jeffersonianum* complex (Amphibia, Caudata). Copeia: 257–300

1041. Valerio CE (1971) Ability of some tropical tadpoles to survive without water. Copeia: 364–365

1042. Van Deusen HM (1969) Feeding habits of *Planigale* (Marsupialia, Dasyuridae). J Mamm 50: 616–618

1043. van Dijk DE (1972) The behavior of southern African anuran tadpoles with particular reference to their ecology and related external morphological features. Zool Africana 7: 49–55

1044. Van Dijk T (1973) A comparative study of hearing in owls of the family Strigidae. Neth J Zool 23: 131–167

1045. van Gelder JR (1973) Ecological observations on Amphibia in the Netherlands II. *Triturus helveticus* Razoumowski: migration, hibernation and neoteny. Neth J Zool 23: 86–108

1046. Van Mierop LHS, Barnard SM (1978) Further observations on thermoregulation in the brooding female *Python molurus bivittatus* (Serpentes: Biodae). Copeia: 615–621

1047. van Zinderen Bakker EM Sr (1976) The evolution of Late-Quaternary palaeoclimates of southern Africa. In: van Zinderen Bakker EM (ed) Palaoecology of Africa and of the surrounding islands and Antarctica. Balkema, Cape Town, Vol 9

1048. Vandenberg JF, Drickamer LC, Colby DR (1971) Social and dietary factors in the sexual maturation of female mice. J Reprod Fertil 28: 397–405

1049. Vaughan TA (1976) Nocturnal behavior of the African false vampire bat *(Cardioderma cor)*. J Mamm 57: 227–248

1050. Verheijen FJ, Reuter JH (1969) The effect of alarm substances on predation among cyprinids. Anim Behav 17: 551–554

1051. Vial JL (1968) The ecology of a tropical salamander. *Bolitoglossa subpalmata,* in Costa Rica. Rev Biol Trop 15: 13–115

1052. Vince MA (1969) Embryonic communication, respiration and the synchronization of hatching. In Hinde RA (ed) Bird vocalizations, their relation to current problems in biology and psychology. Cambridge University Press, London, pp 233–260

1053. Vincent A (1965) The frogs of South Africa, Purnell, Cape Town

1054. Vinegar A (1973) The effects of temperature on the growth and development of embryos of the Indian python, *Python molurus* (Reptilia: Serpentes: Boidae). Copeia: 171–173

1055. Vinegar A, Hutchison VH, Dowling HG (1970) Metabolism, energetics and thermoregulation during brooding of snakes of the genus *Python* (Reptilia: Boidae). Zoologica 55: 19–48

1056. Visser J (1967) First report of ovoviviparity in a southern African amphisbaenid, *Monopeltis c. capensis.* Zool Africana 3: 111–113

1057. Vitt LJ (1973) Reproductive biology of the anguid lizard, *Gerrhonotus coeruleus princeps.* Herpetol 29: 176–184

1058. von Haartman L (1971) Population dynamics. In: Farner DS, King JR (eds) Avian Biology I. Academic, New York, pp 391–439

1059. Voris HK, Bacon JP (1966) Differential predation on tadpoles. Copeia: 594–598

1060. Vuilleumier BS (1971) Pleistocene changes in the fauna and flora of South America. Science 173: 771–780

1061. Vuilleumier F (1967) Phyletic evolution in modern birds of the Patagonian forests. Nature 215: 247–248

1062. Wade O (1930) The behavior of certain spermophiles with special reference to aestivation and hibernation. J Mamm 11: 160–188

1063. Wade O (1950) Soil temperatures, weather conditions, and emergence of ground squirrels from hibernation. J Mamm 31: 158–161
1064. Wade P (1958) Breeding season among mammals in the lowland rain-forest of north Borneo. J Mamm 39: 429–433
1065. Wake DB (1970) Aspects of vertebral evolution in the modern Amphibia. Forma et functio 3: 33–60
1066. Wake MH (1968) Evolutionary morphology of the caecilian urogenital system. I. The gonads and the fat bodies. J Morphol 126: 291–332
1067. Wake MH (1978) The reproductive biology of *Eleutherodactylus jasperi* (Amphibia, Anura, Leptodactylidae), with comments on the evolution of live-bearing systems. J Herpetol 12: 121–133
1068. Walcott C, Gould JL, Kirschvink JL (1979) Pigeons have magnets. Science 205: 1027–1028
1069. Walsberg GE, Campbell GS, King JR (1978) Animal coat color and radiative heat gain: a re-evaluation. J Comp Physiol 126: 211–222
1070. Walls GL (1942) The vertebrate eye and its adaptive radiation. Cranbrook Institute of Science, Bulletin 19. Bloomfield Hills, Michigan
1071. Walters LH, Walters V (1965) Laboratory observations on a cavernicolous poeciliid from Tabasco, Mexico. Copeia: 214–223
1072. Warburg MR (1965a) Studies on the water economy of some Australian frogs. Aust J Zool 13: 317–330
1073. Warburg MR (1965b) Studies of the environmental physiology of some Australian lizards from arid and semi-arid habitats. Aust J Zool 13: 563–575
1074. Warburg MR (1965c) The influence of ambient temperature and humidity on body temperature and water loss from two Australian lizards, *Tiliqua rugosa* (Gray) (Scincidae) and *Amphibolurus barbatus* Cuvier (Agamidae). Aust J Zool 13: 331–350
1075. Warburg MR (1966) On the water economy of several Australian geckos, agamids, and skinks. Copeia: 230–235
1076. Warner JA, Latz MI, Case JF (1979) Cryptic bioluminescence in a midwater shrimp. Science 203: 1109–1110
1077. Warner RR (1975) The adaptive significance of sequential hermaphroditism in animals. Am Nat 109: 61–82
1078. Webb GJW, Messel H, Magnusson W (1977) The nesting of *Crocodylus porosus* in Arnheim Land, Northern Australia. Copeia: 238–249
1079. Webb SD (1976) Mammalian faunal dynamics of the great American interchange. Paleobiol 2: 220–234
1080. Weber E, Werner YL (1977) Vocalization of two snake-lizards (Reptilia: Sauria: Pygopodidae). Herpetol 33: 353–363
1081. Weekes HC (1935) A review of placentation among reptiles with particular regard to the function and evolution of the placenta. Proc Zool Soc London (1935): 625–645
1082. Weir BJ (1974) Reproductive characteristics of hystricomorph rodents. In: Rowlands JW, Weir BJ (eds) The biology of hystricomorph rodents. Symp Zool Soc London, Academic, New York
1083. Weise CM (1963) Annual physiological cycles in captive birds of differing migration habits. Proceeding of the XIII International Ornithological Congress 2: 983–993
1084. Westby GWM (1974) Assessment of the signal value of certain discharge patterns in the electric fish, *Gymnotus carapo*, by means of playback. J Comp Physiol 92: 327–341
1085. Wenzel BM (1973) Chemoreception. In: Farner DS, King JR (eds) Avian biology. Academic, New York, pp 389–415
1086. Westerkov K (1963) Ecological factors affecting distribution of the nesting royal albatross population. Proc XIII International Ornithological Congress (1962) vol 2: 795–811
1087. Westrheim SJ (1975) Reproduction, maturation, and identification of larvae of some *Sebastes* (Scorpaenidae) species in the North Pacific Ocean. J Fisheries Res Board Can 32: 2399–2411
1088. Whitaker AH (1968) The lizards of the Poor Knights Islands, New Zealand. NZ J Sci 11: 623–651

1089. Whitaker AH (1968) *Leiolopisma suteri* (Boulenger), an oviparous skink in New Zealand. NZ J Sci 11: 425–432

1090. Whitaker JO Jr (1963) Food of 120 *Peromyscus leucopus* from Ithaca, New York. J Mamm 44: 418–419

1091. Whitaker JO Jr (1966) Food of *Mus musculus, Peromyscus maniculatus bairdi,* and *Peromyscus leucopus* in Vigo County, Indiana. J Mamm 47: 473–486

1092. White JC Jr, Angelovic JW (1967) Feeding behavior of a young stargazer, *Astrocopus ygraecum.* Copeia: 240–241

1093. Whitford WG, Hutchison VH (1966) Cutaneous and pulmonary gas exchange in ambystomid salamanders. Copeia: 573–577

1094. Whitten WK, Bronson FW, Greenstein JA (1968). Estrus-inducing pheromone of male mice: transport by movement of air. Science 161: 584–585

1095. Wilber CG, Musacchia XJ (1950) Fat metabolism in the arctic ground squirrel. J Mamm 31: 304–309

1096. Wilks BJ (1962) Reingestion in geomyid rodents. J Mamm 43: 267

1097. Williams O (1959) Food habits of the deer mouse. J Mamm 40: 415–419

1098. Willis JL, Moyle DL, Baskett TS (1956) Emergence, breeding, hibernation, movements and transformation of the bullfrog, *Rana catesbeiana,* in Missouri. Copeia: 30–41

1099. Wiltschko W (1975) The interaction of stars and magnetic field in the orientation system of night migrating birds. Z Tierpsychol 37: 337–355

1100. Wiltschko W, Wiltschko R (1972) Magnetic compass of European robins. Science 176: 62–64

1101. Wimsatt WW (1945) Notes on breeding behavior, pregnancy, and parturition in some vespertilionid bats of the eastern United States. J Mamm 26: 23–33

1102. Winn HE (1958) Observations on the reproductive habits of darters (Pisces-Percidae). Am Mid Nat 59: 190–212

1103. Winter JW (1966) Bird predation by the Australian marsupial squirrel glider. J Mamm 47: 530

1104. Wolff JO, Bateman GC (1978) Effects of food availability and ambient temperature on torpor cycles of *Perognathus flavus* (Heteromyidae). J Mamm 59: 707–716

1105. Wood SC, Lenfant CJM (1976) Respiration: mechanics, control and gas exchange. In: Gans C, Dawson WR (eds) Biology of the reptilia. Academic, New York, Vol 5, pp 225–274

1106. Woodruff DS (1976) Courtship, reproductive rates, and mating system in three Australian *Pseudophryne.* J Herpetol 10: 313–318

1107. Woodruff DS (1976) Embryonic mortality in *Pseudophryne* (Anura: Leptodactylidae). Copeia: 445–449

1108. Woodruff DS (1977) Male postmating brooding behavior in three Australian *Pseudophryne* (Anura: Leptodactylidae). Herpetol 33: 296–303

1109. Wourms JP, Cohen DM (1975) Trophotaeniae, embryonic adaptations in the viviparous ophidioid fish, *Oligopus longhursti:* a study of museum specimens. J Morphol 147: 385–402

1110. Wu HW, Liu CK (1940) The bucco-pharyngeal epthelium as the principal respiratory organ in *Monopterus javanicus.* Sinensia 11: 221–239

1111. Wyman RL, Thrall JH (1972) Sound production by the spotted salamander, *Ambystoma maculatum.* Herpetol 28: 210–212

1112. Xavier F, Ozon R (1971) Recherches sur l'activite endocrine de l'ovaire de *Nectoprynoides occidentalis* Angel (Amphibien Anoure vivipare). II. Synthese *in vitro* des steroides. Gen Comp Endocrinol 16: 30–40

1113. Xavier F (1977) An exceptional reproductive strategy in Anura: *Nectophrynoides occidentalis* Angel (Bufonidae), an adaptation to terrestrial life by viviparity. In: Hecht MK et al (eds) Major patterns of vertebrate evolution. Plenum, New York, pp 545–552

1114. Yamamoto Toki-o (1958) Artificial induction of functional sex-reversal in genotypic females of the medaka *(Oryzias latipes).* J Exp Zool 137: 227–260

1115. Yehner RH (1978) Burrow system and home range use by eastern chipmunks, *Tamias striatus:* ecological and behavioral considerations. J Mamm 59: 324–329

1116. Yoshida Hiroichi (1971) Reproduction in the shrew-mole, *Urotrichus talpoides*. J Mamm Soc Jpn 5: 85–90

1117. Yoshida Hiroichi (1972) Reproduction of the long-tailed Japanese field mouse, *Apodemus argenteus*. J Mamm Soc Jpn 5: 170–177

1118. Zarrow MX, Campbell PS, Clark JH (1968) Pregnancy following coital-induced ovulation in a spontaneous ovulator. Science 159: 329–330

1119. Zweifel RG (1954) Adaptation to feeding in the snake *Contia tenuis*. Copeia: 299–230

Index